DIGITAL SYSTEMS
From Gates to Microprocessors

DIGITAL SYSTEMS

From Gates to Microprocessors

Sanjay K. Bose
Department of Electrical Engineering
Indian Institute of Technology
Kanpur, India

JOHN WILEY & SONS
New York Chichester Brisbane Toronto Singapore

First published in 1986
Reprinted 1987
WILEY EASTERN LIMITED
4835/24 Ansari Road, Daryaganj
New Delhi 110 002, India

Distributors:

Australia and New Zealand:
Jacaranda-Wiley Ltd., Jacaranda Press,
JOHN WILEY & SONS, INC.
GPO Box 859, Brisbane, Queensland 4001, Australia

Canada:
JOHN WILEY & SONS CANADA LIMITED
22 Worcester Road, Rexdale, Ontario, Canada

Europe and Africa:
JOHN WILEY & SONS LIMITED
Baffins Lane, Chichester, West Sussex, England

South East Asia:
JOHN WILEY & SONS, INC.
05-05 Block B, Union Industrial Building
37 Jalan Pemimpin, Singapore 2057

Africa and South Asia:
WILEY EASTERN LIMITED
4835/24 Ansari Road, Daryaganj
New Delhi 110 002, India

North and South America and rest of the world:
JOHN WILEY & SONS, INC.
605, Third Avenue, New York, NY 10158, USA

Library of Congress Cataloging in Publication Data

Bose, Sanjay K.
Digital systems—from gates to microprocessors.

1. Digital electronics, I. Title.
TK7868.D5B66 1986 621.3815 85-17922

ISBN 0-470-20252-1 John Wiley & Sons, Inc.
ISBN 0-85226-096-2 Wiley Eastern Limited

Printed in India at Rekha Printers, New Delhi.

To
my parents
and
professors

Preface

This book evolved from the lecture material prepared by the author to teach a course on Digital Systems at I.I.T. Kanpur. This course concentrates on the fundamentals of microprocessors and microcomputer systems as well as tries to familiarize the students with devices that are extensively used in a later laboratory course. Apart from a description of the basic concepts required, the course deals in detail with the Intel 8085A CPU and various common I/O peripherals. This material is treated in depth in Chapters 6–12 and the various associated appendices. We have augmented this with a brief description of single-chip microcomputers in Chapter 13, concentrating in depth on the popular Intel 8048/8748 and 8041A/8741A devices. The Z80 CPU has been presented in some detail in Chapter 14 to familiarize the reader with the features of this powerful microprocessor. Finally, Chapter 15 has been included to acquaint the reader with the aids typically available in designing and developing microprocessor-based systems.

Chapters 6–12 provide the reader with a wealth of detail on the operation of various common LSI integrated circuits. Most of this material is given in the appendices so as to avoid hampering the flow of the chapters themselves. Though these descriptions are given in a fairly detailed fashion, they are not intended to be replacements for the actual data sheets and manuals provided by the manufacturers. Indeed it is hoped that by supporting his or her reading of this text with such manuals, the reader will be able to get a complete picture of the device.

The initial Chapters 1–5 are of an introductory nature, intended to familiarize the reader with the concepts basic to the design of combinatorial and sequential digital circuits. It is hoped that this compact presentation will allow a reader to use this book even if he has no prior knowledge of digital electronics. Moreover, an introductory course on digital circuits may be supported by this text using these chapters along with Chapter 11 dealing with analog interfacing for digital systems. (For the latter, some minimal knowledge of op-amp circuits is however assumed.) With such a course as an introduction, the other chapters may then be used for a second course on digital systems dealing in detail with microprocessors and their peripherals.

The course on Digital Systems taught by the author, as well as this book, concentrates on familiarizing the reader with the operation of 8-bit CPUs providing only a brief mention of 16-bit CPUs. It has been the author's experience that 8-bit CPUs provide the ideal means of introducing students to microprocessors and microcomputers. This is primarily because the simplicity of these devices allow a clear exposition of the underlying concepts as well as of the details of their hardware and their assembly language. The complexity and power of typical 16-bit devices tends to be confusing to the student at the introductory level. Moreover, these devices require a detailed description in terms of their hardware and software, of length sufficient to merit a separate course (and a separate book). The mode of usage typically followed for these devices is also different, concentrating more on off-the-shelf single-board cards and high level language programming rather than on designing one's own hardware and assembly language programs.

Ideally, it is expected that a course following this text will also have an associated lab to allow the students to familiarize themselves with the actual devices dealt with here. For the assembly language programming part, the student may also find it helpful to run and verify his programs using cross-software on a main-frame or on something like a development system. At I.I.T. Kanpur, we provide both these facilities using an 8085A-based workstation developed by our faculty and research engineers. Microprocessor kits may also be used for such a lab even though these tend to be somewhat difficult to use for complex exercises.

SANJAY K. BOSE

Acknowledgements

The author wishes to thank the Quality Improvement Programme, I.I.T. Kanpur for the financial assistance provided by them in the preparation of this manuscript. The help and advice provided by Dr. A. Joshi, Dr. R. Raghuram and Dr. R. N. Biswas is also gratefully acknowledged. Thanks are also due to a large number of undergraduate and graduate students of the E.E. Department, I.I.T. Kanpur who suffered through the lecture notes which preceded this book and helped in correcting numerous errors. The author also wishes to thank Mr. J. S. Rawat and Mr. R. N. Srivastava for their careful typing of the manuscript and Mr. Qasim for drawing the figures.

Preparing a manuscript of this kind is usually a long affair. The author would like to give special thanks to his wife Aditi for her patience and tolerance. Even though her temper got a little frayed at times, she still insisted on helping out with the various little things associated with such a venture.

Contents

Chapter I

Introduction

1.1 THE DIGITAL APPROACH

The basic function of any instrument is to process data presented to it by the external world and then output the results of such processing to the external world. This basic objective is true, regardless of whether the actual processing done is digital or analog in nature. As shown in Fig. 1.1.1, such an instrument is then basically a system for manipulating various data inputs and producing results based on these inputs.

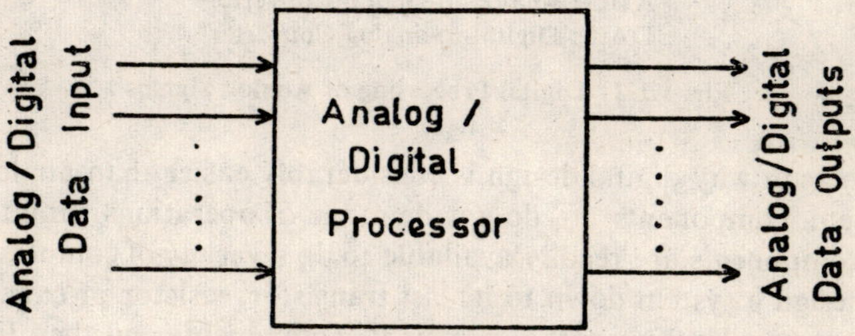

Fig. 1.1.1 Analog/Digital Data Processor

In so far as the world external to the instrument is concerned, it only wants results based on the proper processing of input data. In the absence of other considerations (e.g. cost, speed), it does not care whether the processing is analog or digital. Actually, from a somewhat philosophical view point, analog processing is more direct than digital processing and may indeed appear to be the preferred technique. After all, most of the information generated or used in the real world tends to be analog rather than digital.

Digital processing tends to be more indirect than its analog counterpart. This is because it first requires analog information to be converted to a digital form (i.e. 1s and Øs) and subsequently requires the results of the digital processing to be converted from the digital to the analog form. As shown in Fig. 1.1.2, this implies that in order to do digital processing of analog data, one will always incur the overhead of analog-to-digital and digital-to-analog conversion. It is interesting to note that in spite of this overhead, the current trend in design is more towards the digital approach than the analog one. There are various reasons for this trend. Digital processing tends to be simpler and yet is able to ensure greater processing accuracy. It also allows the designer a considerable degree of freedom in making modifications, and upgrading or redesigning the equipment to incorporate processing changes. Such equipments also tend to be simpler to operate and are more 'user friendly' than their analog counterparts.

Anyone familiar with analog circuit design will testify to the fact that great care must usually be exercised in such a process. This ranges all the way from component stability and component tolerances to impedance matching and signal conditioning. The basic reason for such care is the fact that analog processing operates on continuous variables and the behaviour of the processing circuit must be proper over the entire range of each variable. In contrast, by virtue of quantizing and encoding analog inputs, digital processing need only operate on binary signals, i.e. 1s and Øs. This is one major reason why the design of digital processing circuits is considerably simpler than an equivalent analog design.

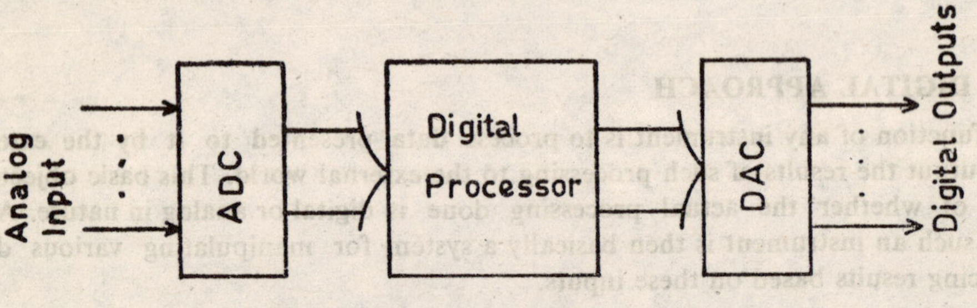

ADC : Analog-to-Digital Converter
DAC : Digital-to-analog Converter

Fig. 1.1.2 Digital Processing of Analog Signals

Another major reason why digital design is considerably easier than analog design is the easy availability of system components to do a wide range of operations. In integrated circuit form (ICs), off-the-shelf components are readily available to do a variety of complex operations. Instead of being forced to design a system down to its last transistor, resistor or capacitor, the designer's job is now reduced to identifying the proper IC components and then interconnecting them (following some simple rules and the manufacturer's specifications) in order to obtain the desired result. The care one had to take in analog design is still needed in the analog-to-digital or the digital-to-analog conversion process but is virtually absent in the design of the actual digital processor. In terms of available digital system components, the industry trend has been to go from Small Scale Integration (SSI) to Medium Scale Integration (MSI) and now to Large Scale Integration (LSI) and Very Large Scale Integration (VLSI) with increasing complexity of function, smaller sizes and greater versatility. The much talked about microprocessor or computer-on-a-chip is a product of this developmental trend which has placed greater computing power in the hands of more people than ever before. Another interesting trend in this direction is the ability to produce custom designed integrated circuits (Custom ICs) to satisfy the needs of a particular user; in other words, if a much needed function is not available in an IC form, it is not unthinkable to go ahead and get the required IC made.

It has been mentioned earlier that whereas analog processing deals with continuous variables which may take on any value in a given range, digital processing has to deal with only two values—that for a 1 and that for a Ø. In the latter, no error will be made unless noise causes enough perturbation to make a 1 look like a Ø and vice versa. This tends to give digital processing an edge over analog processing, in so far as noise immunity is concerned. Because of this, apart from the errors inherent in the analog-to-digital or digital-to-analog conversion, the only other errors present are those caused by the algorithm being used by the digital processor itself. With digital processing, one can therefore obtain great accuracy with only a very minimal effort if one uses a proper processing algorithm.

It is interesting to examine, in this context, the role of the device manufacturer in bringing

digital technology where it is today. In the 'bad old days' when digital systems were first assembled using discrete transistors (and even vacuum tubes), a lot of the design effort was concentrated on circuit minimization, i.e. minimizing the number of components (gates, flip-flops, etc.) in a given system. This was logical since the number of components was directly translatable into cost. The 1960s saw the first changes in this philosophy with the introduction of SSIs by some semiconductor manufacturers. These SSIs typically provided some simple logical functions in one IC package incorporating some 10–15 gates. Minimization techniques were still used but provided cost benefits only if the parts count (number of IC packages) could also be reduced. As IC technology advanced even further, MSIs and eventually LSIs were introduced with greater and greater functional power in one IC package and built using a larger and larger number of gates. With the introduction of these ICs, minimization techniques became only of peripheral interest as systems using MSIs and LSIs did not follow the gate-flip-flop-component count relationship prevalent before. Instead, the considerations of paramount interest were parts count and the cost of the component ICs. Apart from decreasing the parts count, higher degrees of integration also simplified system design and improved their reliability. This has also been a strong contributing factor to the rapid proliferation of digital systems.

The advent of LSIs like microprocessors gave a big boost to the digital approach by providing an enormous amount of cheap computer power in all sorts of applications. For one, it led to low cost microcomputers with capabilities rivalling those of many minicomputers, thereby creating a completely new genre of consumer electronic goods like hobby computers, video games, etc. For another, it revolutionized control applications by providing the computer-on-a-chip as just another system component—very often one of the cheapest components.

The microprocessor gave the system designer another degree of freedom he had not enjoyed before. Previously, a piece of hardware designed for one application could only be used for that application; there was very little chance of it satisfying the needs of some other system. Moreover, upgrading or modifying the design was very often equivalent to redesigning the system from scratch. The advent of the microprocessor changed all that. Now, it is not unusual to find the same hardware (i.e. the same microcomputer board) being used in a variety of different applications and carrying out different functions by running different software. In such a system, incorporating changes is trivial if all that is required is a software change; even if hardware modifications are required, the inherent modularity of these systems makes it easier to incorporate them.

Designing an instrument with microprocessors, also offers some other advantages. The modular nature of such systems usually simplifies maintenance and servicing. As a matter of fact, one can build in a sufficient number of self-tests in the instrument so that the user can do most of the repairs himself. Also, by properly prompting the user, displaying helpful operational information and reducing the complexity of the front panel, the instrument can be made more 'user friendly'.

1.2 OVERVIEW OF THE LATER CHAPTERS

The contents of this text may be roughly divided into two parts. The first part consisting of Chapters 2–5 concentrates on the basics of digital system design, specifically those using SSIs and MSIs. Designing combinatorial and sequential logic using gates and flip-flops is discussed here. A new IEEE/IEC standard representation for digital devices is also described here (in Chapter 5). This subject of 'Dependency Notation' is intended to familiarise the reader with a powerful symbology being projected as the future standard of representation. Since this symbology has been used very rarely for actual circuit diagrams in the text, the reader may skip this if he so desires.

The second part of the text consists of Chapters 6–12 and the associated appendices. This deals with digital system design using LSIs, concentrating specifically on microprocessors and their associated peripheral devices. Starting off with a discussion of general microprocessor architecture and using it to design a microcomputer, it treats the Intel 8085A microprocessor in detail as the one chosen for illustration. Standard peripherals, *I/O* techniques, Analog Interfacing and Interrupts are described in detail. The appendices of this part are intended to provide detailed description and application notes for some common LSI devices.

Chapter 2

Boolean Algebra and Combinatorial Circuit Design

The mathematical basis for digital logic design is provided by the results of Boolean Algebra or the algebra of Boolean variables. Unlike an ordinary variable, a Boolean or binary variable can assume only two values:

(a) TRUE, also denoted by 1 or High (Hi)

(b) FALSE, also denoted by ∅ or Low (Lo)

Other than TRUE/FALSE (or 1/∅), it is not possible for it to take on any other values.

It is possible to define basic functions of these Boolean variables which can then be used to build larger and more complex functional forms and expressions. The results of Boolean Algebra (i.e. its properties, postulates and theorems) can then be used to evaluate such expressions, minimize them and also indicate how they can be realized in circuit form. These basic Boolean functions are given below.

2.1 BASIC BOOLEAN (LOGIC) FUNCTIONS

The basic functions employed in Boolean Algebra are the AND function, the OR function and the NOT function. These are described briefly next.

2.1.1 AND Function

If A, B, and Y are Boolean (Binary, Logic) variables, then

$$Y = A \text{ (AND) } B$$

implies that Y is true (1), only if A and B are both true (1); otherwise, Y is false (∅). This function is also denoted by

$$Y = A \cdot B$$

where the dot (·) denotes the ANDing operation. Moreover, one typically omits the dot (·) and writes A·B as AB.

Fig. 2.1.1 shows a table listing the value of Y for all possible values that A and B can assume. Such a table showing the inputs (A, B in this case) and the resultant outputs is referred to as a 'truth table'.

A	B	Y
∅	∅	∅
∅	1	∅
1	∅	∅
1	1	1

Fig. 2.1.1 Truth Table for the AND Function

The AND function can be easily expanded for more than two variables e.g.

$$Y = A \cdot B \cdot C \cdot D \cdot = A \text{ (AND) } B \text{ (AND) } C \text{ (AND) } D$$

2.1.2 OR Function

If A, B, and Y are Boolean variables then

$$Y = A \text{ (OR) } B$$

implies that Y is FALSE (∅), only if A and B are both FALSE (∅); else, Y is TRUE (1). This is also denoted by

$$Y = A + B$$

Fig. 2.1.2 shows the truth table for the OR function. For more than two variables, the OR function can be expanded as

$$Y = A + B + C + D = A\,(OR)\,B\,(OR)\,C\,(OR)\,D$$

2.1.3 NOT Function

The NOT function applied to a logic variable A, generates its logical inverse denoted by $\bar{A}$. For the case

$$Y = \bar{A}$$

this is shown by the truth table of Fig. 2.1.3.

A	B	Y
Ø	Ø	Ø
Ø	1	1
1	Ø	1
1	1	1

Fig. 2.1.2 Truth Table for the OR Function

A	Y
Ø	1
1	Ø

Fig. 2.1.3 Truth Table for the NOT Function

The AND, OR, and NOT functions should be considered the basic functions of Boolean Algebra. They can be used to create the commonly used NAND (NOT–AND) and NOR (NOT-OR) functions as shown below

NAND : $Y = \overline{A \cdot B}$

NOR : $Y = \overline{A + B}$

The truth tables for the NAND and NOR functions are given in Fig. 2.1.4 and 2.1.5 respectively.

A	B	Y
Ø	Ø	1
Ø	1	1
1	Ø	1
1	1	Ø

Fig. 2.1.4 NAND Truth Table

A	B	Y
Ø	Ø	1
Ø	1	Ø
1	Ø	Ø
1	1	Ø

Fig. 2.1.5 NOR Truth Table

Note that the truth table for the NAND (or NOR) function can also be generated by inverting the output Y in the truth table of the AND (or OR) function.

Logic circuits are available to implement the logic functions described above. A set of standard symbols is used to represent such circuits (referred to as gates). These symbols are shown in Fig. 2.1.6 for the AND, OR, NOT, NAND and NOR functions.

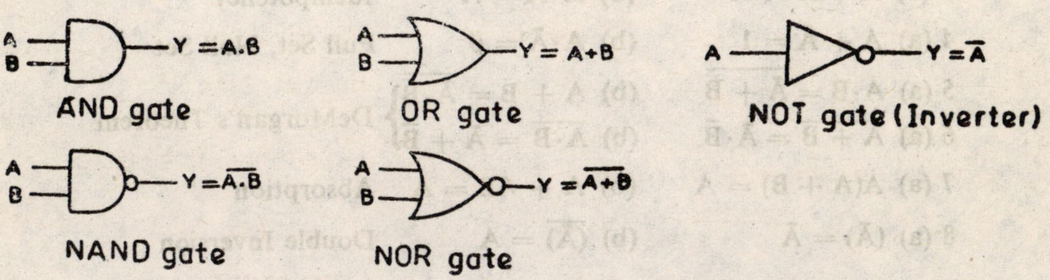

Fig. 2.1.6 Circuit Representations of Elements Implementing Logic Functions

2.2 POSTULATES AND THEOREMS OF BOOLEAN ALGEBRA

In this section, we first present the postulates of Boolean Algebra. It should be noted that these are rules under which this algebra operates; these postulates basically define the operation of the AND, OR and NOT functions discussed in Section 2.1. These postulates may be used to prove the theorems presented in the latter half of this section.

2.2.1 Postulates of Boolean Algebra

The postulates of Boolean Algebra are listed below:

1	(a) $A = 1$ if $A \neq \emptyset$	(b) $A = \emptyset$ if $A \neq 0$	definition of Logic Variable
2	(a) $\emptyset \cdot \emptyset = \emptyset$	(b) $1 + 1 = 1$	
3	(a) $1 \cdot 1 = 1$	(b) $\emptyset + \emptyset = \emptyset$	definition of AND and OR function
4	(a) $1 \cdot \emptyset = \emptyset$	(b) $\emptyset + 1 = 1$	
5	(a) $\bar{\emptyset} = 1$	(b) $\bar{1} - \emptyset$	definition of NOT function

Note that the postulates are not provable since they actually describe the basic properties of Boolean Algebra.

2.2.2 Properties of Boolean Algebra

The Commutative, Associative and Distributive properties of ordinary algebra are also valid for Boolean Algebra. These are summarised below:

Commutative Property : (a) $AB = BA$ (b) $A + B = B + A$

Associative Property : (a) $A(BC) = (AB)C$ multiplicative

(b) $A + (B + C) = (A + B) + C$ additive

Distributive Property : (a) $A(B + C) = AB + AC$ AND-over-OR

(b) $A + BC = (A + B)(A + C)$ OR-over-AND

Special attention should be paid to Distributive Property (b) which is the only property which looks substantially different compared to what one would expect from ordinary algebra.

2.2.3 Some Theorems of Boolean Algebra

Theorems	(a)	(b)	
1	$A + \emptyset = A$	$A \cdot 1 = A$	
2	$A + 1 = 1$	$A \cdot \emptyset = \emptyset$	
3	$A + A = A$	$A \cdot A = A$	Idempotency
4	$A + \bar{A} = 1$	$A \cdot \bar{A} = \emptyset$	Full Set, Null Set
5	$A \cdot B = \overline{\bar{A} + \bar{B}}$	$A + B = \overline{\bar{A} \cdot \bar{B}}$	DeMorgan's Theorem
6	$\overline{A + B} = \bar{A} \cdot \bar{B}$	$\overline{A \cdot B} = \bar{A} + \bar{B}$	DeMorgan's Theorem
7	$A(A + B) = A$	$A + AB = A$	Absorption
8	$(\bar{A}) = \bar{A}$	$\overline{(\bar{A})} = A$	Double Inversion

The theorems can easily be proved by taking the truth table approach. As an example, consider 6(a).

Let $Y = \overline{A + B}$, $Z = \bar{A} \cdot \bar{B}$

For all possible values of A and B, the values of Y and Z may be listed as shown below:

A	B	Y	Z
$\emptyset$	$\emptyset$	1	1
$\emptyset$	1	$\emptyset$	$\emptyset$
1	$\emptyset$	$\emptyset$	$\emptyset$
1	1	$\emptyset$	$\emptyset$

where Y and Z are computed using the postulates and properties of Boolean algebra.

Note that, in the above, Y is identical to Z, . . . for all values of A and B and hence

$$\overline{A + B} = \bar{A} \cdot \bar{B}$$

Algebraic manipulation may also be used to prove a theorem. As an example, consider 7(b).

L.H.S.	$= A + AB$	
	$= A \cdot 1 + A \cdot B$	Theorem 1(b)
	$= A \cdot (1 + B)$	Distributive property
	$= A \cdot 1$	Theorem 2(a)
	$= A$	Theorem 1(b)
	$=$ RHS	

Theorems 5(a), 5(b), 6(a) and 6(b) are especially useful in simplifying logic expressions and in designing digital circuits. They are commonly referred to as DeMorgan's Theorem. It is possible to generalize these theorems for more than two variables as given in the examples below:

(a) $\overline{A + B + C + \ldots} = \bar{A} \cdot \bar{B} \cdot \bar{C} \cdot \ldots$

(b) $\overline{A \cdot B \cdot C \cdot \ldots} = \bar{A} + \bar{B} + C + \ldots$

2.2.4 Duality Property of Boolean Algebra

Boolean Algebra exhibits an interesting duality property. According to this property, every Boolean expression has a dual (expression) where the latter may be found from the former by

(i) changing OR signs to AND signs and AND signs to OR signs,

(ii) changing 1s to Øs and Øs to 1s

in the original expression.

The duality property states that if an equation is true, its dual is also true. Examples of this may be found in Sec. 2.2.1–2.2.3 where the properties, postulates and theorems labelled (b) are the duals of those labelled (a).

The duality property is extremely useful in logic design. This follows from the fact that if an expression $f(\cdot)$ is the dual of another expression $g(\cdot)$ then $f(\cdot) = \overline{[g(\cdot)]}$ provided we also invert all the variables in the expression while finding the dual. As an example of this, consider the following:

$$\text{Let } Y = A\bar{B} + ABC + B\bar{C}$$

$$= A(\bar{B} + BC) + B\bar{C}$$

$$\text{then } \quad \bar{Y} = [\bar{A} + B\cdot(\bar{B} + \bar{C})]\cdot(\bar{B} + C) \quad \text{by duality}$$

$$= [(\bar{A} + B\bar{C})(\bar{B} + C)]$$

$$= \bar{A}\bar{B} + \bar{A}C$$

[It can also be shown that $Y = A + B\cdot\bar{C}$, from which $\bar{Y} = \bar{A}\bar{B} + \bar{A}C$ may be easily obtained by DeMorgan's Theorem].

Finally, it should be noted that DeMorgan's Theorem is really an expression of this duality property.

2.3 LOGIC EXPRESSIONS

Logic variables may be combined using the basic Boolean functions (i.e. AND, OR and NOT) to obtain logic expressions. Two forms of such logic expressions are of special interest. This is because these forms are easily examinable and, as we will see later, can be easily realized using logic gates.

(i) *Sum-of-Product form*: This is really an ORing of ANDed variables such as

$$A\bar{B} + AB\bar{C} + BC$$

(ii) *Product-of-Sum form*: This is an ANDing of ORed variables such as

$$(A + \bar{B})(A + B + \bar{C})(B + C)$$

[Note that using the term Sum to represent ORing and the term Product to represent ANDing is really a hangover from ordinary algebra].

Any given Sum-of-Products expression may be further expanded to obtain another Sum-of-Products form where each AND-term is a unique combination of all the involved variables. This form is referred to as the minterm canonical form or the standard sum, with the ANDed terms

as the minterms. For example,

$$\begin{aligned} Y &= A\bar{B} + AB\bar{C} + BC \\ &= A\bar{B}\cdot 1 + AB\bar{C} + 1\cdot B\cdot C \\ &= A\bar{B}(C + \bar{C}) + AB\bar{C} + (A + \bar{A})BC \\ &= \underbrace{A\bar{B}C + A\bar{B}\bar{C} + AB\bar{C} + ABC + \bar{A}BC}_{\text{minterms}} \text{ — minterm canonical form (standard sum)} \end{aligned}$$

Similarly, one can obtain a Product-of-Sums expression where each ORed term is a unique combination of all the involved variables. This is referred to as the maxterm canonical form or the standard product, with the ORed terms as the maxterms. For example,

$$\begin{aligned} Y &= (A + \bar{B})(A + B + \bar{C})(B + C) \\ &= (A + \bar{B} + C\bar{C})(A + B + \bar{C})(A\bar{A} + B + C) \\ &= \underbrace{(A + \bar{B} + C)(A + \bar{B} + \bar{C})(A + B + \bar{C})(A + B + C)(\bar{A} + B + C)}_{\text{maxterms}} \end{aligned}$$

— maxterm canonical form (standard product)

2.4 KARNAUGH MAPS AND MINIMIZATION OF LOGIC EXPRESSIONS

For a simple and economical realization of a logic expression, it is in the designer's interest to minimize the expression and reduce it to a simple form. For logic expressions having up to four variables, Karnaugh Maps present a simple way of doing so. (For expressions with five or more variables, an alternative method referred to as the Quinn-McClusky technique should be used.)

A Karnaugh Map is really a graphical representation of an expression written in the minterm canonical form. It is a collection of cells where every possible minterm is represented by a cell with the added restriction that adjacent cells (vertically and horizontally) represent minterms which differ by only one variable. The arrangement of the cells and the minterms they represent, are shown in Fig. 2.4.1 for two, three and four variables.

Consider the Karnaugh Map of four variables—D, C, B, A—shown in Fig. 2.4.1. The notation DC on the left of the diagonal line indicates that the binary digits below it have been written in the same sequence e.g. Ø1 implies D = Ø, C = 1. Similarly, the notation BA to the right of the diagonal indicates that the binary digits to its right are also written in the same sequence. The particular combinations of D and C are thereby indicated for each row of cells; the combinations of B and A are similarly indicated for each column of cells.

Each minterm is represented by a cell in the Karnaugh Map. In order to find the minterm for a cell or the cell for a minterm one should realize that, in a Karnaugh Map, the assignments for DC and BA are such that (a) a Ø represents a complemented variable, and (b) a 1 represents an uncomplemented variable. Using this, one can find the minterm for each cell or vice versa; in Fig. 2.4.1 we have indicated the minterm for each cell of the Karnaugh Map.

As mentioned before, the arrangement of cells in a Karnaugh Map was done so that the horizontal (or vertical) neighbour of each cell differed from it in only variable (i.e. the minterm of one has a complemented variable which is uncomplemented in the other, otherwise the two are similar). Extra care is needed when examining the neighbours of cells which lie along the

periphery of the Karnaugh Map. Here one should assume a 'wrap-around' arrangement of the Karnaugh Map such that

(a) cells in the last row are the neighbours of the corresponding cells in the first row, and

(b) cells in the right-most column are the neighbours of the corresponding cells in the left-most column.

For examples of the above, consider the following in the four variable Karnaugh Map of Fig. 2.4.1.

(i) cell $D\bar{C}BA$ is a neighbour of cell $\bar{D}\bar{C}BA$,

(ii) cell $DC\bar{B}\bar{A}$ is a neighbour of cell $DCB\bar{A}$,

(iii) cells $\bar{D}\bar{C}B\bar{A}$ and $D\bar{C}\bar{B}\bar{A}$ are neighbours of cell $\bar{D}\bar{C}\bar{B}\bar{A}$.

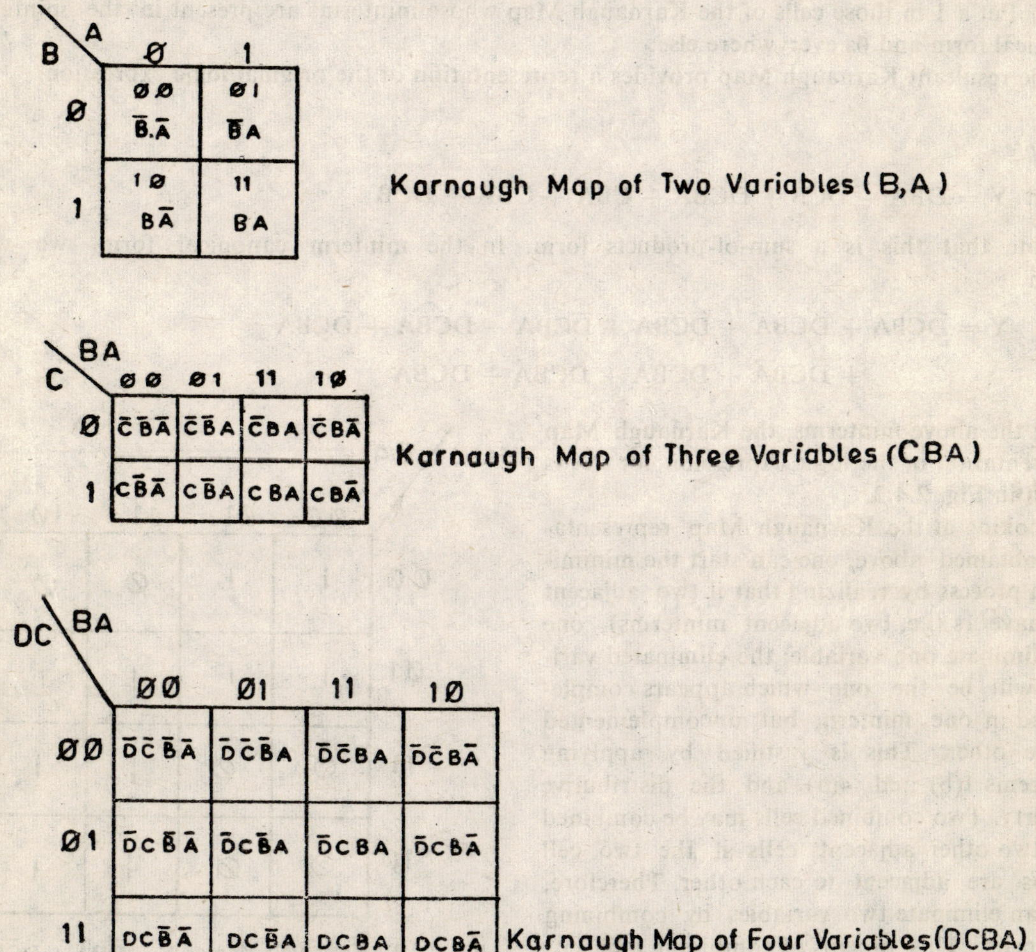

Fig. 2.4.1 Karnaugh Maps with Minterm Designation

The neighbours of a particular cell are referred to as being *adjacent* cells. Note once again that adjacent cells differ in only one variable of their respective minterms, i.e. one minterm has a variable uncomplemented whereas the other one has that same variable complemented. Note also that the wrap around feature shown above regarding cell adjacency is true for both two, three or four variable Karnaugh Maps.

In order to use Karnaugh Maps for minimizing logic expressions, one has to first specify a method of representing a sum of products form of a logic expression on such a map. (For expressions not already in such a form, one has to use Boolean Algebra principles to obtain a sum-of-products form.) Once such an expression is available, one does the following:

(i) Obtain the minterm canonical form by using Theorems 1(b) and 5(a) and the distributive property.

(ii) Put a 1 in those cells of the Karnaugh Map whose minterms are present in the minterm canonical form and Øs everywhere else.

The resultant Karnaugh Map provides a representation of the original logic expression.

Example

Let $Y = \bar{D}\bar{B}\bar{A} + \bar{D}\bar{C}\bar{B} + \bar{D}C\bar{B}A + CBA + CB\bar{A} + D\bar{C}B$

Note that this is a sum-of-products form. In the minterm canonical form, we will obtain

$$Y = \bar{D}\bar{C}\bar{B}\bar{A} + \bar{D}C\bar{B}\bar{A} + \bar{D}\bar{C}\bar{B}A + \bar{D}C\bar{B}A + D\bar{C}BA + \bar{D}CBA$$
$$+ \bar{D}CB\bar{A} + DCBA + DCB\bar{A} + D\bar{C}B\bar{A}$$

Using the above minterms, the Karnaugh Map representation of the logic expression for Y is shown in Fig. 2.4.2.

DC \ BA	ØØ	Ø1	11	1Ø
ØØ	1	1	Ø	Ø
Ø1	1	1	1	1
11	Ø	Ø	1	1
1Ø	Ø	Ø	1	1

Fig. 2.4.2 Karnaugh Map of $Y = \bar{D}\bar{B}\bar{A} + \bar{D}\bar{C}\bar{B} + \bar{D}C\bar{B}A + CBA + CB\bar{A} + D\bar{C}B$

Looking at the Karnaugh Map representation obtained above, one can start the minimization process by realizing that if two adjacent cells have 1s (i.e. two adjacent minterms), one can eliminate one variable; the eliminated variable will be the one which appears complemented in one minterm but uncomplemented in the other. This is justified by applying Theorems 1(b) and 4(a) and the distributive property. Two combined cells may be combined with two other adjacent cells if the two cell groups are adjacent to each other. Therefore, one can eliminate two variables by combining four cells. This process of combining adjacent groups of combined cells is continued as far as possible in order to eliminate the maximum number of variables.

In general, one can say that by combining 2^N adjacent cells (with 1s in them) in the Karnaugh Map representation of the logic expression, N variables may be eliminated. The process of minimization requires that—

(a) combine 2^N cells with N as large as possible in order to obtain one or more large groups of combined cells, and

(b) do (a) such that all the cells with 1s in them may be combined into as few groups as possible where each group has the maximum number of cells that can be grouped together.

The minimized expression can now be expressed in the sum-of-products form where each group of combined cells contributes one product term. The product term for a particular group may be found by omitting from their minterms those variables which got eliminated by the grouping process.

For the Karnaugh Map of Fig. 2.4.2, this minimization process gives rise to three groups of combined cells as shown in Fig. 2.4.3. The product terms for these groups are

Group I $\bar{D}\bar{B}$

Group II $\bar{D}C$

Group III $D\bar{B}$

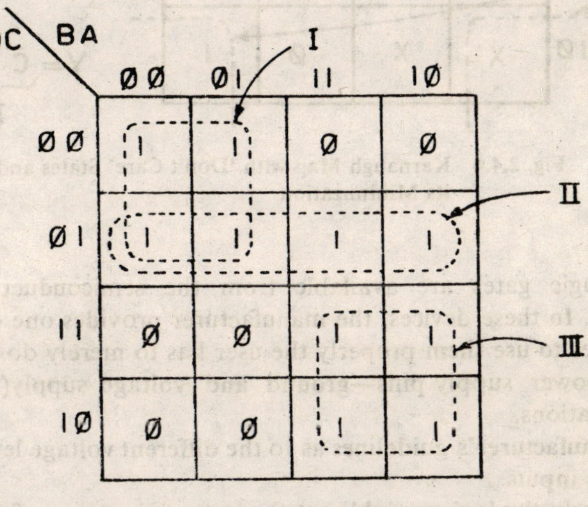

Fig. 2.4.3 Grouping for Minimization

This yields

$$Y = \bar{D}\bar{B} + \bar{D}C + D\bar{B}$$

as the minimized sum-of-products expression for Y.

In all of the previous discussions on minimization, we have assumed that in the minterm canonical form of the expression, a minterm is either present or absent. If a minterm is present, the corresponding cell in the Karnaugh Map has 1; if it is absent, the corresponding cell has a $\emptyset$. Actually a third situation is also possible. This happens when there are 'don't care' conditions.

A particular minterm has a 'don't care' state if the value of the expression for the corresponding inputs could be either a $\emptyset$ or a 1, *as desired*. This is denoted by putting an 'X' in the corresponding cell of the Karnaugh Map. As per his convenience, and in the interests of minimization, the user may choose to put either a 1 or a $\emptyset$ (as desired) in a cell marked with an 'X' and then proceed with minimization as before. 'Don't-care' states and minimization with them are illustrated in Fig. 2.4.4. Note that in this figure the 'don't-care' states $D\bar{C}\ \bar{B}\bar{A}$ and $\bar{D}\bar{C}\ B\bar{A}$ have been assumed to be 1s while the other two ($D\bar{C}\bar{B}A$ and $DCB\bar{A}$) have been taken to be Øs by the process of minimization.

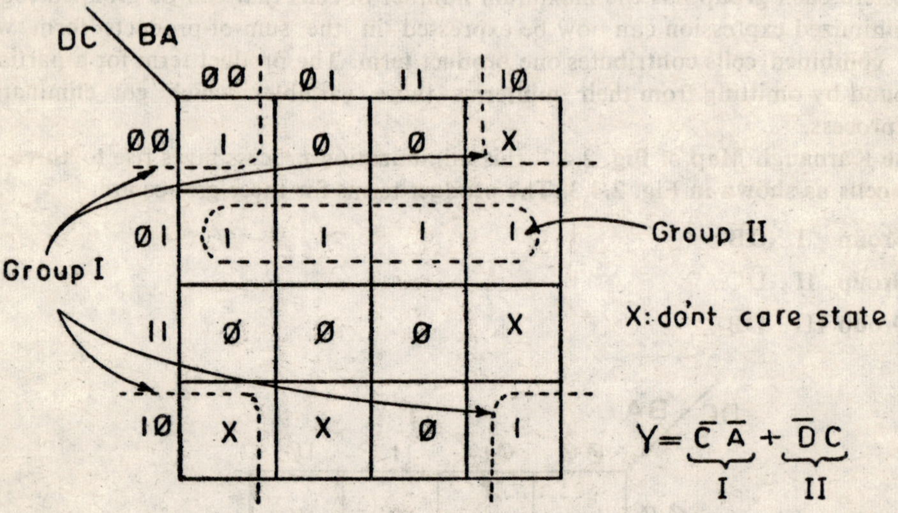

Fig. 2.4.4 Karnaugh Map with 'Don't Care' States and its Minimization

2.5 LOGIC GATES

A wide variety of logic gates are available from the semiconductor manufacturers in an integrated circuit form. In these devices, the manufacturer provides one or more logic gates in a single package. In order to use them properly the user has to merely do the following:

(a) Connect the power supply pins—ground and voltage supply(ies)—properly as per the manufacturer's specifications.

(b) Follow the manufacturer's guidelines as to the different voltage levels required to represent a 1 and a Ø at the gate inputs.

(c) Interpret properly the logic variables at the outputs in terms of the voltage levels appearing there.

(d) Ensure proper loading, i.e. that the output of a gate does not drive more inputs than the number specified by the manufacturer.

Note that in a circuit realization, different voltage levels (actually a range of voltage levels) will be used to represent the logic levels 1 and Ø. The manufacturer's specifications regarding this are—

Input:
- V_{iH} the *minimum* voltage that will be recognized as a 1 at a gate input
- V_{iL} the *maximum* voltage that will be recognized as a Ø at a gate input

Output:
- V_{oH} the *minimum* voltage that will be provided for a gate output of 1 as long as the manufacturer's loading rules are satisfied
- V_{oL} the *maximum* voltage that will be provided for a gate output of Ø as long as the manufacturer's loading rules are satisfied.

Typically $V_{oH} > V_{iH}$ and $V_{oL} < V_{iL}$, in order to provide the user a *noise margin* ($=V_{oH} - V_{iH}$ for 1, $V_{iL} - V_{oL}$ for $\emptyset$) in systems where the output of one gate is driving the inputs of other gates.

The manufacturer also provides specifications regarding the current requirements at the inputs and outputs of such gates. These are

Input: $\begin{cases} I_{iH} \\ I_{iL} \end{cases}$
- I_{iH} maximum current *sinked* by a gate input when the input is at logic 1
- I_{iL} maximum current *sourced* by a gate input when the input is at logic $\emptyset$

Output: $\begin{cases} I_{oH} \\ I_{oL} \end{cases}$
- I_{oH} maximum current *sourced* by a gate output when the output is at logic 1
- I_{oL} maximum current sinked by a gate output when the output is at logic $\emptyset$.

Typically $I_{oH} > I_{iH}$ and $I_{iL} > I_{oL}$, in order to allow a gate output to drive one or more gate inputs. The maximum number of gate inputs that a single gate output can drive (when all the gates are of the same type) is referred to as the *Fan-Out* (of that type of gates). From the above, it can be concluded that

$$\text{Fan Out} = \min\left(\frac{I_{oH}}{I_{iH}}, \frac{I_{iL}}{I_{oL}}\right)$$

There are several different semiconductor techniques available today for implementing logic gates. One would typically use such gates to design circuits for realizing particular logic expressions. For gates using a particular technology, the manufacturer would typically provide the current and voltage specifications mentioned earlier (e.g. I_{iH}, V_{oH} etc.). He also generally provides a Fan-Out figure which is applicable as long as one confines oneself to gates of the same type.

If for some reason, different types of gates (different families using different technologies) are mixed in a circuit, greater care is required (e.g. when a gate of one family is driving one or more gates of another family). The first thing to consider is that the voltages representing the logic levels should be compatible for the two families. If it is not, special interface circuitry is required for doing whatever level shifting is needed. (This information may be obtained using the specifications for V_{iH}, V_{iL}, V_{oH} and V_{oL}). Even if the required voltage compatibility is there, the user still has to consider the loading aspects by examining the I_{iH}, I_{iL}, I_{oH} and I_{oL} specifications.

As an example, consider the situation where we require a TTL 74L series gate to drive TTL 74 series gates.

For the TTL 74L series,

$$I_{oH} = 100\ \mu a,\ I_{oL} = 2\ ma$$

For the TTL 74 series,

$$I_{iH} = 40\ \mu a,\ I_{iL} = 1.6\ ma$$

For this situation,

$$\text{Fan Out} = \min\left[\frac{100}{40}, \frac{2}{1.6}\right] = \min\,[2.5, 1.25]$$
$$= 1.25$$

From the above, one can see that a 74L series gate output can only drive one 74 series gate input.

Some of the more important technologies available today for logic gates are Transistor Transistor Logic (TTL) families and the Metal Oxide Semiconductor (MOS) families. Examples of TTL series gates are 74, 74L, 74LS, 74ALS, etc. All of them are mutually compatible in terms of voltages—the manufacturer specifies $V_{iH} = 2.0v$, $V_{iL} = 0.8v$, $V_{oH} = 2.4v$ and $V_{oL} = 0.4v$ for all these families. Their current requirements are however different and special care is needed (as mentioned earlier) when interfacing gates of different families. For the most popular TTL series, the 74 series, the current specifications are—

$$I_{iH} = 40\ \mu a,\ I_{iL} = 1.6\ ma,\ I_{oH} = 800\ \mu a,\ I_{oL} = 16\ ma$$

A number of different MOS technologies are also in use today. Examples are P-channel MOS (PMOS), N channel MOS (NMOS), Complementary MOS (CMOS), Silicon on Sapphire (SOS) etc. The trend amongst manufacturers now is to ensure voltage compatibility between these devices and the TTL series specifications. Once again, the current compatibility of each device should be checked to ensure proper loading and to calculate the Fan-Out. These calculations are done in the same fashion as shown in the earlier example of a TTL 74L series gate output driving TTL 74 series gate inputs.

2.6 IMPLEMENTING LOGIC EXPRESSIONS WITH GATES

A variety of AND, OR, NAND, NOR and Inverter gates are available in integrated circuit form. These can be used by the circuit designer to implement a given logic expression. In most cases, this is a straightforward implementation following directly from the expression itself. In this section, we will consider a few examples of such implementations.

Example 1

$$Y = A\bar{B} + \bar{A}B$$

This function is used commonly enough to justify a name of its own

$$Y = A \oplus B = A\ (\text{Ex-OR})B$$
$$= A\bar{B} + \bar{A}B$$

Gates performing this function are available in integrated circuit form. The symbol for such a gate is shown in Fig. 2.6.1(a). Figs. 2.6.1(b) and (c) show two ways of implementing the Ex-OR

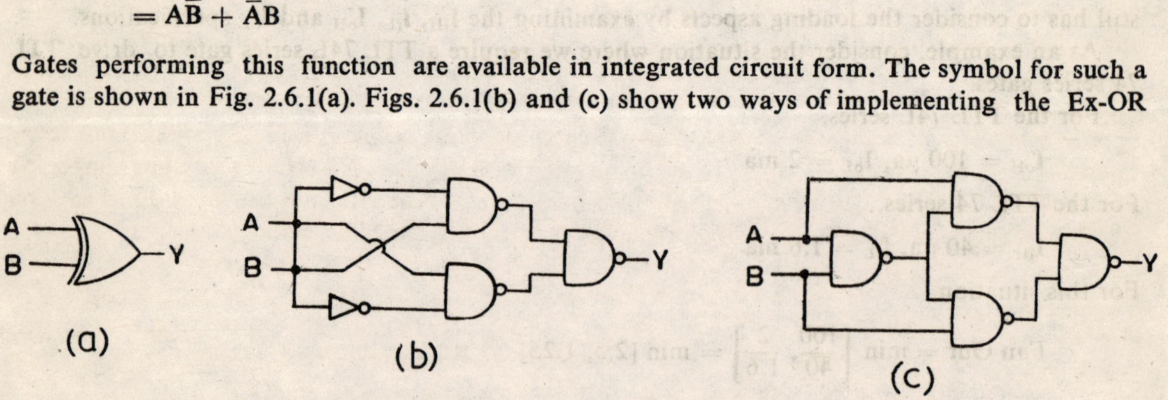

Fig. 2.6.1 Realizations of the Ex-OR Function $Y = A \oplus B = A\bar{B} + \bar{A}B$

function using other gates. The realization of Fig. 2.6.1(b) is relatively easy to see; it is a direct implementation. The realization of Fig. 2.6.1(c) is not that obvious but has the advantage of using a lesser number of gates.

The Ex-NOR function is also defined. Its definition is given below:

$$Y = \overline{A \oplus B} = A \text{ (Ex-NOR) } B$$
$$= AB + \bar{A}\bar{B}$$

Example 2

$$Y = \bar{D}B + \bar{D}\bar{C} + D\bar{B}$$

Two circuits for realizing this function are given in Fig. 2.6.2. Note that the first realization uses only 2-input NAND gates (An inverter may be realized as a NAND gate with all inputs tied together or as a NAND gate with all but one input tied to 1). The other realization uses a 3-input NAND gate.

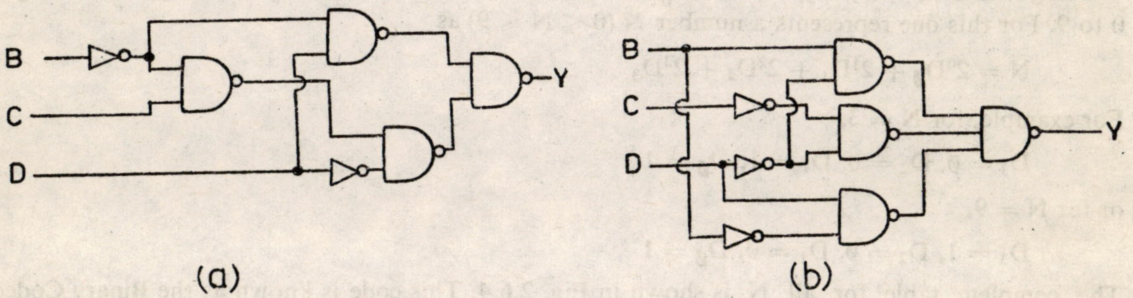

Fig. 2.6.2 Realizations of $Y = \bar{D}B + \bar{D}\bar{C} + D\bar{B}$

It would be appropriate to mention here an important historical point regarding the realization of logic expressions using gates. Previously, the most easily available logic gates were NAND gates with two or more inputs. From theory, one can show that any logic function may be implemented using NAND gates alone. A lot of design effort, therefore, went in to find the most efficient realization (least number of gates) using NAND gates. This is the reason why the realizations using only NAND gates have been given in Fig. 2.6.1 and Fig. 2.6.2.

On the contrary, a variety of logic functions are available today in an integrated circuit form. Therefore, the designer is no longer required to restrict himself to NAND-realizations alone. As a matter of fact, the best approach today is to try for the realization in the most direct fashion possible. A few more examples of logic function realizations are given below.

Example 3

$$Y = \bar{D}CA + DCA + \bar{D}\bar{C}\,\bar{B}\bar{A} + D\bar{C}\,\bar{B}\bar{A} + \bar{D}\bar{C}\,B\bar{A} + D\bar{C}\,B\bar{A} + \bar{D}\bar{C}\bar{B}A$$

By using the Karnaugh Map technique, this may be reduced to

$$Y = CA + \bar{C}\bar{A} + \bar{D}\bar{C}\bar{B}$$
$$= \overline{C \oplus A} + \bar{D}\bar{C}\bar{B}$$
$$= \overline{(C \oplus A)\cdot(D + C + B)}$$

A possible realization is shown in Fig. 2.6.3.

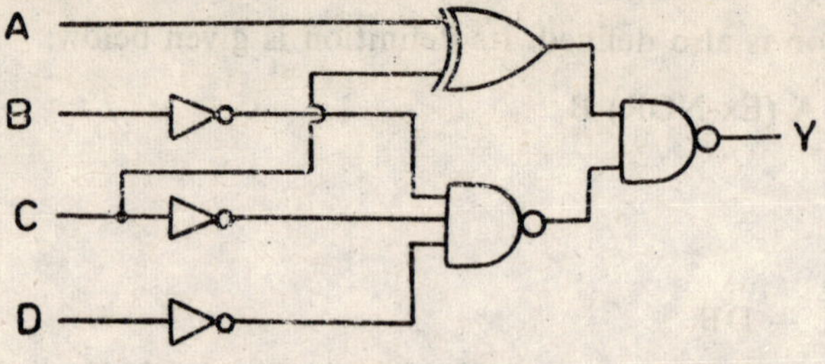

Fig. 2.6.3 Realization of $Y = \bar{D}CA + DCA + \bar{D}\bar{C}\bar{B}\bar{A} + D\bar{C}\bar{B}\bar{A} + \bar{D}\bar{C}B\bar{A} + D\bar{C}B\bar{A} + \bar{D}\bar{C}\bar{B}A$

Example 4: Excess 3-to-BCD Conversion.

Using four bits, $-D_3$, D_2, D_1 and $D_Ø$—a code can be evolved for representing decimal digits 0 to 9. For this one represents a number N ($0 \leqslant N \leqslant 9$) as

$$N = 2^0 D_Ø + 2^1 D_1 + 2^2 D_2 + 2^3 D_3$$

For example, for $N = 3$,

$$D_3 = Ø,\ D_2 = Ø,\ D_1 = 1,\ D_Ø = 1$$

or for $N = 9$,

$$D_3 = 1,\ D_2 = Ø,\ D_1 = Ø,\ D_Ø = 1$$

The complete table for all N is shown in Fig. 2.6.4. This code is known as the Binary Coded Data (BCD) representation of decimal digits.

No.	Excess 3				BCD			
N	E_3	E_2	E_1	$E_Ø$	D_3	D_2	D_1	$D_Ø$
0	Ø	Ø	1	1	Ø	Ø	Ø	Ø
1	Ø	1	Ø	Ø	Ø	Ø	Ø	1
2	Ø	1	Ø	1	Ø	Ø	1	Ø
3	Ø	1	1	Ø	Ø	Ø	1	1
4	0	1	1	1	Ø	1	Ø	Ø
5	1	Ø	Ø	Ø	Ø	1	Ø	1
6	1	Ø	Ø	1	Ø	1	1	Ø
7	1	Ø	1	Ø	Ø	1	1	1
8	1	Ø	1	1	1	Ø	Ø	Ø
9	1	1	Ø	Ø	1	Ø	Ø	1

Note: $D_3 = D_2 = D_1 = D_Ø = X$ (don't cares) for E_3, E_2, E_1, $E_Ø$ = (Ø, Ø, Ø, Ø) (Ø, ØØ1)(ØØ1Ø)(11Ø1)(111Ø)(1111)

Fig. 2.6.4 Representation of Decimal Digits using Excess 3 and BCD Codes

Another representation for decimal digits is given by the Excess 3 code. This also uses four bits—E_3, E_2, E_1, $E_\emptyset$—and represents a number N ($0 \leqslant N \leqslant 9$) according to the following rule:

$$N = 2^0E_\emptyset + 2^1E_1 + 2^2E_2 + 2^2E_2 + 2^3E_3 - 3$$

(The reason for the name Excess 3 should be obvious).
The table for this code is also given in Fig. 2.6.4.

In this example, we want to design a digital circuit which will take the inputs $E_\emptyset - E_3$ and output the BCD code, $D_\emptyset - D_3$ for that number. Using the don't care conditions noted in Fig. 2.6.4, the Karnaugh Maps for D_3, D_2, D_1 and $D_\emptyset$ can be drawn. These K-Maps and the resultant expressions are shown in Fig. 2.6.5. The circuit realization is shown in Fig. 2.6.6.

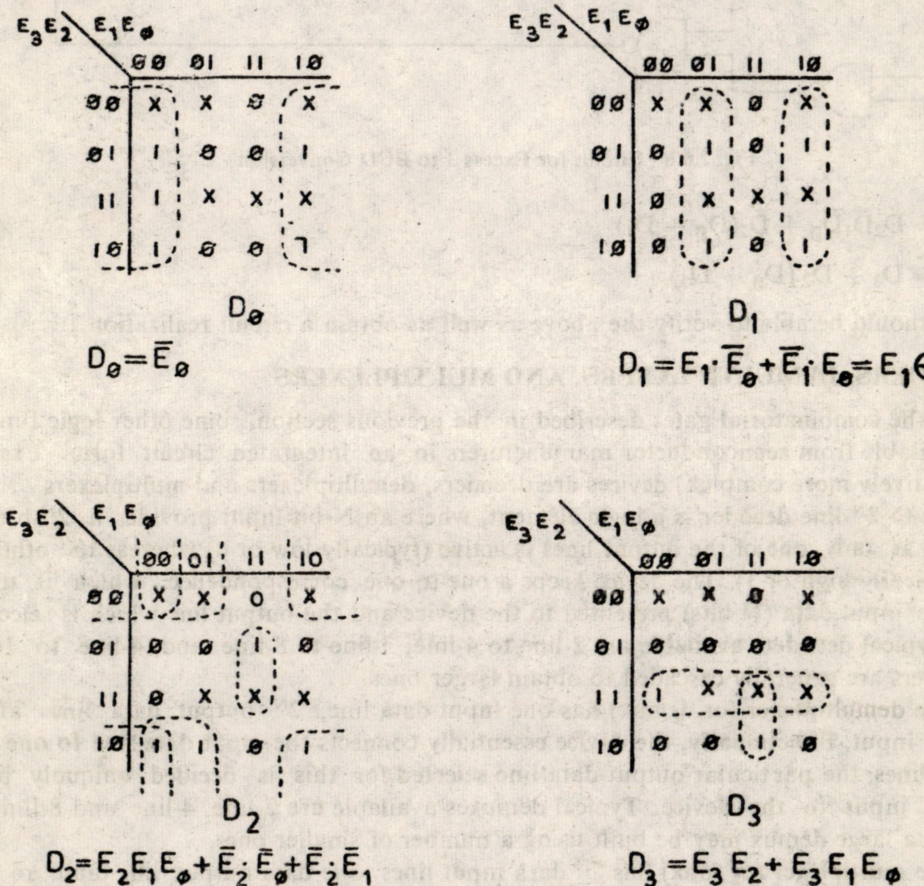

$D_\emptyset = \bar{E}_\emptyset$

$D_1 = E_1 \cdot \bar{E}_\emptyset + \bar{E}_1 \cdot E_\emptyset = E_1 \oplus E_\emptyset$

$D_2 = E_2 E_1 E_\emptyset + \bar{E}_2 \bar{E}_\emptyset + \bar{E}_2 \bar{E}_1$

$D_3 = E_3 E_2 + E_3 E_1 E_\emptyset$

Fig. 2.6.5 Karnaugh Maps for BCD data $D_\emptyset - D_3$ in terms of Excess 3 code $E_3E_2E_1E_\emptyset$

[*Note*: For BCD to Excess 3 conversion, the following expressions may be used.

$$E_\emptyset = \bar{D}_\emptyset$$

$$E_1 = \overline{D_\emptyset + D_1}$$

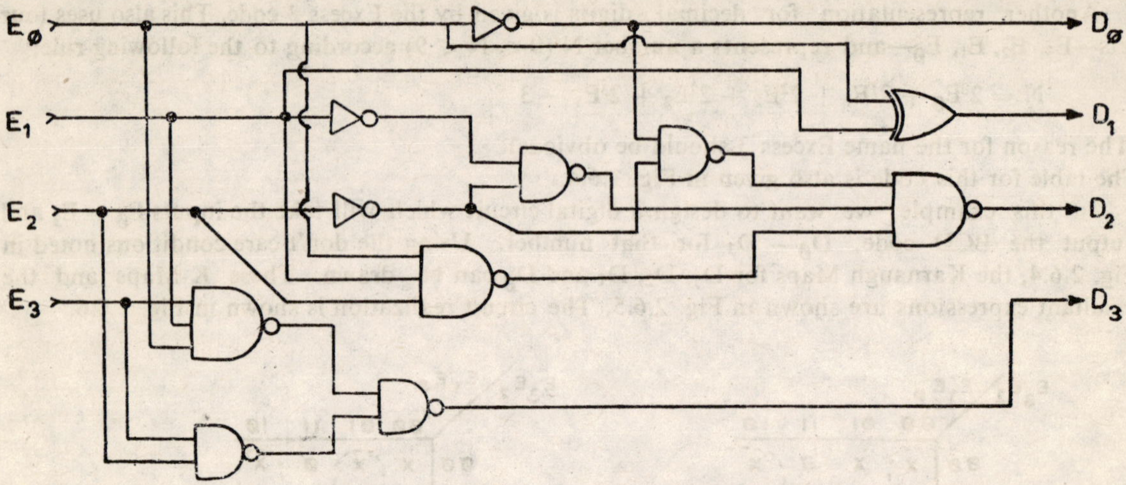

Fig. 2.6.6 Circuit for Excess 3 to BCD Conversion

$$E_2 = D_2\overline{D}_1\overline{D}_\emptyset + \overline{D}_2(D_\emptyset + D_1)$$

$$E_3 = D_3 + D_2(D_\emptyset + D_1)$$

The reader should be able to verify the above as well as obtain a circuit realization.]

2.7 DECODERS, DEMULTIPLEXERS, AND MULTIPLEXERS

Apart from the combinatorial gates described in the previous section, some other logic functions are also available from semiconductor manufacturers in an integrated circuit form. Examples of such (relatively more complex) devices are decoders, demultiplexers and multiplexers.

A N-line to 2^N-line decoder is a logic element, where an N-bit input provides a 2^N-bit output such that only one of the output lines is active (typically low or ∅) whereas the others are inactive (typically high or 1). The device keeps a one-to-one correspondence, which is unique, between the input data (N bits) presented to the device and the output line which is selected to be active. Typical decoders available are 2-line to 4-line, 3-line to 8-line and 4-line to 16-line. Small decoders are generally cascaded to obtain larger ones.

A 2^N-line demultiplexer (or demux) has one input data line, 2^N output data lines and an N-bit select input. Functionally, the device essentially connects the input data line to one of the output data lines; the particular output data line selected for this is decided uniquely by the N-bit select input to the device. Typical demuxes available are 2-line, 4-line and 8-line ones. Once again, a large demux may be built using a number of smaller ones.

A 2^N-line multiplexer (or mux) has 2^N data input lines, one data output line and an N-bit select input. Functionally, this device connects a selected input data line to the output data line; the particular input data line selected is uniquely decided by the N-bit select input. Typical muxes available are 2-line, 4-line and 8-line ones. Larger muxes may be built by cascading smaller ones. Note that from a functional view point, a mux does exactly the opposite of what the demux does and vice versa.

Examples of integrated-circuit decoders, demultiplexers and multiplexers can be found in Chapter 5.

2.8 REPRESENTATION OF NUMBERS AND ARITHMETIC OPERATIONS

In Example 4 of Sec. 2.6 we showed the examples of codes—BCD and Excess 3—that may be used to represent decimal digits (0 to 9) using four binary bits. In general a number N lying between 0 and $(2^K - 1)$ may be represented using K bits $D_{K-1}, \ldots, D_{\emptyset}$ such that

$$N = \sum_{i=0}^{K-1} 2^i D_i \text{ for } 0 \leqslant N \leqslant (2^K - 1)$$

The sequence of bits $D_{K-1} \ldots D_{\emptyset}$ is then referred to as the binary representation of the number N.

A special case of such representation occurs when K = 4. In that case, the four bits $D_3 \ldots D_{\emptyset}$ represent the numbers 0 to 15 as shown in Fig. 2.8.1. This Hexadecimal (or HEX) representation of numbers 0–15 (or HEX digits Ø, 1, 2, 3, 4, 5, 6, 7, 8, 9, A, B, C, D, E, F) is often used in microcomputer terminology as a compact representation of four binary bits (or a nibble). A sequence of eight binary bits (or a byte) may be similarly represented using a sequence

Decimal Number	HEX Represen-tation	D_3	D_2	D_1	$D_{\emptyset}$
0	Ø	Ø	Ø	Ø	Ø
1	1	Ø	Ø	Ø	1
2	2	Ø	Ø	1	Ø
3	3	Ø	Ø	1	1
4	4	Ø	1	Ø	Ø
5	5	Ø	1	Ø	1
6	6	Ø	1	1	Ø
7	7	Ø	1	1	1
8	8	1	Ø	Ø	Ø
9	9	1	Ø	Ø	1
10	A	1	Ø	1	0
11	B	1	Ø	1	1
12	C	1	1	Ø	Ø
13	D	1	1	Ø	1
14	E	1	1	1	Ø
15	F	1	1	1	1

Fig. 2.8.1 Representation of a 4-bit Hexadecimal (HEX) Number

of two Hexadecimal digits.

Coming back to the representation of a general number N lying between 0 and $(2^K - 1)$, recall that K bits were needed to represent N. This is true, if N is always positive—as it is, if it is constrained to lie between 0 and $(2^K - 1)$. A somewhat different approach needs to be taken for representing numbers which are negative.

Consider the situation where the number N is to be represented using binary bits such that

$$-(2^K - 1) \leqslant N \leqslant (2^K - 1)$$

Let $D_K, D_{K-1} \ldots D_1, D_\emptyset$ be the bit sequence needed for representation. There are three common ways in which the signed number N can be represented.

2.8.1 Sign Magnitude Representation

In this representation, the most significant bit D_K represents the sign, i.e.

$$\begin{aligned} D_K &= \emptyset \quad \text{if } N \geqslant 0 \\ &= 1 \quad \text{if } N < 0 \end{aligned}$$

The bit sequence $D_{K-1} \ldots D_\emptyset$ represents the magnitude of N, i.e.

$$|N| = \sum_{i=0}^{K-1} D_i 2^i$$

For example, if $K = 7$, then

$$+73 = \emptyset 1 \emptyset\emptyset 1 \emptyset\emptyset 1$$

$$-73 = 11 \emptyset\emptyset 1 \emptyset\emptyset 1$$

With $K + 1$ bits $(D_K \ldots D_\emptyset)$, one can then represent numbers between $-(2^K - 1)$ and $+(2^K - 1)$.

2.8.2 1's Complement Representation

Representation of a positive number is the same in both the Sign Magnitude and 1's complement form. A negative number in 1's complement form is the exact complement of the corresponding positive number as shown in the example given below:

$$+73 = \emptyset 1 \emptyset\emptyset 1 \emptyset\emptyset 1$$

$$-73 = 1 \emptyset 11 \emptyset 11 \emptyset$$

Note that the most significant bit carries information regarding the sign of the number, i.e. it is Ø for +ive numbers and 1 for −ive numbers. It should also be noted that (including the sign bit) this representation requires $(K + 1)$ bits to represent numbers in the range of $-(2^K - 1)$ to $+(2^K - 1)$.

Consider a number +N which (including the sign bit) requires $(K + 1)$ bits—$D_K \ldots D_\emptyset$—for representation in the 1's complement form. Let $D'_K \ldots D'_\emptyset$ be the representation of the number −N in this form. Defining binary addition of two bits A and B with a Carry In bit (C_{in}) as generating the result Y with a carry Out (C_{out}) bit as shown in Fig. 2.8.2, we can show that

$$(D_K, \ldots, D_\emptyset) + (D'_K \ldots D'_\emptyset) = M_K, \ldots M_\emptyset$$

where

$$M_i = 1 \text{ for } i = 0, \ldots K$$

In other words, adding, a number and its 1's complement sign inverse gives rise to an all 1s number.

2.8.3 2's Complement Representation

Representation of a positive number in 2's Complement Representation is the same as in the 1's Complement and Sign Magnitude representations. However, the representation of a negative number is different and is given below.

Consider a situation where K + 1 bits—$D_K \ldots D_\emptyset$—are available for representing a number N in the range of $-(2^K - 1)$ to $+(2^K - 1)$ (If N is positive the representation is the same as in the earlier representations, i.e. $D_K = \emptyset$, and $D_{K-1} \ldots D_\emptyset$ such that $N = \sum_{i=0}^{K-1} 2^i D_i$). In order to find the representation of a negative number, i.e. $-N$, we proceed as follows.

(i) Find the representation of the number +N.

(ii) Complement all bits in the representation.

(iii) Add 1 to the resulting representation obtained in (ii), following the rules for unsigned addition given in Appendix A2.1.

The result is the representation of −N.

Example (a) What is the representation of −73? ($k = 7$)

Step (i) +73 = ∅1∅∅1∅∅1

(ii) 1∅11∅11∅

(iii) −73 = 1∅11∅111

(b) What is the representation of −22? ($k = 7$)

Step (i) +22 = ∅∅∅1∅11∅

(ii) 111∅1∅∅1

(iii) −22 = 111∅1∅1∅

The algorithm given above for finding representation of a negative number in 2's complement form is the one usually followed by computers. It is also possible to find the 2's complement form of a negative number −N in the following way:

(i) Find the representation of +N.

(ii) Starting with the Least Significant Bit (LSB) do the following. If it is a ∅, keep it as it is and go to the next bit. Continue till a 1 is encountered; then complement the next and all succeeding bits.

The algorithm given above is generally used if one wants to evaluate the 2's Complement Representation of a number by inspection. The algorithm given earlier is most commonly used when one wants to implement a 2's Complementer by hardware or software.

2.8.4 Addition and Subtraction of Signed Binary Numbers

Addition of unsigned (or +ive) binary numbers is a simple operation explained with examples in Appendix A2.1. As shown in the appendix, this operation essentially consists of adding bit i of one number to bit i of the other number along with the Carry In to bit i thereby generating bit i of the result and a Carry Out to bit $(i + 1)$; this process is repeated for each bit position starting from the Least Significant Bits (LSBs) of the two numbers and ending with the Most Significant Bits (MSBs). Addition of two k bit numbers in this case, will generate at the most a $(k + 1)$ bit result. The Carry Out of the addition of the MSBs constitutes the MSB, or the $(k + 1)$th bit, of the result. Note that if the addition requires that the result should also be representable by k bits, then a Carry Out of 1 from the final bit position indicates an overflow condition.

For signed (+ive or −ive) binary numbers, the processes of addition or subtraction of two numbers are very similar to each other. This is because

$$A + (-B) = A - B$$

Therefore, without any loss of generality we will concentrate next on the addition of two signed binary numbers. For purposes of discussion we will assume that each of the operands is represented by $(k + 1)$ bits, i.e. $D_K \ldots D_0$, where the MSB D_K is the sign bit. We further assume that the result also has $(k + 1)$ bits, i.e. $Y_k \ldots Y_\emptyset$, with Y_k as the sign bit of the result.

Consider the operations A + B or A − C, where A, B and C are binary numbers using *1's Complement Representation.* The following algorithm outlines how the above operations may be executed.

(i) If (A + B), go to (ii). If (A − C), find the 1's complement representation of −C and let that be represented by B, i.e. B = −C

(ii) Add A and B (including the sign bit) just as one would add unsigned binary numbers. Let Y′ be the $(k + 1)$ bit result and CY the Carry Out of the sign bit position.

(iii) If CY = ∅, Y = Y′ and go to step (iv). If CY = 1 add CY to Y′ (i.e. add ∅∅ . . . ∅1 to Y′) and let Y be the result [EAC or End Around Carry operation].

(iv) Overflow if (a) $A_k = \emptyset$, $B_k = \emptyset$ but $Y_k = 1$

or

(b) $A_k = 1$, $B_k = 1$ but $Y_k = \emptyset$. Else, Y is the signed result in 1's complement form with Y_k = Sign Bit.

Examples $k = 4$

(A) 12 + 14

```
  ∅11∅∅  (12)
 +∅111∅  (14)
 ------
  11∅1∅
OVERFLOW
```

(B) 12 + 3

```
  ∅11∅∅  (12)
 +∅∅∅11  (3)
 ------
  ∅1111  (+15)
```

(C) 12 − 8   Ø11ØØ (12)
+1Ø111 (−8)
ØØØ11
1
ØØ1ØØ (+4)

(D) 8 − 12   Ø1ØØØ (8)
+1ØØ11 (−12)
11Ø11 (−4)

When numbers A, B, and C are represented in *2's Complement Form*, (A + B) or (A − C) may be found by a process similar to that followed above (for 1's Complement) except that the EAC operation is not needed. For the 2's Complement form, the above operations may be performed by following the algorithm given below.

(i) If (A + B), go to step (ii). If (A − C), find −C in 2's Complement form and denote it by B; i.e. B = −C.

(ii) Add A and B including the sign bit. Let Y be the $(k + 1)$ bit result and ignore the Carry Out of the sign bit position.

(iii) Overflow if (a) $A_k = Ø$, $B_k = Ø$ but $Y_k = 1$

or (b) $A_k = 1$, $B_k = 1$ but $Y_k = Ø$

else Y is the result with Y_k = Sign Bit.

Examples $k = 4$

(A) 4 + 5   00100 (+4)
+ØØ1Ø1 (+5)
Ø1ØØ1 (+9)

(B) −4 − 5   111ØØ (−3)
+11Ø11 (−5)
Carry Out = 1 (Ignore)   1Ø111 (−9)

(C) −14 + 12   1ØØ1Ø (−14)
Ø11ØØ (+12)
1111Ø (−2)

(D) 14 − 12   Ø111Ø (14)
1Ø1ØØ (−12)
ØØØ1Ø (+2)

The 2's Complement form is preferred in arithmetic operations for computers. This is because the extra EAC operation in 1's complement form slows down the execution speed. The algorithm needed for signed binary addition may also be derived as above, for the Sign Magnitude Representation. This is, however, much more complex which is the reason why it is not very popular for use in arithmetic operations.

2.8.5 Implementing Binary Arithmetic by Hardware

The basic piece of hardware needed to implement signed binary addition or subtraction is an adder. This is because we can implement subtraction as 'addition with changed signs' and because inverting a signed number requires only a logical complementation (1's complement) or logical complementation and addition (2's Complement).

Consider the 1-bit addition table shown in Appendix A2.1 for generating a 1-bit result Y_i and the Carry Out $C_{out\,(i)}$ by adding two 1-bit numbers A_i and B_i. One can show that

$$Y_i = A_i \oplus B_i = A_i\bar{B}_i + \bar{A}_iB_i$$

$$C_{out\,(i)} = A_i \cdot B_i$$

A circuit which takes A_i and B_i as inputs and generates Y_i and $C_{out\,(i)}$ is referred to as a *Half-Adder*. The internal circuit of such a half adder is fairly simple and is shown in Fig. 2.8.2.

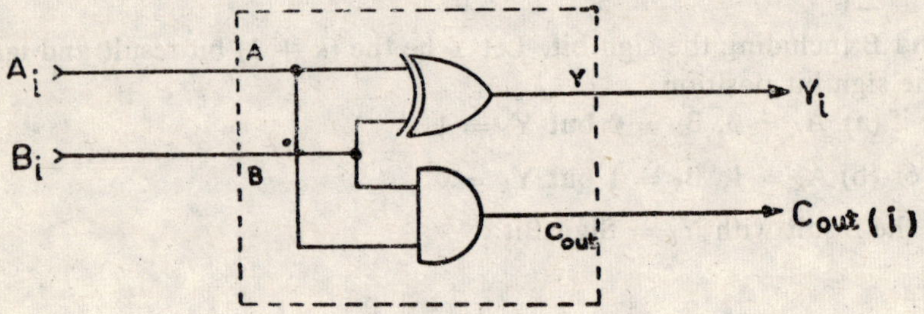

Fig. 2.8.2 Realizing a Half Adder

A Half-Adder does not provide a Carry In $C_{in\,(i)}$ input. A circuit which does that is referred to as a Full-Adder. This can be realized using two half-adders as shown in Fig. 2.8.3. Note

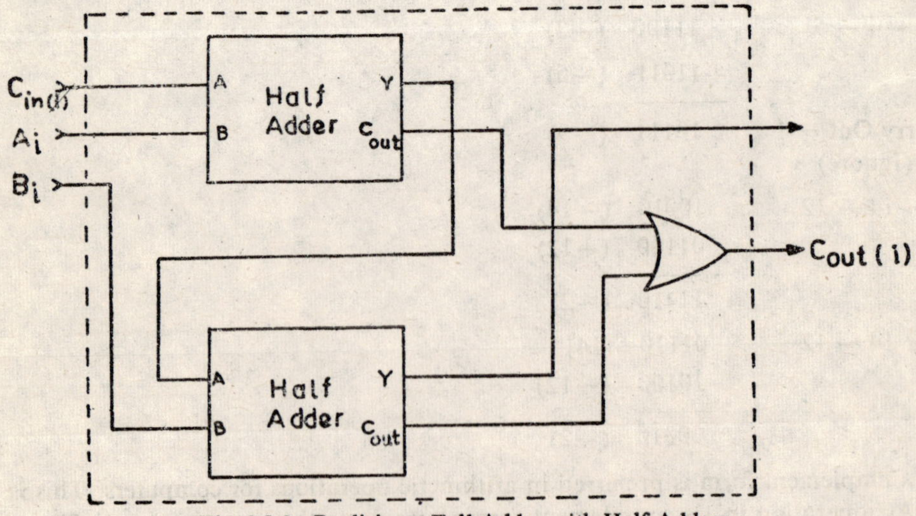

Fig. 2.8.3 Realizing a Full-Adder with Half-Adders

that the inputs to a 1-bit full adder are the two data inputs A_i and B_i and the Carry In $C_{in(i)}$. It, in turn, generates the result bit Y_i and a Carry Out. A number of such Full Adders i.e. N, may be cascaded to obtain a N-bit Adder. This arrangement is shown in Fig. 2.8.4. In this circuit,

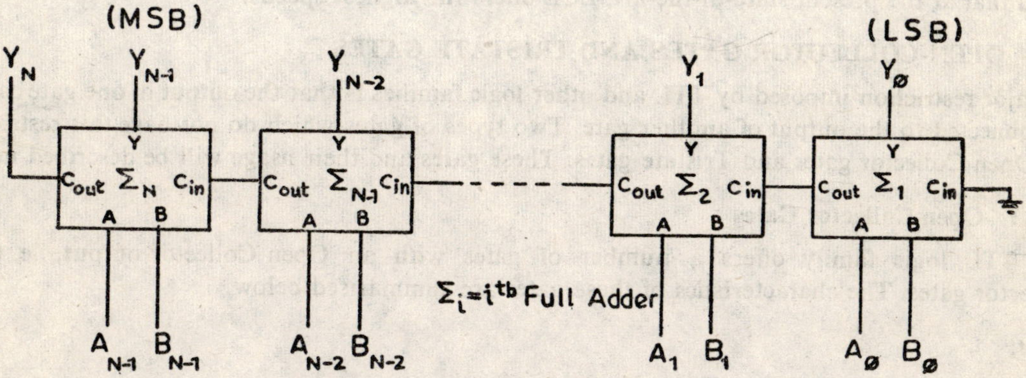

Fig. 2.8.4 N-bit Adder adding N-bit numbers $(A_{N-1} \ldots A_Ø)$, $(B_{N-1} \ldots B_Ø)$ and generating (N + 1)-bit result $(Y_N \ldots Y_Ø)$

note that the Carry In to the LSB stage is grounded and that the Carry Out of the MSB stage is the (N + 1)th bit—the actual MSB—of the result. A number of such adders are available as integrated circuits in the TTL family.

2.9 SPEED CONSTRAINTS ON COMBINATORIAL LOGIC DESIGN

It is unfortunately true that the response of logic circuits to various inputs is not instantaneous. Manufacturers typically provide data on the worst case propagation delay 't_{pd}' for a change in inputs to propagate to the output. When one realizes a logic function using such gates, one has to use these specifications to determine the amount of time after which the output will be valid, once valid inputs have been provided. Though this figure will really vary from gate to gate in the same logic family, typically a thumbrule figure for each gate is used to estimate the response time of a circuit using such gates. For TTL 74 series logic circuits, this typical gate delay is taken to be about 20–30 ns.

Using this Thumb-rule figure for the gate delay, it is fairly easy to examine a combinatorial circuit to find out the worst case response time one may expect. If a maximum of N gates lie between an input and an output, then the worst case response time is Nt_{pd}. As an example, consider the circuit of Fig. 2.6.2(a). The value of N for this case is 4 for the signal to propagate from input to output (i.e. from the B input to the Y output). The worst case response time for this logic circuit is then $4t_{pd}$. If TTL 74 series gates are used, one would then expect a worst case response time of 120 ns. This will be the time taken for a change in the inputs to propagate to the output.

It should be noted that provided one allows enough time for the transients to settle down (i.e. the worst case response time) the speed constraint does not appear for slowly varying inputs. (Of course, one might still get a wrong result if one examines the output during that settling down period). The speed constraint becomes important when the input changes rapidly. In such situations the worst case response time puts a limit (i.e. $1/Nt_{pd}$ per second) on how fast such inputs can change. The designer should take these into consideration when designing logic circuits. Speed improvements may be obtained by minimizing the logic function itself as well as by minimizing the 'cascading of gates' in the realization of the logic function. If still higher speeds are required, there may be no choice but to go for a logic family which provides higher speeds e.g. TTL 74S, 74LS or 74ALS series or the ECL (Emitter Coupled Logic) family. It should be noted that at the present state-of-the-art, ECL offers the highest speeds.

2.10 OPEN-COLLECTOR GATES AND TRISTATE GATES

A major restriction imposed by TTL and other logic families is that the output of one gate cannot be connected to the output of another gate. Two types of gates which do not have this restriction are Open Collector gates and Tristate gates. These gates and their usage will be described next.

2.10.1 Open Collector Gates

The TTL logic family offers a number of gates with an Open Collector output, i.e. Open Collector gates. The characteristics of these gates are summarised below.

Input:

$I_{iH} = 40\ \mu a$

$I_{iL} = 1.6$ ma

Input logic 1: $V_{iH} = 2.0$ V

Input logic Ø: $V_{iL} = 0.8$ V

These are identical to the specifications for normal 74 series TTL

Output:

For logic Ø: $V_{oL} = 0.4$ V

$I_{oL} = 16$ ma

For logic 1: Output is floating with a high impedance

$I_{oH} = 250\ \mu a$ (sinked by the gate output)

Notice that the only situation where these gates differ from normal 74 series TTL is that the logic 1 output is a high impedance floating state rather than a low impedance driver outputting a voltage of at least 2.4 V.

Using these gates in a logic circuit, requires the designer to place a resistance R_L at its output point connecting it to V_{CC} (=5 V). The value of R_L has to be properly chosen depending on the circuit being used. This is explained for various situations below.

Consider an Open Collector TTL inverter by itself, i.e. not driving any other gate, as shown in Fig. 2.10.1. Here the choice of R_L is determined by the current $I_L = \dfrac{5 - V_o}{R_L}$ that this gate output will sink when the output is a logic Ø. For this situation, the specifications state $I_{oL} = 16$ ma. Therefore

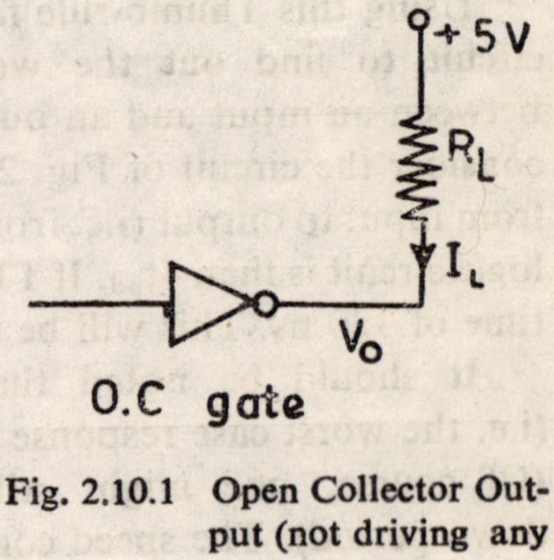

Fig. 2.10.1 Open Collector Output (not driving any other circuit)

$$\frac{5 - V_{oL}}{R_L} < 16 \text{ ma}$$

or $\quad R_L > 290\ \Omega$

[For $R_L < 290\ \Omega$, the I_{oL} specification is violated.]

Consider now the case of Fig. 2.10.2 where N open collector outputs have been tied together through a resistor R_L to +5 V with the tied output driving η TTL inputs. This kind of a connection with open collector outputs—i.e. outputs tied together and connected to +5 V through a pull up resistor R_L—is known as *Wire-AND*ing.

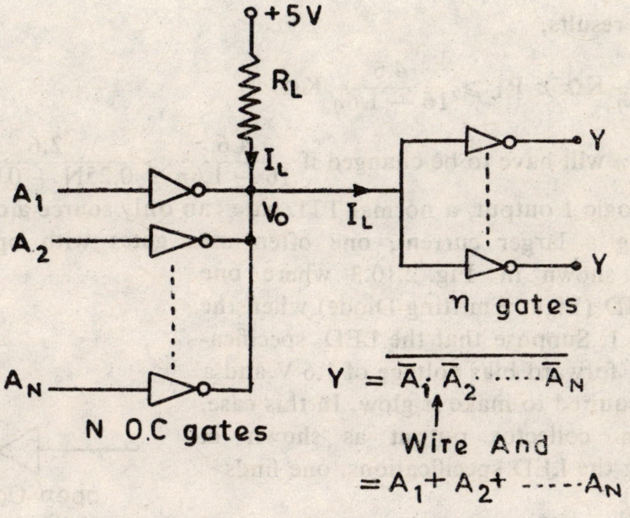

Fig. 2.10.2 Wire Anding Open Collector Outputs

Consider a situation where one or more of the open collector outputs are at logic Ø whereas the others are at logic 1. The resulting V_o will still be Low implying that the tied output will be a logic Ø. This is essentially an ANDing operation on the output logic levels of the open collector gates. This is the reason why such a connection is known as Wire AND. (Note that this can be done only with open collector outputs).

The value of R_L in Fig. 2.10.2 has to be correctly chosen to ensure that V_o falls in the proper range, i.e. 0–0.4V for Ø and 2.4V–5V for 1. As the calculations given below show, this indicates the range of R_L which can be chosen.

Consider V_o Low with one open collector output (say output *i*) at Ø and the other outputs at 1. (This is the worst case.) Output *i* can only sink 16 ma. Therefore

$$I_L + \eta I_{iL} \leqslant 16 \text{ ma}$$

$$I_L = \frac{5 - V_{oL}}{R_L} = \frac{4.6}{R_L}$$

$$\frac{4.6}{R_L} + 1.6\eta \leqslant 16$$

$$R_L \geqslant \frac{4.6}{16 - 1.6\eta} \text{ K}\Omega$$

Consider the case where V_o is High, implying that all the open collector outputs are at their logical 1 level. Now

$$I_L \geqslant NI_{oH} + \eta I_{iH}$$

but $$I_L = \frac{V_{CC} - V_{oH}}{R_L} = \frac{5 - 2.4}{R_L} = \frac{2.6}{R_L}$$

Hence $\frac{2.6}{R_L} \geqslant 0.25N + 0.04\eta$

$$R_L \leqslant \frac{2.6}{0.25N + 0.04\eta} \text{ K}\Omega$$

Combining the above results,

$$\frac{2.6}{0.25N + 0.04\eta} \text{ K}\Omega \geqslant R_L \geqslant \frac{4.6}{16 - 1.6\eta} \text{ K}\Omega$$

[Obviously N and/or η will have to be changed if $\frac{4.6}{16 - 1.6\eta} > \frac{2.6}{0.25N + 0.04\eta}$]

Recall that for a logic 1 output, a normal TTL gate can only source a current of 400 μa. In applications requiring a larger current, one often uses gates with open collector outputs. Consider the circuit shown in Fig. 2.10.3 where one wants to light an LED (Light Emitting Diode) when the gate output is a logic 1. Suppose that the LED specifications are such that a forward bias voltage of 1.6 V and a current of 10 ma is required to make it glow. In this case, one needs an open collector output as shown in Fig. 2.10.3. Following the LED specifications, one finds

$$R_L = \frac{5 - 1.6}{10} = 340 \ \Omega$$

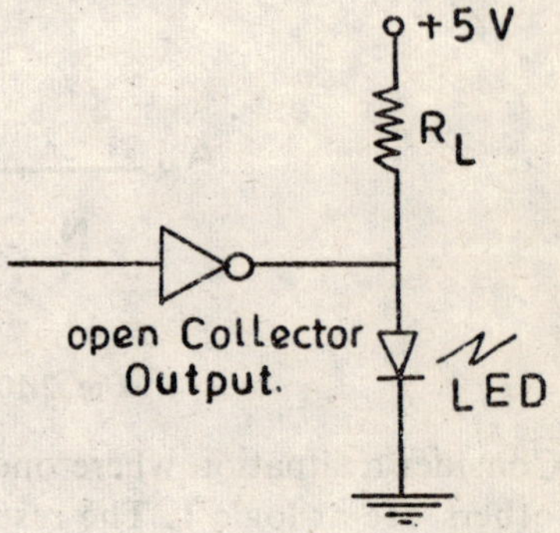

Fig. 2.10.3 Open Collector Output Driving a LED

2.10.2 Tristate Gates

A tristate gate (or more properly a gate with tristate output) essentially provides three output states. Two of these are the normal TTL logic 1 and Ø outputs; the third is a high impedance state where the output is essentially floating with a high impedance. Apart from the inputs defining the logic output of the gate, a tristate gate also requires a Select Input. This Select Input decides whether the gate output will be a true logic level (1 or Ø, depending on the inputs) or it will be in the high impedance state.

A number of tristate gate outputs can be connected together (*without a pull-up resistor*) to provide a common output point 'Y' as shown in Fig. 2.10.4. As shown in this figure, Y can be made to reflect the logic output of a selected gate by driving its select input with '1' and deselecting (select input Ø) the other gates. Y will stay in the High Impedance state if none of the gates are selected. As shown in Fig. 2.10.4, the user should ensure that more than one tristate outputs

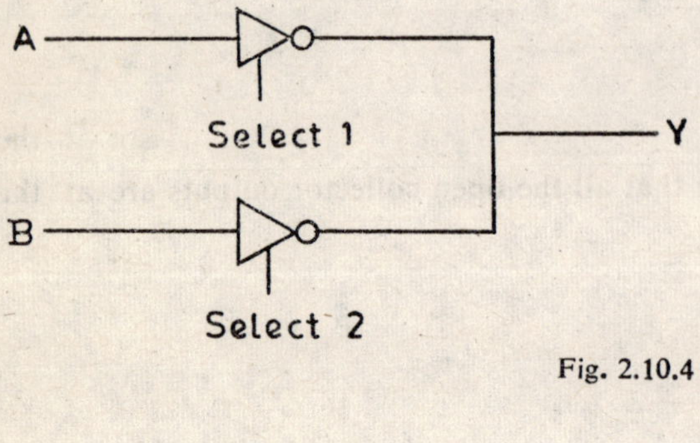

$Y = \overline{A}$ if Select 1 = 1 Select 2 = Ø

$= \overline{B}$ if Select 1 = Ø Select 2 = 1

= High Impedance if Select 1 = Select 2 = Ø

Select 1 = Select 2 = 1 ILLEGAL

Fig. 2.10.4 Tristate Gates

are not simultaneously enabled. This leads to 'conflicts' and is declared to be illegal by the manufacturers of such gates. As long as this condition is followed, the outputs of a number of tristate gates may be connected together.

PROBLEMS

1. For each of the logic expressions given below find its dual

   (a) $Y = A\bar{B} + \bar{A}B$ (b) $Y = AB + \bar{A}\bar{B}$

   (c) $Y = ABC + B\bar{C} + \bar{A}\bar{B}\bar{C}$ (d) $Y = A + \bar{B}\bar{C} + C$

   Also write the truth tables for the above.

2. Find the Minterm Canonical Forms for the following:

   (a) $Y = ABC + \bar{B}\bar{C} + \bar{A}\bar{B}\bar{C}$ (b) $Y = A + \bar{B}\bar{C} + C$

   (c) $Y = AB + BC$ (d) $Y = AB\bar{C} + AB + BC + \bar{A}BC$

3. Find the Maxterm Canonical Forms for the following:

   (a) $Y = (A + B)(\bar{A} + \bar{B} + \bar{C})$ (b) $Y = (\bar{A} + B)(\bar{B} + C)(\bar{C} + A)$

   (c) $Y = (A + \bar{B})(A + B + C)(\bar{A} + C)$ (d) $Y = (A + C)(B + C)(\bar{B} + \bar{C})$

4. Minimize the following using Karnaugh Maps:

   (a) $Y = \bar{D}\bar{C}\bar{B} + D\bar{C}\bar{B} + C\bar{B}A + \bar{C}B\bar{A}$ (b) $Y = \bar{C}A + C\bar{B}\bar{A} + CB\bar{A}$

   (c) $Y = \bar{D}\bar{C}\bar{B}\bar{A} + \bar{D}C\bar{B} + DC\bar{B} + D\bar{C}\bar{B}\bar{A} + CB\bar{A}$

5. Repeat (4) using 'don't cares' for the minterms $\bar{D}CBA$ and DCBA.

6. Design logic circuits to implement the expressions obtained in (4) and (5). First try a design using NAND, NOR, AND, OR, and NOT gates and then repeat the design using only NAND gates.

7. Using 8 bits ($D_7 \ldots D_\emptyset$) obtain a representation of the following numbers in Sign Magnitude, One's complement and Two's complement representation, +70, −110, +25, −13, +96, −65, +115, −5.

8. Carry out the following arithmetic operations using a 8-bit 1's complement and 2's complement representations:

   (a) 70 − 110 (b) −110 + 25

   (c) −25 − 13 (d) 25 + 96

   (e) −13 − 110 (f) 115 − 65

   (g) 5 + 10 (h) −110 + 96

APPENDIX A2.1

Addition of Unsigned Binary Numbers

Adding two one bit numbers A_i and B_i generates a two bit result. Let Y_i be the Least Significant Bit (LSB) of the result and $C_{out\,(i)}$ be the Most Significant Bit. The rules for this addition process are summarized below:

A_i	B_i	Y_i	$C_{out\,(i)}$
Ø	Ø	Ø	Ø
Ø	1	1	Ø
1	Ø	1	Ø
1	1	Ø	1

It is usual to refer to Y_i as the result of the adddition and $C_{out\,(i)}$ as the Carry Out.

A more general case than the above is one where two bits A_i and B_i are added with a Carry In bit $C_{in\,(i)}$ generating a result bit Y_i and a Carry Out bit $C_{out\,(i)}$. The rules for this addition process are given below

$C_{in\,(i)}$	A_i	B_i	Y_i	$C_{out\,(i)}$
Ø	Ø	Ø	Ø	Ø
Ø	Ø	1	1	Ø
Ø	1	Ø	1	Ø
Ø	1	1	Ø	1
1	Ø	Ø	1	Ø
1	Ø	1	Ø	1
1	1	Ø	Ø	1
1	1	1	1	1

(Note that the above table corresponds to that of adding three one bit numbers $C_{in\,(i)}$, A_i, B_i generating a two-bit result $C_{out\,(i)}$, Y_i with Y_i as the LSB and $C_{out\,(i)}$ as the MSB.)

Using the above, one can now present the technique of adding two N bit numbers A and B where A and B are represented as $A_{N-1} \ldots A_Ø$, and $B_{N-1} \ldots B_Ø$ respectively. Let Y be the $(N + 1)$ bit result represented as $Y_N \ldots Y_Ø$. The algorithm for carrying out this addition is given below

(i) $C_{in\,(Ø)} = Ø$, $i = 0$

(ii) Add A_i, B_i and $C_{in\,(i)}$ to obtain Y_i and $C_{out\,(i)}$

(iii) $i = i + 1$

(iv) If $i = N$, go to step (vii)

(v) $C_{in\,(i)} = C_{out\,(i-1)}$

(vi) Go to step (ii)

(vii) $Y_N = C_{out\,(i-1)}$

Consider the following examples.

Examples

1.	A	ØØ1Ø11Ø1
	B	1Ø11Ø111
	$C_{out\,(i)}$	ØØ111111
	Y	111ØØ1ØØ
2.	A	1Ø11Ø11Ø
	B	11ØØ1ØØ1
	$C_{out\,(i)}$	1ØØØØØØØ
	Y	Ø1111111

Note that in this case there is a carry out of the MSB location; this could be an overflow error condition.

Chapter 3

Semiconductor Memory Elements

In the previous chapter, we have described one important building block in digital systems, i.e. combinatorial gates. We have demonstrated how complex logic expressions generating one or more outputs from a number of inputs may be realized using such gates. A common feature of all these circuits was that the outputs were valid only when all inputs were simultaneously valid (may be with some delay from the inputs to the outputs because of the various gate delays). None of the circuits used (or required) memory elements, i.e. an element which will remember the previous value of a logic variable (input, output or an intermediate variable) even though its current value is different. Circuits using such memory elements are very often referred to as Sequential Circuits. This chapter discusses the basic digital building blocks available for building sequential circuits.

3.1 FLIP-FLOPS AND THEIR FUNCTIONS

The basic memory element in a digital system is a Flip-Flop (FF). A flip-flop is a digital circuit with two states which satisfies the following properties:

(a) If left 'undisturbed', it will stay in whatever state it originally had.
(b) It has inputs which may be used to switch it from one state to the other.
(c) It has one or more outputs which are indicative of the state the flip-flop is in.

As we shall see later (Sec. 3.3), there are various kinds of flip-flops. These differ from each other basically in the way in which the input (or inputs) affects its state. In terms of their functions, however, they show a lot of common features.

The basic function of flip-flops is to store data. They can be used to keep a record of what value a variable (input, output or intermediate) had in the past as well as store the current value of a variable for later use. Flip-flops are also used to 'sequence' data, i.e. make a large amount of data appear in a predetermined bit by bit sequence, as desired. Flip-flops are also used to exercise control over the functionality of a digital circuit, i.e. change the operation of a circuit depending on the states of one or more flip-flops. These devices are mainly used in situations which require one or more of these three operations—storage, sequencing or functional alteration of circuit behaviour.

3.2 THE CLOCK INPUT IN SEQUENTIAL CIRCUITS

A general digital system will consist of a number of combinatorial gates (and/or other combinatorial circuits) and a number of flip-flops. Such a system will have (random) propagation delays between the various input, output and internal variables. In a purely combinatorial circuit, this does not pose a problem as long as the user is willing to wait for the outputs to settle down once the inputs have been changed. This is not true for a sequential circuit built using flip-flops. Here erroneous results may occur if a flip-flop is made to store data (or perform some other function) before its inputs have become really valid; this could easily happen because of (random) propagation delays in the system and explicit design steps must be taken to prevent its occurrence.

The problem mentioned above is easily taken care of by incorporating a 'clock' input in the system as shown in Fig. 3.2.1. This is a 1∅1∅ . . . (or . . . Hi, Lo, Hi, Lo . . .) kind of digital

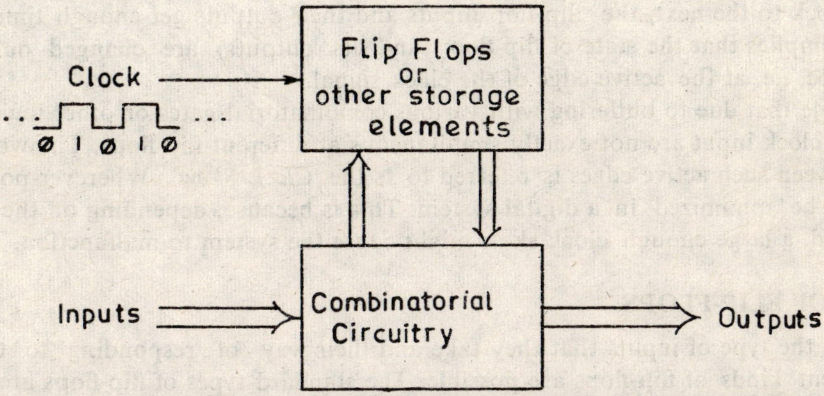

Fig. 3.2.1 Digital System with Combinatorial Gates and Clocked Flip-Flops.

sequence where the duration of a Hi-Lo pair is generally fixed and is referred to as the Clock Period t_{CLK}. The ratio of the Hi-level duration t_{Hi} to the duration t_{CLK} is referred to as the Duty Cycle of the clock waveform.

$$t_{CLK} = t_{Hi} + t_{Lo}$$
$$\text{Duty Cycle} = t_{Hi}/t_{CLK}, \quad f_{CLK} = 1/t_{CLK}$$

Note that a 50% duty cycle clock will have $t_{Hi} = t_{Lo}$. The rise time t_r of a clock input is the time needed for the Lo → Hi transition of the clock waveform, once the signal starts increasing from the Lo state. Similarly, the fall time t_f of the clock waveform is the time taken for the Hi → Lo transition, once the signal starts decreasing from the Hi state. Generally, digital systems prefer clock waveforms with as low a value of rise times and fall times as possible. As a matter of fact, given a clock with large rise and fall times one frequently has to use some special circuitry (e.g. Schmitt triggers). These timing specifications are illustrated in Fig. 3.2.2.

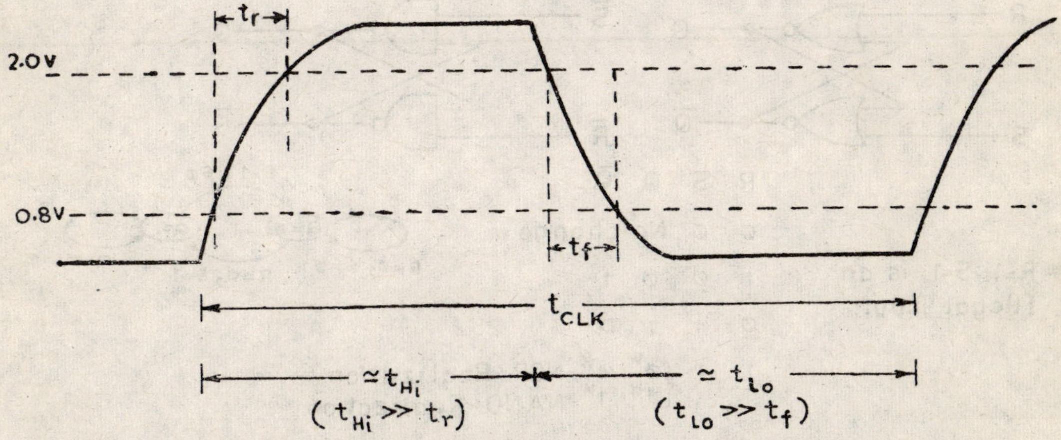

Fig. 3.2.2 Timing Specifications of the Clock Waveform.

Functionally, the clock signal provides a way of synchronizing the operation of the various flip-flops in the system. A typical system is designed such that changes in the states of the flip-flops is only possible at a particular edge (Lo → Hi or Hi → Lo) of the clock input; otherwise, the flip-flop retains its previous state. This procedure has the advantage that from one active edge of the clock to the next, the flip-flop inputs and their outputs get enough time to stabilize. Effectively, it implies that the state of flip-flop (and its outputs) are changed only when the inputs are valid, i.e. at the active edge of the clock signal.

It is possible that due to buffering with various combinatorial gates or other logic, the active edges of the clock input are not exactly simultaneous at different flip-flops. The worst case time difference between such active edges is referred to as the '*Clock Skew*'. Wherever possible, clock skew should be minimized in a digital system. This is because, depending on the propagation delays involved, a large enough clock skew might cause the system to malfunction.

3.3 TYPES OF FLIP-FLOPS

Depending on the type of inputs that they take and their way of responding to these inputs, several different kinds of flip-flops are possible. The standard types of flip-flops are the RS, JK, D and T type flip-flops; these will be described in this section.

One may also classify flip-flops according to the type of clock input required and the way this affects the flip-flop's state. A flip-flop which regards a Lo → Hi transition as the active edge is a *Positive Edge Triggered* flip-flop; conversely, a *Negative Edge Triggered* flip-flop is one where the active edge is a Hi → Lo transition. One can also get a flip-flop with *Master-Slave Triggering*; in such a flip-flop, the inputs are accepted on one edge of the clock signal but the flip-flop state and its outputs change only on the other (complementary) edge.

It is also possible to have a flip-flop without a clock signal. Such a flip-flop is commonly referred to as a *Latch*. Such a circuit is described next since it is the basic building block used to make different types of flip-flops.

3.3.1 A Latch Circuit (RS Latch)

The simplest flip-flop circuit is a latch such as the one shown in Fig. 3.3.1. This consists of two

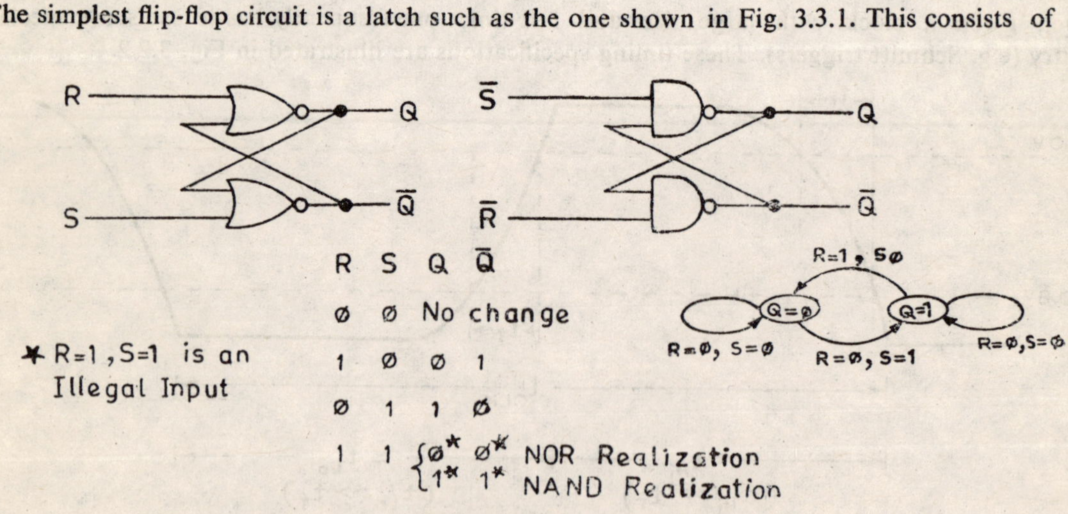

Fig. 3.3.1 R-S Latch, NOR and NAND Realization

gates (either two NORs or NANDs) connected back to back as shown in the figure. This flip-flop has two inputs R, and S, i.e. the Reset and the Set inputs respectively. (Note the inverted inputs for the NAND-latch). The flip-flop has two *stable* states, 1 and Ø. These correspond to the state of the Q output. When the flip-flop is in the 1-state, its Q output is 1; when it is in the Ø-state, its Q output is Ø. Note that in normal operation, the $\overline{Q}$ output is the inverse of the Q output.

Figure 3.3.1 shows a table relating the Q and $\overline{Q}$ output to the inputs R and S. This can be easily derived from the logic of the gates used. Ignoring the R = S = 1 input combination for the moment, one finds that an input of R = 1, S = Ø will put the latch in the Ø-state (Q = Ø, $\overline{Q}$ = 1) regardless of the state it might have been in earlier; similarly an input of R = Ø, S = 1 will put the latch in the 1-state (Q = 1, $\overline{Q}$ = Ø). The input combination R = Ø, S = Ø is essentially a HOLD command to the latch which maintains the state of the latch to whatever state was there before.

In order to store data using such a latch, one can adopt the following procedure. Normally the latch inputs should be kept at R = Ø, S = Ø to hold whatever data was previously stored. If one wants to store a '1', one would apply the combination R = Ø, S = 1 for a short while in order to set the latch, returning the inputs to R = Ø, S = Ø once the latch has been set. One would apply the inputs R = 1, S = Ø for a short while in a similar fashion in order to store a Ø in the latch. This procedure is illustrated by the transition diagram of Fig 3.3.1.

The input combination R = 1, S = 1 is an *illegal input* combination which in normal usage should not be applied to the latch. This is because of the following reasons.

(i) While R = 1, S = 1, the latch outputs are Q = 1, $\overline{Q}$ = 1 (NAND) or Q = Ø, $\overline{Q}$ = Ø (NOR). This however is not a stable state. If one now returns the inputs to R = Ø, S = Ø the output is not held to its previous value; the outputs then become either Q = Ø $\overline{Q}$ = 1 or *Q* = 1 $\overline{Q}$ = Ø depending on the relative propagation delays of the two gates.

(ii) The transition from the R = 1 S = 1 to the R = Ø S = Ø inputs does not produce a deterministic output at Q and $\overline{Q}$. On such a transition, the flip-flop might (randomly) end up in either the 1 or the Ø state. (All other transitions lead to a known and deterministic output). Such an indeterminacy is generally unacceptable in a digital system.

(iii) For normal operation, the Q and $\overline{Q}$ outputs should be the logical inverse of each other. This is violated when the latch inputs are R = 1, S = 1.

Because of the above reasons, the designer is usually required to ensure that the R = 1, S = 1 input is never applied to the R-S latch of Fig. 3.3.1.

The basic latch circuit (specifically the R-S latch) of Fig. 3.3.1 is simple but constitutes the basic element for other, more complicated, flip-flops. It actually has three equilibrium conditions. Two of these, the 1 and Ø states of the latch, are stable equilibria. The other is an unstable equilibrium condition, which is encountered when the latch transits from one state to the other. The time spent in this unstable equilibrium is a function of the propagation delays of the gates used in the circuit (i.e. larger for large gate propagation delays). It is interesting to note that because of this feature, the rise (or fall) time of the latch output is governed by the gate propagation delays, rather than the rise (or fall) time of the input.

3.3.2 The R-S Flip-Flop

The R-S latch of Sec. 3.3.1 required the direct inputs R and S but no clock. The R-S flip-flop is functionally similar to the R-S latch except that it incorporates a clock input. The symbol for an R-S flip-flop is shown in Fig. 3.3.2 which also shows a table summarising the behaviour of the flip-flop.

Referring to Fig. 3.3.2, the circle on the clock input (CLK) implies that the flip-flop output can change only at the $1 \rightarrow \emptyset$ transition of the clock. Absence of the circle would have implied that any change in the flip-flop outputs is possible only at the $\emptyset \rightarrow 1$ transition. Assuming that the flip-flop is edge triggered, therefore implies that the flip-flop of Fig. 3.3.2 is a negative edge triggered flip-flop. (Note that it is also possible to have a positive edge triggered or a master-slave flip-flop).

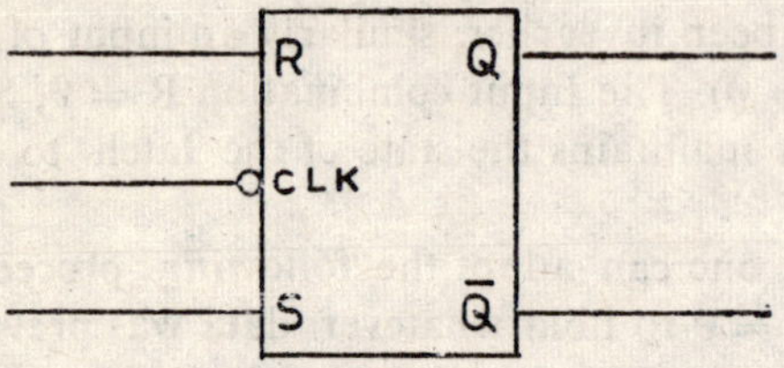

Clock Activ Edge ⌉_

R_n	S_n	Q_{n+1}
Ø	Ø	Q_n
Ø	1	1
1	Ø	Ø
1	1	★

*Ambiguous Output

Fig. 3.3.2 R-S flip-flop (negative edge triggered)

Let Q_n denote the flip-flop output in the nth clock period measured from one active edge to the next, and let S_n and R_n be the respective S and R inputs just prior to the active clock edge at the end of this period. The state of the flip-flop and its output Q_{n+1} in the next clock period is then given by the table shown in Fig. 3.3.2. Note that the operation of a negative or positive edge triggered flip-flop would be identical except that the active edges will be different ($1 \rightarrow \emptyset$ and $\emptyset \rightarrow 1$ respectively).

Figure 3.3.3 shows a circuit for obtaining a negative edge triggered R-S flip-flop using an R-S latch. The associated timing diagram explains the operation of the input circuitry added to the R-S latch to obtain negative edge triggering. Note that the R and S inputs to the latch are normally both at Ø except for a short duration (NAND propagation delay) following the $1 \rightarrow \emptyset$ edge of the clock input. During this short duration the R and S inputs to the latch follow the R and S inputs of the flip-flop resulting in a negative edge triggering action.

Figure 3.3.3 can also be used to point out the relevance of an important specification regarding flip-flops—that of *Setup Time*. This is the minimum time duration just preceding the active clock edge when the inputs have to be held at their valid levels in order to have proper operation of the flip-flop. It is easy to see that because of gate propagation delays, the S and R inputs must be stable some time before the active edge in order to provide the proper latch inputs following the active edge. A change in the S and R inputs just prior to the active edge and very close to it may not get properly reflected in the S and R inputs to the latch immediately following this edge. For flip-flops, one also specifies a 'Settling Time' which is the time taken by the flip-flop to reach the desired state, once the proper inputs have been applied. Figure 3.3.3 also illustrates the maximum value of this time for the arrangement shown.

A circuit very similar to that of Fig. 3.3.3 may be suggested to obtain a positive edge triggered flip-flop from an R-S latch. Apart from the fact that its active clock edge is different, the operation of a positive edge triggered R-S flip-flop will be very similar to its negative edge triggered counterpart.

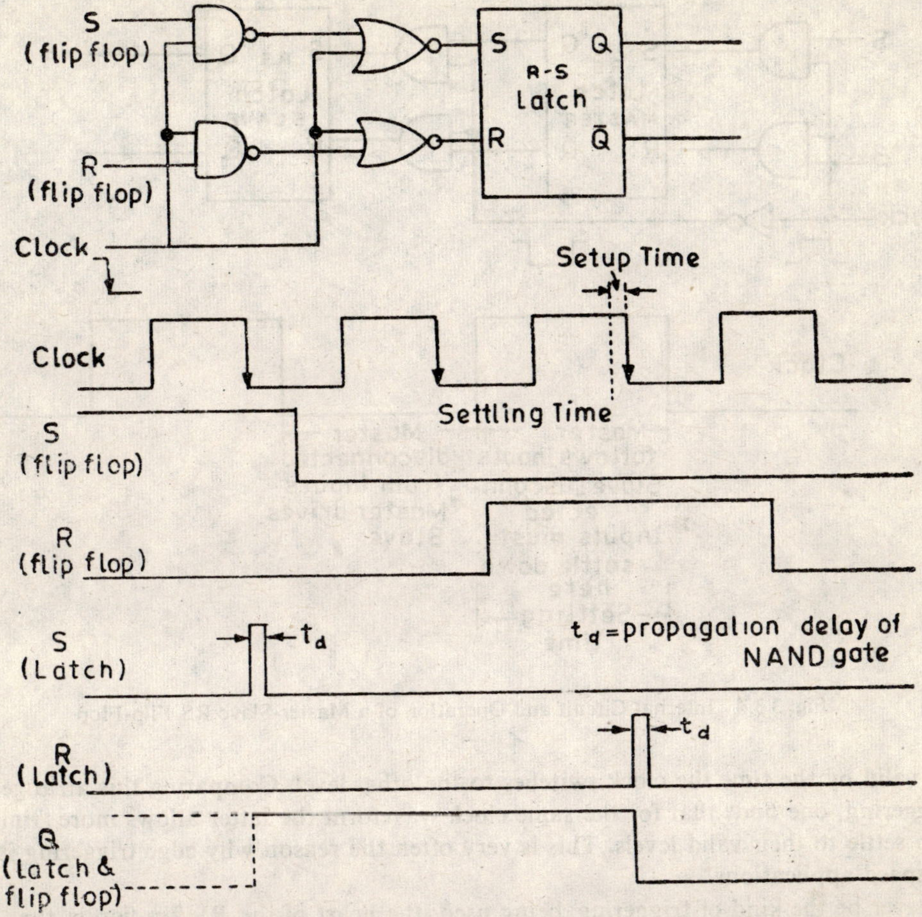

Fig. 3.3.3 Internal Circuit and Operation of a Negative Edge Triggered R-S Flip-Flop

A master slave R-S flip-flop, however, uses a completely different approach to the triggering problem. As shown in Fig. 3.3.4, it contains two cascaded R-S latches. The inputs affect the master latch for one level of the clock (Hi in Fig. 3.3.4). During this time, the slave latch which drives the output is effectively disconnected from the master, i.e. it has R = Ø, S = Ø. For the other level of the clock, the master is disconnected from the inputs and retains its previous contents (induced by the other clock level). During this time, the master also drives the slave latch and hence the outputs of the flip-flop. This operation may be understood by critically examining the circuit shown in Fig. 3.3.4. Note that by using two latches in the master-slave arrangement, one has effectively achieved the following:

(a) The output of the flip-flop does not change during the time that inputs affect the master latch.

(b) The output of the flip-flop changes only when the inputs have been properly latched by the master latch.

The concept of setup time is strictly speaking not applicable to the master slave arrangement. This is because the master latch follows the inputs for one level of the clock. Here the inputs

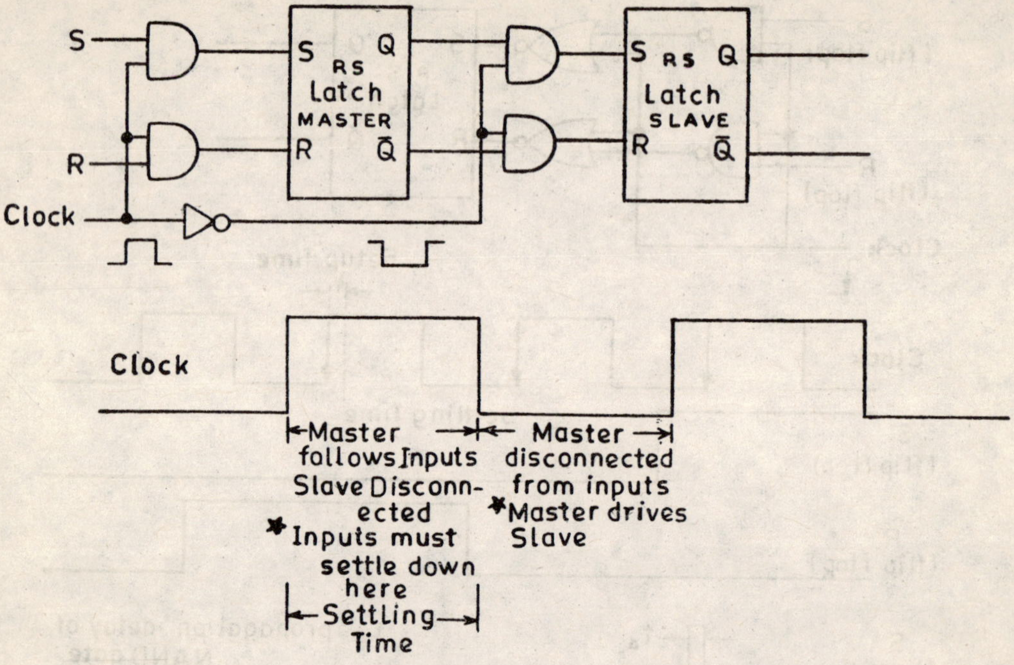

Fig. 3.3.4 Internal Circuit and Operation of a Master-Slave RS Flip-Flop

must be valid by the time the clock switches to the other level. Comparing this arrangement with edge triggering, one finds that for the same clock waveform the latter allows more time for the inputs to settle to their valid levels. This is very often the reason why edge triggering is preferred in high speed applications.

Whatever be the kind of triggering being used, the heart of the RS flip-flop is the RS latch. Just like the latter, the former also responds ambiguously to a R = 1, S = 1 input. For this reason, the designer should take care that this input combination is never applied to a RS flip-flop.

In the remaining portions of this section, we will describe three other types of flip-flops, D, J-K and T flip-flops. Note that all these flip-flops can have either negative or positive edge triggering or a master slave operation. In each case, the considerations mentioned here regarding the triggering of RS flip-flops are also applicable.

3.3.3 The D Flip-Flop

A D flip-flop has a single data input (D) and a clock input. It may have either or both Q and $\overline{Q}$ outputs. The symbol for the D flip-flop is shown in Fig. 3.3.5. This figure also shows the details of its operation. Note that apart from the synchronous D input which depends on the clock, the D flip-flop of Fig. 3.3.5 also has two asynchronous inputs—Preset and Clear—which affect the flip-flop irrespective of the D and the clock inputs.

The Preset input puts the flip-flop in the 1 state when Preset = Ø. Similarly, a Clear input of Ø puts the flip-flop in the Ø state. The D input is effective only when Preset = Clear = 1. Under these conditions, the data at the D input gets shifted into the flip-flop and appears at its output Q. If both Preset and Clear are Ø, the state of the flip-flop is determined by whether the flip-

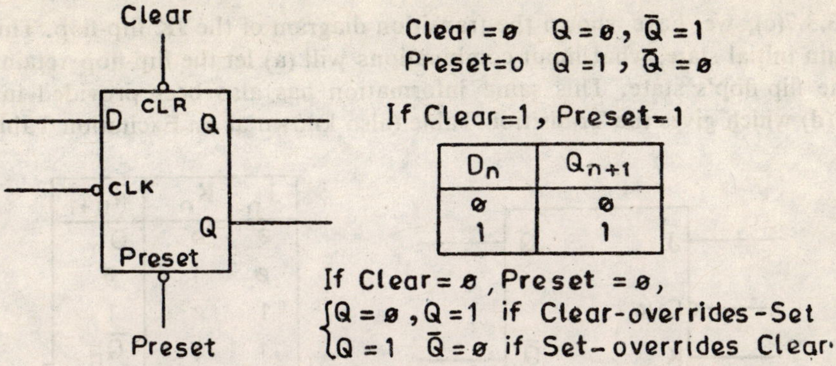

Fig. 3.3.5 D Flip-Flop (Negative Edge Triggered)

flop has 'Set-overrides-Clear or Clear-overrides-Set. Such asynchronous Preset and Clear inputs may or may not be present in all D flip-flops. Such inputs may also be provided in an R-S flip-flop or in the other types of flip-flops discussed in this section. [The circle on Preset and Clear indicate that they are active when Ø].

It is possible to obtain a D flip-flop from an R-S flip-flop as shown in Fig. 3.3.6. Note that in this figure we have shown an R-S flip-flop which has both the asynchronous Preset and Clear

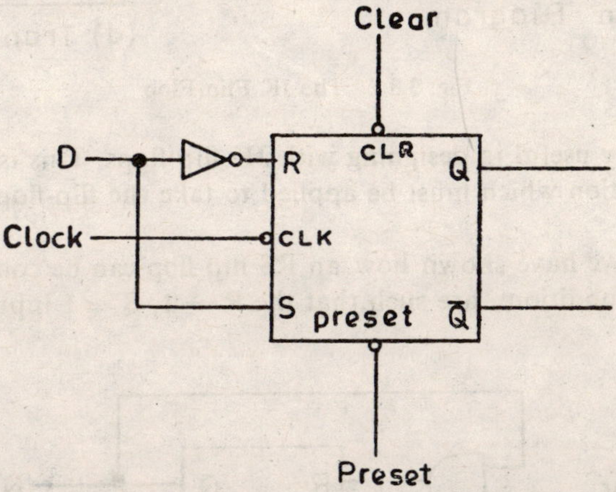

Fig. 3.3.6 Converting RS Flip-Flop to D Flip-Flop

inputs. Also, note the fact that the circuit conditions are such that the illegal RS input R = 1, S = 1 will never appear. (Actually R = Ø, S = Ø will also never appear).

3.3.4 The JK Flip-Flop

The JK flip-flop is very similar to the RS flip-flop (with J corresponding to S and K corresponding to the R input) except in its treatment of the (1, 1) input combination. For the RS flip-flop, this was an illegal combination giving ambiguous results and hence was something to be avoided in the design. In the JK flip-flop, the (1, 1) input combination toggles the flip-flop's state and hence its output. The symbol for the JK flip-flop and its truth table are shown in Figs. 3.3.7(a) and 3.3.7(b) respectively.

In Fig. 3.3.7(c), we have shown the transition diagram of the JK flip-flop. This shows how, given a certain initial state, what input combinations will (a) let the flip-flop retain its state or (b) toggle the flip-flop's state. This same information has also been provided in tabular form in Fig. 3.3.7(d) which gives the Transition Table (also known as an Excitation Table) of the JK

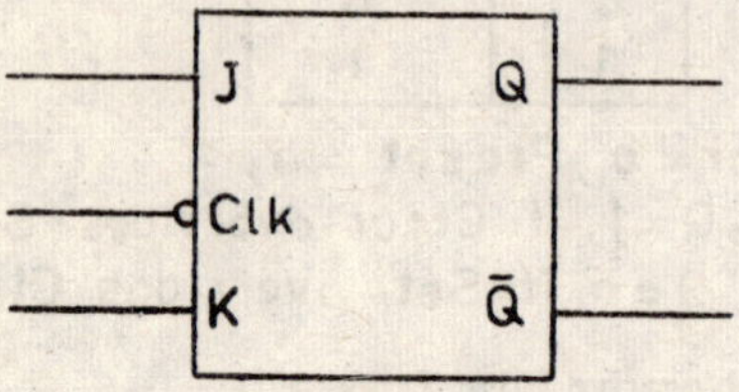

(a) Symbol *may also have Preset & Clear inputs

J_n	K_n	Q_{n+1}
Ø	Ø	Q_n
Ø	1	Ø
1	Ø	1
1	1	$\bar{Q}_n$

(b) Truth Table

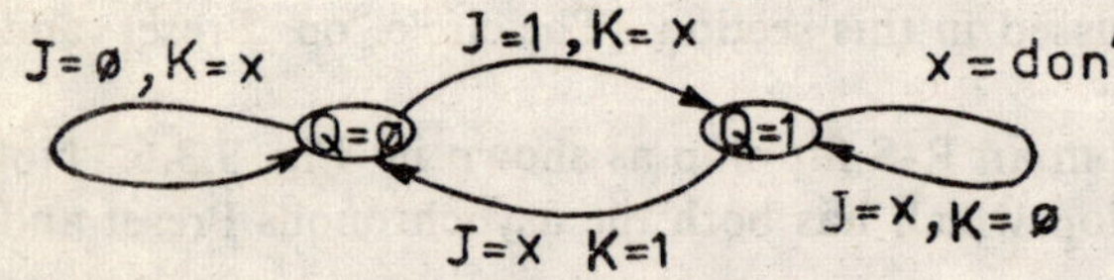

(c) Transition Diagram

x = don't care

$Q_n \rightarrow Q_{n+1}$		J_n	K_n
Ø	Ø	Ø	x
Ø	1	1	x
1	Ø	x	1
1	1	x	Ø

(d) Transition Table

Fig. 3.3.7 The JK Flip-Flop

flip-flop. This table is very useful in designing with JK flip-flops. This is because it specifies the J and K input combination which must be applied to take the flip-flop from a given state Q_n to a desired state Q_{n+1}.

Finally in Fig. 3.3.8, we have shown how an RS flip-flop can be converted to a JK flip-flop. Note that the circuit conditions are such that the R = 1, S = 1 input combination will never arise at the RS flip-flop.

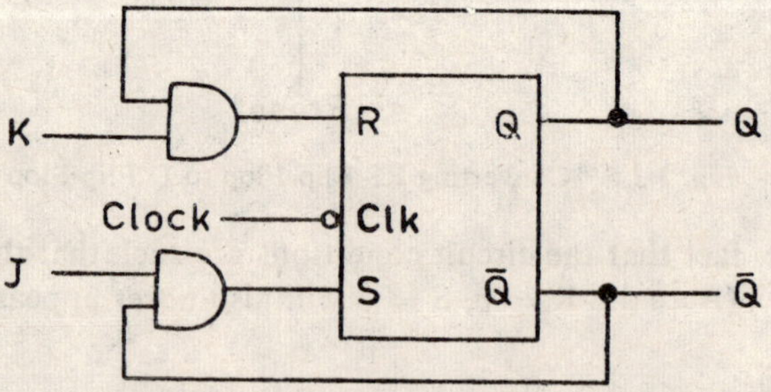

Fig. 3.3.8 RS to JK Flip Flop Conversion

3.3.5 The T Flip-Flop

The T flip-flop is not as commonly used as the JK and D flip-flops. If needed, it is usually built from one of these flip-flops. Its unique feature is a 'toggle' property. When the T input is 1, a clock pulse toggles the flip-flop state; the flip-flop's state remains unchanged if the T input is Ø.

Figure 3.3.9 shows the symbol and the truth table for the T flip-flop. Figure 3.3.10 shows how a T flip-flop can be constructed from a JK flip-flop by simply connecting the J and the K inputs together and providing that as the T input. By comparing Fig. 3.3.10 with Fig. 3.3.8, the reader should be able to see how one can obtain a T flip-flop from an RS flip-flop.

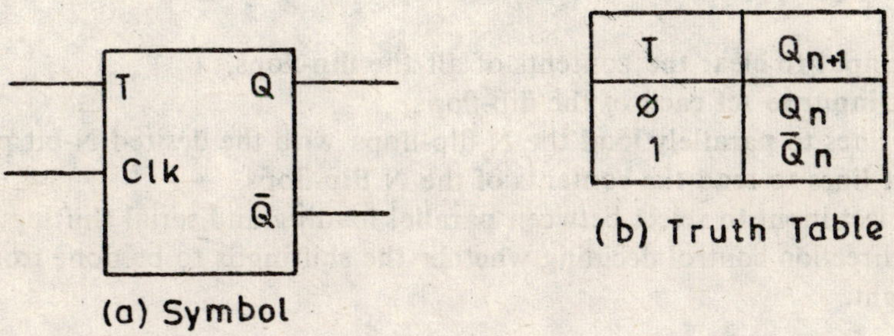

Fig. 3.3.9 T Flip-Flop

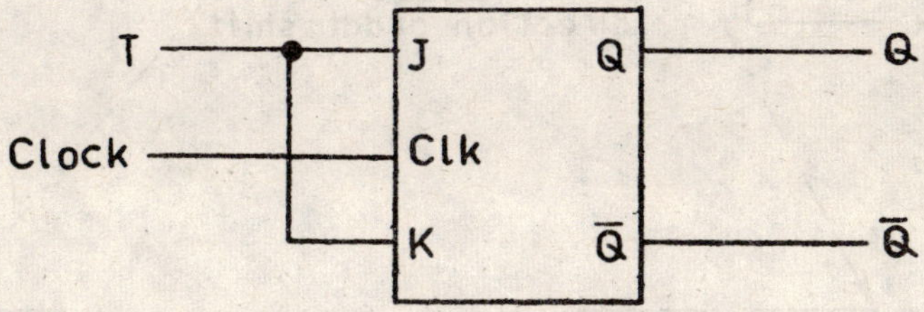

Fig. 3.3.10 JK to T Flip-Flop Conversion

3.4 SHIFT REGISTERS

The flip-flops (especially the D flip-flop) described in Sec. 3.3 serve as the typical data storage units in a memory. These single storage units may be arranged in various ways to obtain a memory which can store more than one data bit. One common way of doing so is a Shift Register, such as the one shown in Fig. 3.4.1. A N-bit shift register basically consists of a series of N D type flip-flops with the output of one D flip-flop feeding the input of the next. The user generally has access to the D input of the first stage and the output of the Nth stage. Data presented at the Data In terminal gets clocked into Flip-Flop 1 following an active clock edge. Simultaneously, the previous contents of the shift register get shifted to the right by one bit; i.e. the Q_n of the ith flip-flop becomes Q_{n+1} of the $(i + 1)$th flip-flop. The data stored in the Nth flip-flop gets shifted out following the clock edge and hence gets lost.

A shift register, such as the one shown as being constructed with D flip-flops in Fig. 3.4.1, will not lose its contents if the clock is stopped. If the clock is stopped, the previous data will be retained and can be clocked once again when the clock starts. Such a shift register whose storage operation is independent of the clock signal's speed (or on its presence or absence) is referred to as a *Static Shift Register*. In the MOS family of logic devices, one also gets a *Dynamic Shift Register*. This is exactly like the static version except that the shift register contents are lost (or at least partially corrupted) unless it is clocked at a certain minimum rate.

Long shift registers are typically available in the MOS family of devices. Such registers are available in both dynamic or static form. Such an N-bit shift register may be used to store N-bits of data. Note that only the data out from the Nth stage is available outside the shift register. This implies that a data bit introduced in the first stage of the shift register becomes available outside only after (N — 1) additional clock edges have gone through.

Short shift registers (4 or 8 bits) are typically available in the TTL family of devices. These may have more inputs or outputs than the ones shown in Fig. 3.4.1. Some of these are given below:

(i) A Clear input to clear the contents of all the flip-flops,
(ii) A Preset input to set each of the flip-flops,
(iii) N input lines to parallely load the N flip-flops with the desired N-bit pattern,
(iv) N output lines to read the contents of the N flip-flops,
(v) Mode Select input to select between parallel loading and serial shifting,
(vi) A shift direction control deciding whether the shifting is to be done from right to left or left to right.

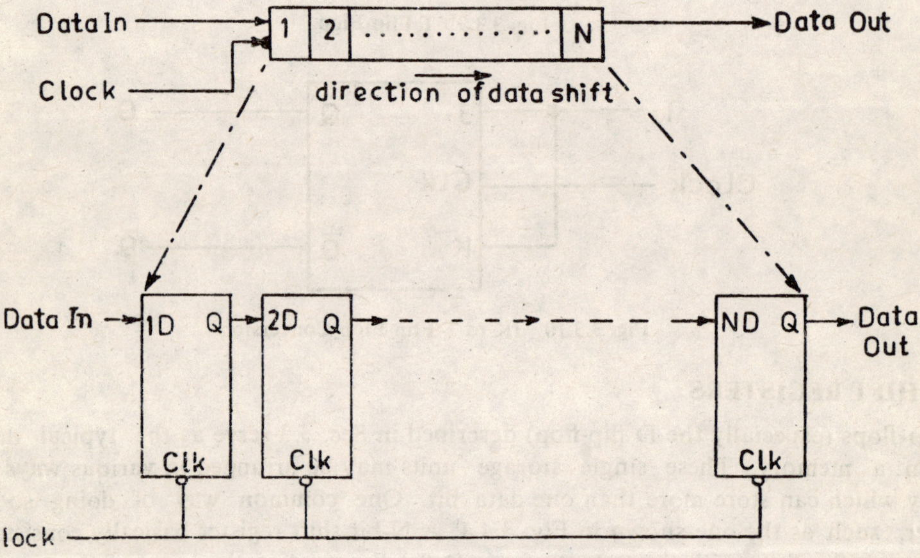

Fig. 3.4.1 A N-bit Shift Register (Static)

A shift register satisfying (iii) may be used to convert parallel data into a serial form. The parallel data is loaded into the shift register and then shifted out serially by the clock. Similarly, a shift register satisfying (iv) may be used to convert a serially received data into a parallel form. Note that a particular IC shift register may only have some of the above mentioned inputs and outputs.

It can be easily seen that a shift register is a serial access memory. Consider the circuit of Fig. 3.4.2 where one has connected a N-bit SR in a circulating configuration. In a typical SR of this type (especially, for N large) one can only read the data bit stored in the Nth stage and change the data bit stored in the 1st stage. In order to READ or WRITE from any other location, the shift register will have to be clocked till that location is in the Nth stage for a READ or the 1st stage for a WRITE. Obviously, one would require more time for a READ or WRITE operation than if one had the ability to access any bit in the SR directly. A further disadvantage is that the access time is random (depending on the current state of the SR and the position of the desired bit); it will be uniformly distributed between 0 and (N − 1) clock periods.

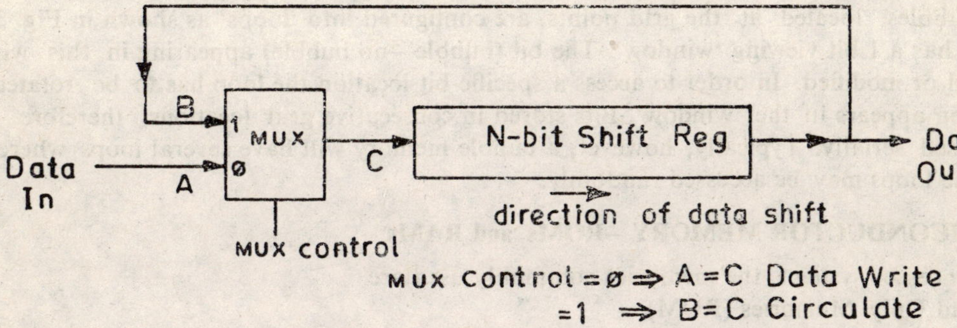

Fig. 3.4.2 N-bit Shift Register in Circulating Mode for Data Storage and Retrieval

Now suppose one wants to store a K bit word in this SR (N > K). We can do it in two ways. We can store this word in K consecutive locations of the SR. In this case a READ operation should rotate the loop to the proper position and then output K bits serially; for WRITE, the K bits will be input serially. The support circuitry for such a memory will also have to keep track of the locations where a K bit word will start (or end).

An alternative method for achieving the purpose will be to place K, N-bit SRs, in parallel. If the READ, WRITE, CLOCK and the LOOP COUNTER (indicating current position of SR loop) of the K SRs are common, the operation of READing or WRITing a K bit word will be similar to that for 1 bit.

It will be appropriate to introduce Bubble Memories (Fig. 3.4.3) at this point. This is because, functionally, they are devices which are partly serial and partly random access. In order to see

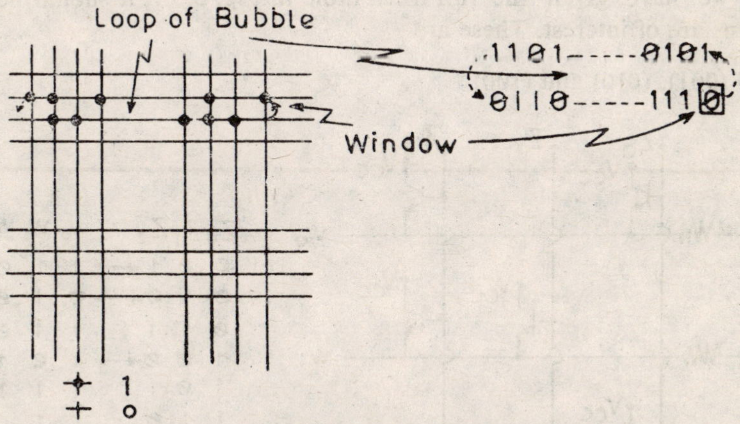

Fig. 3.4.3 Organization of Bubble Memory

this we have to examine the internal structure of these memories. These memories have a thin film of magnetic material grown on a non-magnetized substrate. Microscopic magnetized zones (bubbles) can be created at discrete grid points on the film where these zones have a polarity opposite that of rest of the magnetic film. A 1 (or Ø) at any grid point is represented as the presence (or absence) of a bubble at that point. Unlike a Shift Register, a bubble memory is nonvolatile—the data in memory is retained even when the power supply is turned OFF.

The bubbles, located at the grid points, are configured into 'loops' as shown in Fig. 3.4.3. Each loop has a 1 bit viewing 'window'. The bit (bubble—no bubble) appearing in this window can be read or modified. In order to access a specific bit location the loop has to be rotated till that location appears in the 'window'. Bits stored in consecutive grid locations, therefore, have to be accessed serially. Typically, however, a bubble memory will have several loops where any one of these loops may be accessed randomly.

3.5 SEMICONDUCTOR MEMORY—ROMs and RAMs

In microprocessor systems, the memories commonly used are

(a) Read Only Memories (ROM)
(b) Programmable Read Only Memories (PROM)
(c) Erasable PROMs (EPROM)
(d) Electrically Alterable ROMs (EAROM)
(e) Random Access Memories (RAM) or Read Write Memories (RWM)

It should be noted that all the above memories are random access memories. Any bit stored in these memories may be accessed directly. In this sense, calling (e) a RAM is a misnomer. Since the unique feature of (e) is that 'Read a bit from it' or 'Write a bit into it' may be done with equal facility, a more correct name for this memory would be RWM; however, the name RAM is commonly used and will also be used here.

3.5.1 Read Only Memory (ROM)

A ROM is programmed by the manufacturer and its contents can never be altered. A typical ROM structure using bipolar transistors and MOS is shown in Fig. 3.5.1 and in Fig. 3.5.2 respectively.

Even though we have given the full truth table in Fig. 3.5.1, it should be noted that only three input patterns are of interest. These are

$$Z_\emptyset Z_1 Z_2 = (\emptyset\emptyset 1), (\emptyset 1\emptyset) \text{ and } (1\emptyset\emptyset)$$

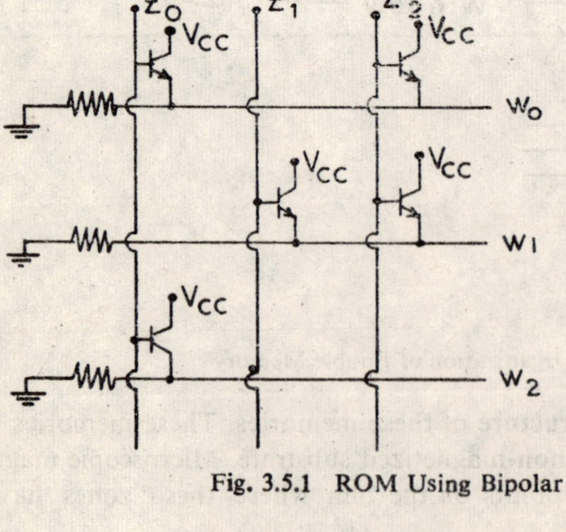

Z_0	Z_1	Z_2		W_0	W_1	W_2
Ø	Ø	1	↔	1	1	Ø
Ø	1	Ø	↔	Ø	1	Ø
Ø	1	1		1	1	Ø
1	Ø	Ø	↔	1	Ø	1
1	Ø	1		1	1	1
1	1	Ø		1	1	1
1	1	1		1	1	1

Fig. 3.5.1 ROM Using Bipolar Transistors

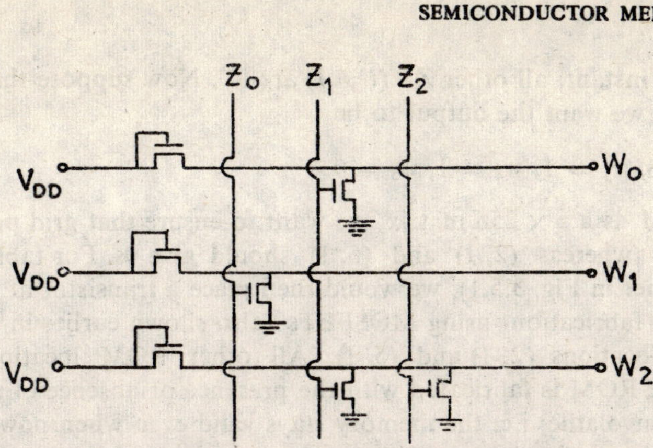

Fig. 3.5.2 ROM Using MOS (same truth table as 3.5.1)

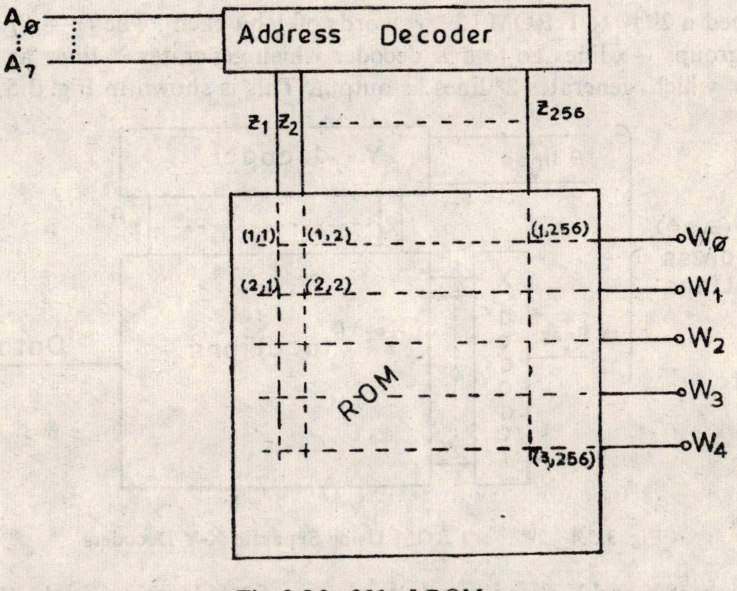

Fig. 3.5.3 256×5 ROM

In order to see this consider how one might build a 256×5 ROM (256 words of 5 bits each) as in Fig. 3.5.3. The address decoder accepts address lines ($A_\emptyset - A_7$) and generates 256 outputs $Z_1 \ldots Z_{256}$ as follows

A_7	A_6	A_5	A_4	A_3	A_2	A_1	A_0	
∅	∅	∅	∅	∅	∅	∅	∅	$Z_1 = 1, Z_2 = \ldots Z_{256} = \emptyset$
∅	∅	∅	∅	∅	∅	∅	1	$Z_1 = \emptyset, Z_2 = 1, Z_3 = \ldots Z_{256} = \emptyset$
∅	∅	∅	∅	∅	∅	1	∅	$Z_1 = \emptyset, Z_2 = \emptyset, Z_3 = 1, Z_4 = \ldots Z_{256} = \emptyset$

and so on.

Considering the ROM, the inputs Z_i, $i = 1, \ldots 256$ to it are such that only one of them (say

with $i = I$) is 1 at any instant; all other Z_is ($i \neq I$) are '∅'. Now suppose that for the inputs such that $Z_I = 1$, $Z_{i \neq I} = \emptyset$, we want the output to be

$$w_1 = 1,\ w_2 = \emptyset,\ w_3 = 1,\ w_4 = 1,\ w_5 = \emptyset.$$

Considering the ROM as a 5 × 256 matrix, we want to ensure that grid points (1, I), (3, I) and (4, I) should yield 1s whereas (2, I) and (5, I) should give ∅s. For fabrication using Bipolar transistors (shown earlier in Fig. 3.5.1), we would then place a transistor at locations (1, I), (3, I) and (4, I) each. For fabrication using MOSFETs (also shown earlier in Fig. 3.5.2), we would place MOS devices at locations (2, I) and (5, I). All other ROM locations can be similarly programmed. Since the ROM is fabricated with the presence or absence of a device signifying 1 or ∅, the memory is nonvolatile; i.e. the memory stays valid even when power is removed.

If the ROM size is large, the straight decoding structure (e.g. 8 Address to 256) shown above, may require too complex a decoder. In this case, a somewhat different organization is very often used.

Suppose we need a $2^{(\alpha+\beta)} \times 1$ ROM [$2^{(\alpha+\beta)}$ words of 1 bit each]. The ($\alpha + \beta$) address lines are divided into two groups —α lines go to a X decoder which generates 2^α lines as output, β lines go to a Y decoder which generates 2^β lines as output. This is shown in Fig. 3.5.4.

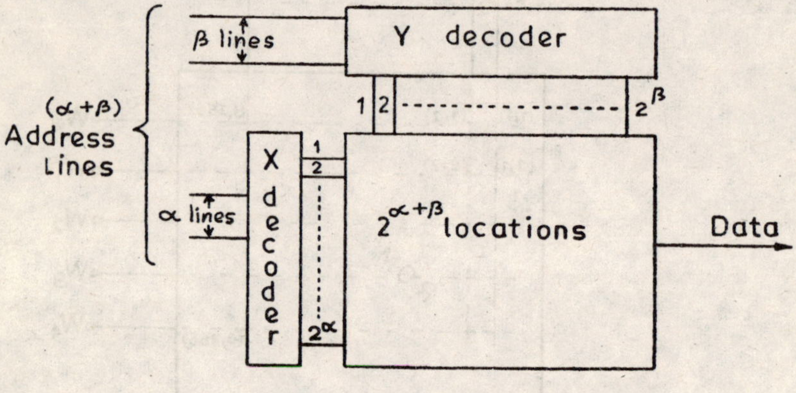

Fig. 3.5.4 $2^{(\alpha+\beta)} \times 1$ ROM Using Separate X-Y Decoders

The combination of X and Y decoders select one out of $2^{\alpha+\beta}$ locations in the ROM, the content of which is then output as data. By putting K such units in parallel, a K-bit data word can be obtained. In our previous example, we needed a 8-to-256 decoder. Using the X-Y structure, we can now get by with two much simpler 4-to-16 decoders.

ROMs are available with the decoders included as part of the IC chip. These are referred to as fully decodable ROMs.

As seen above, the contents of a ROM are fixed at the time of fabrication—the user does not have the freedom of specifying the contents once the ROM has been manufactured. Programmable ROMs (PROM) allow the user the flexibility of programming the contents of the ROM himself. In these memories, the manufacturer places a connecting device at each grid point (corresponding to the intersection of the decoder outputs Z_i and the bit outputs W_i). A fusible link is provided in series with each device. Using a PROM programmer, the user can selectively fuse these links leaving transistors connected only at locations necessary to establish the desired memory content.

Programming a PROM is a one-shot affair—once programmed, its contents cannot be changed. Erasable PROMs (EPROM) allow the user the flexibility of reprogramming the PROM. This is achieved by using 'floating gate transistors (PMOS)'. At normal operating voltage, the gate is entirely isolated from the rest of the chip circuitry. By applying a high enough voltage between its source and drain, a negative charge can be established at the gate thereby putting the transistor in the conducting state. This is done in the process of programming the EPROM chip. Exposure to ultraviolet light drains the negative charge at the gates, thereby erasing the contents of the memory. This can now be reprogrammed as necessary.

It is evident that in order to reprogram a EPROM the whole contents of the EPROM will first have to be erased. Bits in an EPROMs cannot be selectively altered. This feature is allowed by Electrically Alterable ROMs (EAROMs). Using special circuitry, contents of specific bit location can be selectively altered. Till recently, the problem with EAROMs was that they slowly lose their information depending on the number of read operations done on a particular bit location. This problem is not present in ROMs, PROMs or EPROMs.

Let us now consider the number of pins necessary for a ROM IC chip. As an example, we will consider a 4096×8 ($4K \times 8$) EPROM.

Apart from V_{CC} and GND, the chip will need 8 pins for output and 12 pins to address the full 4K or 4096 address space of the EPROM. It will also have a Chip Enable (CE) input. This input serves two purposes.

(i) A typical microprocessor system will have a bussed structure where a large number of IC chips will share the data and address bus. The CE input is useful to avoid contention on the data bus so that the EPROM outputs data to the data bus only when it is selected, i.e. CE = 1 (External logic must ensure that no other device outputs data on the data bus at this time). In order to achieve this, the outputs of the EPROM should use a tristate output stage.

[Tristate outputs are absolutely essential in any device which interfaces to a bus.]

(ii) The Chip Enable (CE) input is also used to decide where a particular memory chip would reside in the address space of the microprocessor and to expand the memory to provide for a larger memory than that which can be provided by one chip. This is discussed in Sec. 7.3.2.

From the discussion of ROM, EPROM etc., it should be evident that the memory output will always be available if the output tristate stage is enabled. As such, a specific READ (RD) signal may not be necessary. EPROMs may, however, have a separate RD or OE (Output Enable) pin. This is generally nothing but another chip enable input. However, in some cases the RD pulse may actually be used to latch the ROM output into a buffer inside the ROM. (The latch output will then have a tristate stage with a chip enable input.) This input is also useful in microprocessor systems because the microprocessor itself will generate a RD pulse when it wants to access data from memory. This input to the memory chip is then useful to ensure that the data is fetched *only when it is needed.*

Our hypothetical $4K \times 8$ EPROM will therefore have 1 V_{CC}, 1 GND, 12 address, 8 Output, 1 CE and 1 RD (or OE) pins. This is exactly what is available in the INTEL 2732 $4K \times 8$ EPROM. (Except that, in contrast to our hypothetical chip, both CE and OE are active Low. This is the reason why the chip uses the notation $\overline{CE}$ and $\overline{OE}$). Let us consider briefly the typical guarantees regarding the timing of various inputs to the 2732A-4 (Fig. 3.5.5).

t_{ACC} = Address to Output Delay < 450 ns

$t_{CE} = \overline{CE}$ to Output Delay < 450 ns

$t_{OE} = \overline{OE}$ to Output Delay < 150 ns

t_{DF} = Output Enable High to Output Float $0 < t_{DF} < 130$ ns

t_{OH} = Output Hold from Addresses, $\overline{CE} \simeq 0$ ns or $\overline{OE}$ whichever occurs first.

Note that there are roughly two kinds of timing specifications. The first kind specifies how long after a valid address, $\overline{CE}$ and $\overline{OE}$ does the output become valid. The device accessing the

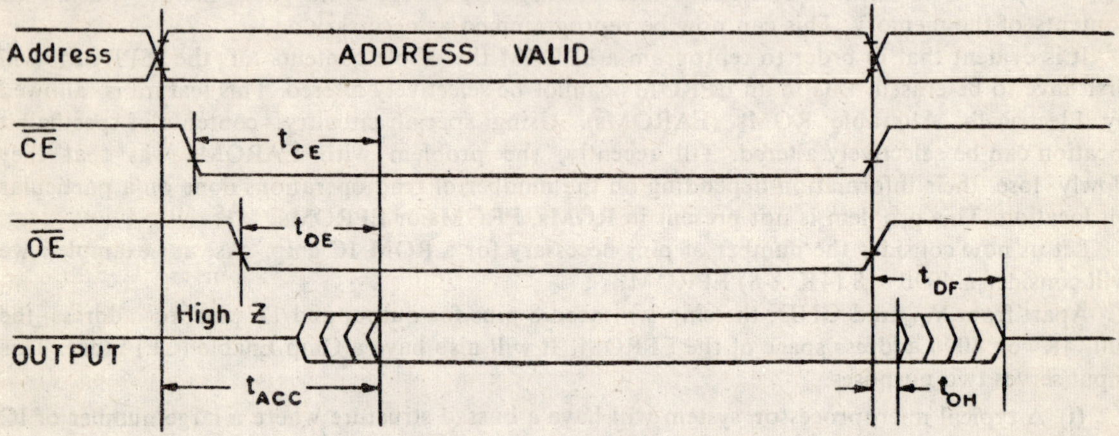

Fig. 3.5.5 Timing for Intel 2732 ROM

EPROM should not try to accept data before this. The second kind specifies how long after a valid address, $\overline{CE}$ or $\overline{OE} = 1$ is rendered invalid, does the output revert to a high Z state (high impedance). During this time, other devices accessing the bus should not output their data onto the bus.

3.5.2 Random Access Memory (RAM)

A typical microprocessor system will generally have two kinds of memory—program memory and data memory. The program memory will generally be implemented using ROMs, PROMs or EPROMs. This is convenient, because the program code will normally not be subjected to changes. We would want to change the contents of this memory only if we want to change the program itself. (This can then be achieved by changing the appropriate ROM and PROM chips or by erasing and reprogramming EPROMs). Data memory contents will, on the other hand, require change in the middle of a program's execution. It is therefore mandatory that the memory used here is of a type that allows the microprocessor (or other appropriate support devices) to change its contents easily. In other words, this memory, implemented using RAM (or RWM) should not only allow a READ function (reading the contents of specific memory locations) but also a WRITE function (changing the contents of specific memory locations) with equal ease.

Functionally a RAM memory chip is organized in a fashion very similar to ROMs, PROMs etc. Individual memory cells are coupled such that they can be accessed via the outputs of a decoder. As in ROMs, the outputs have tristate buffers so that the RAM can be used in a bus architecture, These tristate buffers are enabled via a Chip Enable (CE) and READ (RD) inputs. A RAM also needs a Write (WR) input which would allow the user to address a specific memory location and change its contents. Note that in the case of a ROM, the only possible function is a READ and that the RD input (if present) in a ROM was basically another Chip Select (or Chip Enable) input provided so that the output data will be valid only when the user (microprocessor) wants to read it. This is not the case with RAMs where we may want to do either a RD or WR on a selected RAM. Functionally, the RD input here may be implemented just as in a ROM but, logically, the two inputs are needed to distinguish between the Read and Write operations. It should also be noted that instead of using separate RD and WR inputs we can get by with one RD/$\overline{\text{WR}}$ input. The former, however, is more common since most microprocessor systems will provide RD and WR signals separately. Finally, it should be noted that CE for RAMs can be used (just as in ROMs) to place a particular RAM chip in a specific block in the total available address space of the system as well as to combine more than one RAM chips in order to create a larger RAM block in memory, e.g. combine two 1K×8 RAMs to generate a 2K×8 RAM (see Sec. 7.3.2).

Just as in shift registers, RAMs can be either static or dynamic. Static RAMs can store their contents as long as power is available. Dynamic RAMs need to be refreshed periodically; otherwise their contents are lost. The former is usually implemented using flip-flops whereas the storage element in the latter is generally a capacitor. It should be noted that a RAM data output will also be a tristate output similar to that shown earlier for ROMs.

Like shift registers, RAMs can be either static or dynamic in nature. From the point of view of internal construction, the storage unit for each bit in a static RAM is a flip-flop which may be set or reset to store a 1 or a Ø value. As long as power to the static RAM device is maintained, the flip-flop contents, and hence the memory contents, stay unchanged. The contents do not change on reading and can only be changed by writing into the RAM. In dynamic RAM, the storage element is merely a capacitor. Presence or absence of charge on a capacitor is used to store a 1 or a Ø value. As the charge stored on a capacitor may dissipate with time, the contents of a dynamic RAM device left to itself, will tend to get corrupted because of the discharging of these capacitors. Apart from the normal READ and WRITE operations that one can do with all RAMs,—dynamic or static-dynamic RAMs also require a REFRESH operation. This REFRESH operation renews the charge on the already charged capacitors. This refreshing operation must be done often enough so that the RAM contents stay valid. The manufacturer of a dynamic RAM device usually specifies how the READ, WRITE and REFRESH operations are to be done as well as the minimum rate at which this refreshing action is needed for proper operation of the dynamic RAM. (Proper operation implies that as long as power to the device is maintained, the contents of the RAM can only be changed by an appropriate WRITE operation.)

Compared to Static RAM devices, dynamic RAM devices are traditionally more compact (greater capacity for data storage) and consume less power. This is the reason why dynamic RAMs were popular in applications where a large amount of RAM space is required. In such situations, implementation of the needed memory is more cost-effective through dynamic RAM devices rather than static ones. Actually, with the advent of CMOS Static RAM devices (e.g. Hitachi's 6116), this is no longer completely true. For medium size memory requirements, CMOS static RAMs seem to be the logical choice today from considerations of cost, compactness and power consumption. (Note that, being static devices, they do not require refreshing to maintain their contents.) Usage of dynamic RAMs today, is limited to applications where the size of the required memory is so large that even the CMOS static RAMs would be uneconomical [Note that in terms of compactness, i.e. more memory on one chip, dynamic RAMs are the densest memory devices available today.]

In order to use dynamic RAM devices in a system, one typically needs a Dynamic RAM Controller to take care of the REFRESH requirements. This controller (in IC form or as a circuit card by itself) makes the refreshing action transparent to the user and ensures proper refreshing rate to keep the RAM contents valid. Used in conjunction with the controller, the dynamic RAM looks just like a static RAM to the user. (The only difference, usually experienced, is that if a READ or WRITE cycle clashes with a REFRESH cycle, the latter takes priority. Actually, in that case, the controller will usually ask the user to wait until the refreshing action is over). The availability of Dynamic RAM Controllers in IC form has simplified the job of using dynamic RAMs in a system. However, the cost of these controllers is typically quite high, implying that the size of the memory must be large enough to make it cost-effective.

Timing Diagram for RAM (Read and Write Cycles)

Just as for ROMs, the manufacturer of a RAM will also specify timing constraints that must be followed for proper operation. To see this, we will consider the typical timing requirements of the Intel 2114A (1024×4) static RAM.

READ Cycle (Fig. 3.5.6):

t_{RC} = Read Cycle time > 100 ns

t_A = Access time < 100 ns

t_{CO} = $\overline{CS}$ to output valid < 70 ns

t_{CX} = $\overline{CS}$ to output active > 10 ns

t_{OTD} = Output 3 state from deselection < 30 ns

t_{OHA} = Output hold from address change > 15 ns

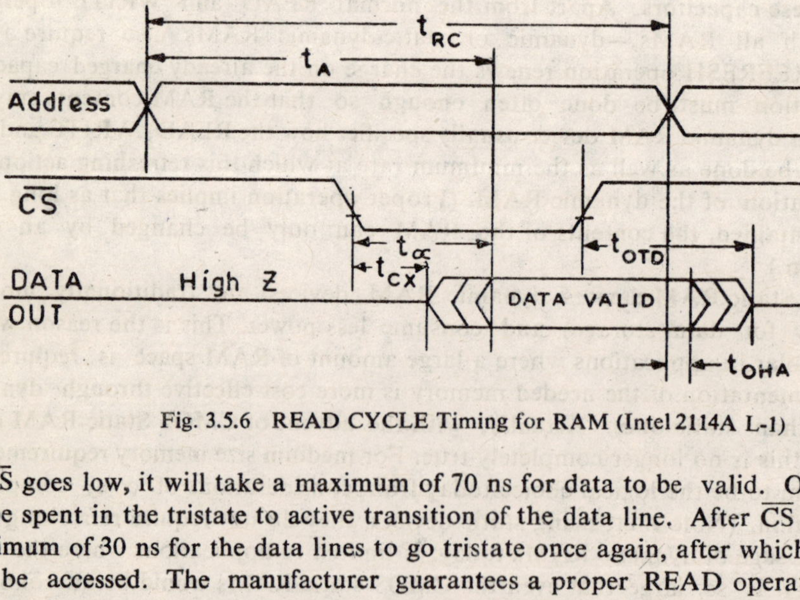

Fig. 3.5.6 READ CYCLE Timing for RAM (Intel 2114A L-1)

After $\overline{CS}$ goes low, it will take a maximum of 70 ns for data to be valid. Of this, 10 ns or more will be spent in the tristate to active transition of the data line. After $\overline{CS}$ goes high, it will take a maximum of 30 ns for the data lines to go tristate once again, after which a new memory word can be accessed. The manufacturer guarantees a proper READ operation if the READ cycle (address change to address change) is 100 ns or more. It may take up to 100 ns for the data to be valid once the proper address has been chosen.

WRITE Cycle (Fig. 3.5.7)

t_{WC} = Write Cycle Time > 100 ns

t_W = Write Time > 75 ns

t_{WR} = Write Release Time $\simeq$ 0

t_{OTW} = Output tristate from write < 30 ns

t_{DW} = Data to write time overlap > 70 ns

t_{DH} = Data Hold from write time $\simeq$ 0 ns

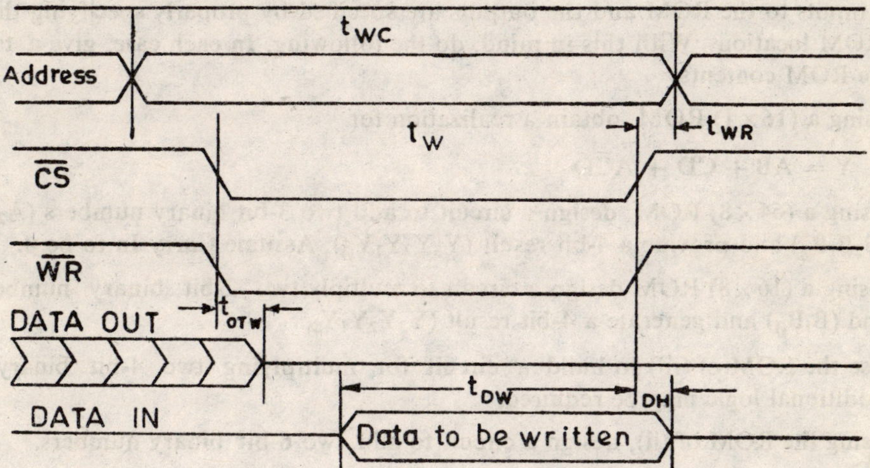

Fig. 3.5.7 WRITE CYCLE Timing for RAM (Intel 2114A L-1)

After $\overline{CS}$ goes low, it takes a maximum of 30 ns for the data out to go to tri-state. The valid data and $\overline{WR} = 0$ overlap needs to be only 70 ns or more. The data does not have to be held valid once $\overline{WR}$ goes to 1. The total Write Cycle time (address change to address change) should be at least 100 ns of which the total time $\overline{CS}$ and $\overline{WR}$ stay low should be at least 75 ns.

PROBLEMS

1. For the R-S latch of Fig. 3.3.1, consider the transition R = 1, S = 1 to R = Ø, S = Ø. Assuming one gate in the latch to be faster than the other, what will be the final state of the latch. Repeat this assuming now that the other gate is faster.

2. A designer wants a circuit with two inputs HOLD and DATA IN and an output DATA OUT to achieve the following:

   When HOLD = 1, DATA OUT should not change and should be equal to the DATA IN value when HOLD was Ø the last time

   When HOLD = Ø, DATA OUT = DATA IN

   Implement the circuit using a R-S latch with whatever other logic is required, ensuring that the input combination R = S = 1 never appears. (This circuit is essentially a D-type latch).

3. Very often one needs to build a desired type of flip-flop using a different flip-flop and additional logic. With this in mind, show how the following conversions may be done

   (a) J-K to D (b) D to J-K

   (c) D to T (d) T to D

   (e) T to J-K (f) R-S to T

4. Using a shift register and additional logic, design a serial-to-parallel and a parallel-to-serial converter. List whatever assumptions you need to make.

5. Strictly speaking a ROM is really a combinatorial logic block where the inputs go as the address inputs to the ROM and the outputs are specified by properly specifying the contents of the ROM locations. With this in mind, do the following. In each case, give a table specifying the ROM contents.

   (i) Using a (16×1) ROM, obtain a realization for

   $$Y = AB + C\bar{D} + A\bar{C}D$$

   (ii) Using a (64×8) ROM, design a circuit to add two 3-bit binary numbers ($A_2A_1A_\emptyset$) and ($B_2B_1B_\emptyset$) and provide a 4-bit result ($Y_3Y_2Y_1Y_\emptyset$). Assume Carry In to be Ø.

   (iii) Using a (16×8) ROM design a circuit to multiply two 2-bit binary numbers ($A_1A_\emptyset$) and ($B_1B_\emptyset$) and generate a 4-bit result ($Y_3Y_2Y_1Y_\emptyset$.

   (iv) Use the ROM of (iii) to build a circuit for multiplying two 4-bit binary numbers. Additional logic may be required.

   (v) Using the ROM of (ii), design a circuit to add two 6-bit binary numbers.

   (vi) Use a (8×8) ROM to build a 3 line-to-8 line decoder.

   (vii) Use a ROM of the required size ($2^K \times N$) to build a 4 line—1 line MUX with two select inputs.

   (viii) Use a ROM of the required size ($2^K \times N$) to build a 1 line to 4 line DEMUX with two select inputs.

Chapter 4

Sequential Circuit Design

We saw in our discussion of Combinatorial Circuits in Chapter 2, that for these circuits the outputs are a function of the current inputs alone. In contrast, a Sequential Circuit is defined in Chapter 3 to be a circuit whose outputs not only depend on the current inputs but also on past ones. Such a circuit obviously requires memory elements, like flip flops, for 'remembering' earlier values of logic variables. In this chapter, we will discuss briefly, the design of various simple sequential circuits like counters as well as indicate how more complex sequential circuits may be designed.

4.1 RIPPLE COUNTERS

The ripple counter is a very basic counter design which is easy to implement with off-the-shelf integrated circuit components. Its simplicity and ease of fabrication makes it a very popular counter. As we shall see next, it also suffers from some very important disadvantages—its speed of operation is limited and decoding its output might cause problems. These should be kept in mind when deciding whether a simple ripple counter would be suitable for a given application.

Consider the simple 3-bit ripple counter shown in Fig. 4.1.1(a) which has been designed using negative edge triggered JK flip-flops. The JK flip-flops have been configured as T flip-flops which toggle when a Hi → Lo edge is applied at their clock inputs. The clock to the counter is connected directly to the clock input of the first flip-flop. The Q-output of this flip-flop is connected to the clock input of the next flip-flop. This connection strategy is repeated for the next flip-flop stage. The clear inputs of all the flip-flops are connected to a common Clear. A Ø at this input clears all the flip-flops and puts the counter in the ØØØ state (i.e. C = Ø, B = Ø, A = Ø).

Referring to the timing diagram of Fig 4.1.1(b), consider a situation where the flip-flop is in the C = Ø, B = Ø, A = Ø state (ØØØ). In Fig. 4.1.1(b), we have shown this being achieved by a Clear = Ø signal. For counting, the Clear input is now taken to 1. Now a Hi → Lo transition of the clock input toggles A to 1; at this point, B and C are equal to Ø. Another Hi → Lo edge transition of the clock signal toggles A to Ø. This causes a Hi → Lo edge at the clock input of the B-flip-flop making that toggle. Now A is Ø and B is 1; the C output is still Ø. Denoting the counter state by the binary number equivalent of CBA, the counter steps from state ØØØ to 111 in steps of 1. The next Hi → Lo edge on the clock input returns the counter to state ØØØ. [This is because the Hi → Lo of the clock toggles A to Ø; this causes a Hi → Lo edge on A which toggles B to Ø and finally the resultant Hi → Lo transition of B toggles C to Ø. Therefore a Hi → Lo of the clock input when the counter is at 111 takes it to ØØØ]. The counter has 8 states from ØØØ(Ø) to 111 (7). This is the reason why the counter is referred to as a Modulo-8 (or Mod-8) counter. Since the counter has three output bits A, B and C, it is also called a 3-bit binary counter.

From the above discussion, it appears that on successive Hi → Lo edges of the clock input, the counter counts up sequentially from ØØØ to 111 (and then to ØØØ, thereafter repeating the sequence). This unfortunately, is not strictly true; as the counter changes state from state N to

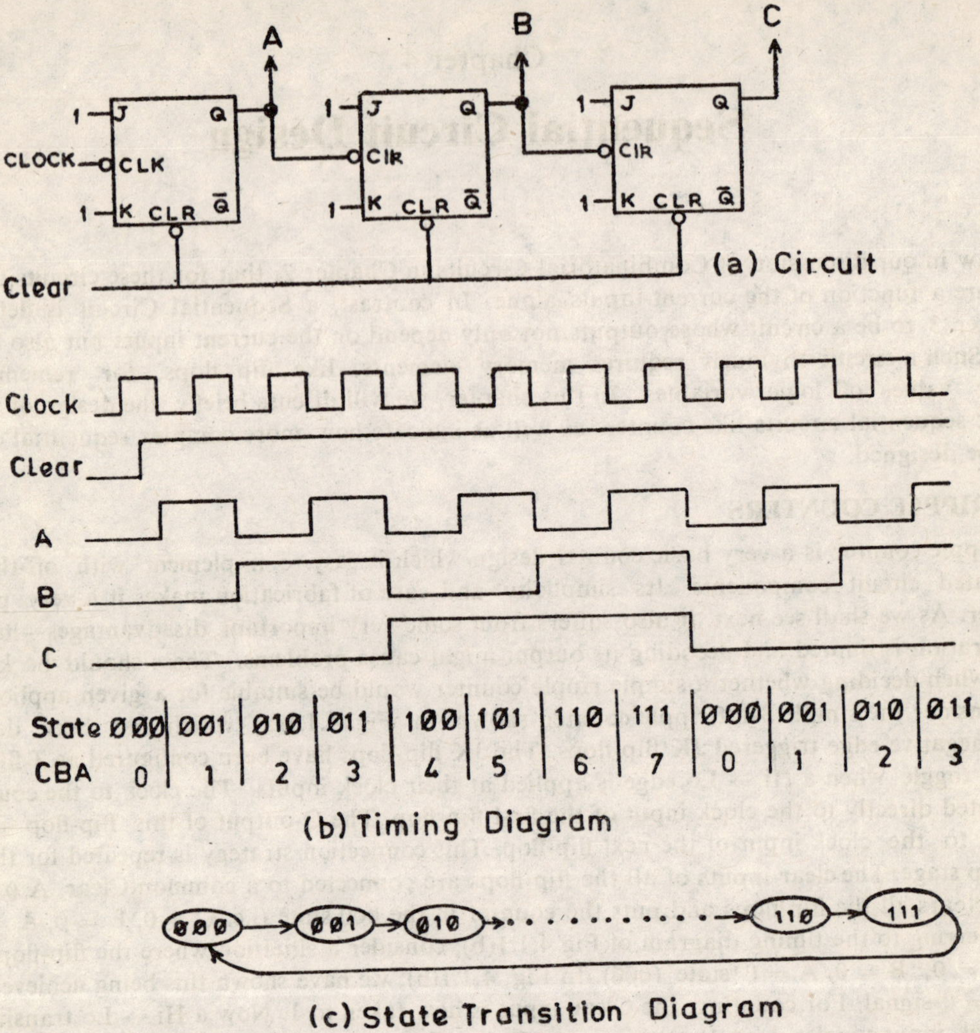

Fig. 4.1.1 3-Bit Modulo-8 Ripple Counter

state (N + 1) because of a clock edge, it might go through some other states (≠N, ≠(N + 1)) in a transient fashion. This arises in the ripple counter because of the time it takes for a signal in the input side to travel to the output, i.e., the time difference between a Hi → Lo edge at the clock input and the change in the Q-output.

Consider the situation where a Hi → Lo edge of the clock input occurs at time T when the counter is in state Ø11, and let t_{pf} be the flip-flop propagation delay. The state of the counter will not change until time (T + t_{pf}); at this instant the counter will go to state Ø1Ø because A will toggle. The toggling of A (1 → Ø) will propagate to the B output at time (T + $2t_{pf}$) when the counter goes to state ØØØ. Finally the toggling of B (1 → Ø) will toggle C (Ø → 1) at time (T + $3t_{pf}$) at which time instant the counter will reach the desired final state of 1ØØ. Note that the desired state transition is from Ø11 to 1ØØ but that the counter goes through the transient states Ø1Ø and ØØØ in the process. Fig. 4.1.2 lists the various transient states that the modulo-8 counter of Fig. 4.1.1 will go through. (Note the curious fact that as the counter goes through count sequence from ØØØ to 111, the transient states appear only in the odd-to-even transitions and that the transient states appearing correspond to *some* of the even states which have appeared earlier).

Most counter applications are such that one has to decode the state of the counter; typically, the system using the counter is required to take some specific action on decoding specific states. For example, we may have a decoding circuitry with our modulo-8 counter which provides eight output lines Q_0-Q_7 such that Q_n is active only when the counter is in state *n*. With a ripple counter, these decoded outputs (actually only those for the even states) will have '*decoding spikes*' corresponding to the transient states. These spikes may cause problems in a system using this counter.

One way of avoiding decoding spikes in the decoded outputs is to 'strobe' the decoder only after the transient counter states have disappeared. By examining the counter after a time delay sufficient for the transient states to disappear, one can completely avoid the 'decoding spikes' problem. This will, however, be done at the cost of a reduced operating speed of the counter-decoder combination.

In order to see this speed problem, consider a N-bit ripple counter constructed like the counter of Fig. 4.1.1. Let t_{pf} be the propagation delay of one flip-flop and let t_{dec} be the time taken by the decoder. The total time taken for a change in the counter's clock input to propagate to the output of the Nth flip-flop is Nt_{pf} in the worst case. A further delay of t_{dec} is required for the decoder. Let f_{CLK} be the frequency of the clock signal. In this case

$$\frac{1}{f_{CLK}} \geqslant Nt_{pf} + t_{dec}$$

or

$$f_{CLK} \leqslant \frac{1}{Nt_{pf} + t_{dec}}$$

gives the upper limit of the clock frequency as $1/(Nt_{pf} + t_{dec})$. For a 3-bit ripple counter with $t_{pf} = 50$ ns and $t_{dec} = 100$ ns, we get that the maximum clock operating frequency is 4 MHz. (For higher operating speeds, synchronous designs must be used. These will be discussed in Sec. 4.2).

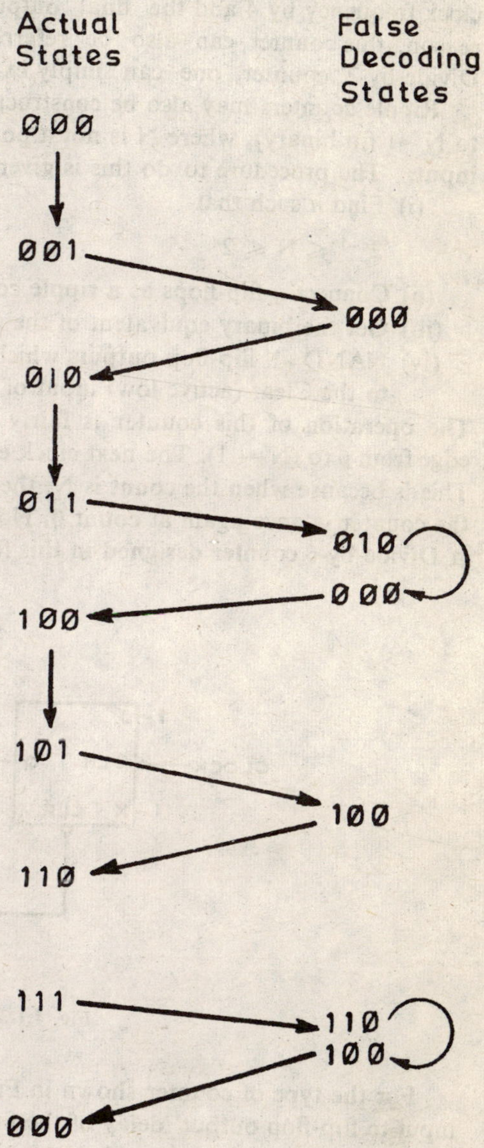

Fig. 4.1.2 False Decoding States in a 3-bit Module-8 Ripple Counter

Consider once again, the 3-bit modulo-8 ripple counter of Fig. 4.1.1. As can be seen from the timing diagram, the output of A divides the clock frequency by 2, the B output divides the clock frequency by 4 and the final output at C divides the input clock frequency by 8. For this reason, this counter can also be referred to as Divide-by-8 counter. In order to obtain a Divide-by-2^n counter, one can simply extend the circuit of Fig. 4.1.1 to one with *n* flip-flops.

Ripple counters may also be constructed to obtain a Divide-by-N counter counting from ϕ to $N-1$ (in binary), where N is not a power of 2. For this, one requires flip-flops with a Clear input. The procedure to do this is given below:

(i) Find *n* such that

$$2^{n-1} \leqslant N \leqslant 2^n$$

(ii) Connect *n* flip-flops as a ripple counter (as for the 3-bit case of Fig. 4.1.1).

(iii) Get the binary equivalent of the number N.

(iv) NAND all flip-flop outputs which would be 1 for a count of N. Connect NAND output to the Clear (active low) input of all the flip-flops.

The operation of this counter is fairly straight-forward. It counts up by 1 for each active clock edge from Ø to (N − 1). The next clock edge makes the flip-flop reach a count of N momentarily. This is because when the count is N, the NAND output is Ø clearing all the flip-flops such that the counter is once again at count Ø. Now the counter resumes counting up to N − 1 once again. A Divide-by-6 counter designed in this fashion is shown in Fig. 4.1.3.

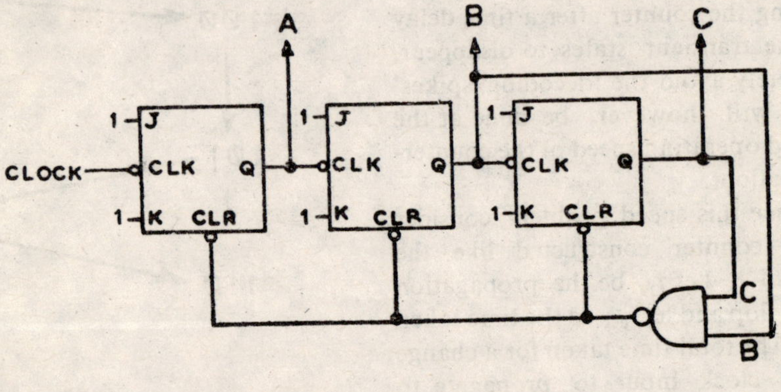

Fig. 4.1.3 Divide-by-6 Ripple Counter

For the type of counter shown in Fig. 4.1.3, the width of the reset pulse is equal to the 'reset input to flip-flop output' delay of the fastest flip-flop. Care has to be taken to ensure that a Clear signal of this duration can indeed reset all the flip-flops. If it cannot, the NAND gate should be replaced by a latch circuit such as the one shown in Fig. 4.1.4.

A ripple counter is an asynchronous counter because apart from the first flip-flop, the outputs of the other flip-flops do not change synchronously with the clock. As the previous discussion showed, most of the problems associated with the ripple counter arises from this asynchronous behaviour. We discuss next, synchronous counters which eliminate the decoding spikes and speed limitations of the ripple counter design by ensuring that all the flip-flops change state synchronously with the clock.

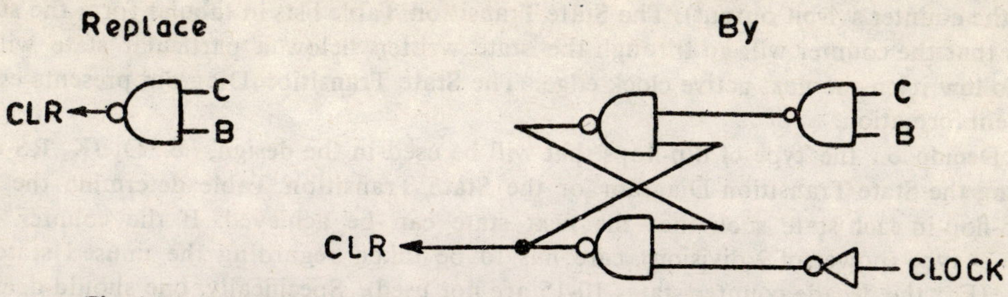

Fig. 4.1.4 Replacing NAND of Divide-by-6 Ripple Counter by Latch Circuit

4.2 SYNCHRONOUS COUNTERS

In a Synchronous Counter, all the constituent flip-flops are under the control of the same clock signal. The inputs to any particular flip-flop (e.g. J-K inputs) are determined as a logical combination of the counter's outputs (i.e. the flip-flop outputs). Therefore, in Synchronous Counters, the speed of clocking is limited only by the delay of one flip-flop and the worst case delays of the combinatorial gating being used to generate the flip-flop inputs. If all the flip-flops have the same propagation delay, the decoded counter output will be free from the 'spikes' encountered in the ripple counter. In any case, the duration of any decoding spike will be at worst equal to the difference in the propagation delays of the slowest and fastest flip-flops.

The general approach to the design of a divide-by-N synchronous counter proceeds as follows. We will illustrate the procedure for a Synchronous Decade (Divide-by-10) Counter.

(i) Determine the number of flip-flops-*n*-needed to implement the counter by using

$$2^{n-1} \leqslant N \leqslant 2^n$$

For the decade counter, e.g. $n = 4$.

(ii) Write the State Transition table or draw the State Transition Diagram of the counter. For the decade counter, these are shown in Fig. 4.2.1. (Note that Q_3 represents the MSB and $Q_Ø$ the

	Q_3	Q_2	Q_1	$Q_Ø$
0	Ø	Ø	Ø	Ø
1	Ø	Ø	Ø	1
2	Ø	Ø	1	Ø
3	Ø	Ø	1	1
4	Ø	1	Ø	Ø
5	Ø	1	Ø	1
6	Ø	1	1	Ø
7	Ø	1	1	1
8	1	Ø	Ø	Ø
9	1	Ø	Ø	1
0	Ø	Ø	Ø	Ø
⋮		⋮		

(a) State Transition Table

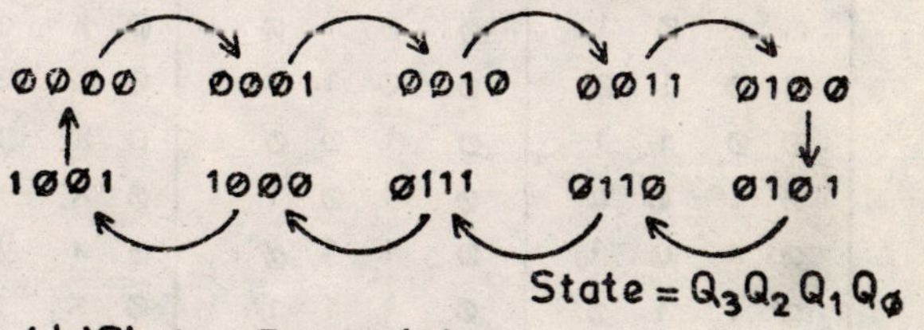

(b) State Transition Diagram

Fig. 4.2.1 State Transition Table and Diagram for Decade Counter

LSB of the counter's 4-bit output). The State Transition Table lists in tabular form the succession of states that the counter will go through the state written below a particular state will be the one to follow it on the next active clock edge. The State Transition Diagram presents essentially the same information.

(iii) Decide on the type of flip-flops that will be used in the design, i.e., D, JK, RS or T. By examining the State Transition Diagram or the State Transition Table determine the input of each flip-flop in each state such that the next state can be achieved. If the counter is not a Modulo-2^k type (power of 2 division), care has to be taken regarding the unused states of the counter. (For the decade counter states 10-15 are not used). Specifically, one should decide what the next state will be if the current state is an unused state. This is important because on power-up, the counter may land up in an unused state. In a pathological situation, the counter may keep switching from one unused state to another without ever entering the desired counting sequence.

Two approaches are recommended for this 'unused-state' problem.

(a) Use the preset and clear inputs of the flip-flops to ensure that the counter can be initialized to a particular state in the count sequence. (It should thereafter stay in the count sequence) or (b) Extend the state transition diagram or the state transition table in such a way that even if the counter starts from an unused state, it will enter the proper counting sequence within a finite number of clock cycles. Determine the inputs of each flip-flop in each state-used or unused-for this extended state transition table (or diagram).

Approach (a) is simple but requires the use of flip-flops having the preset and/or clear inputs. Approach (b) is more complex (more complex combinatorial logic) but does not place such a requirement. Note that in Approach (a), the flip-flop inputs in the unused states will be *don't cares*.

For our decade counter, we will use Approach (a). Fig. 4.2.2 gives the inputs that will be

State				D Flip Flop				JK Flip Flop								T FlipFlop			
Q_3	Q_2	Q_1	Q_0	D_3	D_2	D_1	D_0	J_3	K_3	J_2	K_2	J_1	K_1	J_0	K_0	T_3	T_2	T_1	T_0
0	0	0	0	0	0	0	1	0	X	0	X	0	X	1	X	0	0	0	1
0	0	0	1	0	0	1	0	0	X	0	X	1	X	X	1	0	0	1	1
0	0	1	0	0	0	1	1	0	X	0	X	X	0	1	X	0	0	0	1
0	0	1	1	0	1	0	0	0	X	1	X	X	1	X	1	0	1	1	1
0	1	0	0	0	1	0	1	0	X	X	0	0	X	1	X	0	0	0	1
0	1	0	1	0	1	1	0	0	X	X	0	1	X	X	1	0	0	1	1
0	1	1	0	0	1	1	1	0	X	X	0	X	0	1	X	0	0	0	1
0	1	1	1	1	0	0	0	1	X	X	1	X	1	X	1	1	1	1	1
1	0	0	0	1	0	0	1	X	0	0	X	0	X	1	X	0	0	0	1
1	0	0	1	0	0	0	0	X	1	0	X	0	X	X	1	1	0	0	1
others				X	X	X	X	X	X	X	X	X	X	X	X	X	X	X	X

$D_3 = Q_3 + Q_2Q_1Q_0,\ D_2 = Q_2\bar{Q}_1 + Q_2\bar{Q}_0 + \bar{Q}_2 \cdot Q_1 \cdot Q_0,\ D_1 = \bar{Q}_3 \cdot \bar{Q}_1 \cdot Q_0 + Q_1 \cdot \bar{Q}_0,\ D_0 = \bar{Q}_0$

$J_3 = Q_2Q_1Q_0,\ K_3 = Q_0,\ J_2 = K_2 = Q_1 \cdot Q_0,\ J_1 = K_1 = \bar{Q}_3 \cdot Q_0,\ J_0 = K_0 = 1$

$T_3 = Q_3Q_0 + Q_2Q_1Q_0,\ T_2 = Q_1 \cdot Q_0,\ T_1 = \bar{Q}_3 \cdot Q_0,\ T_0 = 1$

Fig. 4.2.2 Flip-Flop Inputs to Generate the Next State in a Decade Counter

required if D, JK or T flip-flops are being used. (Refer to Fig. 3.3.5, 3.3.7 and 3.3.9).

(iv) Use the Karnaugh Map minimization procedure to obtain a logic expression of the input required at each flip-flop, as a function of the current state outputs Q_3, Q_2, Q_1, $Q_\emptyset$. For the decade counter, these expressions are given in Fig. 4.2.2 for D, JK and T flip-flops.

(Note that if one uses the D flip-flop realization with the expressions given for $D_3D_2D_1$ and $D_\emptyset$ in Fig. 4.2.2, then a counter initialized to 11ØØ or 11Ø1 at power-up will keep toggling back and forth between the states. This should be viewed in context with the unused state problem mentioned earlier).

(v) Use the minimized logic expressions for the flip-flops' inputs to obtain the circuit realization of the counter. For the decade counter, a circuit using JK flip-flops is shown in Fig. 4.2.3.

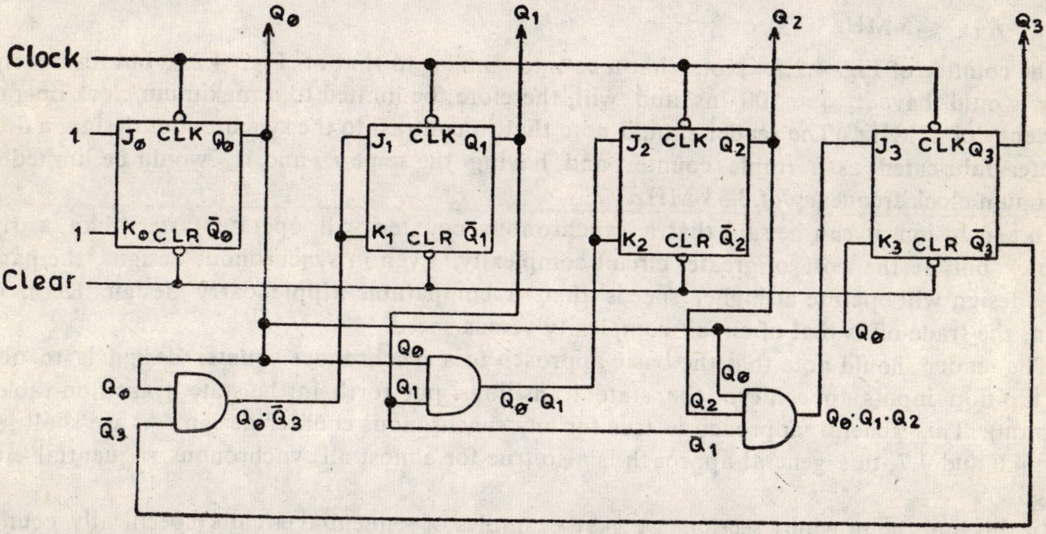

Fig. 4.2.3 Synchronous Decade Counter with Parallel Carry

The realization of the decade counter shown in Fig. 4.2.3 is known as a synchronous decade counter with parallel carry. To see why it is referred to as a counter with parallel carry, consider how the flip-flop inputs are generated. Specifically, note that $J_2 = K_2 = Q_1 \cdot Q_\emptyset$ and that $J_3 = Q_2 \cdot Q_1 . Q_\emptyset$. In the circuit of Fig. 4.2.3, J_3 is directly generated using a 3-input NAND gate. All other inputs are similarly directly generated from the flip-flop outputs using combinatorial logic. Since the propagation delay from the flip-flop outputs to the flip-flop inputs are roughly equal (one gate delay, in this case), valid flip-flop inputs will appear in parallel, i.e. roughly at the same time. This is the reason why the circuit of Fig. 4.2.3 is said to be using parallel carry.

In Fig. 4.2.3, it is also possible to generate J_3 as $J_2 \cdot Q_2$ using a 2-input NAND gate rather than a 3-input one. This kind of realization is referred to as one with *ripple carry* as the J_3 input will now be valid only one gate delay after J_2 is valid. Because of this increased propagation delay, a ripple carry synchronous counter cannot work as fast as one using parallel carry.

Consider the synchronous decade counter with parallel carry shown in Fig. 4.2.3. The total delay between an active edge of the clock and the decoded output of the counter comprises of the following

(a) t_{pf}—the time delay of one flip-flop and

(b) t_{dec}—the time delay of the decoder

Combined with t_{com}—the worst case time delay of the combinatorial logic generating the flip-flop inputs, one gets

$$t_{CLK} \geqslant t_{pf} + t_{dec} + t_{com}$$

Assuming $t_{pf} = 50$ ns, $t_{dec} = 100$ ns and that a logic gate has a propagation delay of 50 ns

$$t_{com} = 50 \text{ ns}$$

and

$$t_{CLK} \geqslant 200 \text{ ns}$$

or

$$f_{CLK} \leqslant 5 \text{ MHz}$$

for the counter of Fig. 4.2.3. Note that a counter similar to that of Fig. 4.2.3 but using ripple carry would have $t_{com} = 100$ ns and will, therefore, be limited to a maximum clock operating frequency of 4 MHz. [The reader should note that in contrast to the synchronous design, a decade counter fabricated as a ripple counter and having the same t_{pf} and t_{dec} would be limited to a maximum clock frequency of 3.33 MHz.]

In conclusion, it can be said that a synchronous counter will operate faster than a ripple counter but at the cost of greater circuit complexity. Even in synchronous designs, the parallel carry design will operate at higher speeds than a comparable ripple carry design. Here, once again, the trade-off is that of circuit complexity versus speed.

The reader should note that the basic approach to a synchronous counter design is to obtain the flip-flop inputs to achieve the state transitions put forth in the state transition table (or diagram). This general approach is true for any synchronous counter design. As we shall see in Secs. 4.6 and 4.7, this general approach is also true for almost all synchronous sequential circuit design.

In this and the previous section, we gave examples of sequential circuits (specifically counters) which used discrete flip-flops (RS, D, JK or T) in their design. In the next few sections, Secs. 4.3–4.5, we will describe briefly some sequential circuits which use Shift Registers in their design. These are the Ring Counters, Johnson Counters and the Linear Shift Register Counters.

4.3 RING COUNTERS

The ring counter is the simplest example of a shift register counter. The basic ring counter circuit merely has a feedback arrangement connecting the serial output of the shift register to its serial input. This basic circuit is shown for the 3-bit case in Fig. 4.3.1 (a).

As can be seen from the transition diagram of the 3-bit ring counter given in Fig. 4.3.1 (b), the count sequence of a ring counter depends very much on the starting state. This situation will also be true for a N-bit ring counter employing basic feedback. In this case also, a number of different counting sequences are possible where—

(a) which count sequence is followed will depend on the initial state and

(b) once the counter is in a particular count sequence, it will stay forever in that sequence without ever reaching a state outside that sequence.

For the 3-bit ring counter of Fig. 4.3.1 (a), four such count sequences are possible. These have been shown in Fig. 4.3.1 (b).

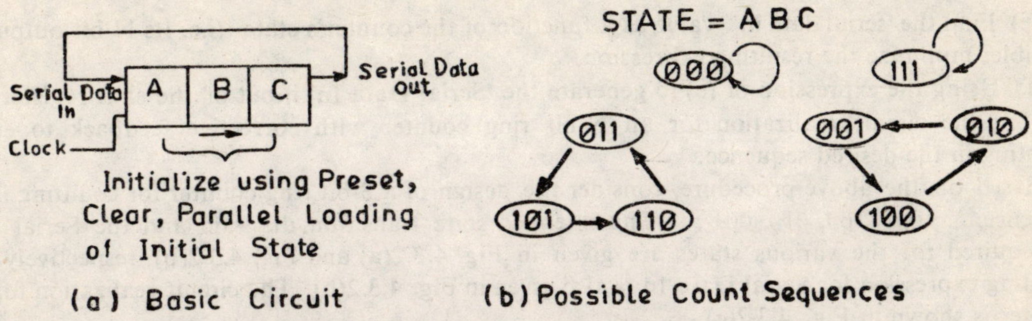

Fig. 4.3.1 3-bit Ring Counter (with Basic Feedback)

Given a ring counter of the basic feedback type, it is therefore essential that one provides an arrangement to force the counter into a 'desirable' state before counting can begin. (In the present context, a desirable state is any one of the states lying in the count sequence one wants executed. It is of course possible for one to be more specific and designate a particular state as the 'desirable state' to be introduced on initialization). If one has access to the Presets and Clears of each of the N-D flip-flops in the shift register this can be easily achieved by selectively setting or resetting these flip-flops. If the N-bit shift register allows parallel loading, one can put the ring counter in the desired initial state simply by loading this into the shift register. One can also break the feedback loop, shift in the desired state in the shift register bit-by-bit and then restore the feedback loop; the ring counter is then ready to count in the desired counting sequence. Note that for the latter, all of these can be done by using a multiplexer at the serial data input, just as we did previously in Sec. 3.4 (Fig. 3.4.2).

Another way of ensuring that a N-bit ring counter goes in the desired counting sequence is by using Corrective Feedback. The following procedure may be used to achieve this:

(a) Identify the desired counting sequence.

(b) For states not in the counting sequence, identify a sequence of states that will finally force the counter into the counting sequence.

The idea here is to ensure that even if the counter starts off in a state outside the counting sequence (due to a random state at power-up or some other reason), it will finally reach a state in the desired counting sequence after a finite number of clock cycles. Thereafter, it should stay in the desired counting sequence.

(c) Using (b), draw a complete state transition diagram. For every possible counter state, this diagram should indicate the successor state. This diagram should be cross-checked to verify that what was indicated in (b) is indeed being done, i.e. that a succession of states exist which allow the counter to enter the desired count sequence even when it starts off in a state outside this sequence.

[Note that in a N-bit ring counter designed with a N-bit shift register, one may only be able to modify the LSB of the previous contents of the shift register. For example, for the 3-bit ring counter, if we want to go from state 111 to ØØ1 we will have to do this in two steps:

$$111 \rightarrow \varnothing 11 \rightarrow \varnothing\varnothing 1$$

A direct transfer from one state to another may not always be possible. This aspect should be kept in mind while doing steps (b) and (c).]

(d) Find the 'serial data in' input required to implement the transition diagram of (c).

(e) Find the 'serial data in' value as a function of the counter's state (i.e. its N-bit output). If possible, minimize the resulting expression.

(f) Using the expression of (e) to generate the 'Serial Data In' input of the shift register, one can find the circuit realization for an N-bit ring counter with corrective feedback to ensure counting in the desired sequence.

Based on the above procedure, consider the design of a 3-bit ring counter for counting in the sequence. . . , ØØ1, 1ØØ, Ø1Ø, ØØ1 . . . The extended state transition diagram and the Serial Data In required for the various states are given in Fig. 4.3.2(a) and Fig. 4.3.2(b), respectively. The resulting expression for Serial Data In is also given in Fig. 4.3.2(b). The circuit realization for this counter is shown in Fig. 4.3.2(c).

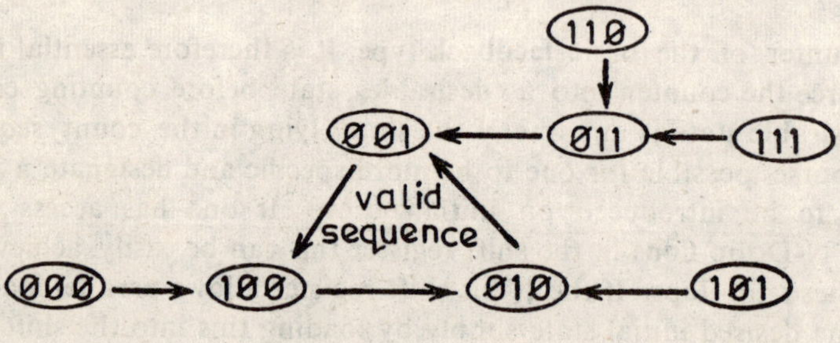

(a) Extended State Transition Diagram

State	D_{in}
ØØØ	1
ØØ1	1
Ø1Ø	Ø
Ø11	Ø
1ØØ	Ø
1Ø1	Ø
11Ø	Ø
111	Ø

D_{in} = Serial Data In

State = A, B, C

$$D_{in} = \overline{A} \cdot \overline{B} = \overline{A + B}$$

(b) Evaluation of D_{in}

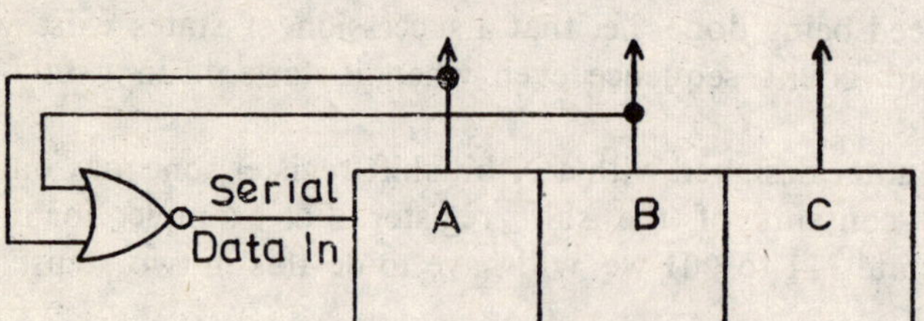

(c) Circuit Realization

Fig. 4.3.2 3-Bit Ring Counter with Corrective Feedback for Self-Start

Note that the most common use of the ring counter is to circulate either a single 1 or a single Ø. The advantage then is that the state of the counter can be directly determined without additional decoding. (In this case, the counter state is directly determined by the position of the circulating 1 or Ø). For such a use, one has to ensure that the counter counts (at least eventually) in the counting sequence circulating a single 1 (or a single Ø).

4.4 JOHNSON COUNTER

The basic structure of a Johnson Counter (also called Twisted Ring or Mobius Counter) is very similar to that of a Ring Counter. The primary difference between the two is that whereas the Ring Counter feeds in the data shifted out (Serial Data Out) of the 'shift register's last stage into its first stage (Serial Data In), the Johnson Counter shifts in the complemented value of that data. Figure 4.4.1 shows such an N-bit counter with the basic feedback arrangement.

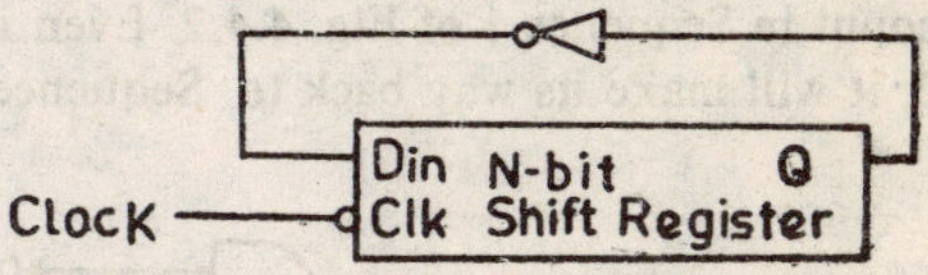

Fig. 4.4.1 N-bit Johnson Counter with Basic Feedback

The properties of a N-bit Johnson Counter with basic feedback are summarised below.

(i) There will be at least one counting sequence with 2N states (length of sequence = 2N).

(ii) There may be other counting sequences but the length of these cannot be greater than 2N.

(iii) The counting sequence followed depends on the initial state of the counter.

(iv) If one selects a counting sequence of 2N states as the desired counting sequence, then the counter will have $(2^N - 2N)$ undesirable states.

Figure 4.4.2 shows the counting sequences produced by a 4-bit Johnson Counter. Note that in this case, both the sequences have 8 states each and one can choose any one of the two to be the desired count sequence. If nothing else is done, then just as for the Ring Counter, the 4-bit Johnson Counter may start-up in any one of the two sequences of Fig. 4.4.2 on power-up. (It may also transfer from the desired sequence to the undesirable sequence because of noise effects).

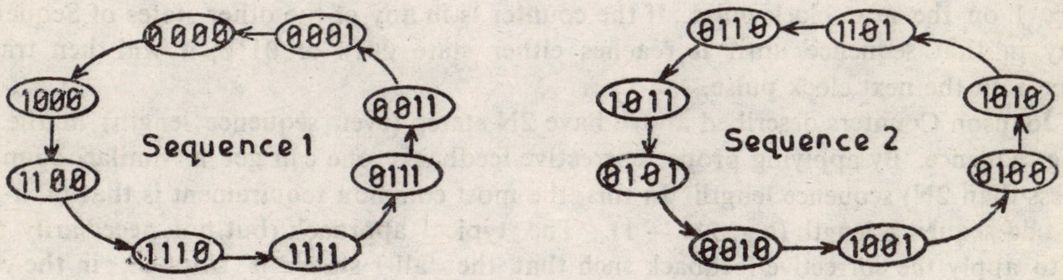

Fig. 4.4.2 State Transition Diagram for 4-bit Johnson Counter

To place a Johnson Counter in the desired counting sequence, one can adopt the same strategies as those suggested for the Ring Counter in Sec. 4.3. The Counter can be placed in the desired initial state (a state in the desired sequence) by

(i) initializing the flip-flops of the counter to the desired state using the Preset and Clear inputs of the flip-flops.

(ii) parallel loading of the desired state.

(iii) serial loading of the desired state (by shifting in the desired state with the feedback broken).

One can also use corrective feedback to ensure that even if the counter lands up in an undesired state, it will end up in the desired counting sequence after a finite number of clock cycles. This procedure is similar to the corrective feedback procedure employed for the Ring Counter (e.g. the one illustrated in Fig. 4.3.2) and involves generating an appropriate Serial Data In signal which is a function of the state of the counter.

Figure 4.4.3 shows a 4-bit Johnson Counter with modified (corrective) feedback designed to count in Sequence 1 of Fig. 4.4.2. Even if the counter, somehow, lands up in a state in Sequence 2, it will make its way back to Sequence 1 because of the corrective feedback. The circuit of

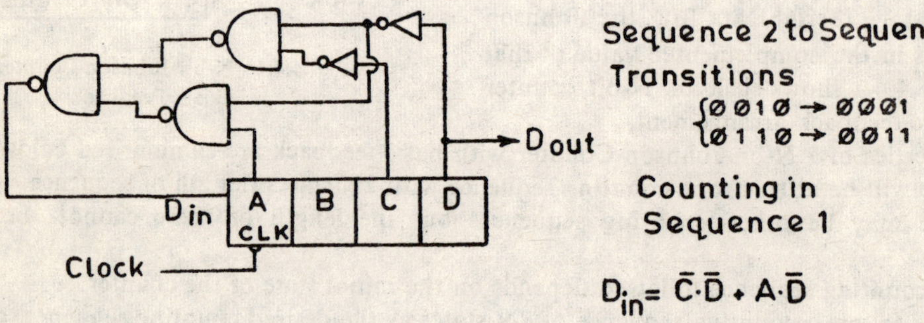

Fig. 4.4.3 4-bit Johnson Counter with Modified Feedback

Fig. 4.4.3 provides two exit points from Sequence 2 to Sequence 1. These are

(i) ØØ1Ø (Sequence 2) to ØØØ1 (Sequence 1) and

(ii) Ø11Ø (Sequence 2) to ØØ11 (Sequence 1)

If the counter is in states ØØ1Ø or Ø11Ø, it will directly transfer to the corresponding state of Sequence 1 on the next clock pulse. If the counter is in any of the other states of Sequence 2, it will stay in that sequence until it reaches either state ØØ1Ø or Ø11Ø; it will then transfer to Sequence 1 on the next clock pulse.

The Johnson Counters described above have 2N states (even sequence length) in the desired counting sequence. By applying proper corrective feedback, one can get a similar counter with other (less than 2N) sequence length. In this, the most common requirement is that of a counter with an odd sequence length (e.g. 2N − 1). The typical approach (but not necessarily the only one) is to apply the corrective feedback such that the 'all-1 state' is skipped, in the counting sequence. For the 4-bit Johnson Counter case, one can then apply the following feedback

$$D_{in} = \bar{C}.\bar{D} + A.\bar{B}.\bar{D}$$

With this feedback, all the properties of the counter of Fig. 4.4.3 will be maintained except that

(i) in the counting sequence, state Ø111 will follow state 111Ø (state 1111 is skipped) and

(ii) a counter starting up in state 1111 will transfer to state Ø111 (and hence to the counting sequence) in the next clock pulse.

4.5 LINEAR SHIFT REGISTER COUNTERS

Counters of this type are also constructed using N-bit shift registers but typically yield much longer cycle lengths than the Johnson (cycle length = 2N) or the Ring (cycle length N) Counters. With proper feedback connections, these linear shift register counters can yield cycle lengths of $(2^N - 1)$. This is comparable to the 2^N cycle length achievable with a N-bit synchronous binary counter. Compared to the latter, however, they are easier to build and are very economical in situations where a long cycle length is required. Note that these counters are also synchronous in nature and have all the associated advantages of synchronous counters. A Linear Shift Register Counter of cycle length $(2^N - 1)$ is referred to as a Maximum Length Sequence (MLS) Generator/Counter. A Linear Shift Register Counter with a cycle length less than $(2^N - 1)$ is referred to as a Non-Maximum Length Sequence (Non-MLS) Generator/Counter.

4.5.1 MLS Counters

The basic circuit of a N-bit MLS Counter is given in Fig. 4.5.1. This figure also shows the unique feature of all MLS Counters—the fact that the only logic required to derive D_{in} (of the Shift Register) from the current state of the counter (State = $Q_{\emptyset} \ldots Q_{N-1}$) is an Ex-OR operation involving some of the counter's outputs. Figure 4.5.2 lists some of the logic expressions that can be used for this purpose.

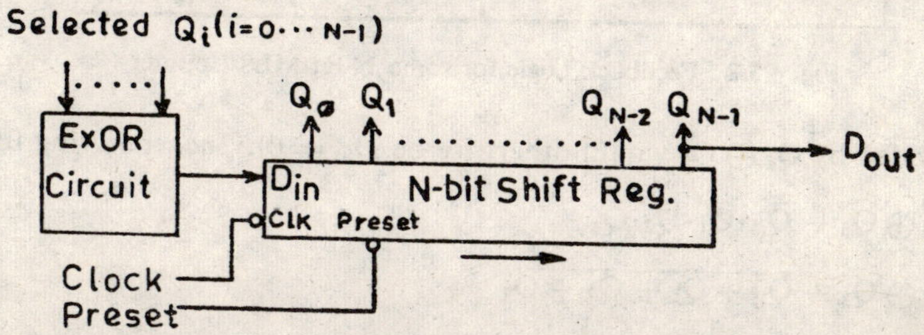

Fig. 4.5.1 Basic Circuit for a N-bit MLS Counter

Fig. 4.5.1 also indicates one problem inherent with these MLS Counters. In their basic form these counters are not self-starting. Note that the Ex-OR operation on k-Øs will generate the result Ø. This implies that the state ØØ . . . Ø is not a state in the MLS Counter's counting sequence since a MLS counter accidentally in this state will always remain there. A MLS counter can be made to be self-starting by incorporating additional (corrective) logic. The purpose of such additional logic will be to force the counter to go to state 1Ø . . . Ø from the state Ø. . . Ø and thus enter the desired counting sequence. It can also be made to start in state 1 . . . 1 (valid state in counting sequence) by using the Preset input (if available) as shown in Fig. 4.5.1.

Consider the design of a 4-bit MLS Counter (Cycle Length = 15). From Fig. 4.5.2, the Serial Data In (D_{in}) to the shift register is given by

$$D_{in} = Q_2 \oplus Q_3$$

Note that with D_{in} calculated as above, the counter will not be self-starting. In order to make it self-starting, we have to ensure that $D_{in} = 1$ when the counter is in state ØØØØ. (For state ØØØØ,

N No. of Shift Reg. Stages	D_{in} Feed back Expression
3	$Q_1 \oplus Q_2$
4	$Q_2 \oplus Q_3$
5	$Q_2 \oplus Q_4$
6	$Q_4 \oplus Q_5$
7	$Q_5 \oplus Q_6$
8	$Q_3 \oplus Q_4 \oplus Q_5 \oplus Q_7$
9	$Q_4 \oplus Q_8$
10	$Q_6 \oplus Q_9$
11	$Q_8 \oplus Q_{10}$
12	$Q_5 \oplus Q_7 \oplus Q_{10} \oplus Q_{11}$

Fig. 4.5.2 Feedback Logic for some N-bit MLS Counters

the feedback provided by $Q_2 \oplus Q_3$ would otherwise be $D_{in} = \emptyset$). Incorporating this

$$D_{in} = Q_2 \oplus Q_3 + \bar{Q}_\emptyset \cdot \bar{Q}_1 \cdot \bar{Q}_2 \cdot \bar{Q}_3$$

$$= Q_2 \oplus Q_3 + \overline{Q_\emptyset + Q_1 + Q_2 + Q_3}$$

for a 4-bit self-starting MLS counter. The circuit for this counter is given in Fig. 4.5.3(a). Fig. 4.5.3(b) gives the state transition diagram of this counter. Note the ØØØØ to 1ØØØ transition shown there.

The sequence of data bits appearing at the Serial Data Out (D_{out}) of a N-bit MLS Counter is referred to as a Maximal Length Sequence or a Pseudo Random Sequence of length $2^N - 1$. (This sequence also appears at the output of the other D-flip-flops of the shift register in an appropriately shifted form). For a 4-bit MLS counter, this sequence is given in Fig. 4.5.3(c).

The MLS Sequence has some very interesting properties. For example, an MLS sequence of cycle length $(2^N - 1)$ will always have (2^{N-1}) 1s and $(2^{N-1} - 1)$ Øs, i.e. one more 1 than Øs. Moreover, the distribution of 1s and Øs in the sequence satisfies some randomness properties rather closely, especially if the cycle length is sufficiently long. This is the reason why they are also referred to as a Pseudo Random Sequence or Pseudo Noise Sequence. MLS sequences are very often used in applications where an (approximately) random sequence is needed; the states of the MLS Counter are similarly used if one needs a sequence of (approximately) random numbers. Another important property of MLS sequence is that Ex-ORing a sequence with a shifted version of itself generates a time-shifted version of the original sequence. This is shown

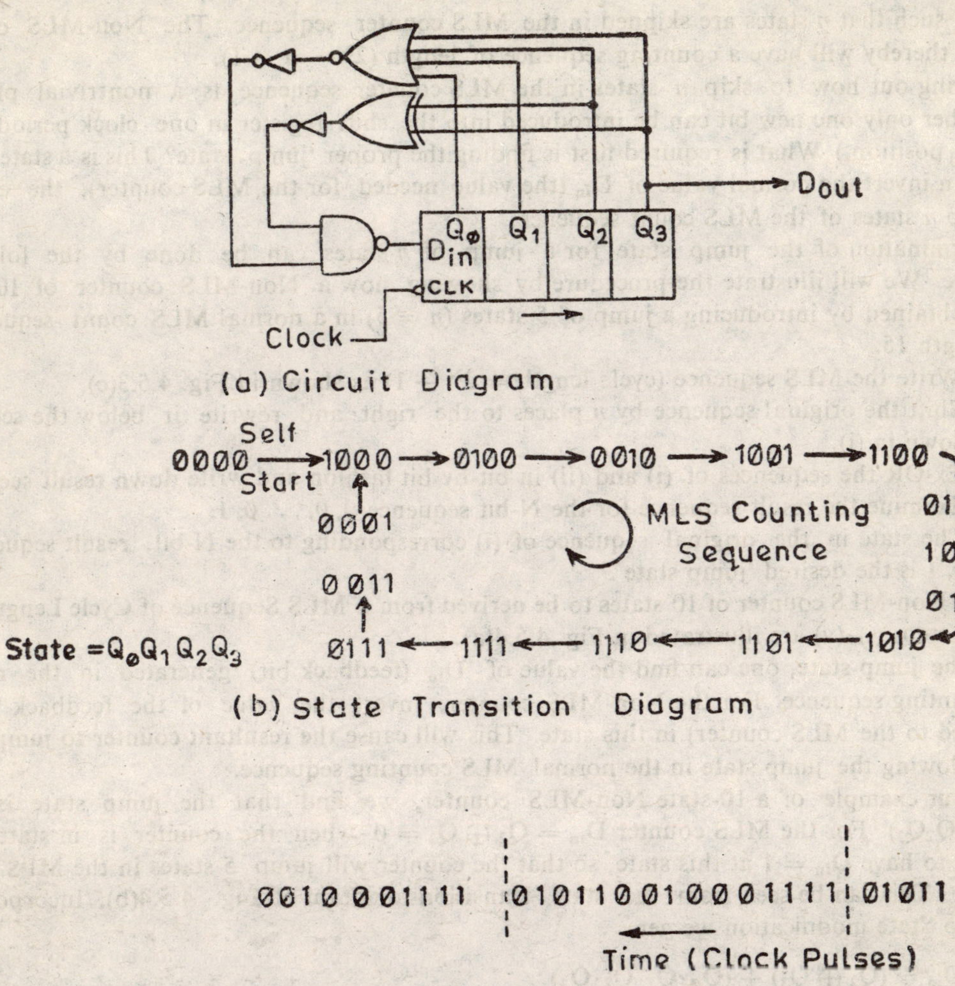

(a) Circuit Diagram

(b) State Transition Diagram

(c) Maximum Length Sequence of Length 15

Fig. 4.5.3 4-Bit Maximum Length Sequence Counter with Self-Start Feature

below for a MLS sequence of cycle length = 15

	. . . ØØ1ØØØ1111Ø1Ø11ØØ1Ø . . .
Shift left by 2	. . . 1ØØØ1111Ø1Ø11ØØ1ØØØ . . .
Ex-OR (bit by bit)	. . . 1Ø1Ø11ØØ1ØØØ1111Ø1Ø . . .

Note that in the above, the Ex-ORed result is the original sequence shifted left by 9 places.

4.5.2 Non-MLS Counters

If a N-bit shift register is being used, one can obtain a Non-MLS counter by the simple expedient of first building a MLS counter and then incorporating appropriate feedback in addition, such that n states are skipped in the MLS counter sequence. The Non-MLS counter obtained thereby will have a counting sequence of length $(2^N - n - 1)$.

Figuring out how to skip n states in the MLS counter sequence is a nontrivial problem (Remember only one new bit can be introduced into the shift register in one clock period—that in the $Q_\emptyset$ position.) What is required first is finding the proper 'jump' state? This is a state where if one can invert the normal value of D_{in} (the value needed for the MLS counter), the counter will jump n states of the MLS count sequence.

Determination of the jump state for a jump of n states can be done by the following procedure. We will illustrate the procedure by showing how a Non-MLS counter of 10 states may be obtained by introducing a jump of 5 states ($n = 5$) in a normal MLS count sequence of cycle length 15.

(i) Write the MLS sequence (cycle length $= 2^N - 1$) as shown in Fig. 4.5.3(c).

(ii) Shift the original sequence by n places to the right and rewrite it below the sequence written down in (i).

(iii) Ex-OR the sequences of (i) and (ii) in bit-by-bit fashion and write down result sequence.

(iv) Examine the result sequence for the N-bit sequence Ø, Ø, . . . Ø, 1.

(v) The state in the original sequence of (i) corresponding to the N-bit, result sequence of Ø, Ø . . . Ø, 1 is the desired 'jump state'.

For a Non-MLS counter of 10 states to be derived from a MLS Sequence of Cycle Length = 15 ($N = 4$), steps (i)–(v) are illustrated in Fig. 4.5.4(a).

For the jump state, one can find the value of D_{in} (feedback bit) generated in the normal MLS counting sequence. For the Non-MLS counter, invert the value of the feedback bit D_{in} (compared to the MLS counter) in this state. This will cause the resultant counter to jump the n states following the jump state in the normal MLS counting sequence.

For our example of a 10-state Non-MLS counter, we find that the jump state is 11ØØ. ($= Q_\emptyset Q_1 Q_2 Q_3$). For the MLS counter $D_{in} = Q_2 \oplus Q_3 = \emptyset$ when the counter is in state 11ØØ. We want to have $D_{in} = 1$ at this state, so that the counter will jump 5 states in the MLS Count Sequence. This can be seen from the state transition diagram of Fig. 4.5.4(b). Incorporating this Jump State modification we get

$$D_{in} = (Q_2 \oplus Q_3) + (Q_\emptyset \cdot Q_1 \cdot \overline{Q_2} \cdot \overline{Q_3})$$

If we include the self-start feature (exit the ØØØØ state to ØØØ1 state), we get

$$D_{in} = Q_2 \oplus Q_3 + Q_\emptyset Q_1 \overline{Q}_2 \overline{Q}_3 + \overline{Q}_0 \overline{Q}_1 \overline{Q}_2 \overline{Q}_3$$

$$= (\overline{Q_\emptyset \oplus Q_1} \cdot \overline{Q}_2 \cdot \overline{Q}_3) + (Q_2 \oplus Q_3)$$

The resultant circuit for a 10-state Non-MLS Counter is given in Fig. 4.5.4(c).

4.6 COUNTERS WITH AN ARBITRARY COUNT SEQUENCE

There are applications where a system designer is asked to generate a N-bit counter counting in some arbitrary counting sequence. This, for example, would be the case if the counter outputs are being directly used to enable/disable some other subsystems. An example of such an arbitrary state sequence is given in Fig. 4.6.1 for $N = 3$. Note that in Fig. 4.6.1, the self-start feature has been incorporated.

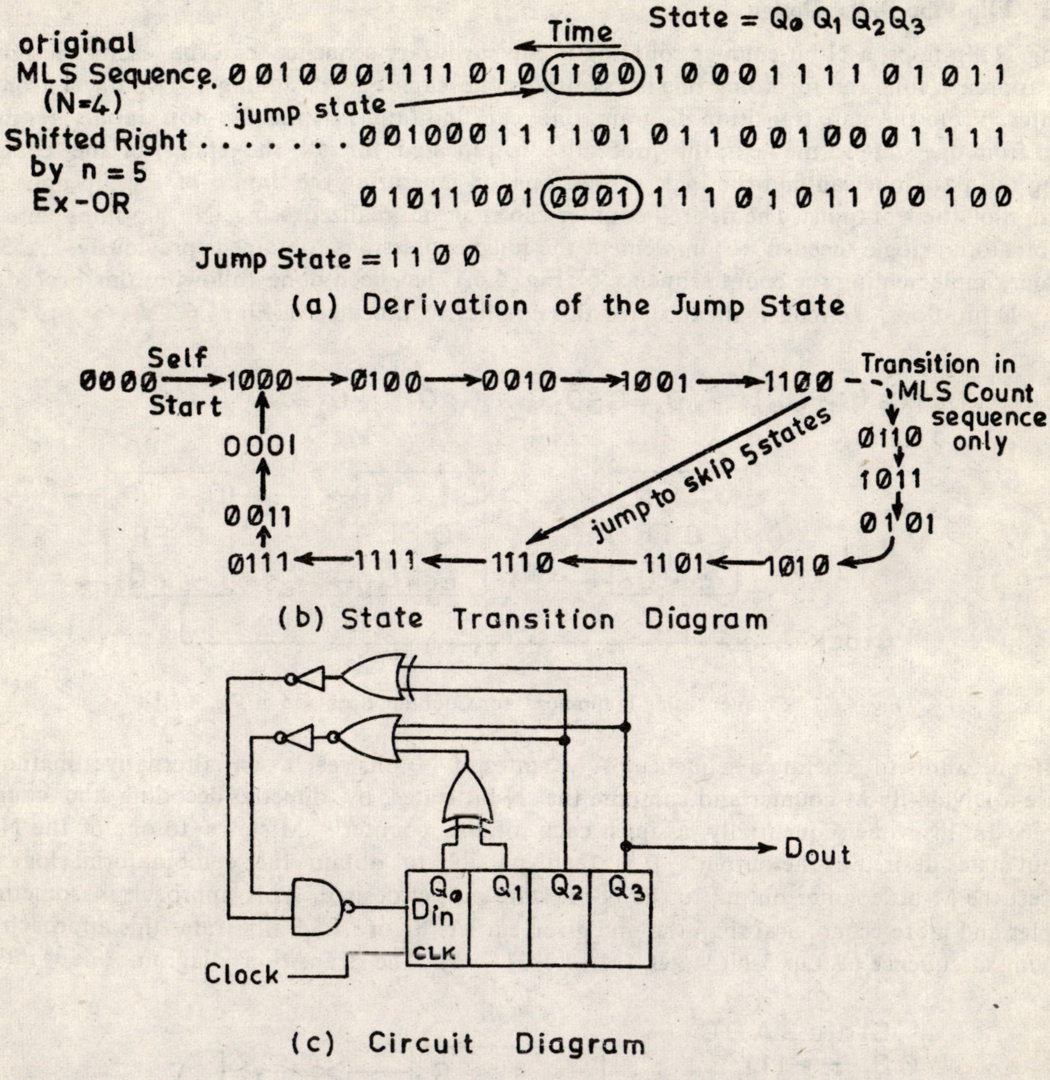

Fig. 4.5.4 10-State Non-MLS Counter Using a 4-Bit Shift Register

There are two approaches to the designing of such counters—one using flip-flops and the other using Read-Only-Memory (ROM). These two approaches are described next.

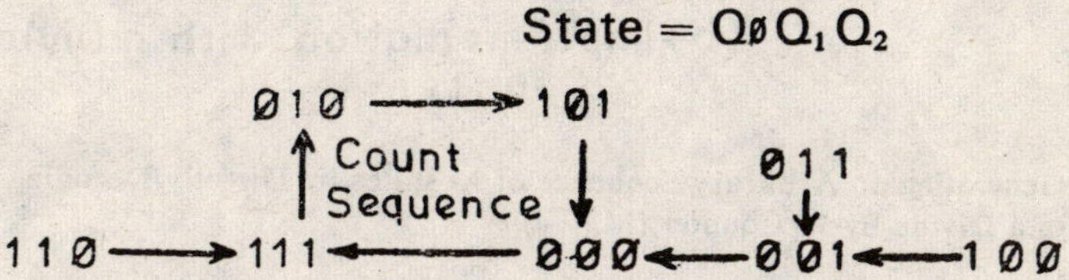

Fig. 4.6.1 State Transition Diagram of a 3-bit Counter Counting in a Prespecified Count Sequence (Self-Starting)

4.6.1 Flip-Flop Based Design

Using N flip-flops, a N-bit counter counting in any arbitrary sequence can be easily designed. The approach followed for doing this is essentially the same as designing a N-bit synchronous counter. From the state transition diagram, one can find (and list) the flip-flop inputs required to go from one state to the next; this procedure is repeated for all the states of the counter. Using this data, one can find the logic expressions for generating the inputs of the flip-flops as a function of their outputs. The desired counter can then be realized using N flip-flops and the combinatorial logic needed to implement the logic expressions obtained previously. A 3-bit counter, implementing the count sequence of Fig. 4.6.1, has been done following this procedure. Using D flip-flops, a circuit realization for this counter is indicated in Fig. 4.6.2.

$$D_\emptyset = \bar{Q}_2 (Q_1 + \bar{Q}_\emptyset) \quad , D_1 = \bar{Q}_\emptyset \bar{Q}_1 \bar{Q}_2 + Q_\emptyset Q_1 \quad , D_2 = \bar{Q}_2 + \bar{Q}_\emptyset Q_1$$

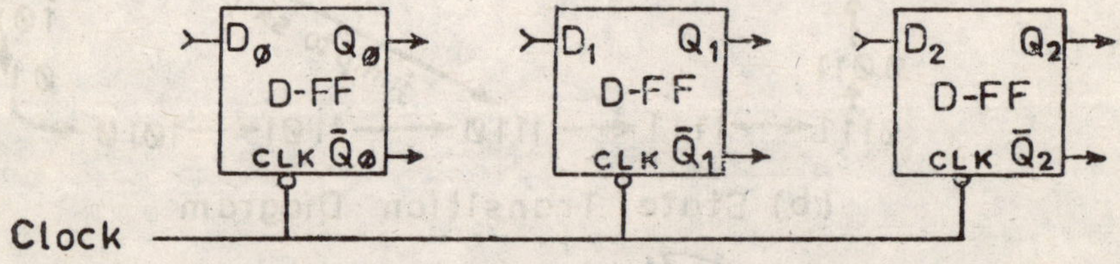

Fig. 4.6.2 Counter (using D flip-flops) for Counting Sequence of Fig. 4.6.1

If one wants to generate a sequence of M states of N bits each, an alternative method is to use a Divide-By-M counter and compute the N-bit states by directly decoding the counter output. In this, one sequentially assigns each of the counter's M-states to one of the N-bit output states desired at the output. It is then possible to obtain the combinatorial logic to convert the M-bit counter output to the N-bit state output desired. This approach is sometimes simpler and more economical than the one given earlier. Figure 4.6.3 illustrates this approach for deriving a sequence of four 3-bit states (M = 4, N = 3). The transition diagram one wants to

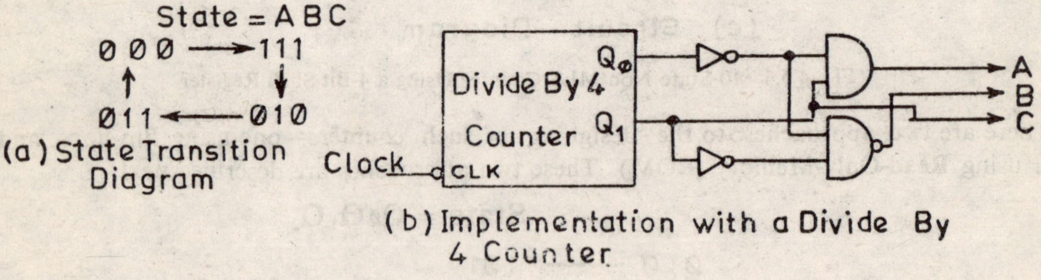

Fig. 4.6.3 Generating an Arbitrary Sequence of M states by Directly Decoding the Output of a Divide-By-M Counter (M = 4)

realize in this case is given in Fig. 4.6.3(a). The following assignment of states is done between the Divide-By-4 counter outputs and the desired outputs.

Divide-By-M Counter Outputs	Desired Outputs
$Q_\emptyset$ Q_1	A B C
∅ ∅	∅ ∅ ∅
∅ 1	1 1 1
1 ∅	∅ 1 ∅
1 1	∅ 1 1

Obviously with the appropriate combinatorial logic, the outputs ABC will step through the desired sequence as the Divide-By-M counter steps through states ∅∅ to 11 (and back to ∅∅). The combinatorial logic needed is given by—

$$A = \bar{Q}_\emptyset \cdot Q_1,\ B = Q_\emptyset + Q_1,\ C = Q_1$$

The circuit realization for this counter is given in Fig. 4.6.3(b). Not that, in this approach, no start-up problems will be encountered, provided the Divide-By-M counter itself is free from start-up problems.

4.6.2 ROM Based Design

A counter counting in an arbitrary state sequence can also be generated using a ROM and a shift register with a parallel input/output feature. In order to generate a sequence of M states with N-bits for each state one requires a $2^K \times L$ ROM and a N-bit parallel input/output register where $L \geqslant N$ and $K \geqslant \text{Max}(\log_2 M, N)$. The general circuit for doing this is shown in Fig. 4.6.4 where $A_\emptyset - A_{K-1}$ are the address inputs of the ROM and $D_\emptyset - D_{L-1}$ are its data outputs. By connecting $\overline{OE}$ to ∅ (output enable), the ROM's data outputs are kept permanently enabled. An active edge of the clock latches the lower N data outputs ($D_\emptyset - D_{N-1}$) of the ROM into the register. The register's outputs are once again fedback as the lower N address inputs ($A_\emptyset - A_{N-1}$) of the ROM.

It is vital that the ROM be properly programmed in this case. The reader should be able to see that the current state (i.e. the register outputs) act, in some form, as the address of the ROM location which contains, in some form the next state desired. The programming of the ROM should be such that this can be achieved with the circuit of Fig. 4.6.4 (or some variation thereof).

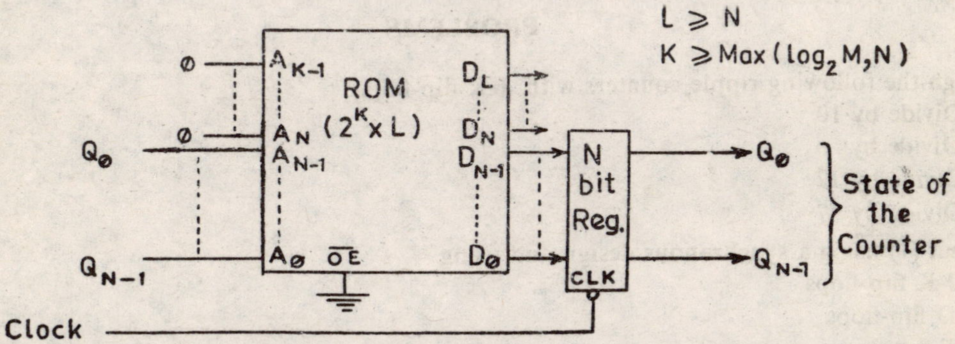

Fig. 4.6.4 ROM Based Design of a Counter for Generating an Arbitrary Sequence of M states with N-bits for Each State

Note that this circuit may have start-up problems if the register starts up in a state outside the desired sequence. Having the ROM programmed to force the register to some state in the desired sequence, will eliminate such start-up problems. Figure 4.6.5 shows the contents of 16 × 8

A_3	A_2	A_1	$A_Ø$	D_3	D_2	D_1	$D_Ø$
Ø	Ø	Ø	Ø	X	1	1	1
Ø	Ø	Ø	1	X	Ø	Ø	Ø
Ø	Ø	1	Ø	X	1	Ø	1
Ø	Ø	1	1	X	Ø	Ø	1
Ø	1	Ø	Ø	X	Ø	Ø	1
Ø	1	Ø	1	X	Ø	Ø	Ø
Ø	1	1	Ø	X	1	1	1
Ø	1	1	1	X	Ø	1	Ø
1	Ø	Ø	Ø	X	X	X	X
⋮	⋮	⋮	⋮				
1	1	1	1				

X = Don't Care

Fig. 4.6.5 Contents of 16×4 ROM to be Used in Fig. 4.6.4 to Obtain a Circuit for Realizing the State Transition Diagram of Fig. 4.6.1

ROM used in the circuit of Fig. 4.6.4 to obtain the state transition diagram shown in Fig. 4.6.1. Note that this counter will be self-starting and that locations 1ØØØ to 1111 of the ROM and the bit D_3 of locations ØØØØ to Ø111 are not used; this is the reason why they have been put as Don't Cares (X).

The scope of this text does not allow a detailed discussion of generalized sequential circuit design. The reader should however be aware of the fact that such a design essentially boils down to designing a sequential circuit for generating an arbitrary state sequence. In that respect, the approaches discussed in this section will be found useful.

PROBLEMS

1. Design the following ripple counters with J-K flip-flops
   (a) Divide by 10
   (b) Divide by 5
   (c) Divide by 12
   (d) Divide by 7
2. Repeat (1) using a synchronous design and using
   (i) J-K flip-flops
   (ii) D flip-flops
   (iii) T flip-flops
3. (a) Obtain the state transition diagram for a 4-bit Ring Counter.
   (b) Choose a particular counting sequence in (a) as the valid counting sequence. Modify the feedback to provide the self-start feature.
4. Repeat (3) for a 4-bit Johnson Counter

5. Design non-MLS Linear Shift Register counters counting in a counting sequence of the following lengths:
(a) 9, (b) 11, (c) 12, (d) 13, (e) 14
6. A designer wants a counter to count in the following sequence:

   ...., Ø, 13, 8, 7, 9, 4, 2, 3, 15, 11, 1, Ø, 13, 8 ...

   Design a counter to do this. You may adopt the following strategies
   (a) Ripple counter with decoding
   (b) Synchronous counter with decoding
   (c) Direct realization using J-K, D or T flip-flops
   (d) MLS counter with decoding
   (e) ROM based design.

Chapter 5

Dependency Notation

This chapter is devoted to introducing a new Symbology—Dependency Notation—suggested for representing digital circuits. This notation fills a long-standing need for ways of representing MSI devices like decoders, multiplexers, counters etc. The power of this notation lies in the fact that the symbol used to represent the device is generally sufficient to describe its operation. Given a circuit schematic which uses dependency notation, one does not have to keep referring to the manufacturer's data sheets for devices in order to see how the circuit operates. Moreover, this notation also simplifies considerably the job of drawing circuit schematics and understanding them. The objective of the following sections is to introduce this notation to an extent which will allow the reader to both use and understand this notation. It is also suggested that the reader refer to recent editions of the data books of TTL device-manufacturers, which uses this notation, in order to achieve greater familiarity. This chapter discusses the use of this notation for specific circuits, e.g. counters, multiplexers, shift registers etc. and introduces some of the formal rules of dependency.

5.1 LOGIC CONVENTION AND INDICATOR SYSTEMS

The 'logic convention' being used decides the method of representing a logical (binary) quantity. These are described below.

(a) Positive Logic Convention:

Logic 1 = positive voltage (H)

Logic Ø = more negative voltage (L)

This convention is the one commonly used today. In this convention, a small circle (○) placed at inputs or outputs inverts the corresponding logical variable. Some gates using this notation are shown in Fig. 5.1.1.

(b) Negative Logic Convention:

Logic 1 = negative voltage (L)

Logic Ø = more positive voltage (H)

This convention is rarely used today and will not be discussed any further. One should note, however, that it uses the same negation indicator as the Positive Logic System.

(c) Mixed-Logic Convention: This convention allows the designer to redefine the meaning of logic value '1' (as either H or L) at different places in the circuit. This convention uses a polarity indicator (—◺) to define voltage levels at any particular point in the circuit, i.e. presence of —◺ implies L = 1 and H = Ø, absence of —◺ implies L = Ø, and H = 1. The representation of some gates using this convention is shown in Fig. 5.1.2. Comparing these with the positive logic convention representations shown in Fig. 5.1.1, it is obvious that (at least for these cases) one can consider, the polarity indicator —▷ as functionally equivalent to the negation indicator. Though this is true, the reader is advised not to view it in this fashion but try to go by the

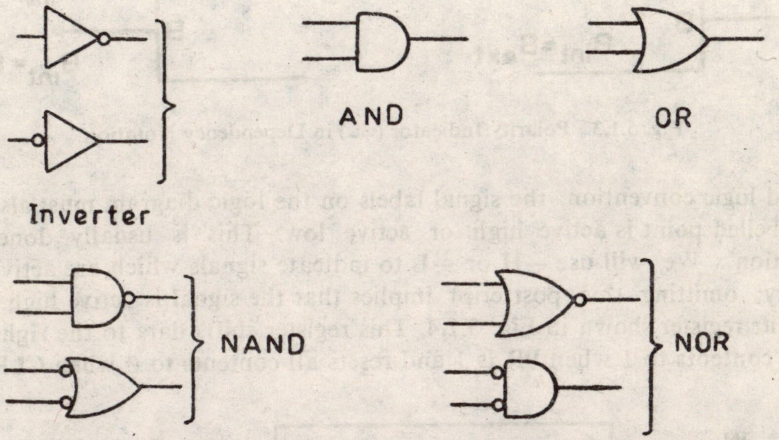

Fig. 5.1.1 Positive Logic with Negation Indicator (○) Representation for Some Common Gates

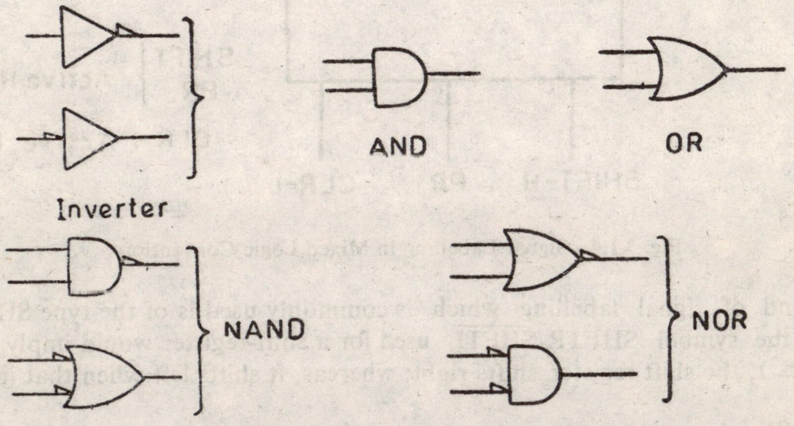

Fig. 5.1.2 Mixed Logic with Polarity Indicator (—▷) Representation for Some Common Gates

definition of the polarity indicator. For example, the proper interpretation of a NAND gate in this convention is either 'an AND gate with active-low output' or 'a NOR gate with active-low inputs'.

The use of the polarity or negation indicators at the inputs or outputs of a device provides a way of distinguishing between the internal and external logic states of the device. This is illustrated in Fig. 5.1.3 for the mixed logic convention using polarity indicators. (Note that the same thing could have been done with negation indicators for the positive logic convention). As an example, for the NAND realization with 'an AND gate with active low inputs'—an external logic state of 1 corresponds to an internal logic level of L (or a ∅ by positive logic convention).

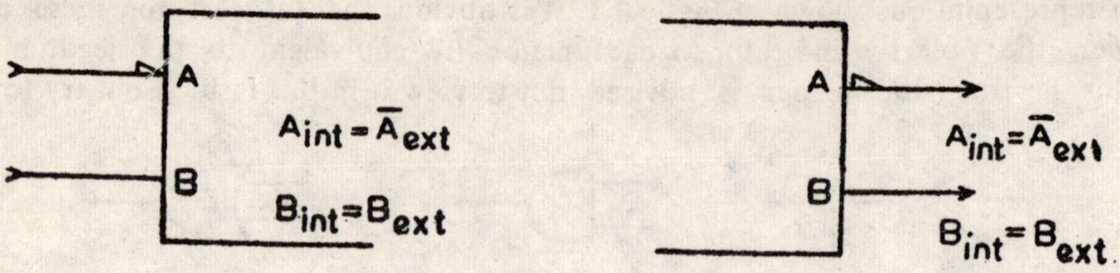

Fig. 5.1.3 Polarity Indicator (—◺) in Dependency Notation

In the mixed logic convention, the signal labels on the logic diagram must also identify if the signal at the labelled point is active high or active low. This is usually done by a 'Signal Name Convention'. We will use —H or —L to indicate signals which are active high or active low respectively; omitting this postscript implies that the signal is active high. As an example consider the shift register shown in Fig. 5.1.4. This register shifts data to the right when SHIFT is 1, presets all contents to 1 when PR is 1 and resets all contents to Ø when CLR is Ø.

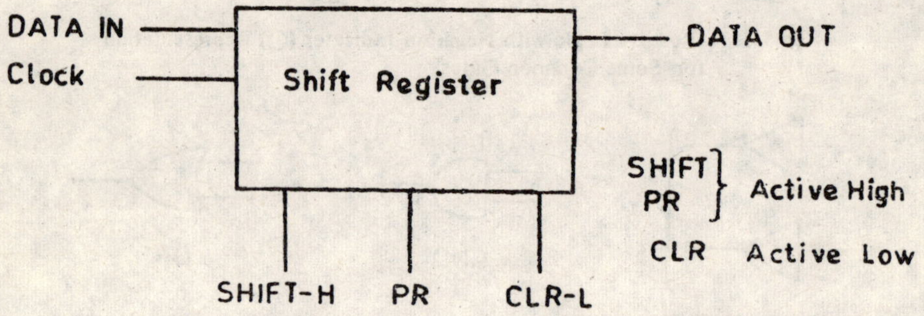

Fig. 5.1.4 Signal Labelling in Mixed Logic Convention

Another kind of signal labelling which is commonly used is of the type SHFTR/$\overline{\text{SHFTL}}$. For example, the symbol SHFTR/$\overline{\text{SHFTL}}$ used for a shift-register would imply that when the labelled input = 1, the shift register shifts right; whereas, it shifts left when that input = Ø.

5.2 DECODERS

Dependency notation uses the label X/Y to denote a decoder block. Inputs are labelled (inside the block) with their weights. The X/Y symbol implies that only that output whose weight is equal to the sum of the active inputs, properly weighted, will be the one which will be active. (The reader should note that with a binary weighting sequence, i.e. 1, 2, 4, . . . , etc., one will actually get a decoder action from the above).

This decoder representation is illustrated in Fig. 5.2.1 which shows the representation for the 74LS42, 4-line to 10-line decoder. (Note that in this figure, signal-flow is from left to right and the numbers outside the block show the actual pin numbers on the IC package. This convention will be used in all figures throughout this section). As per our earlier description, if the inputs to this device are pins 12 = Ø, 13 = 1, 14 = 1, 15 = Ø, then the output labelled 6 (pin 7) is Low and all other outputs are high.

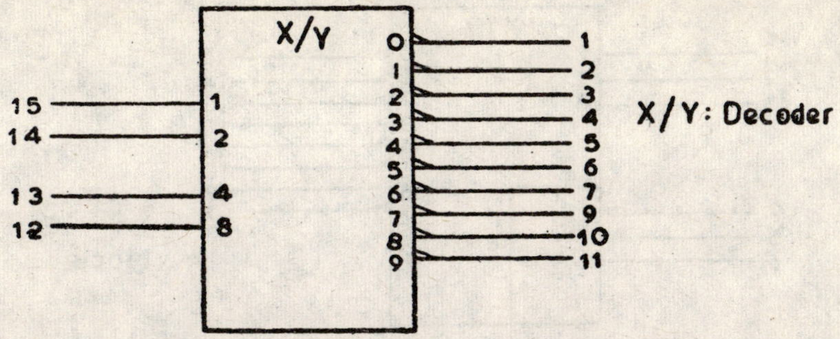

Fig. 5.2.1 74LS42, 4-Line to 10-Line Decoder

Another decoder, the 74LS139 dual 2-line to 4-line decoder is shown in Fig. 5.2.2. The horizontal line through the middle of this block denotes the fact that the IC package contains two

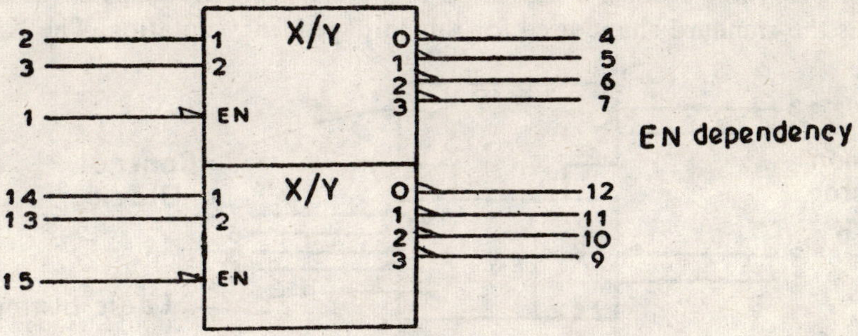

Fig. 5.2.2 74LS139, Dual 2-Line to 4-Line Decoder

completely independent decoders (i.e. the inputs and outputs of one do not affect the outputs of the other). The EN input illustrates the *EN* (*enable*) *dependency* of the dependency notation. The EN dependency symbol is reserved for inputs that enable outputs. In Fig. 5.2.2, if the EN input of a particular decoder is active (Low, in this case—note the polarity indicator), then only does the decoder action make one of its outputs active (low); if EN is not active, all the outputs of that decoder will be inactive. For example, in Fig. 5.2.2, if the inputs are pins 1 = Ø, 2 = Ø, 3 = 1, 15 = 1, 13 = X, 14 = X (X = don't care) then the output at pin 6 = Ø and other output pins will output 1s. The reader should note that, in some sense, the EN dependency is a way of ANDing the EN input with what would have been the output of the corresponding decoder without such an EN input. We will find this to be true even when the EN dependency is shown for circuits other than decoders.

A 74LS138 3-line to 8-line decoder with multiple (three) enable inputs is shown in Fig. 5.2.3. In order to show multiple enable inputs, dependency notation uses an '&-block', with an EN outside, within the dependency symbol of the device. All the inputs to the &-block must be active (i.e. pins 4 = Ø, 5 = Ø, 6 = 1) in order to have EN active. Note that a decoding action is present only when EN is active; if EN is inactive, all outputs will be inactive (i.e. all outputs will be at 1). Note that the 'internal inputs' of the &-block are ANDed in order to produce the EN value internally. One may also label an input as ENm. This implies that the input or output labelled 'm' will be active only if the input labelled ENm is also active.

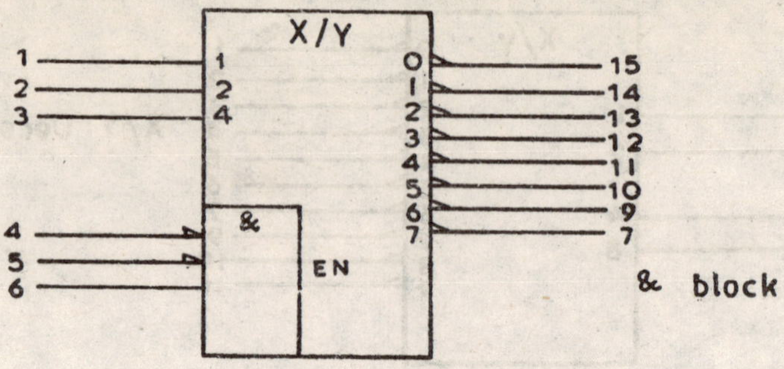

Fig. 5.2.3 74LS138, 3-Line to 8-Line Decoder

It is quite common to find two or more logic elements controlled by a common set of control inputs. In dependency notation, one can denote this by using a common control block as shown for the dual 2-line to 4-line decoder 74LS155 in Fig. 5.2.4. Note the shape of the common control block, which is the standard shape used for this in dependency notation. The X/Y label on this

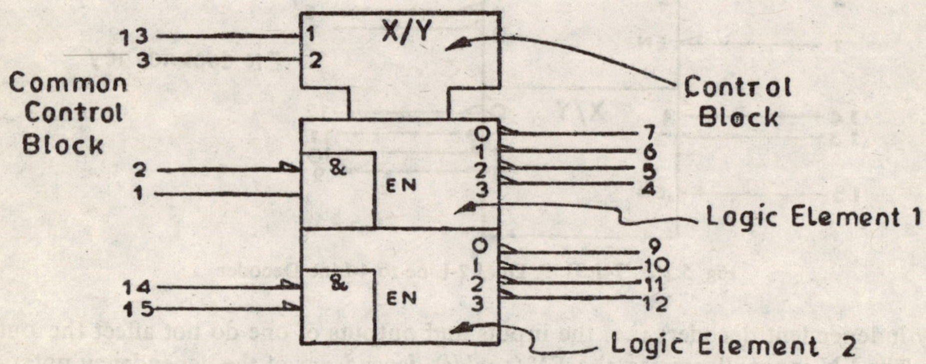

Fig. 5.2.4 74LS155 Dual 2-Line to 4-Line Decoder

block indicates that the decoding actions of the controlled logic elements are based on the inputs to this block, i.e. pin 13 = weight 1, pin 3 = weight 2. Also note that the enable inputs—2 for each—of the two decoders are separate. The decoded output of the upper element is active only

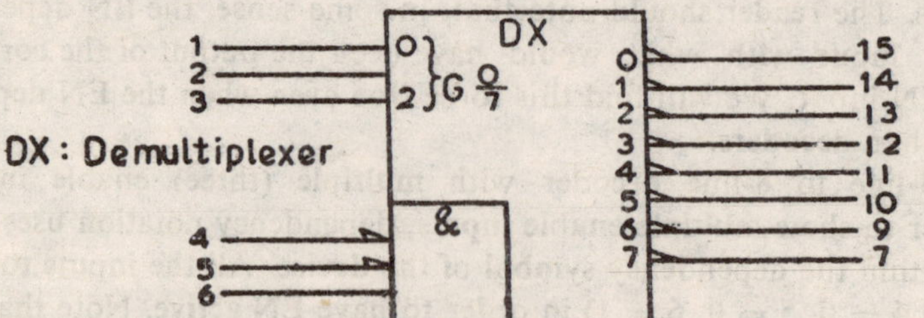

Fig. 5.2.5 Alternative Representation of 74LS138, 3-Line to 8-Line Decoder (As a Demultiplexer Using the DX Label, AND Dependency and an &-block for Input Data)

when pin 2 = Ø and pin 1 = 1 and that of the lower element is active only when pin 14 = Ø and pin 15 = Ø.

A decoder with one or more enable-inputs may also be viewed as a demultiplexer with the data input to the demultiplexer being provided as the logical combination (in some way) of the enable-inputs. Viewed in this fashion, a decoder (with enable)/demultiplexer may also be represented in an alternative way; this is shown in Fig. 5.2.5 for the 74LS138 device (shown earlier in Fig. 5.2.3). This uses the symbol DX (for demultiplexer) and an *AND dependency* indicated by the letter G. This AND dependency is discussed next in relation to the dependency notation for multiplexers.

5.3 MULTIPLEXERS

In dependency notation, a logic block with the symbol MUX represents a multiplexer. A fundamental dependency notation mechanism referred to as AND-dependency is needed in order to symbolically represent a multiplexer. This AND dependency is denoted by the letter G and is always followed by a number, e.g. G2. Using Gn (n is a number) to label an input (or some other logical entity), implies that any other input or output (in this same logic element) labelled 'n' will be ANDed with the logical entity labelled Gn.

As an example, consider the 74LS157 Quad 2-Input Multiplexer shown in Fig. 5.3.1. Consider the first logic element, with inputs at pins 2 and 3 and output at pin 4. (Note that the G1

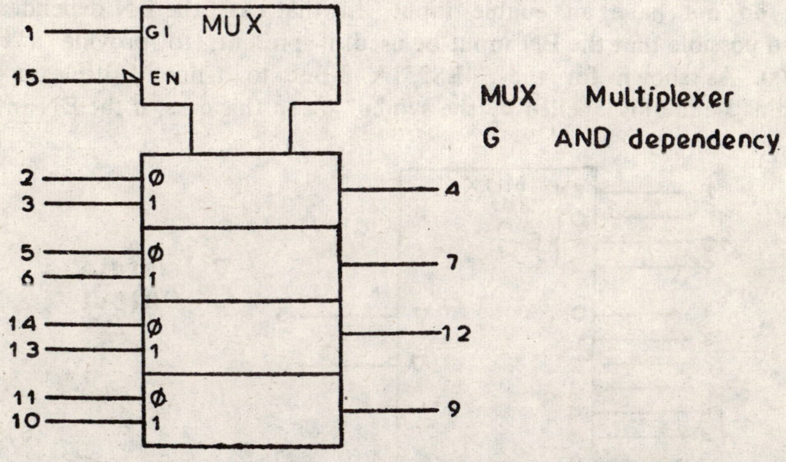

Fig. 5.3.1 74LS157, Quad 2-Input Noninverting Multiplexer

and EN inputs to this device reside in the common control block and will therefore affect all the four logic elements). If the EN input at pin 15 is Ø, the G1 input will decide whether the output at pin 4 will be connected to input pin 2 or 3; specifically, if G1 = Ø; pin 4 = pin 2; if G1 = 1; pin 4 = pin 3. If pin 15 is 1 (EN inactive) the output at pin 4 will be Ø (Ø is pin 4's inactive state, note the absence of the polarity indicator at 4). The other three logic elements in this IC package will also behave in an analogous fashion.

A somewhat different form of the AND dependency notation is shown by the label $G\frac{0}{7}$ in the representation of the 74LS151 8-line to 1-line Multiplexer of Fig. 5.3.2. In this case, the label $G\frac{0}{7}$ is really a shorthand notation for the labels G0, G1, G2, G3, G4, G5, G6 and G7. One of these is selected by the binary grouping of inputs 11, 10 and 9 with weights of successive powers of 2, i.e. $2^0 = 1$, $2^1 = 2$, $2^2 = 4$, respectively. If the net weight thus calculated is n the AND dependency symbol Gn becomes active. If the enable input is active (pin 7 = Ø), then the outputs become active if the input labelled n is also active. (If the enable input is inactive, pin 7 = 1, then all the outputs will be inactive regardless of which Gn is active). Functionally, this implies that pin 5 follows the input at input n, whereas pin 6 will provide the inverse of this. For this particular IC, consider the situation where the input pins 7 and 9 are Ø whereas pins 11 and 10 are 1. The chip is enabled (since the enable input (7) is active) and G3 symbol also becomes active. In this case, pin 5 output follows the input at pin 1 (input 3) whereas pin 6 provides the logical inverse of this.

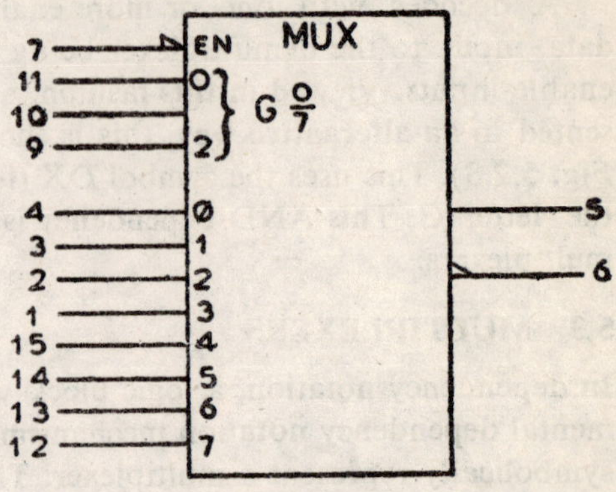

Fig. 5.3.2 74LS151, 8-Line to 1-Line Multiplexer

[*Note*: Looking back at Fig. 5.3.1, the reader should realize that we should have used the notation G$\frac{0}{1}$ there instead of G1. Since there are only two AND dependency symbols in this case, using G1 alone also implies the presence of G0 even though this is not explicitly stated.]

It is not necessary for a MUX to have an enable input. Several multiplexers exist, e.g. 74LS152 which do not have an enable input. In that case, the EN dependency will not be present. It is also possible that the EN input be used, if present, to provide a tristate control for the output(s). As shown for the 74LS251A 8-line to 1-line Multiplexer of Fig. 5.3.3, a tristate output (pins 5 and 6) is labelled by the symbol ▽. In this case, if the EN input is inactive,

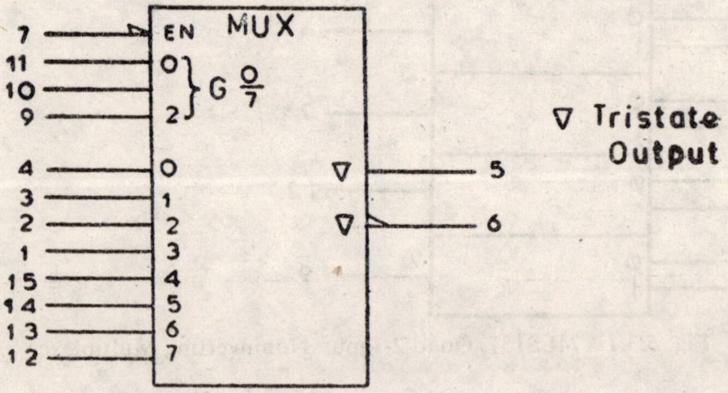

Fig. 5.3.3 74LS251A 8-Line to 1-Line Multiplexer with Tristate Output

the outputs are also inactive, that is they are in the tristate condition. If the EN input is active (pin 7 = Ø), the operation of this device follows the operation of the 74LS151 (Fig. 5.3.2) device.

Parenthetically, one should emphasize that the inactive state of a tristate output (labelled ▽) is the tristate, or High-Z, condition; in the active state, this output is Ø or 1 depending on the inputs. One should also mention here that dependency notation uses the symbol $\underline{\Diamond}$ to label Open-Collector outputs. Similarly the symbol ▷ inside the logic block indicates a circuit with unusual drive capability, i.e. a buffer. The use of these symbols is shown in Fig. 5.3.4 for the 7406 Hex Inverter Buffer/Driver with Open Collector outputs.

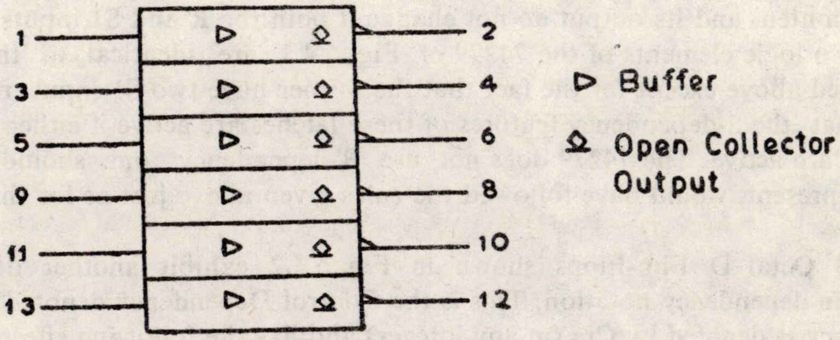

Fig. 5.3.4 7406 Hex Inverter Buffer Driver (Open Collector Outputs)

5.4 FLIP-FLOPS, SHIFT REGISTERS AND COUNTERS

The Quad Set-Reset Latch, 74279, shown in Fig. 5.4.1 illustrates the use of the R and S symbols in dependency notation. If these symbols are used by themselves (without a number following),

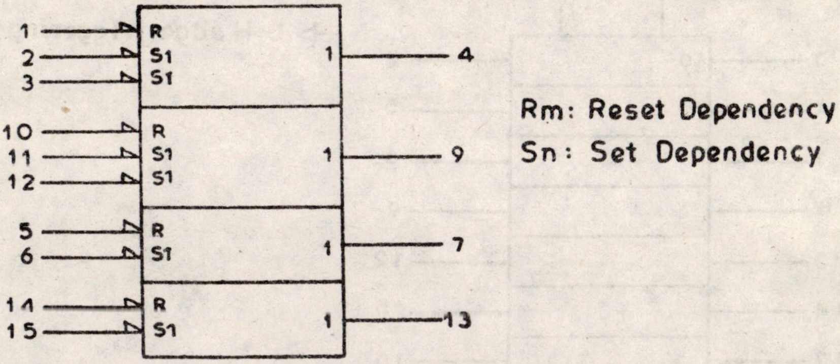

Fig. 5.4.1 74279, Quad Set-Reset Latch

they imply the following:

(a) If the R input is at its internal 1 state, the internal value of the affected outputs is Ø.

(b) If the S input is at its internal 1 state, the internal value of the affected outputs is 1.

Note that (a) and (b) above may not be able to specify the output state for the combination R = S = 1. In this case, Set and Reset dependencies are used; these are denoted by the symbols Rm and Sn respectively, where 'm' and 'n' are numbers. These obey the following rules:

(a) When an Sn input is active (internal 1), outputs affected (i.e. those labelled n) will react just as if they would to the input combination S = 1, R = Ø regardless of the actual input R.

(b) When an Rm input is active, outputs affected will react just as if they would to the input combination S = Ø, R = 1, regardless of the actual input S.

(c) When an Sn or Rm input is inactive (internal Ø), it has no effect.

Based on the above rules, we can now examine the lowest logic element of the 74279 device shown in Fig. 5.4.1. If pin 15 is at Ø (internal value of S1 = 1), then the affected output, labelled

1 at pin 13, is 1 and this will be stored in the latch. (Note that the label 1 at pin 13 implies that S1 affects this output). On the contrary, if the R input is active (pin 14 = Ø) and S1 is inactive (pin 15 = 1), then the output at pin 13 is Ø and this is stored in the latch. Since this is an R-S latch, the latch content and its output do not change if both the R and S1 inputs are inactive.

The upper two logic elements of the 74279 of Fig. 5.4.1 are identical to the lowest logic element described above except for the fact that the former have two S1 inputs rather than one. This implies that the S-dependency features of these latches are active if either one of these S1 inputs (or both) are active. The 74279 does not use R-dependency; one should note that R-dependency, if present, would have followed the rules given above just as for the S-dependency case.

The 74LS273 Octal D Flip-Flops shown in Fig. 5.4.2 exhibit another dependency very commonly used in dependency notation. This is the Control Dependency denoted by the letter C. This C-dependency is denoted by Cm (m any integer) and has the following effects.

'When an input or output labelled Cm is active (i.e. at its internal 1 state), the inputs affected by Cm are enabled; otherwise these inputs are disabled'.

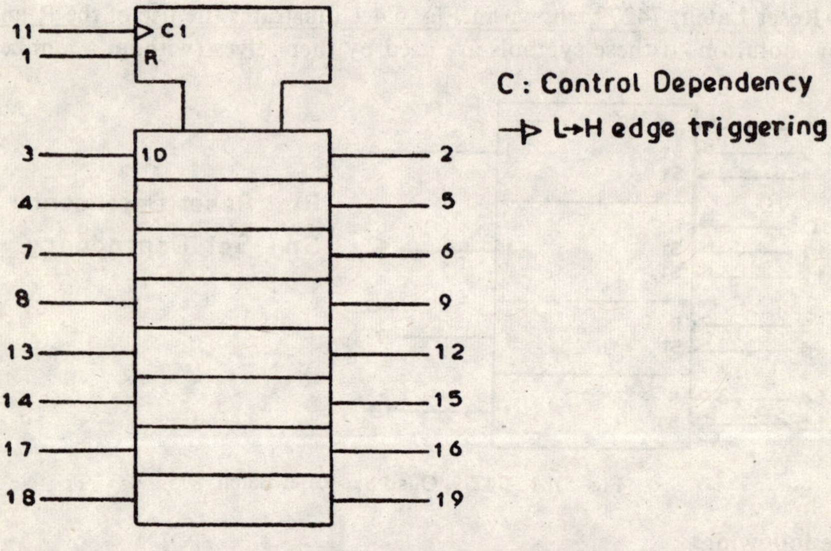

Fig. 5.4.2 74LS273 Octal D Flip-Flops

Fig. 5.4.2 also introduces the Dynamic Indicator symbol '–|>'. The use of this symbol at an input indicates that the logic element senses the Ø → 1 transition of the *internal* logic level and not the internal logic level itself. In Fig. 5.4.2, this implies that the D flip-flops of the 74LS273 device are positive edge triggered (i.e. sensitive to the Ø → 1 transition at input pin 11). One should also note that in Fig. 5.4.2, only one of the logic elements has been labelled. This is commonly done if the other logic elements also use the same labels.

Based on the above description, the operation of the 74LS273 IC is easily understood. If the R-input at pin 1 is active (logic 1), then all the 8D flip-flops are reset and their outputs go to Ø. If the R-input is inactive, the data at the 1D inputs get latched into the flip-flops and appears at the outputs whenever the C1 input at pin 11 is active (i.e. pin 11 goes through a Ø → 1 transition). The flip-flops' contents and their outputs stay unchanged if both the C1 and the R input stay inactive.

The reader should note the way the C1-1D inputs show Control dependency. As per our earlier description of control dependency, the 1D inputs to the flip-flops are active only when C1 is active. The difference between the C1 and R inputs is also worth mentioning once again; the C1 input is edge sensitive (positive edge, i.e. Ø → 1) whereas the R input is level sensitive (logic 1) and exercises an overriding effect on the former. If a clock signal is being applied to the C1 input, we may refer to the 74LS273 as being operated in a synchronous mode with an asynchronous reset input.

[It is worthwhile mentioning another IC, the 74LS374, here. This is absolutely identical to the 74LS273 except that the R label on pin 1 is replaced by an EN label and the polarity indicator symbol and that all the outputs are tristate instead of being active high. The reader should examine the operation of this device].

J-K flip-flops may be represented in very much the same fashion as D flip-flops. Instead of an input labelled with the letter D, the logic element has two inputs, one labelled with J (for the J-input) and the other with K (for the K-input). The 7473 and the 74LS73, Dual J-K flip-flops have been shown in this way in Fig. 5.4.3.

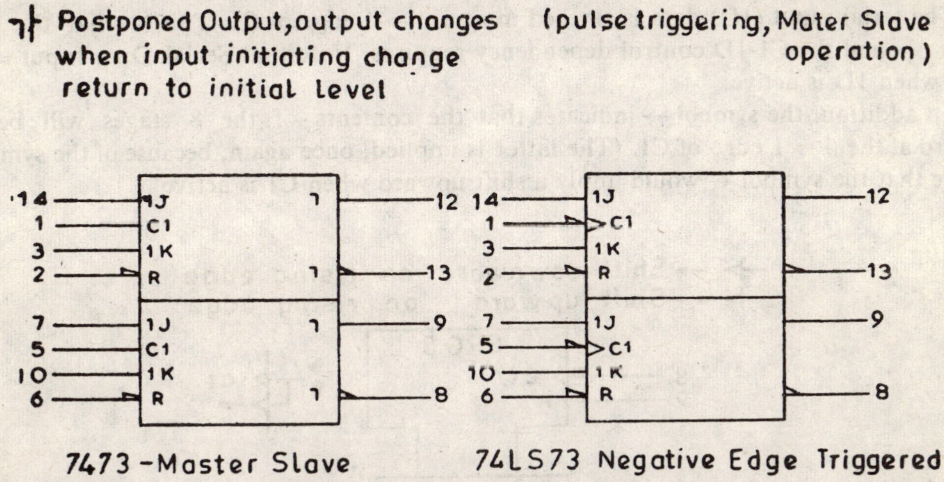

Fig. 5.4.3 7473/74LS73 Dual J-K Flip-Flops

Consider the representation of the 74LS73 first. The control dependency between the C1-1J and C1-1K can be easily seen. The –|> symbol indicates edge triggering and the presence of the polarity indicator on this input implies that the J-K flip-flops are negative edge triggered, i.e. sensitive to a 1 → Ø transition on the input at pins 1 and 5. The 1J and 1K values at this edge are the respective J and K inputs to the flip-flops. The R input is asynchronous. A Ø at these inputs clears the corresponding flip-flops. Note that the flip-flops provide both the Q (pins 12, 9) and the $\bar{Q}$ (pins 13, 8) outputs.

In contrast to the 74LS73 which is negative edge triggered, the 7473 is a Master-Slave J-K flip-flop. As shown in Fig. 5.4.3, this is indicated by the symbol ⏋ at its outputs and by the fact that the C1 inputs no longer have the dynamic indicator symbol –|>. (The presence of –|> implied edge triggering). The definition of the ⌝ symbol has also been given in Fig. 5.4.3.

Fig. 5.4.4, showing the 74LS377 octal D flip-flop with clock enable, illustrates an interesting way of using dependency-notation. Here the AND dependency lies between the G1-1C2 inputs and the Control dependency between the 1C2-2D inputs. Because of these dependencies, the inputs to the D flip-flops are active only at those $\emptyset \rightarrow 1$ transitions of pin 11 which take place while pin 1 = $\emptyset$.

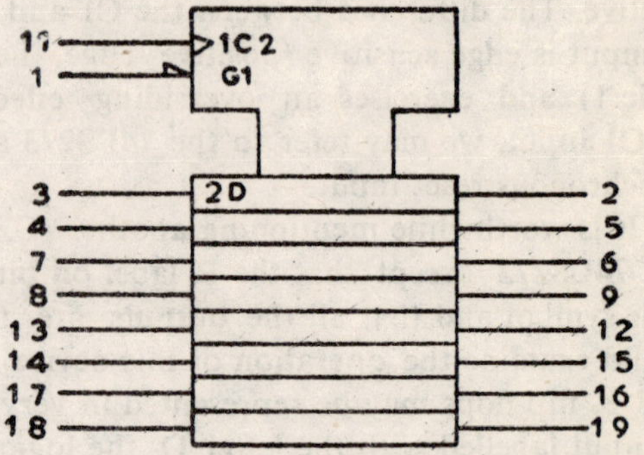

Fig. 5.4.4 74LS377 Octal D Flip-Flop with Clock Enable

In dependency notation, a *n*-bit shift register is indicated by a logic element labelled SRGn. A 8-bit shift register, the 74LS164, is shown in Fig. 5.4.5. The serial input to this shift register is provided by ANDing the inputs at pins 1 and 2; the outputs of all the 8 shift register stages are available in parallel. A low at pin 9 (R input active) resets all the shift register stages. The C1/→ labelling of pin 8 indicates the following:

(a) The serial input (1D) data is clocked in at a $\emptyset \rightarrow 1$ edge of C1. This is what the dynamic indicator –|> and the C1-1D control dependency requires. Note that Serial Data Input = (pin 1) ·(pin 2) when 1D is active.

(b) In addition, the symbol → indicates that the contents of the 8 stages will be shifted downward at the $\emptyset \rightarrow 1$ edge of C1. (The latter is implied, once again, because of the symbol –|>.)

[Note that the symbol ← would imply a shift upward when C1 is active].

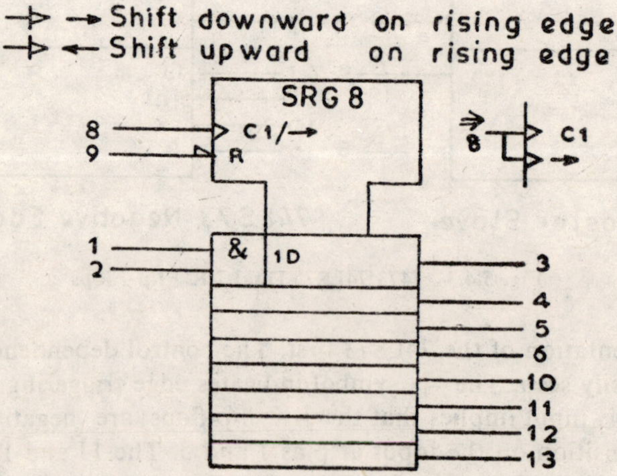

Fig. 5.4.5 74LS164, 8-bit Serial In/Parallel Out Shift Register

The different ways in which a clock input may be active are shown in Fig. 5.4.6. Note that all the examples shown in Fig. 5.4.6 are for rising edge sensitive inputs. Falling edge sensitive inputs may be symbolised by placing a polarity indicator symbol (—▷) on the input line.

Device	Symbol	Action at rising edge
J-K flip-flop	Cn	nJ and nK inputs active
D flip-flop	Cn	nD input active
Shift Register	→ (←)	Shift downward (upward) by one
Counter	+ (−)	count up (down) by one

Fig. 5.4.6 Examples of Rising Edge Sensitive Inputs

A somewhat more complicated use of dependency notation ideas is shown in Fig. 5.4.7 which shows one possible representation of the 74LS166 8-bit Parallel/Serial In, Serial Out Shift Register. This introduces some new symbols which are described next.

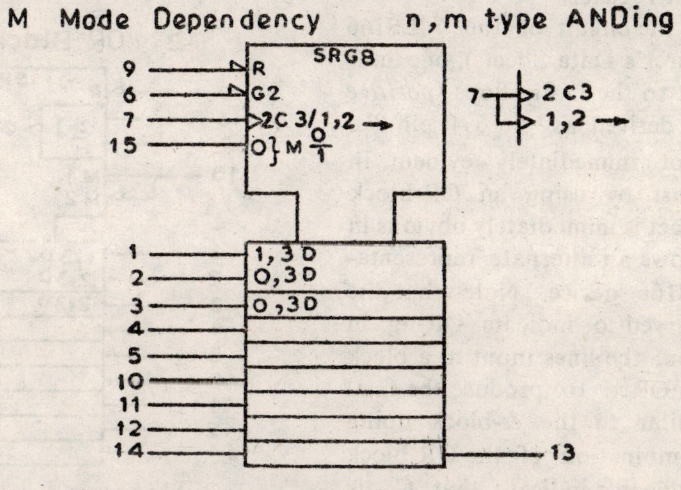

Fig. 5.4.7 74LS166, 8-bit Parallel/Serial In, Serial Out Shift Register

The letter M represents Mode Dependency. One way of showing this dependency is by applying a symbol like $M\frac{0}{m}$ to the group of inputs, which decide the mode of operation of the logic element. The inputs are decoded (by applying binary weights to them) in order to calculate the net weight n (say). The symbol $M\frac{0}{m}$ is actually a shorthand notation for M0, M1 . . . Mm. If the mode inputs decide a net weight n, then Mn will be active, i.e. the logic element will be working in Mode n. (Note that this decoding of mode is done in the same way as it is done for G-dependency.) The inputs or outputs affected in Mode n will have the number n appearing in their labels. In Fig. 5.4.7, the 74LS166 operates in Mode 0 (Parallel In or Load) if pin 15 = 0; if pin 15 = 1, the device works in Mode 1 (Serial Shift).

As before, the G2-2C3 dependency shown in Fig. 5.4.7 implies that C3 will be active if pin 6 = Ø and a Ø → 1 transition occurs on pin 7. The 'n, m' type labelling appearing in the figure implies that both n and m labels should be active for that input or output to be affected. (This is another kind of ANDing terminology used in dependency notation). In this particular case, these labels imply the following.

(i) 1, 3D (pin 1)-input to the D flip-flop will take place from pin 1 when M1 and C3 are both active or specifically pin 6 = Ø, pin 15 = 1, pin 7 = 0 → 1 edge.

(ii) Ø, 3D (pins 2, 3, . . . etc.)-input to the eight D flip-flops will be active from these labelled pins when M0 and C3 are both active.

(iii) 1, 2 → —the rising edge of C3 will cause data to be shifted downwards when M1 and G2 are both active.

Based on the above description, the reader should examine the operation of the 74LS166 and satisfy himself that it indeed operates in the Parallel In or Load mode in Mode 1 and in the Serial Shift mode in Mode 0. It should also be noted that the output of the 74LS166 (pin 13) is not affected by the M, C or G dependencies and always reflects the content of the lowest D flip-flop in the 8-bit shift register.

If one examines the circuit of the 74LS166 (refer to Manufacturer's Data Sheets), one finds that the clock input to the D flip-flops (*positive edge triggered*) is derived as 'pin 6 + pin 7's. This ORing is not immediately evident in Fig. 5.4.7. In contrast by using an OR-block labelled '⩾ 1' this fact is immediately obvious in Fig. 5.4.8 which shows an alternate representation for the 74LS166 device. Note that the symbol '⩾1' is reserved to indicate ORing in dependency notation; the lines input to a block labelled '⩾1' are ORed to produce the final result. (This is similar to the &-block doing ANDing). The combination of the OR-block and the dynamic indicator indicate that C3 is active when the variable (pin 6 + pin 7) make a positive transition. The M-dependency shown in Fig. 5.4.8 is also different from the previous representation. Here the mode dependency is explicitly shown as M1 and M2 where M1 is active (Serial Shift) if pin 15 = 1 and M2 is active (Load) if pin 15 = Ø. Specifically the following should be noted by the reader in this representation.

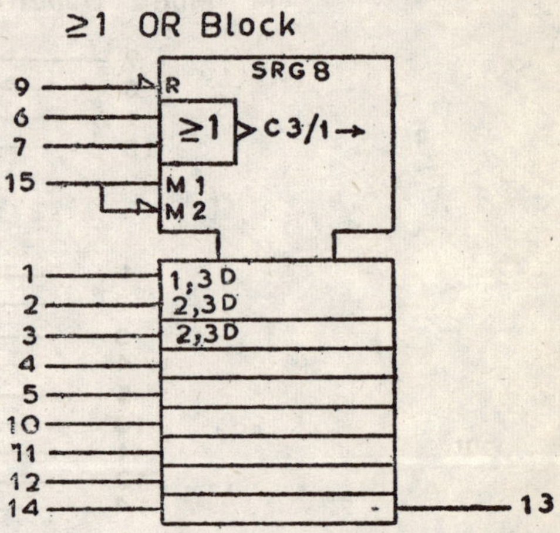

Fig. 5.4.8 Alternative Representation of the 74LS166

(i) –|> C3/1 → implies 'shift downward' on rising edge of C3 if M1 is active,

(ii) 1, 3D input to the first stage is active when M1 is active and C3 is active (i.e. rising edge of C3),

(iii) 2, 3D input to the eight stages (Load) is active when M2 and C3 are active.

Figure 5.4.7 and Fig. 5.4.8 illustrate the fact that a representation using dependency notation is not necessarily unique. Qualitatively, the representation of the 74LS166 device using Fig. 5.4.8 is probably more complete since it emphasizes the interchangeable nature of the pins 6 and 7 inputs. Figure 5.4.7, on the other hand, emphasizes the normal usage of the device with a clock input and another clock-enable input.

Dependency notation uses two different labels to identify counters. These are CTRn and CTRDIVm where n and m are integers. The label CTRn identifies the logic element as *n*-bit divide-by-(2^n) counter whereas CTRDIVm stands for a divide-by-m counter.

The dependency notation for a 4-bit presettable binary counter, 74LS161A, is shown in Fig. 5.4.9. All the counter stages are reset by inputting pin 1 = ∅ (i.e. R input active); pin 1

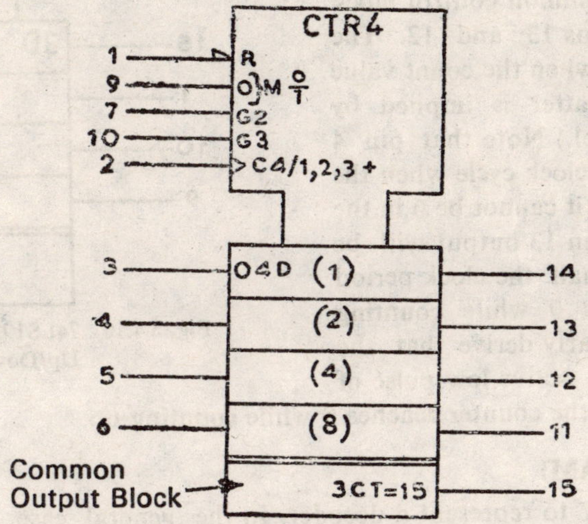

Fig. 5.4.9 74LS161A, 4-Bit Binary Counter

must be 1 in order to make it count. The mode input, pin 9, allows the choice of two modes—M0 and M1. With M0 active, the rising edge of the C4 input at pin 2 presets the counter stages with the inputs at pins 3, 4, 5 and 6; this is shown by the 0, 4D labelling of the preset inputs. In order to count up, the G2, G3 inputs and M1 should be active. In that case, the counter counts up on every rising edge of the C4 input. This is implied by the C4/1, 2, 3+ label on pin 2 (M1 active, G2 active, G3 active). Note that as described in Fig. 5.4.6, the + sign indicates a counting up operation. (This counter cannot count down.)

A new label introduced in Fig. 5.4.9 is the 3 CT = 15 label on output pin 13. In general, a label i, . . . , m, CT = n implies that the labelled output is active if the counter has a count of n and if the dependencies labelled i, . . . , m are all active. Following this, the output at pin 13 in Fig. 5.4.9 is 1 only when the count value is 15 and pin 10 = 1; this is the Terminal Count output. Note that this output is common to all the counter stages. The double-line at the top of the element providing this output is used to indicate a Common Output Block in dependency notation. Alternatively, one could also have placed the 3CT = 15 output in the Common Control Block itself.

A slightly more complicated example of a counter is shown in Fig. 5.4.10. This shows the 74LS192 Presettable BCD Decade Up/Down Counter. This counter is reset by making the R input active, i.e. pin 14 = 1. This counter is preset by inputting pin 11 = ∅ at which instant the counter contents get preset to the logic values at pins 15, 1, 10 and 9. Note that this can be done regardless of the inputs at pins 5 and 4—the clock inputs. The clock for counting up should be connected to pin 5 with pin 4 = 1, at which point the counter will count up on the rising edges of the pin 5 input. Similarly the clock for down-counting should be connected to pin 4 with pin 5 = 1; this will make the counter count down at the rising edges of the pin 4 input. These two facts are denoted by the G2, 2+ and G1, 1− labels. The common control block provides two outputs on pins 13 and 12. The pin 13 output goes low when the count value is ∅ and pin 4 = ∅. (The latter is implied by the $\bar{2}$ in the $\bar{2}$ CT = ∅ label.) Note that pin 4 will be ∅ for only half a clock cycle when the counter is counting down (it cannot be ∅ in the count up operation). The pin 13 output will be an active low pulse of half the clock period when the counter reaches ∅ while counting down. The reader can similarly derive that the pin 12 output will be an active low pulse of half the clock period when the counter reaches 9 while counting up.

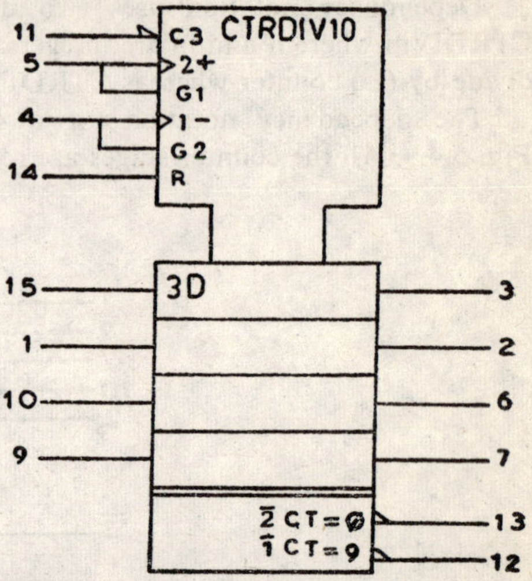

Fig. 5.4.10 74LS192 Presettable BCD Decade Up/Down Counter

5.5 MEMORY (ROM, RAM)

We have used the label X/Y to represent a decoder. In the general case, dependency notation uses this label to represent any general combinatorial circuit e.g. coder, code converters etc. Since a ROM (or PROM) is functionally equivalent to a combinatorial circuit one also uses the X/Y label for these devices. The representation of the INTEL 2716 EPROM is shown in Fig. 5.5.1. This is a 2K×8 EPROM storing 2K 8-bit words; the exact contents of the EPROM will have to be specified by a separate table as indicated in the figure. Based on our previous discussion of decoders, the reader should be able to understand the representation of the 2716 device from Fig. 5.5.1.

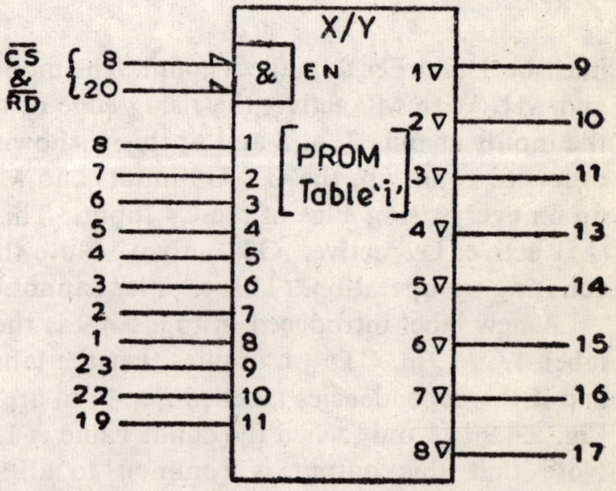

Fig. 5.5.1 Intel 2716, 2K×8 EPROM

In order to represent a RAM, however, one needs to introduce Address Dependency or A-Dependency. This is done for those logic elements, e.g. memories, where address inputs are used to choose specific sections of a large multidimensional array. In the case of memories, this selected section is usually referred to as a word consisting of one or more bits. Using address dependency one only needs to have the symbolic representation of a general section of the array rather than requiring a representation of the entire array. (In the case of memories, this general section will be that needed to represent typically the storage of one word). As for the inputs and outputs to this general section, the following points should be kept in mind.

(i) An input of the array shown for a particular element of the general section is common to the corresponding elements of all selected sections of the array.

(ii) An output of the array shown at a particular element of the general section is actually obtained by ORing the outputs of the corresponding elements of all the selected sections of the array.

(iii) If an output is indicated to be open collector or tristate in nature, then this refers to the output of the array as a whole and not to those of the sections of the array.

(iv) Those inputs of the logic element not affected by an affecting address input have their defined effect on all sections of the array.

(v) Those inputs of the logic element affected by an address input have their defined effect only on those sections selected by the address input.

An input to the logic element which creates address selection is labelled Am where m stands for the address of the particular section of the array which is selected by this input. If more than one input lines are present one can also apply binary weighting to these inputs and label the entire set as $A\frac{0}{m}$. Note that $A\frac{0}{m}$ is really a shorthand notation for A0, A1 . . . Am. The binary weighting to find the active address input is done in the same way as for the $G\frac{0}{m}$ and $M\frac{0}{m}$ representations discussed earlier.

For the general section shown, inputs and outputs controlled by the address inputs are labelled with the label A; other dependencies may also be shown separated by commas.

Based on the above description of the A-dependency representation, consider the Intel 2142 1K × 4 RAM shown in Fig. 5.5.2. Note that there are two control dependency inputs C1 and C2; the former acts as the WRite signal and the latter as a ReaD signal where both signals are active low. For either reading or writing, the RAM must be selected. This is done by inputting pin 9 = ∅, pin 5 = 1 which has the effect of making EN3 active. The ten address inputs to the RAM provide 1024 address dependencies A0, A1 ... A1023; depending on the address input (in binary weighted form), only one of them will be active at any given instant. The logic element shown corresponds to one general memory word; the particular memory word being selected in a given situation will depend on the address control input. The A, 1, 3D label at the inputs imply that the data on pins 15, 14, 13, 12 will get written into the selected address (for which A is active) provided C1 is active ($\overline{WR}$ = pin 11 = ∅) and EN3 is active ($\overline{CS}$ = ∅ ⇒ pin 9 = ∅, pin 5 = 1). Similarly the 2, 3, A▽ label at the outputs imply that the contents of the selected address will be output on pins 15, 14, 13, 12 provided C2 is active and EN3 is active and that

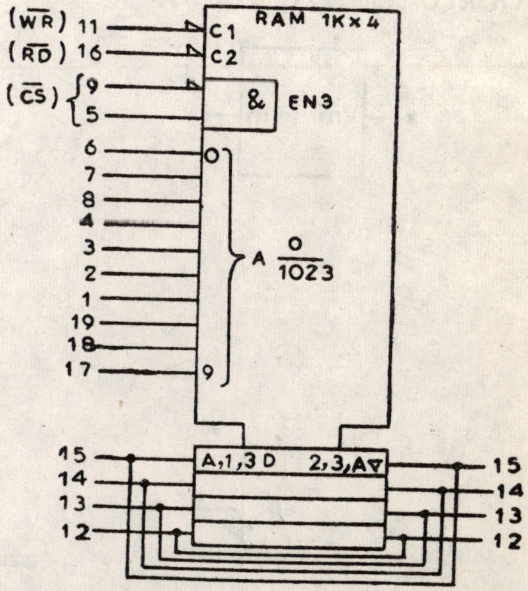

Fig. 5.5.2 Intel 2142, 1K×4-Bit Static RAM

otherwise the output will be in the High-z state. Finally, it should be noted that for this RAM, data input and output occur through the same pins. This is indicated by the external connections shown between the input and the output lines.

5.6 OTHER DEPENDENCIES

In the previous sections we have described the following dependency types

EN Enable
G AND
C Control
S Set
R Reset
M Mode
A Address

Apart from these, dependency notation also specifies the following dependencies

V OR
N Negate, Exclusive — OR
Z Interconnection

A brief description of these is given next.

V(OR) Dependency (Fig. 5.6.1)

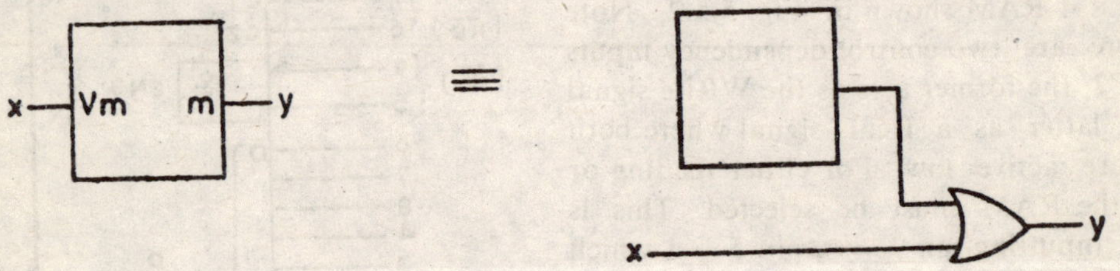

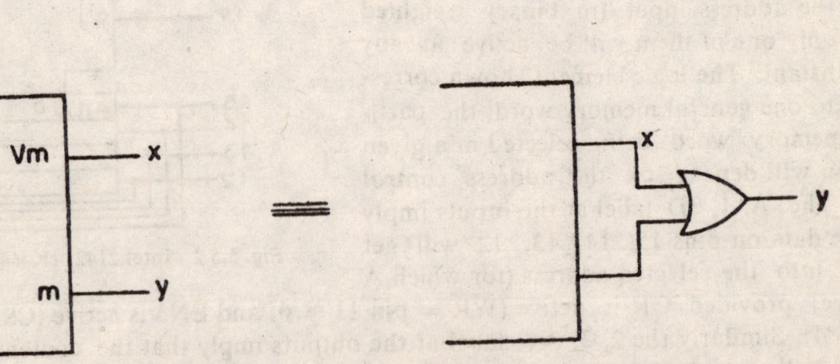

Fig. 5.6.1 V(OR) Dependency

If a Vm (m integer) input or output is at its internal 1 state, those inputs or outputs affected by it also are at their internal 1 states. If a Vm input or output is at its internal Ø state, then the inputs or outputs affected by it are at their otherwise normally defined internal logic states.

N (Negate or X-OR) Dependency (Fig. 5.6.2)

If an Nm (m integer) input or output is at its internal Ø state, inputs or outputs affected by it are at their normally defined internal logic states. If an Nm input or output is at its internal 1 state,

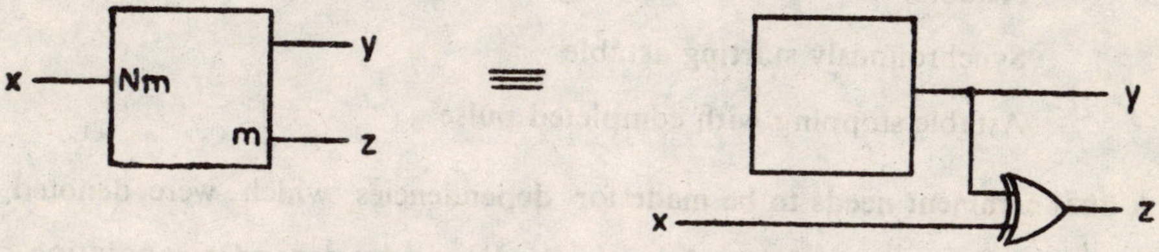

Fig. 5.6.2 N (Negate or XOR) Dependency

inputs or outputs affected by it are at the logical inverse (i.e. negation) of their otherwise normally defined internal logic states.

(Note that $X \oplus 1 = \overline{X}$ and $X \oplus \emptyset = X$. This is the reason why the N-dependency is also referred to as X-OR Dependency).

Z (Interconnection) Dependency

This is used to show internal connections between inputs, outputs, internal inputs and/or internal outputs. The internal logic state of an input or output labelled 'm' is the same as the internal logic state of the input or output labelled as Zm (m integer).

5.7 OTHER QUALIFYING SYMBOLS

A number of qualifying symbols have been introduced in the previous sections, e.g. $\geqslant 1$, X/Y, MUX, DX, ▷, SRGm, CTRm, CTRDIVm, ROM, RAM. Dependency notation also allows for a number of other qualifying symbols listed below:

$= 1$	ExOR
$=$	All inputs at same state
$2k$	Even no. of inputs active
$2k + 1$	Odd no. of inputs active
1	One input active
⎍	Schmitt trigger input, input with hysterisis
Σ	Adder
$P - Q$	Subtracter
CPG	Look-ahead carry generator

π	Multiplier
COMP	Magnitude Comparator
ALU	Arithmetic Logic Unit
$\_\sqcap\_$	Retriggerable Monostable
$1\sqcap\_$	Nonretriggerable Monostable
G	Astable
!G	Synchronously starting astable
G!	Astable stopping with completed pulse

Note: A final comment needs to be made for dependencies which were denoted earlier as M $\frac{0}{m}$, G $\frac{n}{m}$, A $\frac{n}{m}$ etc. The general form of the these, allowed by dependency notation, is of the type M $\frac{n}{m}$, G $\frac{n}{m}$, A $\frac{n}{m}$ etc. ($m > n$). If the latter representation is used, one should add '*n*' to the net weight calculated (by the binary weighting of the inputs) in order to find the particular dependency (Mx, Gx, Ax; value of *x*) which will be active.

Example: Consider an input labelled M$\frac{3}{4}$. If input is Ø, M3 will be active; if input is 1, M4 will be active.

PROBLEMS

The dependency notation symbols for a few devices are shown in Fig. P5.1–P5.8. Describe their operation. You may find it useful to compare your description with the manufacturer's data sheets.

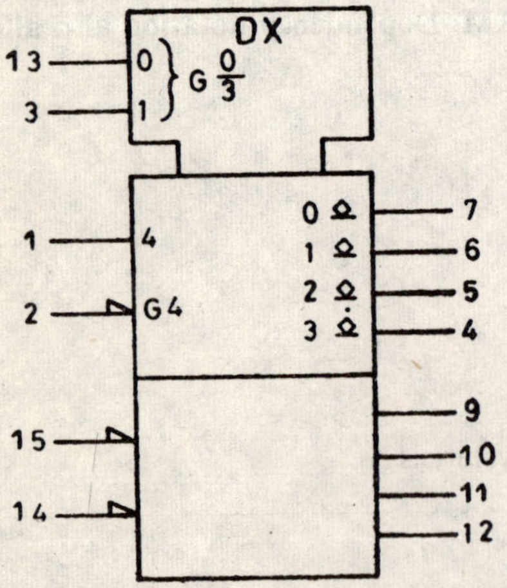

Fig. P5.1 74LS156 Dual 1 of 4 Decoder

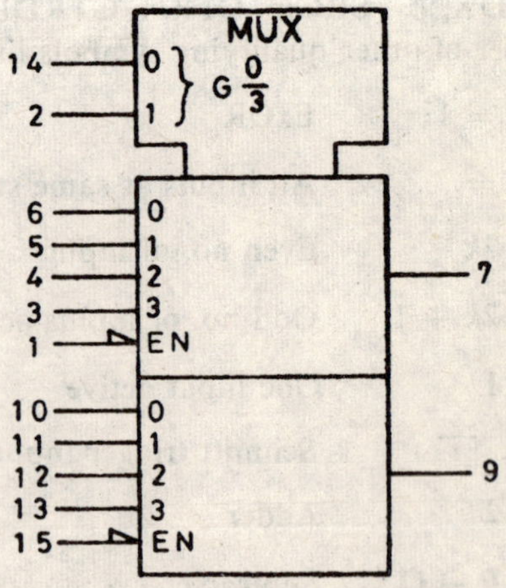

Fig. P5.2 74LS153 Dual 4-Line to 1-Line Multiplexer

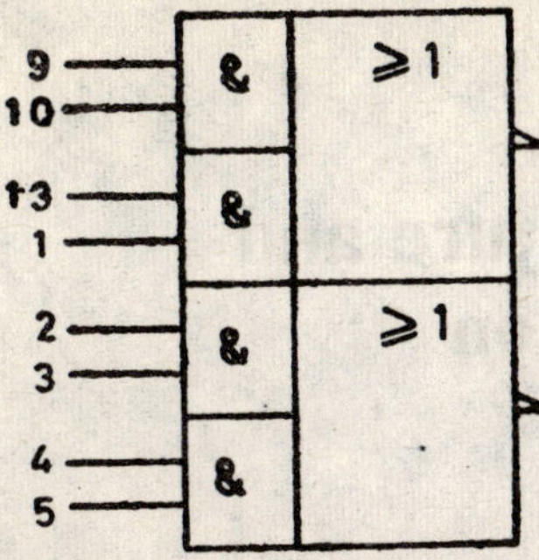

Fig. P5.3 7451 Dual AND-OR-Invert Gates

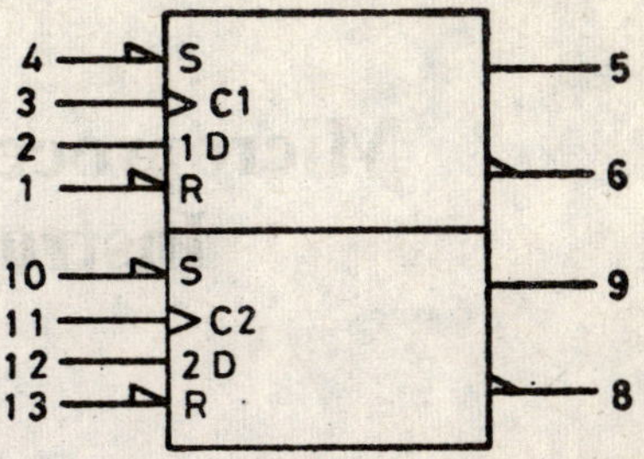

Fig. P5.4 74LS74A Dual D Flip-Flops

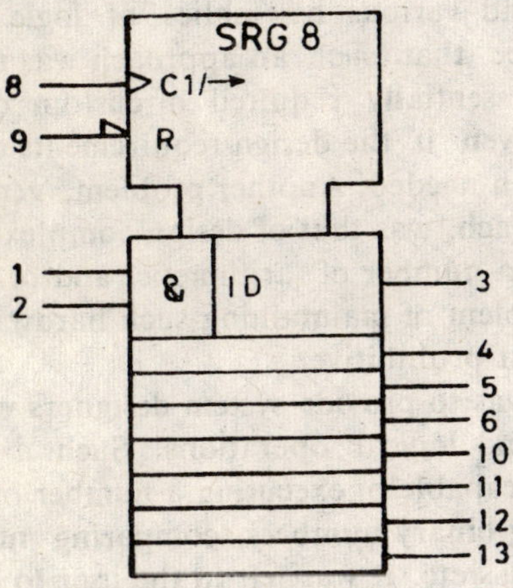

Fig. P5.5 74LS164 8-bit Serial In/Parallel Out Shift Register

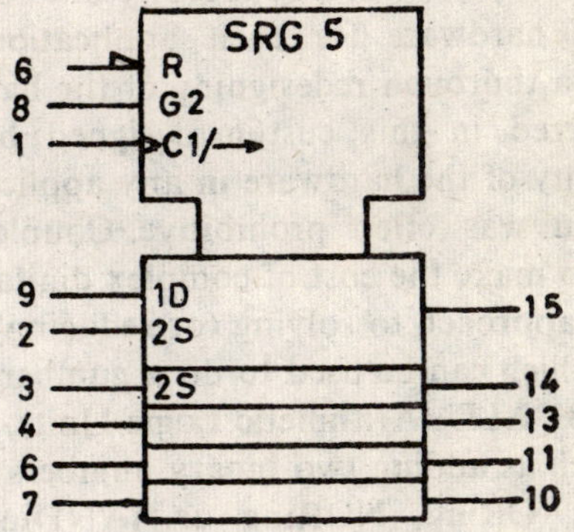

Fig. P5.6 74LS96 5-bit Shift Register

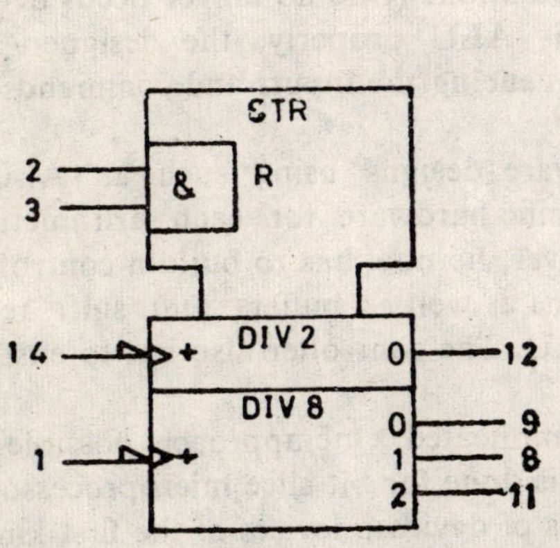

Fig. P5.7 74LS93 4-bit Binary Ripple Counter

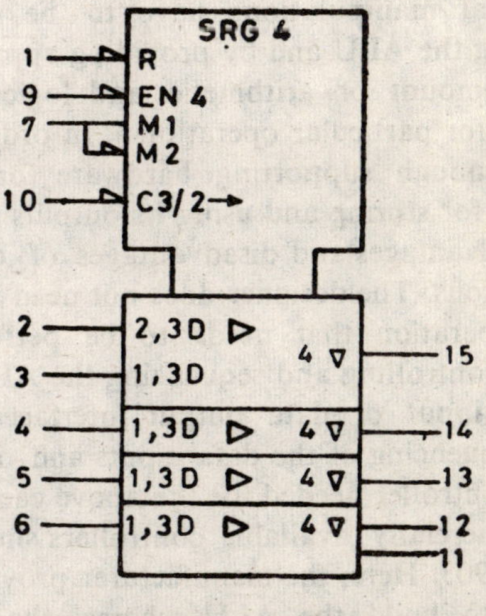

Fig. P5.8 74LS395A 4-bit Cascadable Shift Register with Tristate Outputs

Chapter 6

Microprocessor Architecture and Instruction Execution

Before the advent of microprocessors and other LSI devices, the typical way of doing digital information processing was by means of dedicated hardware. Thus, one used an Adder for additions/subtractions, a Multiplier for multiplications and various assemblies of logic devices for implementing logical operations. It is easy to see that such an approach was not only cumbersome, but was also a costly one to follow. One essentially required a custom designed piece of hardware for each application; moreover, even if the design requirements changed slightly, a thorough redesigning of the hardware was often needed. Another problem, very often encountered in this custom designed hardware approach, was that of design complexity. The complexity of the hardware in any application requiring a number of arithmetic and/or logical operation was often prohibitive. Coupled with the problem of maintaining such hardware, this tended to make the cost of complex digital systems almost prohibitive.

One approach to solving (or reducing) this problem was to provide system designers with one device which can be used to do a number of arithmetic and logical operations. Such a device, e.g. 74181 ALU (Arithmetic Logic Unit), was inherently capable of executing a number of operations such as adding two binary numbers, complementing binary numbers, comparing numbers, ANDing, ORing, NORing or NANDing binary variables etc. It was left to the user to provide the input data and specify the operation to be executed; the chip, in turn, provided the results of the operation to the user.

These multifunction ALU chips are convenient to use in systems where a lot of arithmetic and logical manipulations have to be done on digital data. It is easy to see that by properly controlling the ALU and by providing appropriate data to it, the same device can be used to do a large amount of arithmetic and logical data manipulations (One no longer needs dedicated hardware for particular operations). In order to use this ALU properly, the designer should provide enough supporting hardware for properly sequencing the inputs and commands to the ALU and for storing and using its outputs appropriately.

The advantages and disadvantages of digital hardware designs using such an ALU, are fairly obvious. The designer does not need to design specific hardware for each arithmetic and logical operation that needs to be performed. However, he now has to build a controller for properly controlling and sequencing the ALU's operations as well as buffers and shift registers for data input or data output interfaces with the ALU. The controller also has to ensure the proper sequencing of the data inputs and outputs.

The controller needed for the above can get fairly complicated. One approach possible is to use commercially available controllers similar to what is done for bit slice microprocessors like the AM 2903. Here, the manufacturer provides two kinds of devices; devices of the first kind can be used to build the ALU whereas the other kind of devices can be used as the controller. A more detailed description of such a microprocessor can be found in Sec. 6.5.

In the bit slice approach, a considerable amount of effort is still required to obtain a functional ALU and its corresponding controller. Apart from the hardware, one needs to program the controller properly in order to obtain the desired functions. (This is referred to as microprogramming.) These tasks may be made easier by using microprocessors or CPU-on-a-chip (CPU : Central Processing Unit). Here the ALU, the controller and some data storage registers are made available on a single device chip. The CPU's ALU and controller is designed by the manufacturer to recognize and execute a predetermined repertoire of instructions. In order to realize a particular application, the user now merely provides a proper data input-output interface to the CPU and a set of instructions for it to execute. (This set of instructions constitute the program.) The ALU and the controller inside the CPU ensure the proper execution of each instruction. The internal architecture of such a CPU is described next.

6.1 ARCHITECTURE OF A TYPICAL CPU

In order to describe the architecture of the CPU, it would be worthwhile to discuss what we basically want the CPU to do. From our previous discussion, it should be obvious that the CPU essentially executes a series of instructions (i.e. the program). The bulk of these instructions require the CPU to perform unary or binary operations. (Unary operations require only one operand e.g. inverting a byte of data, inverting signs, shifting a byte left or right etc. Binary operations require two operands e.g. addition, subtraction, AND, OR etc.) Moreover, for each operation (instruction), the CPU will need sources from where to get the operands as well as a destination where it can store the result. These will normally be storage elements which are either inside the CPU itself (registers, latches etc.) or outside it (memory devices—read only or read/write). In order to serve any useful purpose, the CPU must have some means of communicating with the world outside it; in other words, the CPU should provide some means of transferring data from external storage elements (memory) to internal ones (registers) and vice versa. Even the simplest CPU would, therefore, need support memory in order to be functional.

A Simple CPU

The internal architecture of a simple CPU is shown in Fig. 6.1.1. The functional blocks shown for this CPU fall into roughly four categories:

(i) ALU for arithmetic and logic operations
(ii) Temporary storage elements (Temp Reg., ACC, IR, Flag Reg.)
(iii) Memory addressing storage elements (PC, DC)
(iv) Controller for (i) to (iii) (Instruction Decoder and Machine Cycle Encoder)

The heart of the CPU is the 2-input 1-output *ALU* block which is capable of performing a number of unary and binary operations on the operands provided by the *Accumulator* (ACC) and the *Temporary Register*. The actual operation to be performed is specified by the Instruction Decoder block. Note that the ALU is typically a combinatorial logic block whose output can be routed either to be internal storage elements (ACC or Temp Reg.) or to the external storage elements (memory etc.). In addition to getting their data from the ALU, the ACC or the Temp Reg can also be loaded directly from memory.

For proper operation, a CPU needs external memory for three reasons. These are listed below:

(i) Data Source
(ii) Data Destination
(iii) Instruction Source (sequence of instruction forming the program).

As described above, external memory can be both a source (via ACC etc.) of operands to the ALU or a destination for the results generated by it. For proper routing of results to or from external memory, one needs to generate the address for it. This specifies the actual memory location which is required to be the source (or the destination) for the data. External memory is also needed to specify the instruction to be executed. Note that external memory needed as a data source can be implemented using either read-only memory (ROM) or read/write memory (RAM); however, RAM will be needed when external memory is needed as the data destination. External memory acting as the Instruction Source is usually implemented with ROM.

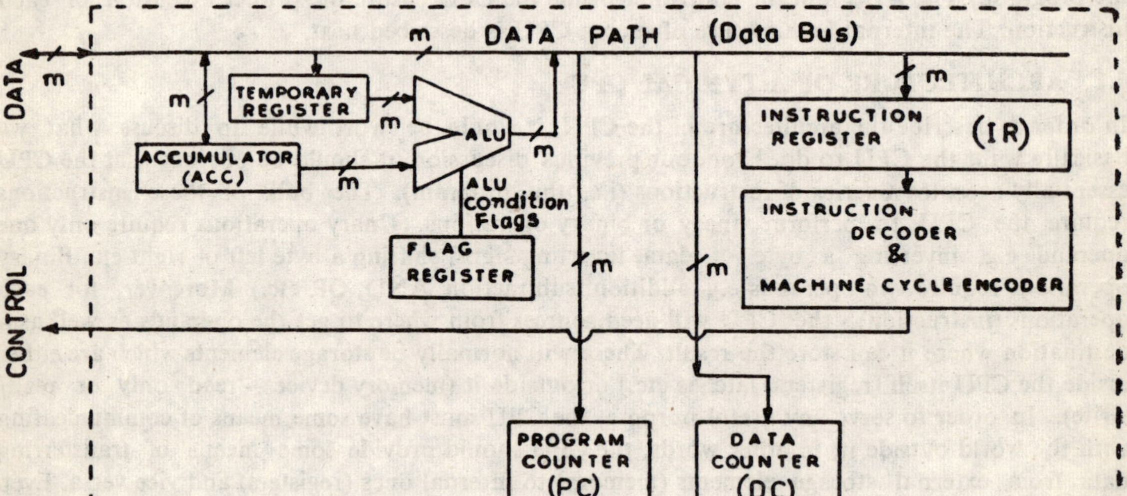

Fig. 6.1.1 Architecture of a Simple CPU (*m*-bit data, *n*-bit address)

The CPU of Fig. 6.1.1 provides two functional blocks. *Program Counter* (PC) and *Data Counter* (DC), for addressing external memory. The Data Counter points to the memory locations needed for data transfers between the CPU and external memory. The Program Counter, on the other hand, points to that memory location which will provide the *next* instruction to be executed. Both PC and DC can be loaded with data from the data bus; the CPU will typically have instructions for doing so. Execution of one instruction will normally increment PC so that it points to the next instruction. If instructions are sequentially arranged in memory, this will guarantee that they will also be sequentially executed. Sometimes, program execution requires that non-sequential instructions be executed e.g. JUMP or CALL type instructions. This requires changing PC directly instead of merely incrementing it. This can be achieved by directly loading PC as mentioned before. Typical CPUs also provide an automatic initialization of PC to a predetermined value whenever it is RESET. This initialization address should have the first instruction of the program to be executed. Apart from being loaded through the Data Bus, the Data Counter (DC) can generally also be directly incremented or decremented. Typical CPUs normally provide instructions to do so.

It should be noted that the distinction regarding the roles of PC and DC is generally not as rigid as implied above. There are situations where PC is also used to address a data source. For example one can have instructions where the data is provided by the instruction itself. In this case the data is part of the program memory; the binary code (*op code*) representing the instruction and the data are placed sequentially in the program memory.

The ALU of Fig. 6.1.1 also produces a FLAG output which is stored in the *FLAG REGISTER*. This consists of bits which indicate some special conditions that were noted in the last operation executed by the ALU. Examples of such FLAG information are given below:

(i) Z = 1 if ALU result is Ø; else Z = Ø (Zero Flag)

(ii) S = 1 if ALU result is —ive; else S = Ø (Sign Flag)

(iii) CY = 1 if ALU operation generated a carry out of the most significant bit; else CY = Ø (Carry Flag)

(iv) P = 1 if ALU result has even parity; else P = Ø (Parity Flag)

These FLAGs may be used to control the operation of the CPU in a manner described later.

The binary code (op-code) representing the instruction to be executed, is fetched from the (program) memory location pointed to by PC and loaded into the *Instruction Register* (IR). The Instruction Register passes this op code to the Instruction Decoder which interprets this op code appropriately in order to decide what operations need to be done for executing this instruction. For example one might have an instruction which requires (a) load data from memory pointed to by DC into Temp Reg, (b) add ACC contents to Temp Register contents, (c) store result in ACC. This instruction would then execute (ACC) ← (ACC) + (Memory Location pointed to by DC). The Machine Cycle Encoder then generates the control signals needed to perform each of the 'micro-operations' that are to be executed in order to execute this instruction. These control signals go to all the other blocks of the CPU shown in Fig. 6.1.1. Some control signals are also needed for components outside the CPU. For example, these external control signals also decide whether one needs to read from external memory or write to it. This functional block also updates PC appropriately (increments it or loads data into it) and fetches the next instruction from memory (into IR) once execution of the current instruction is over. This new instruction is then executed.

Apart from getting data (op-code) from IR, the Instruction Decoder block also monitors the FLAG register contents. This allows the CPU to execute Conditional Instructions, i.e. instructions which may or may not be performed depending on the status of one or more FLAG bits. (If it is not to be executed, the CPU merely moves on to the next instruction.)

Finally, it should be noted that we have shown a CPU handling m-bit data and n-bit addresses in Fig. 6.1.1. Typical values of m are 4, 8, 16, 32 etc. whereas n is normally 16 or 20. The Intel 8085A CPU discussed in Sec. 6.3 has a 8 bit data bus and 16 bit address bus. Another CPU, Intel 8086, discussed in Sec. 6.4 has a 16 bit data bus and a 20 bit address bus.

6.2 INSTRUCTION EXECUTION

In this section we will concentrate on the operation of the CPU of Fig. 6.1.1 in greater detail by examining how some specific instructions will be executed. In the previous section we have indicated that each instruction is further subdivided by the CPU into a number of simpler steps referred to as machine cycles. The CPU requires at least one clock cycle (of an internally or externally generated clock signal) to execute each of these machine cycles. Since an instruction might require one or more machine cycles, the instruction itself would take one or more clock cycles to be executed.

For the sake of convenience (and without much loss of generality), we will assume that $m = 8$ and $n = 16$ in Fig. 6.1.1. This implies that data size is one byte whereas an address consists of two bytes. The reader should keep this in mind when referring to Fig. 6.1.1 for the purposes of this section.

The instructions discussed as examples here require three kinds of Machine Cycles for execution. These are the Op Code Fetch (OF), Memory Read (MR) and Memory Write (MW) machine cycles. The Op Code Fetch Machine cycle is one where the CPU reads the program memory for the op code of the next instruction to be executed and loads it into the Instruction Register. The Memory Read machine cycle is one where the CPU reads external memory (program or data area) to obtain one byte of data which it loads into an internal register (*other than IR*). A Memory Write machine cycle is, similarly, one where the CPU writes one byte of data into external memory.

The following notation will be used in our discussions of instruction execution:

(·)	Contents of indicated register or memory location. For example (ACC) = content of ACC (ØØØØ) = content of location ØØØØH ((PC)) = content of location pointed at by PC
$(PC)_H$	High order byte of PC
$(PC)_L$	Low order byte of PC
$(DC)_H$	High order byte of DC
$(DC)_L$	Low order byte of DC
WW, XX, YY . . .	Consecutive byte storage in program memory with WW = ((PC)) XX = ((PC) + 1) ⋮ etc. (Note that WW should be the op code of the instruction to be executed.)
⟨XYZ⟩	op code of instruction whose mnemonic is XYZ (Note that the mnemonics used will actually be those of the Intel 8085A CPU.)

Address storage convention = Lower order byte in lower memory location,
(memory storage) Higher order byte in higher memory location

6.2.1 EXECUTION OF SOME SIMPLE INSTRUCTIONS

(A) Instruction: LDA ØØØ2H

Operation: (ACC) ← (ØØØ2), Load accumulator direct

Storage: ⟨LDA⟩, Ø2, ØØ

Machine Cycles:

(i) OF: (IR) ← ((PC)) = ⟨LDA⟩

(PC) ← (PC) + 1

(ii) MR: $(DC)_L$ ← ((PC)) = Ø2

(PC) ← (PC) + 1

(iii) MR: $(DC)_H$ ← ((PC)) = ØØ

(PC) ← (PC) + 1

(iv) MR: (ACC) ← ((DC))

The timing diagram for the execution of this instruction is shown in Fig. 6.2.1. It requires four machine cycles—one OF and three MR MCs. The details of the CPU's operation during each machine cycle have been given above.

MC No	1	2	3	4
MC Type	OF	MR	MR	MR
Address Bus	(PC)	(PC)	(PC)	(DC)
Data Bus	((PC)) = <LDA>	((PC)) = Ø2	((PC)) = ØØ	((DC))
Operation	(IR) ← ((PC)) (PC) ← (PC)+1	$(DC)_L$ ← ((PC)) (PC) ← (PC)+1	$(DC)_H$ ← ((PC)) (PC) ← (PC)+1	(ACC) ← ((DC))

—Time→

Fig. 6.2.1 Execution of LDA Addr Instruction

(B) Instruction: ADD M

Operation: (ACC) ← (ACC) + ((DC)) Add accumulator contents to contents of memory location indicated by DC and store result in accumulator

Storage: ⟨ADD M⟩

Machine Cycles:

(i) OF: (IR) ← ((PC)) = ⟨ADD M⟩

(PC) ← (PC) + 1

(ii) MR: (Temp Reg) ← ((DC))

(ACC) ← (ACC) + (Temp Reg)

The timing diagram for the execution of this instruction is shown in Fig. 6.2.2. Note that this instruction requires only two machine cycles with the actual addition being performed by the ALU in the second MC (i.e. the MR-MC) itself.

MC No	1	2
MC Type	OF	MR
Address Bus	(PC)	(DC)
Data Bus	((PC)) = <ADD M>	((DC))
Operation	(IR) ← ((PC))	(Temp.Reg) ← ((DC))
	(PC) ← (PC) + 1	(ACC) ← (ACC) + (Temp Reg)

—— Time ——→

Fig. 6.2.2 Execution of ADD M Instruction

(C) Instruction: STA 1ØFCH

Operation: (1ØFC) ← (ACC) Store accumulator direct

Storage: ⟨STA⟩, FC, 1Ø

Machine Cycles:

(i) OF: (IR) ← ((PC)) = ⟨STA⟩

(PC) ← (PC) + 1

(ii) MR: $(DC)_L$ ← ((PC)) = FC

(PC) ← (PC) + 1

(iii) MR: $(DC)_H$ ← ((PC)) = 1Ø

(PC) ← (PC) + 1

(iv) MW: ((DC)) ← (ACC)

Fig. 6.2.3 shows the timing diagram for the execution of this instruction.

MC No	1	2	3	4
MC Type	OF	MR	MR	MW
Address Bus	(PC)	(PC)	(PC)	(DC)
Data Bus	((PC)) = <STA>	((PC)) = FC	((PC)) = 1Ø	(ACC)
Operation	(IR) ← ((PC))	$(DC)_L$ ← ((PC))	$(DC)_H$ ← ((PC))	((DC)) ← (ACC)
	(PC) ← (PC) + 1	(PC) ← (PC) + 1	(PC) ← (PC) + 1	

—— Time ——→

Fig. 6.2.3 Execution of STA Addr Instruction

(D) Instruction: JMP Ø7EØH

Operation: (PC) ← Ø7EØ Jump direct, continue execution from location Ø7EØ

Storage: ⟨JMP⟩, EØ, Ø7

Machine Cycles:

(i) OF: (IR) ← ((PC)) = ⟨JMP⟩

(PC): ← (PC) + 1

(ii) MR: $(DC)_L$ ← ((PC)) = EØ

(PC) ← (PC) + 1

(iii) MR: $(DC)_H$ ← ((PC)) = Ø7

(PC) ← (DC) = Ø7EØ

Fig. 6.2.4 shows the timing diagram for the execution of this instruction.

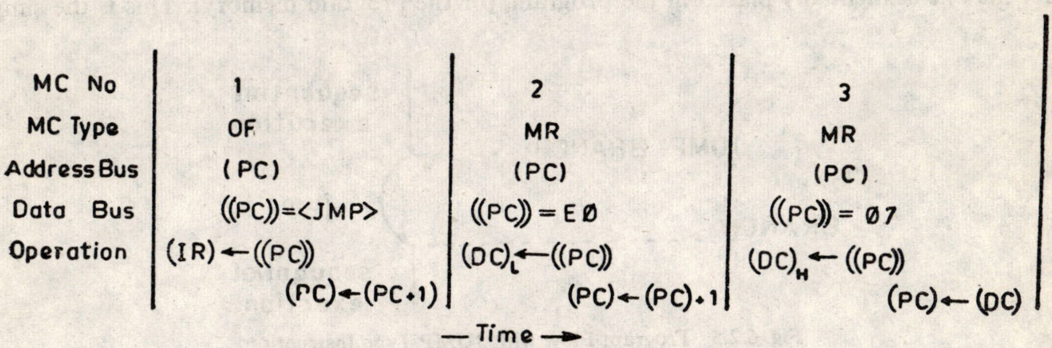

Fig. 6.2.4 Execution of JMP Addr Instruction

(E) Instruction: JZ Ø7EBH

Operation: If Z = 1, (PC) → Ø7EB

If Z = Ø, (PC) ← $(PC)_{old}$ + 3

Storage: ⟨JZ⟩, EB, Ø7

Machine Cycles:

(i) OF: (IR) ← ((PC)) = ⟨JZ⟩

(PC) ← (PC) + 1

(ii) MR: $(DC)_L$ ← ((PC)) = EB

(PC) ← (PC) + 1

(iii) MR: $(DC)_H$ ← ((PC)) = Ø7

Check condition of Z FLAG.

If Z = 1, (PC) ← (DC)

If Z = Ø, (PC) ← (PC) + 1

The timing diagram for this instruction has not been shown. Based on the earlier figures, the reader should be able to draw it himself.

6.2.2 Modifications to the Simple CPU of Fig. 6.1.1

The CPU of Fig. 6.1.1 is not capable of executing an important class of instructions. These are the CALL and RETurn type of instructions which provide subroutine linkage. Before discussing the modifications needed to provide such linkages, it would be useful to review the subject of program flow.

In a CPU such as ours, PC is initialized to a predetermined address value (decided by the manufacturer) whenever it is reset. The first instruction of whatever program is being executed must lie at this address. Here onwards, program execution is mostly sequential with PC being appropriately incremented by each instruction (so that it points to the next sequential instruction in memory which is also the next instruction to be executed). There are, however, some important exceptions to this sequential program flow.

The first exception occurs when the program incorporates a JUMP type instruction as shown in Fig. 6.2.5. In this case, once the JUMP instruction is executed, the next instruction executed is not the one sequentially placed in the program (or the program memory). This is the simplest

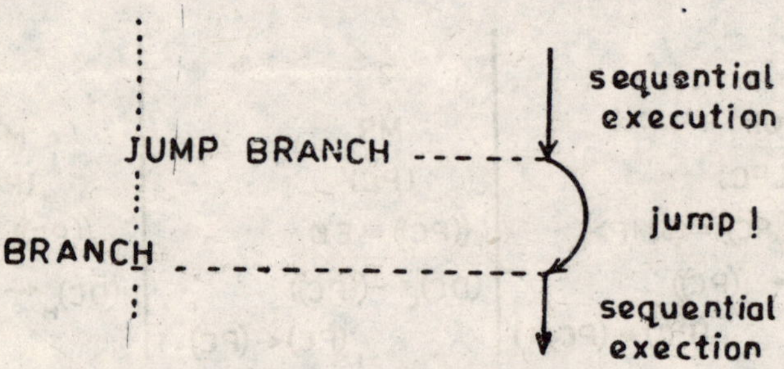

Fig. 6.2.5 Program Flow with JUMP Type Instruction

kind of branching instruction one finds in a microprocessor CPU. As shown by an example in Sec. 6.2.1 a 'JMP Addr' type of instruction can be easily executed by our simple CPU by first getting 'Addr' from program memory and then loading that into PC.

It is also possible to have Conditional Jump instructions in a program e.g. 'J condition Addr'. When such an instruction is encountered, one of two things can happen. If the specified condition (usually a FLAG bit) is true (the FLAG is set), program execution continues from 'Addr'; otherwise, the next instruction in the program memory is executed. This type of branching is illustrated in Fig. 6.2.6. In Sec. 6.2.1 we had discussed how our simple CPU would execute such an instruction. Note that whether the JUMP is conditional or unconditional, the actual JUMP

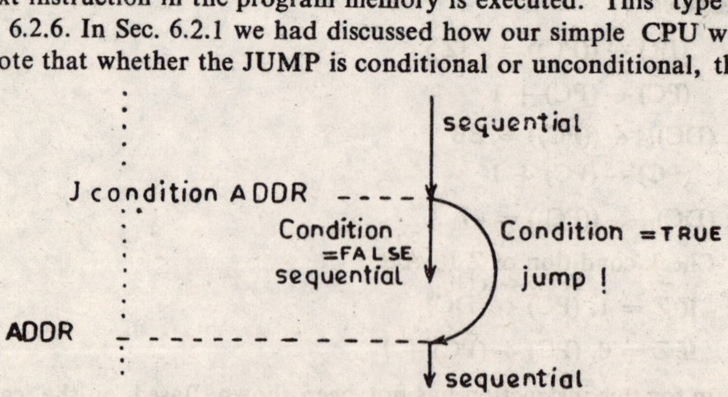

Fig. 6.2.6 Program Flow with a Conditional Jump Instruction

is executed by directly loading PC with the address of the instruction one wants to jump to; this jump-address is, in turn, provided by the appropriate data following the instruction's op code. Note that conditional jump instructions are also examples of conditional instructions (promised earlier!). As a matter of fact, conditional jumps, conditional calls and returns (described next) are usually the only conditional instructions one encounters in the instruction sets of most microprocessors.

Two very important instructions that we have still not provided for in the instruction set of our CPU are CALL and RETurn (and conditional forms of these instructions). In order to see what is involved in these instructions, we will briefly examine the concept of subroutines itself since CALL and RETURN are only used to implement subroutines—CALL is used to branch program flow to a subroutine and RETURN is used to return from the subroutine and resume the previous program flow.

In most programs, one finds that there exist a number of frequently used instruction sequences. A typical instruction sequence of this kind may be inserted physically in the program wherever it is needed. An alternative approach would be to store this instruction sequence, referred to as a subroutine, only once in memory and arrange for this routine to be executed whenever the program calls for execution of this sequence. It is obvious that the latter approach will need less memory than the former. Consider the following example of a program where the Nth instruction is a call to a subroutine MULT.

MULT
⋮
N − 1 —
N Execute Subroutine MULT CALL MULT ---------- Execute MULT
N + 1 — ----------
Return (from MULT to N + 1)

The CALL MULT instruction must do the following:

(i) Store the address of instruction (N + 1) somewhere
(ii) Jump to the starting address of MULT

The RETURN instruction placed at the end of MULT must now do the following:

(i) Recover the address of instruction (N + 1)
(ii) Jump to instruction (N + 1)

It is easy to see that both the CALL and RETURN instructions are very similar to the JUMP instruction with some modifications. The CALL instruction not only executes a JUMP but also stores the return address to which the program must return once the execution of the subroutine is over. The RETURN address is also a JUMP where the address to jump to is recovered from the storage area where it was stored by the corresponding CALL instruction.

One can have a simple program structure where there is one main program and a number of subroutines. The subroutines are always called from the main program and that a return from a subroutine always brings us back to the main program. One can also have a *Nested Subroutine* structure where one subroutine can call another. It is also possible to have a *Recursive Subroutine* where the subroutine can call itself. (The 'depth' of nesting or recursion may be fixed or unlimited depending on the implementation.) Ideally, our CPU should be able to implement both Nested and Recursive Subroutines. We will now examine the method by which the CPU can handle subroutines.

It is obvious from our earlier discussion, that in order to implement a CALL instruction in our CPU, we need some kind of storage in the CPU which will store the return address for the subroutine, (i.e. the address of the instruction which follows in the program memory). If we are prepared to forego nested and recursive subroutines, we can easily provide this storage by providing an extra register in the CPU where PC contents are stored before PC is changed to the starting address of the subroutine. Another approach will be to require that the first memory word or words of the subroutine should store the return address of the subroutine when it is called. (The RETURN instruction merely changes PC to the contents of these words.) This latter approach, common in older minicomputers, will not work with microprocessors because the microprocessor's program is generally implemented in read-only memory (ROM) rather than read/write memory (RAM). One can use a modified form of this approach by providing RAM locations to store return addresses for each subroutine whenever it is called. Apart from having implementation problems (proper initialization, ensuring some predetermined address space is always implemented in RAM etc.), this structure will not allow recursion. (On recursion, previous return addresses will get destroyed.)

All the above problems can be avoided by providing a LIFO Stack (Last In First Out) shown in Fig. 6.2.7. Whenever a CALL is executed, the PC contents are 'pushed into' the stack.

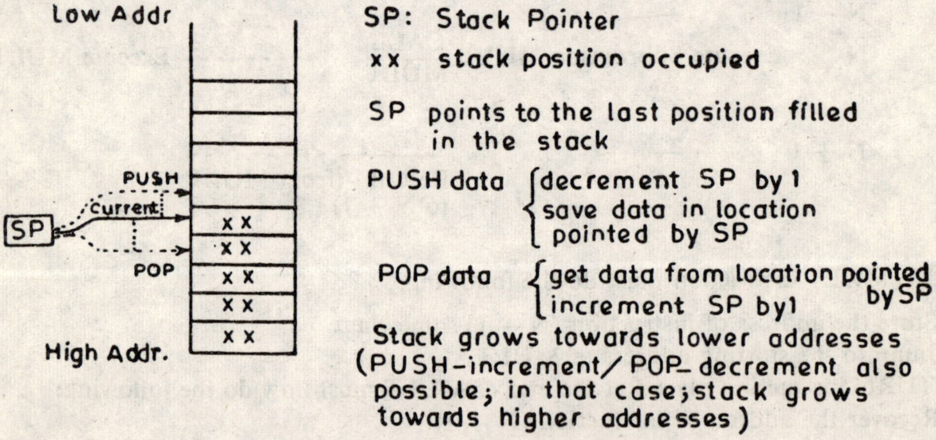

Fig. 6.2.7 Organization of a LIFO Stack

A RETURN instruction then merely 'pops out' the content off the top of the stack and loads them into PC (Note that the stack will have to be of the Last In First Out or LIFO type.) It can be shown that this method of calling subroutines or returning from them, will work for any number of nested or recursive subroutines provided a stack of sufficient depth is available. It is for these reasons that a Stack is almost routinely available in all microprocessors.

One can of course implement a Stack inside the CPU itself. Apart from the extra storage one would then require within the CPU, this is not preferred for another reason. If the Stack is inside the chip, its 'depth' and hence the extent to which nesting and recursion can be done will be limited. (More importantly, this limitation will essentially be decided by the manufacturer and not the user). It is more common to actually implement the stack in the system's RAM leaving only a Stack Pointer (SP) register in the CPU. The user will then have to take the responsibility of initializing SP so that it points to a location where RAM is present. A PUSH will increment (decrement) SP whereas a POP will correspondingly decrement (increment) SP. The user now has to take care of the following in his program:

(i) SP should always point at a RAM location.

(ii) RAM area designated for Stack should not be (unknowingly) used by other program instructions which can write data into memory.

In order to provide the subroutine linkage instructions in our CPU (Fig. 6.1.1) we will introduce a Stack Pointer (SP) in our CPU and require that the user implement the actual stack in external RAM. The CPU with this modification is shown in Fig. 6.2.8. Note that SP is similar

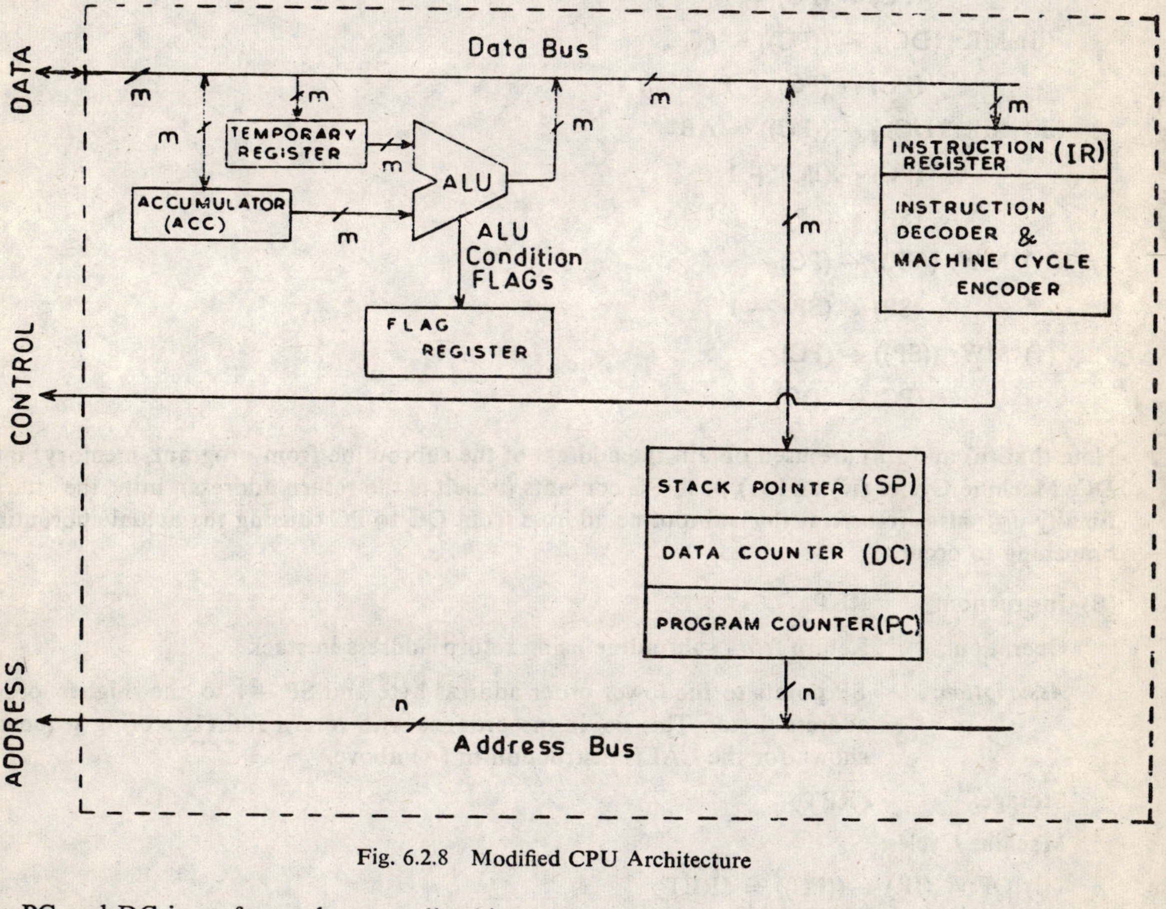

Fig. 6.2.8 Modified CPU Architecture

to PC and DC in so far as they are all n-bit registers taking their inputs from the data bus and outputting an address to the address bus. (For our present discussion, n is 16.) It should be noted that SP, DC, and PC all point to the same 2^n m-word address space. It is up to the user to ensure that the program memory, data memory and stack memory areas do not overlap each other.

6.2.3 Subroutine Linkage Instruction and Other Instructions Using Stack

In this section, we present how subroutine linkage instructions and other instructions using the stack can be implemented using the modified CPU of Fig. 6.2.8. These implementation details are given in terms of the constituent machine cycles using the notation described earlier in Sec. 6.2.1.

(A) Instruction: CALL ABØC

Operation: Jump to ABØC saving return address in stack

Storage: ⟨CALL⟩, ØC, AB

Machine Cycles:

(i) OF: $(IR) \leftarrow ((PC)) = \langle CALL \rangle$

$(PC) \leftarrow (PC) + 1$

(ii) MR: $(DC)_L \leftarrow ((PC)) = ØC$

$(PC) \leftarrow (PC) + 1$

(iii) MR: $(DC)_H \leftarrow ((PC)) = AB$

$(PC) \leftarrow (PC) + 1$

$(SP) \leftarrow (SP) - 1$

(iv) MW: $((SP)) \leftarrow (PC)_H$

$(SP) \leftarrow (SP) - 1$

(v) MW: $((SP)) \leftarrow (PC)_L$

$(PC) \leftarrow (DC)$

Note that (ii) and (iii) are used obtain the address of the subroutine from program memory into DC. Machine Cycles (iv) and (v) push PC contents (which is the return address) into the stack. Finally (v) also transfers the subroutine address from DC to PC causing the actual subroutine branching to occur.

(B) Instruction: RET

Operation: Return from subroutine using return address in stack

Assumption: SP points to the lower order address byte and SP +1 to the higher order address byte. This is in accordance with return address storing sequence shown for the CALL instruction in (A) above.

Storage: ⟨RET⟩

Machine Cycles:

(i) OF: $(IR) \leftarrow ((PC)) = \langle RET \rangle$

$(PC) \leftarrow (PC) + 1$

(ii) MR: $(PC)_L \leftarrow ((SP))$

$(SP) \leftarrow (SP) + 1$

(iii) MR: $(PC)_H \leftarrow ((SP))$

$(SP) \leftarrow (SP) + 1$

Note that the CALL and RET instructions as described above may be used as a pair to enter and exit a subroutine. Moreover, the implementation will allow nesting and recursion up to any depth, limited only by the maximum stack size that can be allowed.

A side benefit of using the stack is that it allows easy storage and retrieval of data. One can push data into the stack and pop it out as required. Note that unlike instructions like LDA Addr or STA Addr (Sec. 6.2.1) one does not have to specify a memory address for this; the stack area is itself used for storage. This can be a considerable advantage in many programming situations. We describe next examples of two instructions which should be used as a pair to store and retrieve data using the stack.

(C) Instruction: PUSH PSW

Operation: Store ACC and FLAG Register in stack

Storage: ⟨PUSH PSW⟩

Machine Cycles:

(i) OF: $(IR) \leftarrow ((PC)) = \langle \text{PUSH PSW} \rangle$

$(PC) \leftarrow (PC) + 1$

$(SP) \leftarrow (SP) - 1$

(ii) MW: $((SP)) \leftarrow (ACC)$

$(SP) \leftarrow (SP) - 1$

(iii) MW: $((SP)) \leftarrow (\text{FLAG Reg.})$

(D) Instruction: POP PSW

Operation: Restore ACC and FLAG Register using data stored in stack by PUSH PSW

Storage: ⟨POP PSW⟩

Machine Cycles:

(i) OF: $(IR) \leftarrow ((PC)) = \langle \text{POP PSW} \rangle$

$(PC) \leftarrow (PC) + 1$

(ii) MR: $(\text{FLAG Reg.}) \leftarrow ((SP))$

$(SP) \leftarrow (SP) + 1$

(iii) MR: $(ACC) \leftarrow ((SP))$

$(SP) \leftarrow (SP) + 1$

A CPU such as that of Fig. 6.2.8 will also typically provide instructions to directly manipulate the stack pointer. SP as shown below:

(E) Instruction: LXI SP, FFFEH

Operation: Load FFFE in SP

Storage: ⟨LXI SP⟩, FE, FF

Machine Cycles:

(i) OF: $(IR) \leftarrow ((PC)) = \langle LXI\ SP \rangle$

$(PC) \leftarrow (PC) + 1$

(ii) MR: $(SP)_L \leftarrow ((PC)) = FE$

$(PC) \leftarrow (PC) + 1$

(iii) MR: $(SP)_H \leftarrow ((PC)) = FF$

$(PC) \leftarrow (PC) + 1$

(F) Instruction: INX SP/DCX SP

Operation: Increment/Decrement SP by 1

Storage: $\langle INX\ SP \rangle / \langle DCX\ SP \rangle$

Machine Cycle: (OF): $(IR) \leftarrow ((PC)) = \langle INX\ SP \rangle / \langle DCX\ SP \rangle$

$(PC) \leftarrow (PC) + 1$

$(SP) \leftarrow (SP) + 1 / (SP) \leftarrow (SP) - 1$

6.3 INTERNAL ARCHITECTURE OF THE INTEL 8085A CPU

The Intel 8085A CPU is a reasonably powerful, general purpose microprocessor manufactured by Intel Corp. It is an extremely popular 8-bit microprocessor with excellent software and hardware support both from Intel as well as from others. It is a processor with a 8-bit data bus and a 16-bit address bus and is shown schematically in Fig. 6.3.1. Comparing this with our modified CPU of Fig. 6.2.8, we note the following differences.

(i) Instead of the single general purpose register—the Accumulator—in Fig. 6.1.1 and Fig. 6.2.8, the 8085A CPU has seven such registers. Apart from the Accumulator, these are labelled B, C, D, E, H, and L. These registers are 8-bit registers; however the pairs B-C, D-E, and H-L may also be used as three 16-bit registers B, D and H respectively. These can act as both the source and destination of ALU data.

(ii) Apart from the FLAGs S, Z, P, CY mentioned earlier for our CPU of Fig. 6.1.1, the 8085A also provides an Auxiliary Carry (AC) FLAG. This is set to 1 if the ALU operation results in a carry from bit 3 to bit 4 (lower nibble to higher nibble, accumulator bits are labelled Ø – 7, bit Ø = LSB, bit 7 = MSB) of the accumulator.

(iii) The 8085A does not have a specific register acting as the data counter DC. The 8085A, however, allows the user to specify any of the B-C, D-E or H-L register pairs as providing the data address in certain instructions. It also has instructions where the data address comes from a register pair but where the instruction does not explicitly indicate which register pair. In such situations, it is always the H-L register pair which provides the data address. One can therefore consider the H-L register pair as something like a Data Counter in the 8085A.

Apart from the above differences, the 8085A is very similar to the modified CPU shown in Fig. 6.2.8. Its operation is also reasonably similar to the instruction execution illustrated in Sec. 6.2. (More will be said about this in Chapters 8 and 9). However, the following features of the 8085A are also worth noting.

(i) The presence of a large number of accumulator-like working registers is a very big advantage of designing with the 8085A CPU. Memory accesses for data are minimized by

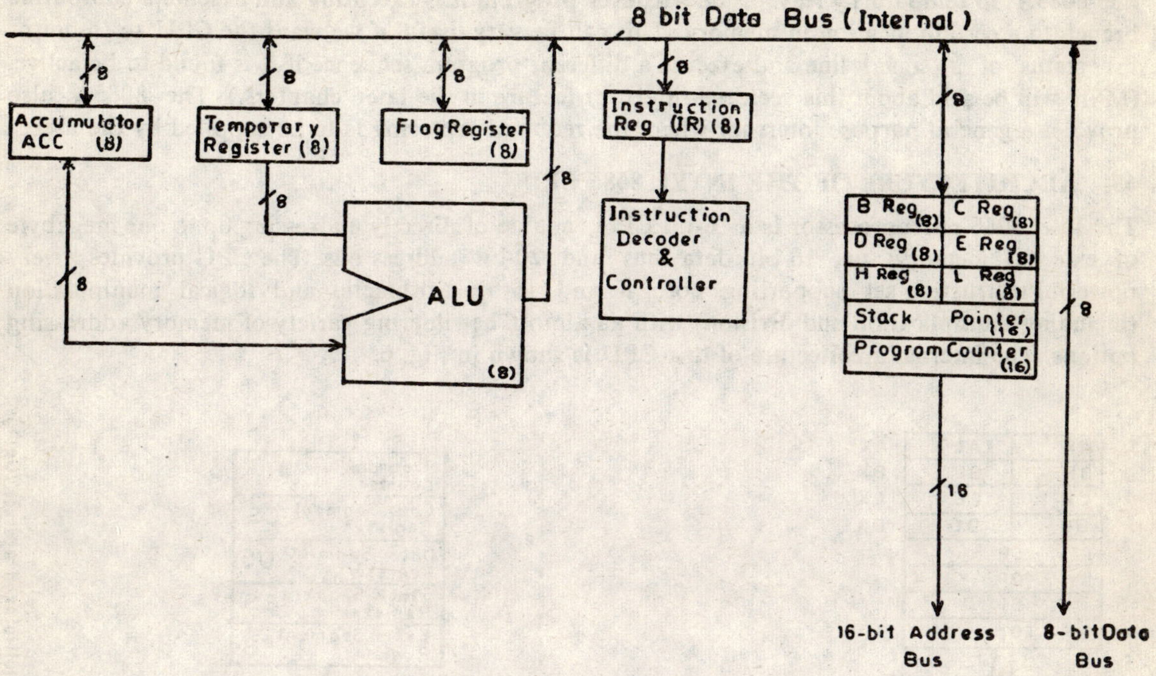

Fig. 6.3.1 Internal Organization of the 8085A CPU

using these registers to store intermediate results in arithmetic and logical manipulations. This typically makes the 8085A a very efficient CPU in terms the number of memory accesses required in a program.

(ii) Even though it is primarily a 8-bit machine, some 16-bit data manipulations can be easily implemented with the 8085A by properly using its B-C, D-E and H-L register pairs.

(iii) The 8085A is primarily intended for binary arithmetic. However, the AC FLAG makes it reasonably easy to implement manipulation of packed BCD data (i.e. two BCD digits represented by one byte). This is further facilitated by a Decimal Adjust instruction in its instruction set.

(iv) It can address a total of 64 K bytes in memory via PC, SP or one of the register pairs. It is up to the user to divide this address range appropriately into program area, data area and stack area. Note that for proper operation, the user should initialize the stack pointer (SP) before instructions using the stack are issued.

(v) On RESET, the 8085A CPU initializes PC to ØØØØH. Therefore, the first instruction in the program should be at ØØØØH in the memory address space of the CPU.

(vi) Most of the 8085A's instructions increment PC to point at the op code of the next instruction. The PC can also be directly loaded in order to execute branch type instructions like JUMP, CALL, RETURN etc.

(vii) Conditional branch instructions are the only type of conditional instructions available in the 8085A CPU.

(viii) The 8085A provides a vectored interrupt feature (not shown in Fig. 6.3.1) which is found to be extremely useful in most applications. Activating a vectored interrupt signal, forces the 8085A to temporarily leave aside whatever program it is executing and execute a subroutine branch to a certain location in memory. This can be very useful if we want the CPU to monitor the status of a signal line and execute a different program sequence if it is found to be active. (More will be said about this vectored interrupt feature in the later chapters.) The 8085A also provides a general purpose interrupt where the required vectoring is to be supplied by the user.

6.4 ARCHITECTURE OF THE INTEL 8086 CPU

The Intel 8086 microprocessor is a 16-bit CPU capable of directly addressing up to one megabyte of external memory; i.e. 16-bit data bus and 20-bit address bus. The CPU provides a very powerful instruction set supporting both 8 and 16-bit arithmetic and logical manipulation (including multiplication and division) with an almost bewildering variety of memory addressing options. The internal architecture of this CPU is shown in Fig. 6.4.1.

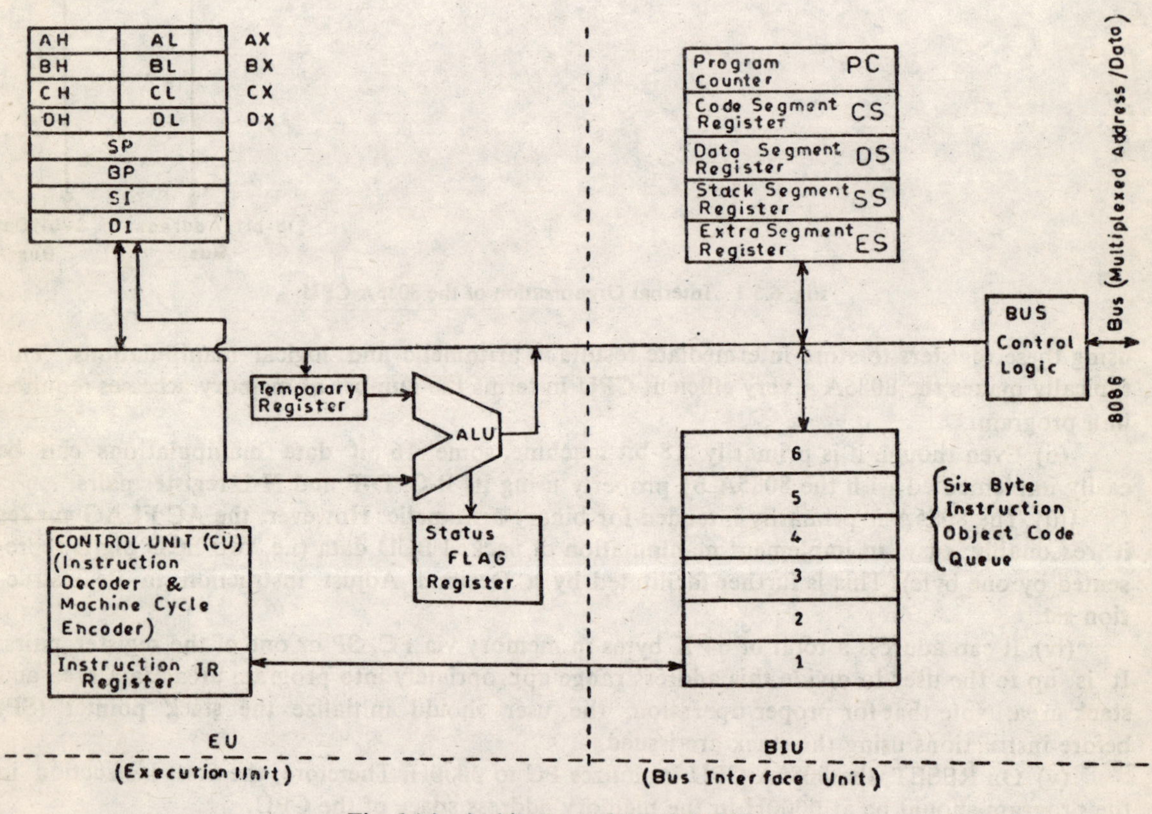

Fig. 6.4.1 Architecture of the 8086 CPU

Examining the 8086 CPU of Fig. 6.4.1, we find that in spite of some similarities, (some of the functional blocks in the 8086 are similar to what we had encountered before) its architecture is substantially different from the CPU architecture shown earlier. A major difference lies in the fact that the 8086 CPU can be viewed as a combination of two units—an Execution Unit (EU) and a Bus Interface Unit (BIU). The Execution Unit is roughly similar to the CPUs discussed before except for the fact that it does not communicate directly (address, data etc.) with the outside world (i.e. external memory); such communication is handled by the BIU. (In other words, if the EU needs data from a particular location, it will actually ask the BIU to activate external logic appropriately and fetch the required data; if it wants to store data in a particular location, it will pass the data to the BIU and ask it to store it in the desired location.)

An interesting feature of the 8086 CPU is its Instruction Object Code Queue. This is really a six byte deep stack which the BIU tries to keep full at all times. Whenever one or more locations in the stack become empty, the BIU fetches bytes sequentially from program memory until it is full (Addressing of program memory described later). Note that the EU usually takes bytes from this queue instead of asking the BIU to directly fetch a byte from program memory. (It is possible that the latter might also take place under exceptional conditions, e.g. when the EU executes a branch instruction.) The presence of the Instruction Object Code Queue greatly enhances the memory accessing efficiency of the 8086. Whenever the external bus is not being used (i.e. by the 8086 to fetch or store data or by circuitry around the 8086), the BIU will fetch bytes from program memory and store it in the queue. Then, when the byte is actually needed, it will already be there in the BIU and the BIU will not need to access memory for it. The efficiency in this kind of arrangement comes from the fact that most of the memory accesses of the 8086 will be done by the BIU concurrently with other processes going on in the EU. An arrangement of this kind is typically referred to as '*pipelining*'. A pipelined architecture like that of the 8086 is frequently used in minicomputers to improve the efficiency of system operation.

In the 8086, the BIU has the responsibility of generating the address needed for the different kinds of memory accesses demanded by the CPU (i.e. program, data or stack). The 20-bit address needed for this are generated using the Program Counter and the four segment registers CS, DS, SS and ES. It should be noted that PC, CS, DS, SS, and ES are all 16-bit registers and that they are used to generate a 20-bit address by computing

$$EA = 2^4(\text{Reg A}) + (\text{Reg B}).$$

With Reg A and B both 16-bit registers, EA would be the 20-bit *effective address*. Particular choices of Reg A and Reg B will depend on what is being accessed (program, data or stack) and the addressing mode. Some typical examples are given below:

(i) Program Reg A = CS

Reg B = PC

(ii) Stack Reg A = SS

Reg B = SP

(iii) Data Almost any combination with

Reg A = One of CS, DS, SS, ES in BIU

Reg B = One of SP, BP, SI, or DI in EU

For data, EA can also be computed as

$$EA = 2^4(\text{Reg A}) + (\text{Reg B}) + X + Y$$

where Reg A, Reg B as above, and

X = ØØØØ or (BX)

Y = ØØØØ or an 8-bit signed offset (sign extended) or 16-bit signed offset specified by the instruction

As one can see from the above, the 8086 allows a very large number of ways in which data may be addressed.

In general one can say that all the addresses generated by the 8086 are done by

(i) Shifting the contents of a BIU segment register four places left (with Øs filled in the 4-LSBs).

(ii) Adding the contents of one or more EU register and/or

(iii) Adding a displacement specified by the instruction itself.

The segmentation concept that the user is now required to follow, introduces some other interesting features of the 8086 CPU. Referring back to the CPU architectures of Fig. 6.2.8 and Fig. 6.3.1 one finds that there the program area, data memory area and stack area were completely overlapping from the CPU's point of view. (The user took care to ensure that there was no actual overlap during execution.) In other words, there was no segmentation of any kind between the three types of memory in these CPUs.

The four segment registers of the 8086 are used to identify four segments in the memory address space of 64 K bytes each. By loading the segment registers with appropriate values, the user has the freedom of deciding whether the four segments should be completely non-overlapping, partially overlapping or completely overlapping. Subject to explicit redefinition by the user, the 8086 CPU normally assumes that the program code will come from the program segment, data will use the data segment and stack operations will use the stack segment. The user still has the freedom to specify in an instruction which segment to use for the data. He can also load new values into CS, DS, SS, and ES in order to define and use new segments for program, data or stack. (ES is an extra segment register which can be used like DS for memory addressing.)

The EU looks very much like the CPU of the 8085A. The only striking difference is the greater number of registers available on the 8086. The 16-bit register AX serves as the primary accumulator of the 8086 CPU. The 8086 also provides three other 16-bit working registers BX, CX, DX. Apart from their accumulator like operation, these registers also do the following:

(i) BX serves as a base register for computing memory addresses

(ii) CX acts as a counter for multi-iteration instructions where execution continues until CX increments or decrements to 0

(iii) DX is used for data transfers between I/O ports (discussed in Chapter 7) and the memory location addressed by DX.

Each of these 16-bit registers is also accessible as two 8-bit registers i.e. AX = AH — AL, BX = BH — BL, CX = CH — CL, DX = DH — DL where H stands for the high order byte and L stands for the lower order byte.

In addition to these general accumulator-type working registers, the EU of the 8086 also contains four other 16-bit registers referred to as Stack Pointer (SP), Base Pointer (BP), Source Index register (SI) and Destination Index register (DI). The Stack Pointer is used in conjunction with the Stack Segment register to generate the 20-bit address needed by the 8086 for stack addressing. The Base Pointer is used for Base Relative Addressing and SI and DI for Indexed Addressing. (These addressing modes are described in Section 7.4.) For certain operations, SI points of the data source and DI points to the data destination; this is the reason for the Source Index and Destination Index nomenclature. Finally it should be pointed out that unlike the accumulator type registers AX, BX, CX and DX, the registers SP, BP, SI and DI are used purely for address generation (either for stack operations or data memory accessing).

The Intel 8086 CPU is a very powerful microprocessor with features reminiscent of some of the more powerful minicomputers. The reader should note that the above discussion is intended to give a flavour of its advanced CPU architecture and should not be considered as an exhaustive description of its features. The reader should refer to the User's Manual for this CPU and other supporting literature from Intel in order to get a complete description of this microprocessor. Intel also provides a 8088 CPU whose internal architecture is virtually identical to the 8086. The 8088 CPU, unlike the 8086, communicates with the external world via a 8-bit data bus rather than a 16-bit bus. This is done even though the 8088 is also a 16-bit CPU.

6.5 BIT SLICE ARCHITECTURES

From our discussion of CPU architectures in Sec. 6.2–6.4, it should be evident that the functional blocks inside a CPU can be divided into two types—(a) the ALU and associated storage registers and (b) the Controller for controlling (a). An instruction (in the CPU's instruction set) is then executed by the Controller asking the ALU (and its associated registers) to perform a number of simple operations. In the type of CPUs discussed in Sec. 6.2–6.4, the manufacturer has pre-programmed the Controller to properly execute the simple operations needed to execute each instruction in its instruction set. These simple operations and their sequence is referred to as the microprogram for a particular instruction.

In typical CPUs such as the 8085, 8086 etc., the user does not have any control over the microprogram or the sequencing of the microinstructions that are performed inside the CPU. Bit Slice microprocessors allow the user this privilege. In a microprocessor of this type, the user has control over the microprogram and can write microprograms to recognize and execute any instruction desired. In effect, the user creates his own instruction set by writing the microprogram for each instruction in this set. At a higher level, his CPU can then execute a program containing these instructions.

The manufacturer of a bit slice processor provides the user with two kinds of components—an ALU and Register Slice and a Microcontroller Slice. The user concatenates the ALU slices to obtain whatever ALU data size is required; the micro-controller slices are also concatenated to obtain the overall microcontroller for this type of CPU. This arrangement is shown schematically in Fig. 6.5.1.

The ALU slices are only able to execute a few microinstructions. (A typical microinstruction is like one of the steps shown for an instruction in Sec. 6.2). The microcontroller has the job of generating the right control signals in the right sequence in order to—

(a) regulate and direct data flow between the ALU registers and between them and memory.

(b) make the ALU execute the right microinstructions in the proper sequence.

A detailed description of a bit-slice architecture is beyond the scope of the present discussion. It is necessary, however, to point out the reasons why such an architecture is sometimes preferred over the CPU-on-a-chip type of products. Two important reasons for preferring the bit-slice approach is given below.

(i) Speed: Typical microprocessors are MOS devices whereas bit-slice processors are bipolar LSIs. This allows the latter to work at higher clock rates and hence with lower execution times.

(ii) Specialized Instruction Set: There are times when the application demands some special instructions which may not be there in the instruction set of a typical CPU chip. Using bit-slice processors, one can design a CPU with the required instructions in its repertoire.

These are the two main reasons why at times the designer may forego the use of a

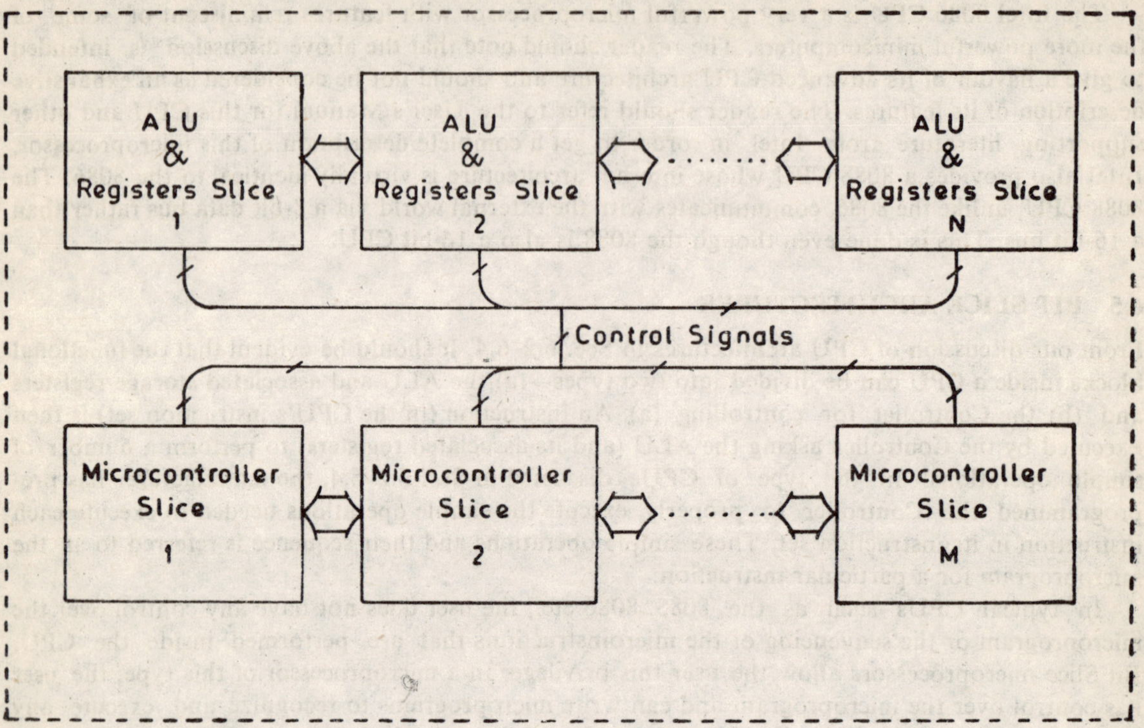

Fig. 6.5.1 Building a Bit-Slice CPU with ALU and Microcontroller Slices

predesigned CPU chip, and take on the somewhat more complex task of designing his own CPU with bit-slice components.

PROBLEMS

1. For our hypothetical CPU (the one with the Stack Pointer SP) describe briefly how a CALL and RET instruction will be executed.
2. Suppose the entire memory space of a CPU were to be occupied by RAM. Can you suggest a method of subroutine linkage which, without using a stack, will allow implementation of nested subroutines (but not recursive ones). Show (Machine Cycle wise) how the subroutine linkage will be executed in a CPU of the kind shown in Fig. 6.1.1 assuming whatever extra registers were needed.
3. In the CPU of Fig. 6.2.8, (with m = 8, n = 16) one wants to implement an instruction XPCSP which will do the following

   XPCSP: $(PC) \leftarrow (PC) + 1$

   $(PC)_L \leftarrow ((SP)),\ (PC)_H \leftarrow ((SP) + 1)$

   Show how such an instruction could be executed.

4. In the CPU of Fig. 6.2.8 (with m = 8, n = 16), show how a POP instruction (popping the stack) will be executed using the following as examples

   POP A; $(A) \leftarrow ((SP))$, $(SP) \leftarrow (SP) + 1$

   POP DC; $(DC)_L \leftarrow ((SP))$, $(DC)_H \leftarrow ((SP) + 1)$

   $(SP) \leftarrow (SP) + 2$

   POP PC; similar to above.

5. Using (3) and (4) show that

   LXI DC, Addr; $(DC) \leftarrow$ Addr

   PUSH DC; opposite of POP DC

   XPCSP

   will be equivalent to a CALL Addr instruction.

6. Show how the following instructions may be executed in the CPU of Fig. 6.2.8 (m = 8, n = 16)

   (i) SPDC; $(SP) \leftrightarrow (DC)$

   (ii) LDAX DC; $(A) \leftarrow ((DC))$

   (iii) STAX DC; $((DC)) \leftarrow (A)$

   (iv) INX SP; $(SP) \leftarrow (SP) \leftarrow + 2$

   (v) NOP; No Operation

   (vi) SUB M; $(A) \leftarrow (A) - ((DC))$

   (vii) ADI Data; $(A) \leftarrow (A) +$ Data (Data = 8 bits)

   (viii) SUI Data; $(A) \leftarrow (A) -$ Data (Data = 8 bits)

   Assume that PC is properly incremented by each of the above.

Chapter 7

Microcomputer Organization

7.1 INTRODUCTION

In a very general sense, a microcomputer is best regarded as a system incorporating a CPU and associated hardware whose purpose is to manipulate data in some fashion. This is exactly what any digital circuit designed using SSIs and MSIs (i.e. gates, flip-flops etc.) will also do. It would therefore seem from this naive definition of a microcomputer, that the present burst of activity in this area is merely selling old wines in new bottles. The fact that this is not true, arises primarily because of the power inherent in the basic component of any microcomputer system—the microprocessor or the 'CPU on a chip'.

As described in the previous chapters, the microprocessor should be regarded as a general purpose logic device. In contrast to standard SSIs and MSIs, where the manufacturer decides what the device will do, with microprocessors it is the user who decides what the device should do by asking it to execute a proper set of instructions. A microcomputer, from this viewpoint, is merely an assembly of devices whose sole task is to ensure that the instructions desired are indeed carried out properly and to allow the microprocessor to communicate with the real world, i.e. the user environment. The power of the microcomputer lies in the fact that if the application changes, the same system can still be used by appropriately modifying the instructions to be executed and probably incorporating some nominal changes in the hardware. In contrast, a digital circuit designed using SSIs and MSIs for some application will probably need to be completely redesigned if the application changes significantly. Our definition of a microcomputer is not at fault; it just fails to mention the reason why it is such a powerful tool for a system designer.

The objective of a microcomputer is to manipulate data in a certain fashion specified by the system designer. A typical microcomputer achieves this objective by getting its component CPU, namely the microprocessor, to execute a number of instructions in the proper sequence. This sequence of instructions comprises the program that is executed by the microcomputer. This program, properly executed, should embody what the system designer wants the microcomputer to do.

A microcomputer which does nothing other than manipulate data present within itself, will not be of much use to anybody. In order to do something meaningful, the data being manipulated should depend in some fashion on inputs provided to the microcomputer. Similarly, the data manipulations being carried out by the microcomputer would be completely senseless unless the results of these manipulations affect things outside the microcomputer itself. A microcomputer should provide outputs which, in some way, depend on its inputs; the way inputs and outputs are related is decided by the program that gets executed.

Summarising, one can say that a microcomputer is an assembly of devices, including a CPU, which manipulates data depending on one or more inputs and according to a program, in order to generate one or more outputs. With the understanding that one can change the data manipulation being done, merely by changing the program, this is probably a better definition of a microcomputer than what was stated earlier. For the purposes of this chapter, this will be the definition followed. This 'black box' definition is illustrated in Fig. 7.1.1.

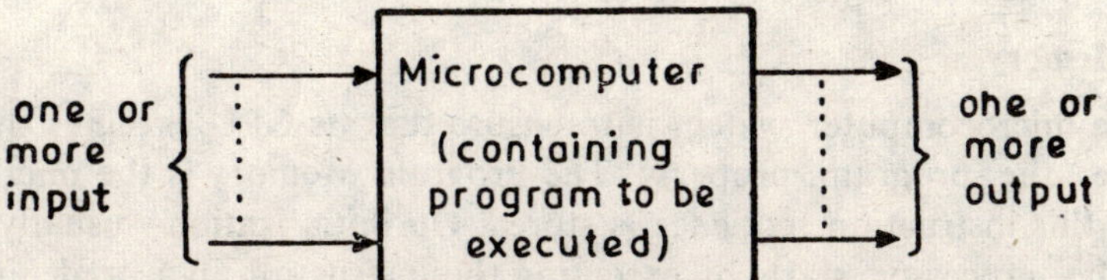

Fig. 7.1.1 Microcomputer as a Black Box

7.2 BASIC COMPONENTS OF A MICROCOMPUTER (MEMORY AND I/O)

The microprocessor is the basic component of a microcomputer system. By itself, however, it is able to achieve very little. Typically, we need to support this device with a number of 'peripheral' or 'support' devices in order for it to function meaningfully. Based on the definition of a microcomputer given in Section 7.1 (its objectives and how it achieves them), we can specify in a general sense the functional blocks that are needed to support the operation of a microprocessor. For our purposes, we have identified five functional blocks that are needed to build a basic microcomputer using a microprocessor. These functional blocks are illustrated in Fig. 7.2.1 and are described briefly in this section.

The functional blocks shown in Fig. 7.2.1 do not necessarily represent physical devices; the division shown has been made more on the basis of functionality. If one looks at the various

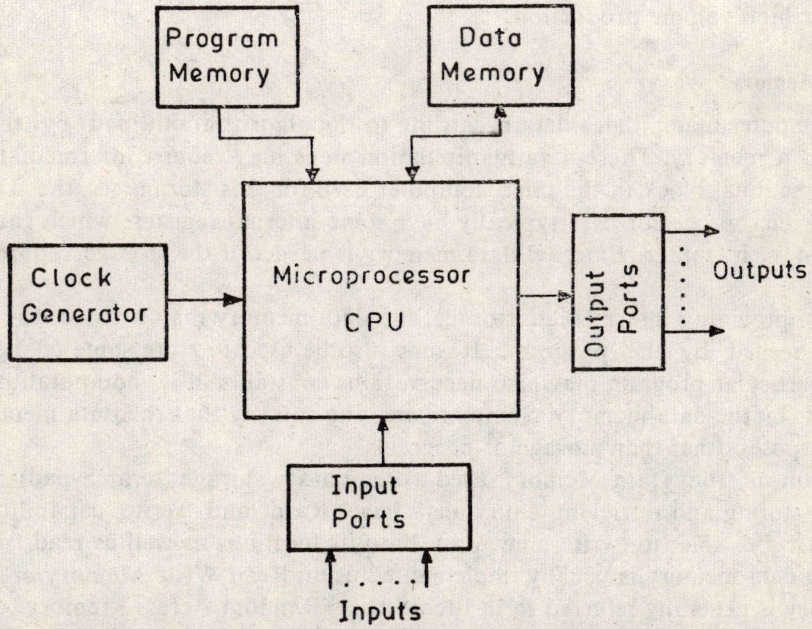

Fig. 7.2.1 Basic Components of a Microcomputer

devices available from different manufacturers, one can easily find examples where two or more of these functional blocks have been combined in one device or combined with the microprocessor. As for the microprocessor itself, this may have an architecture of the type shown in Fig. 6.2.8. (This has been described earlier in Chapter 6.)

7.2.1 Program Memory

The basic task of a microcomputer system is to ensure that its CPU executes the desired instruction sequence, i.e. the program, properly. The Program Memory is the microcomputer system component where this instruction sequence is stored. On initialization—usually a Power-Up or Manual Reset—the processor starts by executing the instruction in a predetermined location in program memory; the first instruction of the program should therefore be in this location. Thereafter, the control of the system passes to this program.

In typical microprocessor systems, the program to be executed is usually a fixed one which does not change. Therefore, one generally finds microprocessor programs stored in Read-Only-Memory or ROMs, i.e. the program memory is implemented in ROM. The contents of a ROM are fixed by the manufacturer during device fabrication. Programmable ROMs (PROMs) are available which allow the user the flexibility of programming the ROM (deciding its contents) himself. However, a PROM once programmed cannot be reprogrammed. Erasable PROMs (EPROMs) allow this feature; their entire contents can be erased by shining ultraviolet light on them, after which they can be programmed once again. The erasure, however, is total so that one cannot change the bit contents of a specific location without erasing and reprogramming the entire EPROM. This feature of selectively changing the contents of specific locations is allowed by Electrically Alterable ROMs (EAROMs). Instead of ROMs, one can design systems with the program memory implemented in PROMs, EPROMs or EAROMs. Typically EPROMs or EAROMs are used during product development; PROMs are used for low volume production and ROMs for high volume production.

7.2.2 Data Memory

A microcomputer manipulates data according to the algorithm outlined by the instructions in the program memory. These data manipulation steps may require intermediate results to be stored. The functional block in the microcomputer used for this storage is the Data Memory. Note that a microprocessor will typically have some internal registers which can also be used, if available, for such storage. External data memory is needed if the storage requirements exceed this resource.

Apart from providing intermediate storage, the data memory may be used to provide some of the data needed by the program. It may also be used to store some of the results of the program. A particular program may also need various constants in its computations. These are also provided by the data memory. Summarising, one can say that the data memory is used for all storage purposes other than storage of programs.

The portion of the Data Memory used as temporary storage (scratch-pad) of intermediate results, or for storing and retrieving data must have Read and Write capability; the microprocessor must be able to write new data into its locations as well as read from them. This portion of the data memory is usually implemented using Read Write Memory or RWM. (This type of memory is generally referred to in literature as Random Access Memory or RAM. We will also use this name here, even though it really is a misnomer.)

Just as for programs, the constants used in the data manipulation process seldom change. The Data Memory segment used for this storage is typically implemented using ROMs, PROMs or EPROMs. This storage is virtually identical to that used for storing programs.

7.2.3 Input Ports and Output Ports

The Input and Output Ports provide the microcomputer the capability to communicate with the outside world. The input ports allow data to pass from the outside world to the microcomputer—data which will be used in the data manipulation being done by the microcomputer. The output ports are used by the microcomputer to send data to devices outside the system. These are, typically, results of the data manipulation process. As an example, consider a microcomputer system which implements a dumb CRT based terminal. The keyboard will interface to this system via an input port. On the other hand, output ports will be used to drive the CRT display, and possibly other LED displays. The terminal to host computer interface will also require two ports; an input port for data from the host to the terminal and an output port for data from the terminal to the host.

7.2.4 Clock Generator

Operations inside the microprocessor, as well as in other parts of the microcomputer, are usually synchronous by nature. This is done so that events in different parts of the system can proceed in a systematic and reliable fashion. The clock needed to perform this synchronous operation is provided by the Clock Generator. Some microprocessors require special clock waveforms (e.g. a 2-phase clock, or a clock with non-TTL levels) for operation. In such cases, the manufacturer of that microprocessor typically provides a special Clock Generator chip which provides the desired clock waveform.

As far as this functional block is concerned, there is a definite trend among microprocessor manufacturers to integrate it on the microprocessor chip itself. The user is merely required to put proper additional timing circuitry, e.g. crystal, R-C or L-C network. The microprocessor, in that case, also provides an output clock signal which can be used by other devices in the microcomputer system for their own timing and synchronization. Keeping this trend of integration in mind, we will assume hereafter that, unless specifically mentioned, the Clock Generator is part of the microprocessor itself.

7.2.5 Integration of Functional Blocks

The functional blocks identified above are the basic requirements of any microcomputer system. Regardless of the application, these blocks will have to be present in some form in order that the microcomputer system may function meaningfully. Typically, LSI chips which realize one or more of these functional blocks are available from various IC manufacturers. As a matter of fact, the manufacturer of a particular microprocessor will also normally supply LSI devices which will implement these functional blocks for that microprocessor. Using this "family" of compatible devices, a simple microcomputer system can be easily configured.

There is also a trend among manufacturers to integrate two or more of these functional blocks in one LSI device. For example, one can get devices which have ROM, RAM and Input/Output Ports on the same chips. For simple systems such integrated devices may be quite attractive since they reduce chip count and circuit size and also simplify the hardware. The extreme limit of such integration would be to put ROM, RAM and Input/Output Ports on the microprocessor chip itself. This is indeed done in the Single Chip Microcomputers available today. These devices are ideal for applications where the size of the circuitry and simplicity of the hardware are the primary design considerations.

7.3 INTERCONNECTING COMPONENTS IN A MICROCOMPUTER

In Sec. 7.2 we have described the functional blocks which are the necessary components of any microprocessor system. In Fig. 7.2.1 we have shown these components, where each one has been shown connected separately to the microprocessor (or CPU). These connections are intended to show the direction of the signal transfer, rather than the actual connections. Due to various reasons, such separate connections between the CPU and other devices are not preferred. Some of these reasons are summarised below:

(i) If we were to follow a connection strategy like that shown in Fig. 7.2.1, we will need a prohibitively large number of signal lines, input and output, from the CPU. To see the magnitude of this problem, one should realize that (a) separate data lines will be needed for each memory block or port, (b) bidirectional data lines—probably two sets of lines—will be needed for RAMs and (c) the memory blocks will also need separate address lines to access specific locations inside each block. Moreover, the CPU must output separate control lines, e.g. READ ($\overline{RD}$) or WRITE ($\overline{WR}$), for each block.

(ii) The complexity required of the CPU will be quite great if one wants each of these blocks to be simultaneously controlled, e.g. reading one data word from memory, and writing another to an output port at the same time. Such a complexity of operation may be difficult to achieve.

(iii) The capability of expanding the system by adding more components (memories, ports or other special purpose devices) will be severely limited. This should be evident from the fact that if a CPU is fabricated on a chip such that it provides interfaces to N devices, adding one more device will at least be difficult, if not impossible

(iv) The complexity of the overall system will also be great. With such a large number of signal lines, laying out the circuit itself may be difficult.

In order to resolve these and other problems, a Bussed Architecture, described next, is used to connect the components in a Microcomputer System.

7.3.1 The Bussed Architecture for Microcomputers

Consider a situation where there are N devices connected to one signal line. Device 'i' can use this line to transfer data to Device 'j' provided:

(a) Device 'i' knows when to output such that Device 'j' is in a position to receive the data. This can be easily ensured if Device 'j' has some means of signalling Device 'i' to output data.

(b) Other than Device 'i', no other device outputs on the signal line during this time. Device 'i' should be the only device driving the signal line at this time.

(c) Other than Device 'j', no other device should interpret the data on the signal line as intended for themselves.

Obviously, if the above conditions can be ensured, the same signal line can be shared by N devices to transfer data between any two of these devices. This is exactly the approach taken to implement a Bussed Architecture. The signal lines that are shared by a number of devices are referred to as the Bus.

Normal TTL gates are not suited for driving the Bus in a Bussed Architecture. This is because such a gate will always be outputting and it will not be possible to satisfy condition (b) above. A special gate, referred to as a Tristate Gate, is required for driving a Bus. This gate, described in Sec. 2.10.2, is illustrated once again in Fig. 7.3.1. It has the feature that when it is

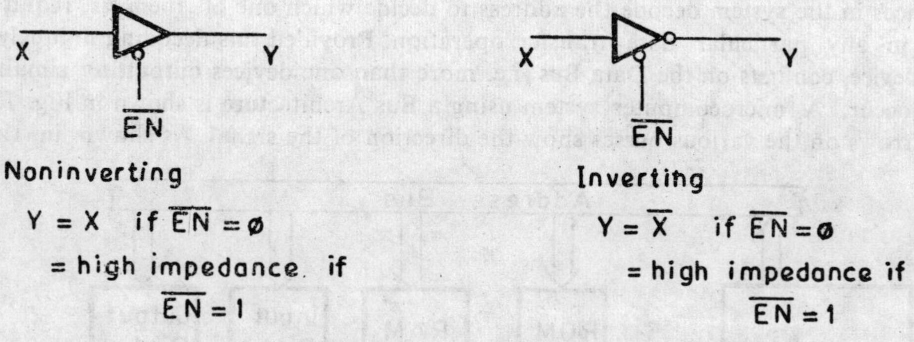

Fig. 7.3.1 Tristate Gates

disabled (i.e. $\overline{EN} = 1$) its output goes to a high impedance (High-Z) state; otherwise, its output merely follows the input if it is a non-inverting gate or is the complement of the input if it is an inverting one. If the outputs of two such gates are tied together, we can decide which device, if any, should drive the output. This can be done by enabling ($\overline{EN} = \emptyset$) only that gate which should be driving the shared output line.

In a bussed architecture, a set of data lines referred to as the Data Bus is shared by a number of devices to transfer data from one device to another. Care must be taken to ensure that at any instant, only one device outputs on the data bus; the other devices which *can* output on the data bus should be restrained from doing so (i.e. their tristate drivers should be in the High-Z state). In a simple microcomputer system, this sharing of the data bus is done under the control of the microprocessor (CPU). Before describing how this is done, it should be noted that in simple microcomputers the data either goes from the CPU to a device or from a device to the CPU. (More complex systems, where this is not true, are described later.) In such a system, 'non-CPU device to another non-CPU device' type of transfers does not take place. This considerably simplifies the problem of properly controlling the data bus. Unless specifically stated, we will always assume that data either goes from the CPU (Write Operation) or comes to the CPU (Read Operation).

The CPU controls the signal flow on the data bus by outputting proper control signals on some other lines. Some of these lines are used to signify if the CPU wants a Read Operation or a Write Operation; these form part of the Control Bus. Two signals needed in the Control Bus (output by the CPU) are $\overline{RD}$ and $\overline{WR}$ where

$\overline{RD} = \emptyset$; Read Operation, data from device to CPU

$\overline{WR} = \emptyset$; Write Operation, data from CPU to device

Note that because of our earlier assumption, $\overline{RD}$ and $\overline{WR}$ cannot be simultaneously active (i.e. Low); however, $\overline{RD}$ and $\overline{WR}$ can be simultaneously inactive (e.g. when the CPU is doing something completely internal to itself).

The Control Bus merely signals the direction of data flow on the Data Bus where the CPU is either the source or the destination of the data. The information regarding the choice of the other (non-CPU) device involved in this data transaction is carried on some other signal lines output by the CPU; these are referred to as the Address Bus. The address appearing on the Address Bus identifies the device the CPU wants to Read from or Write to; in the case of memory devices, this address also identifies the particular location inside the device. The (non-CPU) devices in the system decode the address to decide which one of them is required to be involved in any particular data transfer operation. Provided this decoding uniquely identifies only one device, conflicts on the Data Bus (i.e. more than one devices outputting simultaneously) will not occur. A microcomputer system using a Bus Architecture is shown in Fig. 7.3.2. Note that the arrows on the various busses show the direction of the signal. As shown in this figure,

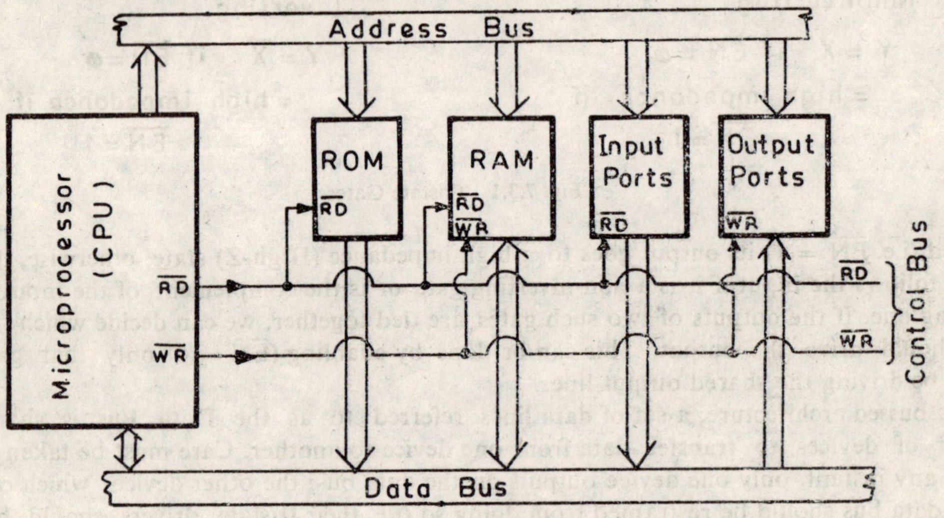

Fig. 7.3.2 Bus Architecture of a Microcomputer

the Address Bus and Control Bus are always driven by the microprocessor. The Data Bus is bidirectional and may be driven either by the CPU or by the other devices. Specifically, Input Ports and ROMs may drive the Data Bus whereas Output Ports only get data from the Data Bus. The RAM is an exception since it can either drive the Data Bus (retrieve data) or get data from it (store data). The techniques used to address Memory devices and Input/Output (I/O) Ports are described next.

7.3.2 Memory Addressing

Consider a memory device having n address inputs ($A_\emptyset, \ldots, A_{n-1}$) and m data lines ($D_\emptyset, \ldots, D_{m-1}$) where $A_\emptyset$ and $D_\emptyset$ are the least significant address and data bits, respectively. Such a memory device will provide 2^n storage locations for m-bit data words. If the memory device is a ROM, it is assumed to have a $\overline{RD}$ input alone and the data lines are outputs with tristate control. If the device is a RAM, it is assumed to have both a $\overline{RD}$ and a $\overline{WR}$ input and the data lines can be either inputs or outputs with tristate control. For both a ROM or a RAM, we will assume the existence of a $\overline{CS}$ input such that $\overline{CS} = \emptyset$ is a prerequisite for both a reading or writing operation; moreover, when $\overline{CS} = 1$, the data lines of the memory device is in the High-Z state. Both a ROM and a RAM of this type are illustrated in Fig. 7.3.3. Note that to read data from such a memory device both $\overline{RD}$ and $\overline{CS}$ should be Low; then the device outputs the data stored in location ($A_{n-1}, \ldots, A_\emptyset$) on the data lines. Similarly, to write data into a RAM of this type, both $\overline{CS}$ and $\overline{WR}$ should be Low; then the data on the data bus is stored in location ($A_{n-1}, \ldots, A_{\emptyset}$) of the device.

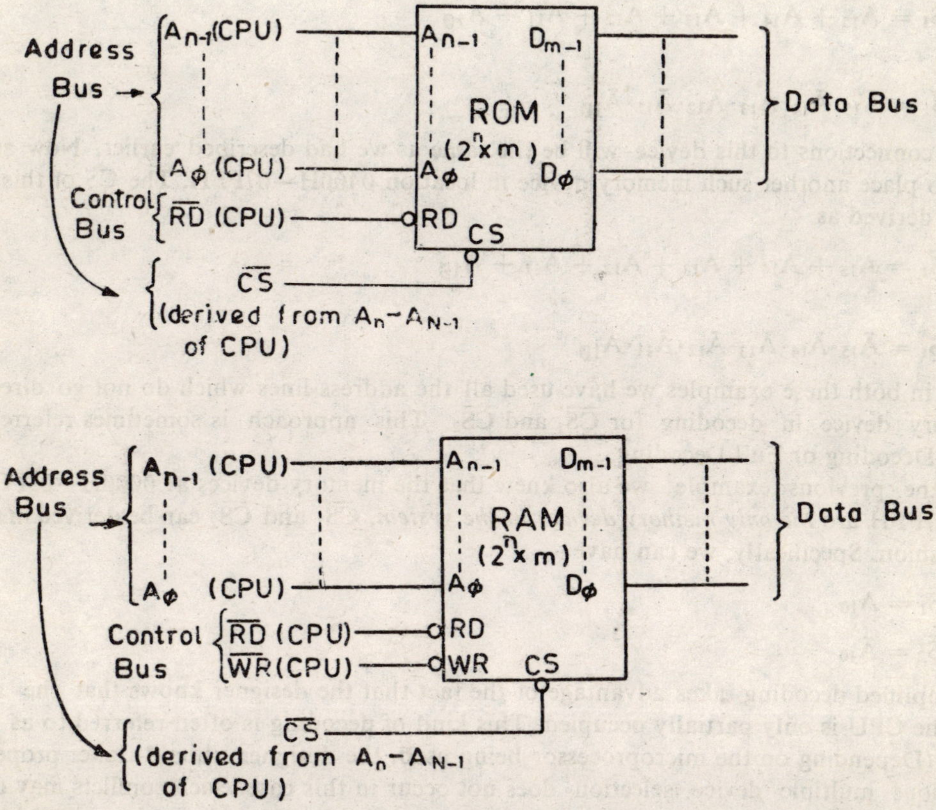

Fig. 7.3.3 Connecting a ($2^n \times m$) ROM and RAM

Consider the situation when a memory device of this type is to be interfaced to a microprocessor with N (N > n) address lines ($A_{\emptyset}, \ldots, A_{N-1}$) and m data lines ($D_{\emptyset}, \ldots, D_{m-1}$). (If the microprocessor has more than m data lines, more such memory devices can be set up in parallel; if it has less, some of the m data bits output by the memory, or input to it, are 'don't cares'.) The data lines of both the CPU and the memory device are connected to the data bus of the system. For the control bus, $\overline{RD}$ will go from the CPU to the device both for ROM or RAM; the $\overline{WR}$ will also be similarly connected if the memory device is a RAM. Out of the N address lines of the CPU, constituting the address bus, the lower n — $A_{\emptyset}, \ldots, A_{n-1}$—will go to the corresponding address inputs of the memory device. The $\overline{CS}$ input of the device will be appropriately derived from the upper order address bits A_n — N_{N-1}. The designer has the freedom to derive $\overline{CS}$ in such a way that the device occupies any consecutive 2^n locations in the 2^N address range of the microprocessor. This connection strategy has also been shown in Fig. 7.3.3.

Let us consider some specific examples to illustrate how the $\overline{CS}$ input is used to place a memory device in a desired memory location in the address space of the microprocessor. Specifically, we assume that the microprocessor has 16 address lines ($A_{\emptyset}$ — A_{15}) and 8 data lines

($D_Ø - D_7$) as in the Intel 8085A processor. The address range in this case is ØØØØH – FFFFH or Ø to 64 K. Now if we want to place a 1Ø24×8 memory device ($n = 10$, $m = 8$) in the address space ØØØØH – Ø3FFH, we can derive $\overline{CS}$ as

$$\overline{CS}_1 = A_{15} + A_{14} + A_{13} + A_{12} + A_{11} + A_{1Ø}$$

or

$$CS_1 = \bar{A}_{15} \cdot \bar{A}_{14} \cdot \bar{A}_{13} \cdot \bar{A}_{12} \cdot \bar{A}_{11} \cdot \bar{A}_{1Ø}$$

The other connections to this device will be the same as we had described earlier. Now suppose we want to place another such memory device in location Ø4ØØH—Ø7FFH. The CS of this device should be derived as

$$\overline{CS}_1 = A_{15} + A_{14} + A_{13} + A_{12} + A_{11} + \bar{A}_{1Ø}$$

or

$$CS_1 = \bar{A}_{15} \cdot \bar{A}_{14} \cdot \bar{A}_{13} \cdot \bar{A}_{12} \cdot \bar{A}_{11} \cdot A_{1Ø}$$

Note that in both these examples we have used all the address lines which do not go directly to the memory device in decoding for $\overline{CS}_1$ and $\overline{CS}_2$. This approach is sometimes referred to as Complete Decoding or Full Decoding.

If in the previous example, we also knew that the memory devices at ØØØØH—Ø3FFH and Ø4ØØH—Ø7FFH are *the only memory devices in the system*, $\overline{CS}_1$ and $\overline{CS}_2$ can be derived in a much simpler fashion. Specifically, we can have

$$\overline{CS}_1 = A_{10}$$

$$\overline{CS}_2 = \bar{A}_{10}$$

Such a simplified decoding takes advantage of the fact that the designer knows that the address range of the CPU is only partially occupied. This kind of decoding is often referred to as Linear Selection. (Depending on the microprocessor being used, the designer should take proper care that spurious multiple device selection does not occur in this case. Such conflicts may damage LSIs connected to the data bus.) A decoding strategy for obtaining the various $\overline{CS}$ which uses only some of the address lines not going directly to the memory device can also be followed. This is sometimes referred to as Partial Decoding. Integrated Circuit decoders like the 8205/3205, 74LS138, 74LS139 etc. are very commonly used to decode the required chip selects ($\overline{CS}$s) from the address bus. The reader is referred to the manufacturers' data sheets for further details regarding these chips.

A microcomputer designer needs to keep one other fact in mind while deciding where to place memory devices in the memory address space of the system. The particular microprocessor being used may have one or more inputs (e.g. RESET) which force the device to execute instructions from a predetermined (decided by the microprocessor designer) location in its address space. For example, the RESET input causes the Intel 8080/8085A to execute instructions from location ØØØØH. The microcomputer designer should make sure that there is a memory device in these locations which contains the desired instructions.

7.3.3 Addressing Input and Output (I/O) Ports

From the point of view of the microprocessor, the distinction made between memory devices and I/O ports is, to some extent, artificial. To the microprocessor, the world outside consists of devices which it can read (select using address bus and output $\overline{RD}$ = Ø) from or devices which it can write to (select using address bus and output $\overline{WR}$ = Ø). It cares very little whether the device thus selected is a memory location or an I/O port.

The primary reason for distinguishing between I/O ports and memory arises from the fact that a number of microprocessors available today do make this distinction. These microprocessors have specific instruction—e.g. IN or OUT—for handling I/O ports and output additional signals on the control bus to distinguish between the situations when they are accessing memory and when they are accessing I/O ports. As a result, the I/O Address Space of such a microprocessor is distinct from its Memory Address Space. An easy way of ensuring this is to output an IO/$\overline{M}$ signal on the control bus where

IO/$\overline{M}$ = Ø; memory being addressed

= 1; I/O ports being addressed

Using this, the $\overline{CS}$ of a memory device should be derived such that it can be Ø only if IO/$\overline{M}$ = Ø and the proper address is present on the address bus. Similarly the $\overline{CS}$ of an I/O port should be Ø when IO/$\overline{M}$ = 1 and the address of this port appears on the address bus. Alternatively, the $\overline{CS}$s of the various devices can be derived from the address bus alone but instead of having just IO/$\overline{M}$, $\overline{RD}$ and $\overline{WR}$, the control bus now outputs

$\overline{MEMR}$; Ø for reading from memory

$\overline{MEMW}$; Ø for writing to memory

$\overline{IOR}$; Ø for reading from input port

$\overline{IOW}$; Ø for writing to output port

As shown in Fig. 7.3.4, these latter signals can be easily derived from IO/$\overline{M}$, $\overline{RD}$ and $\overline{WR}$.

$\overline{MEMW} = (IO/\overline{M}) + \overline{WR}$ $\qquad$ $\overline{IOW} = \overline{(IO/\overline{M})} + \overline{WR}$

$\overline{MEMW} = (IO/\overline{M}) + \overline{RD}$ $\qquad$ $\overline{IOR} = \overline{(IO/\overline{M})} + \overline{RD}$

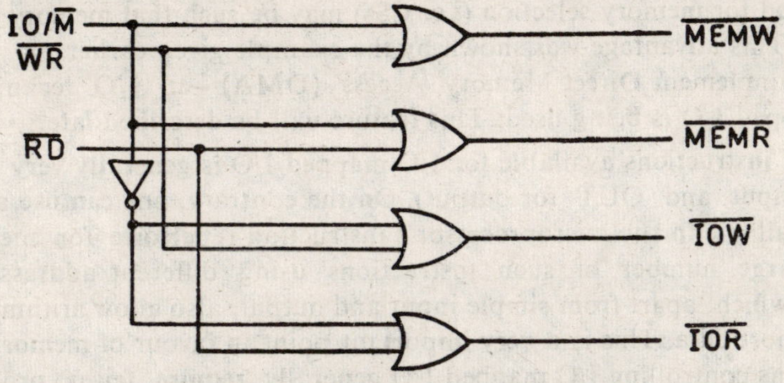

Fig. 7.3.4 Deriving Separate READ and WRITE Signals for I/O and Memory

Addressing I/O ports such that, as above, the I/O Address Space is distinct from the Memory Address Space is referred to as *I/O Mapped I/O Addressing*. The method used for deriving $\overline{CS}$s for the various I/O ports in this case has also been described above. (Sometimes the microprocessor is such that only a limited number of address pins actually carry relevant address information when I/O mapped I/O is being used. This should be taken into account in the design.) The facility of using I/O Mapped I/O is provided by microprocessors like Intel's 8080, 8085, 8086, 8088, Zilog's Z80 etc.

We had mentioned earlier that there is really no need to distinguish between an I/O Address Space and Memory Address Space except, perhaps, from the viewpoint of convenience. It is, therefore, quite viable to address an I/O port exactly as if one were addressing a memory device. In this case, $\overline{RD}$ and $\overline{WR}$ would be the only control signals needed and the $\overline{CS}$s of the various I/O ports are decoded from the address bus in the same manner as addressing memory. This approach to I/O addressing is known as *Memory Mapped I/O Addressing*. This mode of I/O addressing can be used with any microprocessor. As a matter of fact this is the only I/O addressing mode that some microprocessors, e.g. Motorola 6800, will support.

7.3.4 Comparison of I/O Mapped and Memory Mapped I/O

In systems using microprocessors which allow I/O Mapped I/O, the decision to go for one type of I/O addressing or the other is usually a matter of convenience. For example, suppose one is using a decoder like the 8205 to decode chip selects ($\overline{CS}$) for some memory modules. If after providing the required chip selects, one discovers that some 8205 decoded outputs are still unused, then these will generally be used to select I/O ports. These ports will, therefore, be memory mapped. Apart from such considerations, there are some pros and cons to these I/O addressing techniques. These are described below.

(a) Using Memory Mapped I/O we can typically address more I/O ports than when using I/O Mapped I/O. This is because in the latter case only a limited number of the address lines generally carry useful address information. For example, the 8085A can address 64 K memory locations but can only address 256 I/O ports. (Of course, 256 I/O ports is much more than what is usually needed. This is at the most a very dubious advantage in favour of memory mapped I/O.)

(b) Using I/O mapped I/O, we can distinguish between separate I/O and memory address spaces. Conceptually, this is sometimes an advantage in favour of I/O mapped I/O. On the other hand the decoding used for memory selection (i.e. $\overline{CS}$s) may be such that memory mapped I/O is easier to implement. This advantage was shown by the example given earlier.

(c) It is easier to implement Direct Memory Access (DMA)—an I/O technique described later—when I/O mapped I/O is being used. This feature will be described later.

(d) The variety of instructions available for I/O mapped I/O is generally very limited (typically, just IN for input and OUT for output). On the contrary, one can use all the memory access instructions available in the microprocessor's instruction repertoire for memory mapped I/O. Generally, a large number of such instructions using different addressing modes (see Sec. 7.4) is available which, apart from simple input and output, also allow arithmetic and logical manipulation of I/O port data. This is a very important point in favour of memory mapped I/O.

(e) The instructions controlling I/O mapped I/O generally require fewer program memory bytes to encode than the instructions for memory mapped I/O and usually also execute faster. This makes I/O mapped I/O usually more efficient than memory mapped I/O.

In summary, it should be noted that the difference between I/O mapped and memory mapped I/O techniques is generally very small; one really cannot say that one is superior to the other. In the final analysis, which I/O technique is actually used in a design is mostly a matter of convenience and, to some extent, a reflection of the designer's own bias.

7.4 MEMORY ADDRESSING MODES

The CPU of a microcomputer system is a device executing instructions which manipulate one or more data words in some fashion. The data to be manipulated must have a source (memory, internal register or input port) and a destination (memory, internal register or output port). Any data manipulation instruction that the CPU executes should somehow specify the source and destination of the data. In this section, we will describe the way this is done when the source or destination is a memory location where the desired data is to be read from or stored into. These techniques are referred to as Memory Addressing Modes. It should be noted that the CPU itself will have to decode the instruction in order to find out the addressing mode being used. Moreover, if memory mapped I/O is being used, these various addressing techniques can also be used for addressing I/O ports. Finally, in all the examples given below, we will assume that one of the CPUs internal registers (e.g. Accumulator or some other register) is either the source or the destination of the data. This is usually the case with most microprocessors.

Before describing the various memory addressing modes, we introduce the following notation in order to make the description easier.

EA = Effective memory address of the location which is either the source or the destination of the data word

(x) = data word at memory location address 'x' or in register x of the CPU.

Finally, we will use various instruction mnemonics to illustrate the addressing modes. Wherever possible, assembly language mnemonics of the 8085A will be used; where this is not possible, we will introduce our own mnemonics. Mnemonics of the 8085A will be preceded by a '*' in order to identify them as such.

7.4.1 Immediate Addressing

In instructions using Immediate Addressing, the data is provided by the instruction itself. For example

* MVI A, ØFH; move $\emptyset F_H$ into accumulator

Strictly speaking, this is not really a memory addressing mode since one cannot really associate a data memory location with the source data. This kind of instructions is often used to initialize the CPU's internal registers with known values.

7.4.2 Direct Addressing

In Direct Addressing, the instruction itself directly provides the address for data access (source or destination). For example,

* LDA 2ØC1H; load accumulator with contents of location $2\emptyset C1_H$

$$EA = 2\emptyset C1_H,\ (A) \leftarrow (2\emptyset C1_H)$$

*Mnemonics Copyright Intel Co. 1985.

* STA 2ØCØH; store accumulator contents at location $2ØCØ_H$

$$EA = 2ØCØ_H, (2ØCØ_H) \leftarrow (A)$$

This addressing mode is available in most microprocessors. It provides a simple and convenient way of manipulating RAM contents or loading constants from ROM. Note that with this addressing mode, EA is fixed for a given program.

7.4.3 Indirect Addressing

In instructions using Indirect Addressing, the instruction specifies the address of (one or more, as required) memory locations whose contents specify the actual address for the data. For example,

LDI 2ØC1H; Load accumulator indirect using contents of location $2ØC1_H$.

In this particular example, how EA is actually calculated will depend on the microprocessor. Some possibilities are:

(a) If a 16 bit address is to be generated, and if a memory data word contains 16 bits, then EA could be given by

$$EA = (20C1)_H$$

and the instruction LDI 20C1H does the following

$$(A) \leftarrow ((20C1_H))$$

(b) If a 16 bit address is to be generated, and if a memory data word contains 8 bits, then EA could be given by

$$EA = (2ØC2_H)(20C1_H)$$

i.e. concatenation of the contents of locations $2ØC2_H$ and $2ØC1_H$. The instruction LDI 2ØC1H then does the following

$$(A) \leftarrow ((2ØC2_H)(2ØC1_H))$$

As the above examples show, the actual calculation of EA might be done differently by different CPUs. The common feature of all indirect addressing instruction, however, is the fact that the actual address for data access is specified by the contents of some other locations, whose address, in turn, is directly specified by the instruction itself.

Both Indirect and Direct Addressing instructions require similar amount of program memory storage space. In terms of the time required to execute the instruction, direct addressing is usually faster. However, indirect addressing provides a very powerful feature of allowing variable data addresses which cannot be done with direct addressing. This is described in greater detail below.

In Direct Addressing, the EA needed is part of the program memory (i.e. in ROM) and cannot be changed under program control. On the other hand, in Indirect Addressing, the EA is specified by the contents of some other memory locations. If these locations are in RAM, it will be possible to change them under program control and hence change EA. This is the reason why indirect addressing will allow the actual address of the data to be changed. This is a very useful feature in various applications, e.g. a program loop where we would like to access different memory locations (e.g. sequential) everytime the loop is executed.

In spite of its power, indirect addressing instructions are not popular in microprocessors because of their relative complexity. Instead, most of them have another form of indirect addressing known as *Register Indirect Addressing*. In this addressing mode, the instruction specifies one or more of the CPU's internal registers whose contents provide the actual EA for the data. For example

* LDAX B; Load accumulator with contents of memory location whose address is (B)(C)

$$EA = (B)(C),\ (A) \leftarrow ((B)(C))$$

Register Indirect Addressing also allows EA to be changed under program control. Compared to Indirect Addressing, it generally requires less program memory and executes faster. However, since only a limited number of internal registers will be available inside the CPU, it does not have the full power of Indirect Addressing.

In CPUs having a Data Counter (DC), or where a DC can be assumed, a special case of register indirect addressing is usually available. This is referred to as *Implied Addressing*. In this case, the instruction does not specifically indicate a register as containing EA; instead, EA is taken to be the contents of DC.

* MOV B, M; $EA = (H)(L),\ (B) \leftarrow ((H)(L))$ $DC = H - L$ Load B with the contents of memory location pointed at by H-L contents; H-L assumed to be the Data Counter.

7.4.4 Program Relative Addressing

Instructions using this addressing mode usually specify a (signed) shift. This shift is added to the address of the instruction (its location in memory) in order to generate the memory address for data. The EA can then be calculated in either a direct or indirect fashion as shown by the following example.

LDPR Addr; $(A) \leftarrow (EA)$

If this instruction is at memory location N, then—

EA for Dir. Prog. Rel. Addr. $= N + \text{Addr}$

EA for Indir. Prog. Rel. Addr. (possibly) $= (N + \text{Addr} + 1),\ (N + \text{Addr})$

7.4.5 Indexed Addressing

This addressing mode requires the CPU to have an Index Register (IxR) and can be either direct or indirect. In *Indexed Direct Addressing*, the EA is the sum of a direct address and an implied address, where the direct address is provided by the instruction itself and the implied address is taken as the contents of IxR. In this case, an instruction of the form

Instr. Addr

will generate

$$EA = (IxR) + \text{Addr}.$$

This kind of addressing can generate variable EAs with the same instruction just as in Indirect Addressing. It is a popular addressing scheme in most of the powerful microprocessors like Z80, 8086, 8088 and the 6800.

* Mnemonics Copyright Intel Co. 1985.

Indexed addressing can also be done in an indirect fashion. This kind of Indexed Indirect Addressing can be done with either pre-indexing or post-indexing. Examples of this are given below, provided the instruction is of the form

Instr. Addr.

Pre-indexed Indirect Addressing:

$$EA = ((IxR) + Addr + 1)((IxR) + Addr)$$

Post-indexed Indirect Addressing:

$$EA = (Addr + 1)(Addr) + (IxR)$$

7.4.6 Base Relative Addressing

Apart from an Index Register, some microprocessors now also provide a Base Register. The Base Register is virtually indistinguishable from the Index Register. Functionally the only difference is that whereas the Base Register is generally not modified during address calculations, the Index Register may be changed, incremented or decremented.

In Base Relative Addressing, the effective address is computed as follows:

EA = Content of Base Register

+ Content of Index Register

+ Displacement, if any, specified by the instruction

A common way of using Base Relative addressing is to efficiently address blocked data in memory. For example, the Base Register may point to the origin of a large data area in memory to be used for storing different tables. The Displacement (provided by the instruction) might then point to the origin of a table where the Index Register identifies specific entries in that table.

The addressing schemes described above may or may not be all present in a microprocessor. The earlier microprocessors usually had only a limited set of such schemes because of implementation problems. A popular scheme, almost always present in that case, was Implied Addressing. This is because it was simple to implement and is also quite powerful.

In terms of the popular microprocessors available today, one can make quite a distinction in the power and flexibility of their instruction set as regards addressing modes. Intel's 8080 and 8085 CPUs, for example only provide Direct, Register Indirect, and Implied Addressing. Motorola's 6800 and the Z80 on the other hand provide an instruction set which supports a greater variety of addressing options, i.e. base page address and indexed addressing. The 16-bit processors available today, also support a large variety of addressing modes. So much so, that in terms of flexibility of addressing, they almost rival the addressing modes of minicomputers.

7.5 INPUT/OUTPUT (I/O) TECHNIQUES

Apart from memory, input and output ports are the other non-CPU devices in a microprocessor system which act as the source and destination of data. Unlike memory, however, I/O ports are seldom operated under the complete control of the microprocessor. Their operation is basically asynchronous in so far as the system designer cannot usually state in an *a priori* fashion when one of them will require service (i.e. an input port wants the CPU to read its contents, an output port wants data from the CPU). This variability arises primarily from the fact that unlike other devices in the microcomputer, I/O ports are directly influenced by the world external to the microcomputer. The various I/O techniques described in this section are all intended to interface I/O ports to the microprocessor such that their asynchronous behaviour can be withstood while providing them 'service' as and when it is needed. In this context, it should be noted that it may not be possible to respond immediately to an I/O port's request for service; what is usually required is that service be provided within some worst case time period after such service is requested. Fig. 7.5.1 illustrates the typically asynchronous nature of such I/O service requests. As shown there, one or more I/O ports in the system may demand service in an asynchronous and essentially random fashion.

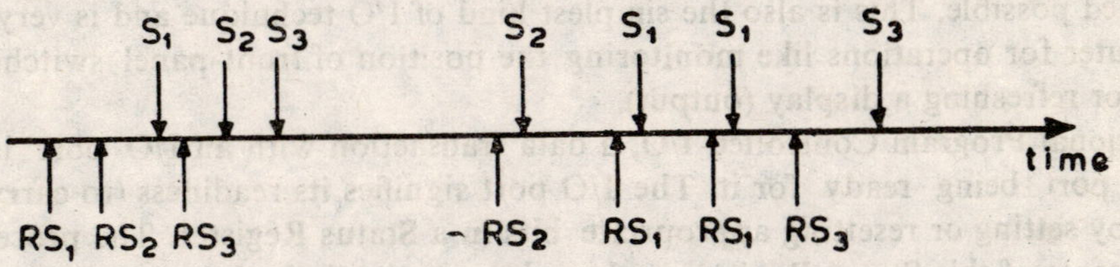

RS_i: Request for Service from I/O port 'i'
S_j: Service granted to I/O port 'j'

Fig. 7.5.1 Asynchronous I/O Timing

Another problem that needs to be solved when interfacing two or more I/O ports in a microcomputer system is priority of service. Consider a situation where two or more I/O ports are demanding service and the processor is busy so that it cannot provide service to any one of these ports. When the processor is finally able to service these ports, it will have to decide which port to service first (e.g. first-come-first-served, one can assign priorities to each port and serve the one with the highest priority first, last-come-first-served etc.). In other words, the system designer should provide some rule which will arbitrate in situations where multiple I/O devices are demanding service. The I/O techniques used should also incorporate this 'priority of service' within themselves.

Before presenting brief descriptions of some of the common I/O techniques used, it should be pointed out that the designer is in no way obliged to use only one of them in his system. It is entirely possible, and actually quite common, to use one technique for some I/O ports and other techniques for other I/O ports. One, however, generally uses one particular I/O technique for a particular I/O port. It is certainly possible to design a system where more than one I/O technique is provided to service a particular I/O port; that, however, is not considered a good design practice (increased complexity!) and not generally done.

The common I/O techniques one usually finds in microcomputer systems are—

Program Controlled I/O

Interrupt Controlled I/O

Hardware Controlled I/O or DMA (Direct Memory Access)

These techniques are described briefly, next. A more detailed description of these techniques will be given later in Chapter 10.

7.5.1 Program Controlled I/O

With Program Controlled I/O, the I/O transaction is done completely under the control of the processor. The processor inputs data from an input port and outputs data to an output port only when it encounters the appropriate instruction in its software. Two kinds of Program Controlled I/O are possible—Unconditional or Conditional Program Controlled I/O.

Unconditional Program Controlled I/O represents the extreme situation where the processor reads data from an input port or writes data to an output port without even checking whether such a transaction can be done. It carries out the data transaction assuming that such a transaction is indeed possible. This is also the simplest kind of I/O technique and is very often used in a microcomputer for operations like monitoring the position of front-panel switches (input) or for updating or refreshing a display (output).

In Conditional Program Controlled I/O, a data transaction with an I/O port is conditioned on the I/O port being ready for it. The I/O port signifies its readiness (to carry out the data transaction) by setting or resetting appropriate bits in a Status Register. The processor, in turn, reads the contents of this Status Register and carries out the desired data transfer only when it finds that particular I/O port to be ready. The Status Register itself is very often read as another input port by an unconditional program controlled I/O method. The algorithmic approach used in this I/O technique is illustrated by the block diagrams of Fig. 7.5.2.

Even though data transactions using Conditional Program Controlled I/O are done strictly under the control of the CPU, I/O ports served by this technique (and external devices connected to these ports) can still exercise some nominal control over these transactions. This can be done by means of the Status Register; a device requesting or refusing service can convey this information to the CPU by properly setting or resetting appropriate bits in the corresponding Status Register. For I/O ports requesting service from the CPU in this fashion, the CPU's software should be such that it examines the Status Registers of these ports as often as possible. This should be done often enough so that an I/O port requesting service is attended to before it generates another service request.

7.5.2 Interrupt Controlled I/O

Most microprocessors provide one or more Interrupt inputs which can be used to interrupt the program being executed by the CPU so that it can provide service to the interrupting device. Once the required service has been provided, the CPU may return to its original (interrupted) task. Most CPU's also have an ENable Interrupt flip-flop which can be set or reset under program control; the ENI F/F is typically also reset on power-up and after an interrupt is acknowledged by the CPU. The CPU can respond to an interrupt signal only when the ENI F/F is set. (In the 8085A, the ENI F/F is referred to as the INTE F/F).

It is desirable to have interrupts disabled at power-up so that the microcomputer system can properly initialize itself without being interrupted. Moreover, the software elsewhere might contain a sequence of instructions which the designer does not want to be interrupted; the ENI F/F should be reset when such instructions are being executed. Finally, the interrupt inputs should also be disabled when the CPU acknowledges an interrupt and should stay disabled until the CPU explicitly enables them.

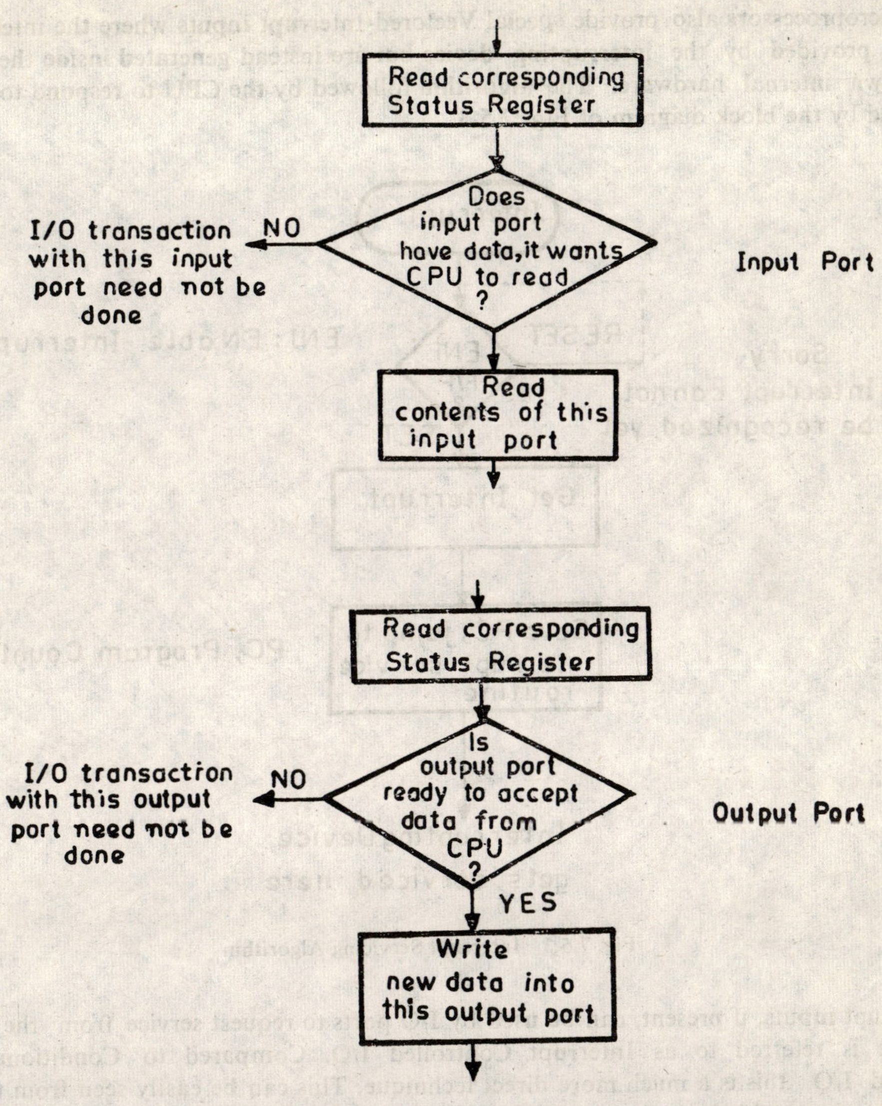

Fig. 7.5.2 Conditional Program Controlled I/O Technique for Input and Output

When the CPU acknowledges an interrupt signal, it is normally the responsibility of the interrupting device to provide the CPU (via the system bus) with an Interrupt Vector. This interrupt vector contains the address in memory where the program servicing that interrupt has been stored. The CPU executes a Subroutine Call to this address, i.e. saves the Program Counter (PC) and jumps to that address. If a Subroutine Return is encountered at the end of this program, the original PC value is restored and the CPU returns to the original (interrupted) routine. It should be evident that as long as different interrupting devices provide different interrupt vectors, the same interrupt input can be shared by a number of interrupting devices. In this case, one needs the additional requirement that either multiple devices should not interrupt at the same time, or if they do, there should be external hardware to arbitrate between them.

Some microprocessors also provide special Vectored-Interrupt inputs where the interrupt vectors are not provided by the interrupting device but are instead generated inside the CPU by the CPUs own internal hardware. The algorithm followed by the CPU to respond to interrupts is illustrated by the block diagram of Fig. 7.5.3.

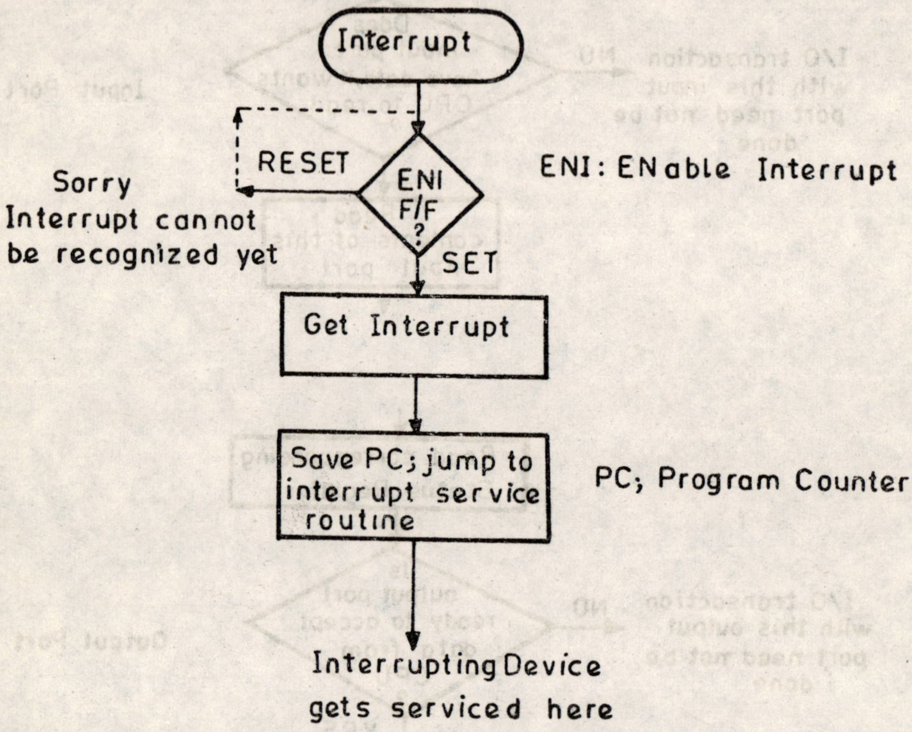

Fig. 7.5.3 Interrupt Servicing Algorithm

Interrupt inputs, if present, can be used by I/O ports to request service from the CPU; this technique is referred to as Interrupt Controlled I/O. Compared to Conditional Program Controlled I/O, this is a much more direct technique. This can be easily seen from the fact that if interrupts are enabled, the interrupting I/O port will immediately get serviced; it will not have to wait for the CPU to examine it to see if service is required. Since the I/O port itself drives the interrupt input, it exercises direct control over when it is serviced even though any service provided is still under the control of the CPU. (Remember that it can deny such service by keeping its interrupts disabled.) Compared to Conditional Program Controlled I/O, a lot of CPU overhead, in terms of time and software, is also saved. This is because the CPU no longer has to repeatedly check to see if service is required. Hardware external to the CPU is more complex for Interrupt Controlled I/O because of the need to provide proper Interrupt Vectors and to arbitrate between multiple I/O devices sharing the same interrupt. Since interrupts are completely asynchronous Inputs to the CPU, debugging a system which uses interrupts is also generally more difficult than debugging a system without interrupts. This I/O technique will be discussed in greater detail in Chapter 10.

7.5.3 Hardware Controlled I/O or DMA (Direct Memory Access)

Consider an application where N data words need to be transferred between the data memory and an I/O port. If either Interrupt Controlled or Conditional Program Controlled techniques were to be used, each one of these N data words will need to be routed *through* the CPU (i.e. the CPU will have to obtain one data word from the source, send it to the destination and then repeat the process for the other data words). Apart from consuming time, this process will require software to handle data routing and will also tend to tie up the CPU while the data transaction is in progress. In an application like this, it would be more efficient if one can design external hardware which can transfer data directly between selected data memory and I/O ports without invoking specific attention from the CPU. This is exactly what is done in Hardware Controlled I/O or Direct Memory Access (DMA).

Proper DMA operation requires the presence of a DMA Controller in the system. This device essentially mimics the CPU during a DMA operation by driving the Address Bus with the proper memory address and by

(a) outputting $\overline{MEMR} = \emptyset$ and $\overline{IOW} = \emptyset$ for 'DMA Read' (Output) operation

(b) outputting $\overline{MEMW} = \emptyset$ and $\overline{IOR} = \emptyset$ for 'DMA Write (Input) operation

(c) ensuring the selection of the proper I/O port by means of separate signal lines (Chip Selects) connected directly to individual I/O ports.

During a DMA operation, the CPU should relinquish the System Bus and allow the DMA Controller to control it.

The CPU is required to properly initialize the DMA Controller (see Chapter 10 for details). Now if one of the I/O ports being served by this DMA Controller requests service, the Controller first asks the CPU to relinquish control of the System Bus. Once this is done the DMA Controller provides the requested service (data from port to memory or memory to port) to the I/O port by carrying out the proper DMA operation, i.e. DMA Read or DMA Write.

Note that once the DMA Controller has been properly initialized, service requests from I/O ports being served by that Controller can be handled without requiring any further attention from the CPU. However, the CPU's own operation is affected by the fact that, in case it wants to access the System Bus while a DMA operation is going on, it will have to wait until the DMA operation is over. In effect, DMA operations create interference and some delays in the CPU's own program execution. In Chapter 10, we describe different kinds of DMA techniques which

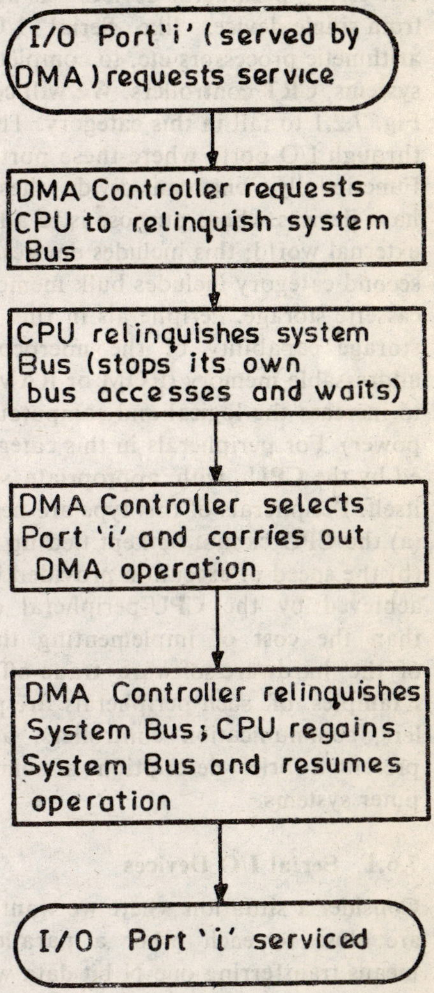

Fig. 7.5.4 Servicing I/O ports using DMA Techniques

trade-off speed and complexity in order to reduce such interference. A simplified block diagram illustrating a DMA operation is shown in Fig. 7.5.4.

It should finally be noted that DMA is a viable approach only if the required data transaction is between memory and I/O ports. DMA techniques will not work for data transactions between I/O ports and the CPU. As a matter of fact, in a DMA transfer, the data itself does not enter the CPU or the DMA Controller. This, in turn, implies that it will be difficult to use DMA in a situation where some logical decision has to be made depending on the data word actually being transferred.

7.6 SOME OTHER MICROCOMPUTER COMPONENTS

The term—peripheral devices—as applied to microcomputers covers a wide range of devices—from single devices like Serial I/O devices, counters, timers, various kinds of controllers, arithmetic processors etc. to complete multi-device systems like Key Boards, Floppy-Disk storage systems, CRT controllers. We will consider any device not in our basic microcomputer system of Fig. 7.2.1 to fall in this category. The microcomputer system communicates with these devices through I/O ports where these ports are either external to the device or integrated into it. Functionally, one can divide these peripherals into three categories. In the first category we have devices whose purpose is to allow easy communication between the microcomputer and the external world; this includes devices such as keyboard, LEDs, CRT displays, switches etc. The second category includes bulk memory or storage systems like floppy-disk storage, hard disk or cassette storage. Peripherals in this category are intended to increase the program or data storage capability of the microcomputer beyond what is available in the systems immediately addressable memory (ROM or RAM). In the third category, we have devices which are intended to increase the logical and computational capabilities of the CPU itself (basically enhance its power). For peripherals in this category, the functions implemented can very often be implemented by the CPU, with appropriate software, thereby eliminating the need for the peripheral itself. Peripherals of this type are nevertheless used when the design considerations are such that (a) the CPU cannot be kept tied up for the length of time needed to execute such software, (b) the speed of execution provided by the CPU-software combination is lower than what can be achieved by the CPU-peripheral combination, or (c) the cost of the peripheral itself is lower than the cost of implementing the required software. These peripherals are good examples of the hardware-software trade-off one often encounters in the design of microcomputers. Examples of such peripherals are programmable counters, serial I/O devices, interrupt controllers, communication controllers, arithmetic processors, multipliers etc. In this section we will present a brief description of some of the peripherals one commonly encounters in microcomputer systems.

7.6.1 Serial I/O Devices

Consider a situation where we want to transfer data between two systems. Unless these systems are close to each other a parallel transfer would not be very economical. (Parallel transfer means transferring one N bit data word at a time using N signal lines.) A more practical solution is to convert data words from a parallel to a serial format and transfer the data one bit at a time using only one signal line; this is referred to as Serial I/O or Serial Data Communications. Various device manufacturers market LSI devices which can provide serial data communication capability to a microcomputer system. These devices are often referred to as

Universal Synchronous/Asynchronous Receiver Transmitters (USARTs) or Universal Asynchronous Receiver Transmitters (UART). Serial I/O techniques, USART/UART etc. are discussed in detail in Chapter 10.

7.6.2 Communication Controllers

In several applications, a number of microcomputer systems often need to communicate with each other or with a Central Controller. In such applications, the data transfer is usually serial. Moreover, each of the component systems are required to follow certain rules regarding the data transaction and data formats. These rules, referred to as data protocols, are intended to ensure the intelligibility of the communication as well as to guarantee the reliability, speed and efficiency of such communications.

Given a particular application, it is certainly possible to design one's own protocol which will satisfy the communication needs of that application. However, very often, it is better to simply use one of the protocols developed by the industry e.g. BISYNC, SDLC, HDLC, ADCMP etc. Using these protocols not only simplify the design of the communication system, but also ensure compatibility with other systems using the same protocol.

Using one of these standard protocols for communications provides one more advantage—there are several LSI devices available today which can act as a Communication Controller for one or more of these protocols. Implementing protocols using the corresponding controller peripherals greatly simplify the design procedure. Examples of such devices are Intel's 8273 and 8274 and Fairchild's 3846. The reader is referred to the corresponding data books for more information on these devices. It should be noted that even though these controllers implement quite a bit of the protocol's rules, they rarely implement the protocol fully. The CPU is expected to implement those protocol rules which the controller itself is unable to support. The reader interested in using these LSI controllers is advised to keep this fact in mind.

7.6.3 Programmable Timers/Counters

In various applications using microcomputer systems, it is sometimes necessary to measure time. There are generally two reasons for doing this—to keep a record of the 'time of day' (absolute time) or to measure the time interval between two events (relative time). Both kinds of measurements can be made using either software or external hardware. In this section we briefly describe LSI devices known as Programmable Timers/Counters which are very often used when one wants to make this time measurement using external hardware.

These Timer/Counters measure time using an external clock signal which must be provided to the device. The basic unit of time is in that case the time interval between successive pulses of this clock. The typical Timers/Counters commercially available can be programmed by the CPU (writing a proper control word into the corresponding output port) to provide a variety of useful functions. Some of these are given below:

(i) Divide-by-N counter where the value of N is software controllable by means of the CPU. One common use for such a counter is for variable baud rate generation under software control for serial data transfers.

(ii) Measure the time interval between two events, say E1 and E2. The counter is initialized with a know value V1 when E1 occurs. It is then read when E2 occurs; let V2 be this value. The difference between the two values V1 and V2 represents the time elapsed between events E1 and E2.

(iii) Suppose the CPU wants to execute a task 'T', t seconds after event E occurs. It should load the Timer/Counter with the value corresponding to t seconds. When event E occurs the counter's output is forced Low and the counter starts decrementing to Ø. When this is reached, the counting is stopped and the counter output goes High. Using the Low-to-High transition of the counter output, one can interrupt the CPU and force it to execute task T, t seconds after event E occurs.

Intel's 8253 is a good example of such a Programmable Timer/Counter. It is a common microprocessor peripheral device which is useful in many applications including those mentioned above. (This device actually contains three Timer/Counters on the same chip.) This peripheral device is described in detail in the Appendix A7.

7.6.4 Bulk Memory/Mass Storage Devices

Bulk memory or mass storage devices are used to store information (program or data) at a lower cost per bit than what is possible with semiconductor devices (ROM or RAM). They are also used in applications where the total memory requirement of the system exceeds the address space of the microcomputer. In such cases, program and data are moved from secondary storage in some bulk memory device (background memory) to semiconductor RAMs (foreground memory) only as they are needed; program or data, no longer immediately needed may similarly be moved from foreground memory to background memory until they are needed once again. This type of 'Swapping' procedure may be adopted even when the address space limit of the microcomputer has not been reached but cost considerations put a limit on the number of semiconductor memory devices. It should be noted that even though the cost per bit of these storage devices are typically lower than that of semiconductor memories, their access times are comparatively much longer. This implies that systems incorporating such storage would be slower in operation.

A wide variety of mass storage mediums are commercially available for use in microcomputer systems. Examples of such storage mediums are paper tape, cassette tapes, floppy disks, hard disks, magnetic bubbles, CCDs (Charge Coupled Devices), etc. Paper tapes are probably the oldest medium but have become almost obsolete today. Cassette tapes—audio or data cartridge types—are extremely popular in small systems and with hobbyists because of their low cost. However, they have the disadvantage of comparatively high access times. Floppy disks (and mini-floppies) can provide much faster access times even though they are costlier than cassette tapes. They are becoming increasingly popular in medium size systems in applications where low access times are imperative. Hard disks are comparatively rare in microcomputer systems. This is because of their cost and the fact that in typical systems it is difficult to justify their use from the viewpoint of faster access or higher storage capacity, their two main advantages.

All the storage mediums, other than magnetic bubbles or CCDs, suffer from one major disadvantage. All of them have rotating mechanical parts which do not work reliably unless care is taken to keep the operating environment clean. Magnetic bubbles, or bubble memories, and CCDs are unique in providing a mass storage medium with no moving parts. Bubble memories are commercially available from a few semiconductor manufacturers. Unfortunately, they are still not as popular as they, perhaps, should have been.

A number of semiconductor device manufacturers also provide a number of controllers (in the form of LSI chips) for interfacing a microcomputer system with these bulk memory devices. Using these peripherals, the design procedure for incorporating such memory in the system gets considerably simplified. Such LSI controllers are currently available for floppy disk and bubble memory storage systems. (No specific LSI controllers are available for cassette tape storage. However, if the cost can be justified, properly programmed single chip microcomputers like the Intel 8048/8041 devices can be used for this purpose.)

7.6.5 Keypads and Keyboards

For manual data entry of a large number of variables into a microcomputer system, keypads and keyboards are commonly used. The term keypad is generally applied to a keyboard containing a small number of keys; otherwise, the two are quite similar.

In a typical keyboard, a unique binary code is associated with each key. From the microcomputer's point of view, 'reading' a keyboard essentially implies inputting (via an input port) the code corresponding to the key that has been pressed. There are two ways in which this 'reading' is commonly done.

If the number of keys are small, one can directly use an IC encoder or combinatorial circuitry to provide the binary code corresponding to the key being pressed. This is illustrated in Fig. 7.6.1. Here we show an encoder which outputs an N-bit code corresponding to which of 2^N keys has been pressed; if more than one key are simultaneously pressed, the N-bit output recognizes only one of them based on some predetermined priority. The encoder provides another output KP which serves to inform the microprocessor that some key has been pressed and that the encoder output should be read. The microprocessor then reads the encoder output through an input port.

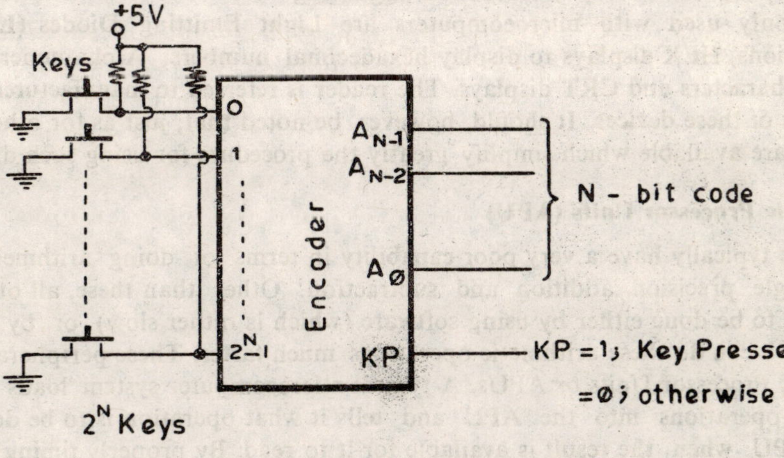

Fig. 7.6.1 Direct Encoding of a 2^N Key Keypad

It can easily be seen that if N is large, the encoder becomes complex and difficult to implement. In such a situation a 'Scanning Method' is commonly employed. Here the keys in the keyboard are periodically scanned to see if any of them is pressed. The microprocessor can do this scanning itself using an appropriate combination of I/O ports and software. It is more common to do this with an LSI Keyboard Controller which continually scans the keyboard for pressed keys, informs the microprocessor when a key is pressed and identifies the pressed key.

(Once again, this is a hardware software trade-off.) One problem in this scanned approach is 'rollover', i.e. what happens if more than one key is pressed either inadvertently or because the operator is entering data too fast. There are two approaches to solving this problem—2-key rollover and N-key rollover. In 2-key rollover, once a key is pressed, any other pressed key is not recognized until the first key is released. In N-key rollover, each key closure is recognized as it is encountered in the keyboard scan, regardless of the actual sequence in which they were pressed.

Keys in any keyboard will have a 'debouncing' problem just like any other switch—i.e. one key closure will actually be seen as a number of key closures. Avoiding this problem by debouncing the key can be done either by software or by hardware. A circuit for doing this is shown in Fig. 7.6.2.

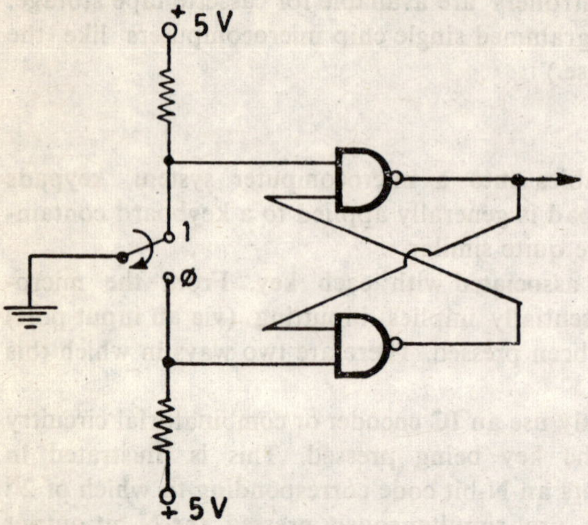

Fig. 7.6.2 Debouncing Circuit using R-S Latch

7.6.6 Display Devices

In order to interact meaningfully with a human operator, any microcomputer system would need display devices to convey information regarding results or its own status. The various displays commonly used with microcomputers are Light Emitting Diodes (LEDs) to signal ON/OFF conditions, HEX displays to display hexadecimal numbers, Alphanumeric displays to display ASCII characters and CRT displays. The reader is referred to manufacturers' data sheets for a description of these devices. It should, however, be noted that, just as for other peripherals, LSI controllers are available which simplify greatly the procedure for using such displays.

7.6.7 Arithmetic Processor Units (APU)

Microprocessors typically have a very poor capability in terms of doing arithmetic operations other than single precision addition and subtraction. Other than these, all other arithmetic operations have to be done either by using software (which is rather slow) or by using special peripherals which can do these arithmetic operations much faster. These peripherals are referred to as Arithmetic Processor Units or APUs. A typical microcomputer system loads the data for the arithmetic operations into the APU and tells it what operation is to be done. The APU informs the CPU when the result is available for it to read. By properly timing the use of the APU, a microcomputer can do complex arithmetic operations with remarkable speed while simultaneously keeping the CPU's own software and time overhead small.

7.7 MULTIPLE PROCESSOR ARCHITECTURE

With the microprocessor CPU very often being the lowest cost LSI in a microcomputer system, it is not surprising to find systems using more than one CPUs. One approach commonly followed is to divide the microcomputer's function into sets of tasks such that each CPU handles a particular set of tasks. Compared to single CPU systems, a multiple processor system will provide higher throughput, faster processing, higher reliability and greater modularity of design.

Architectures using multiple processors typically become more complicated than systems using single processors. In multiple processor systems a number of design problems have to be solved. Examples of these are the most effective partitioning of tasks between multiple processors, sequencing and control of interprocessor data transactions, arbitration and use of shared bus, usage and control of subsystems which are being shared by more than one processor etc. A number of solutions for these have been evolved depending on the particular application. This section introduces the reader to some of the simpler multiple processor architectures.

A simple (relatively!) architecture can be evolved using multiple processors if each processor performs independent tasks on separate data such that they can combine the results of these tasks when needed. This class of architectures, referred to as Multiple Instruction Multiple Data (MIMD) systems can achieve a considerable degree of parallelism in the execution of their operations.

A true Multiprocessing System is an MIMD system where the same tasks can run on different processors. One particular CPU in the system acts as the system controller running an operating system software. This CPU allocates tasks, to other CPUs in the system in a dynamic fashion such that the system's resources are used as efficiently as possible. Such systems tend to be very complex in terms of both their hardware and software and are beyond the purview of this text.

Another type of MIMD system, which is simpler and far more common than multiprocessor systems, is a Distributed System. In a Distributed System, individual processors are assigned tasks in a fixed fashion. The decision as to which task is to be handled by which processor is made at the design stage and is never changed. (Note that unlike multiprocessor systems, the task allocation is not dynamic.) Two different kinds of distributed systems are described next.

7.7.1 Master-Slave Type of Distributed Multiple Processor System

As shown in Fig. 7.7.1, systems of this type typically have one master CPU and one or more slave CPUs. The slaves work with their own program memory, data memory and I/O ports where the system bus of each slave is local only to itself and does not interact with the system

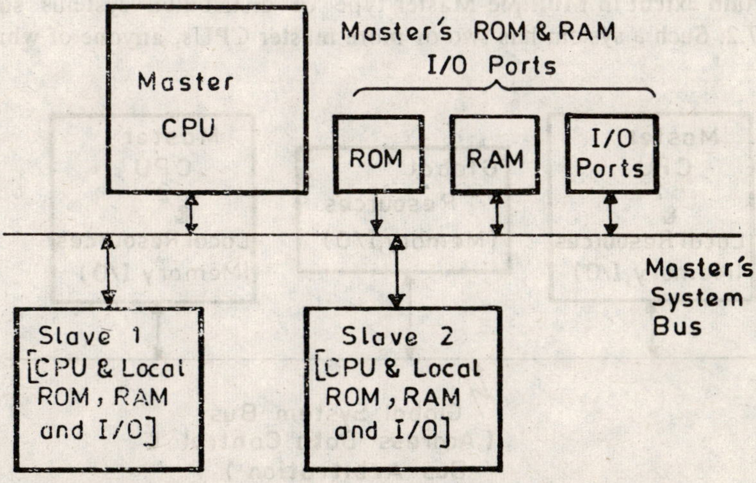

Fig. 7.7.1 Master-Slave Multiple Processor Architecture

busses either of the master or of other slaves. Communication between a slave and the master is through one or more bidirectional I/O ports (or suitable combination of input and output ports) and is done completely under the control of the master. (Note that an input port for a slave will be an output port for the master and vice versa.) Each slave CPU is identified with one or more tasks it should execute on command from the master. When the master wants a particular task to be executed, it appropriately commands the proper slave and also provides it with all the relevant data it needs for this job. If results are expected, it interrogates the slave some time later to find out what results were obtained from that particular task. Several single chip microcomputers are now available which are specifically designed to act as slave CPUs in a Master-Slave type of Distributed Multiple Processor systems, e.g. Intel 8048, 8041 etc. [See Ch. 13].

There are a few points worth noting about the Master-Slave architecture for a distributed system. The system resources (ROM, RAM, I/O etc.) available to the master are not available to the other slave CPUs which, of course, may have similar resources of their own. The reason for this is that, as far as its own system bus (address, data, control) is concerned, the master CPU is the only bus-master (i.e. the only controller regulating bus accesses). From the viewpoint of bus accesses, the master regards the slaves as being I/O ports which drive the bus only when they are asked to input data to the master CPU. Since all data transactions between the master CPU and its bus and the slave CPU and their local busses are under the control of the master, situations involving contentions between various CPUs do not arise. This leads to a considerable simplification in the design of these systems (compared to general MIMD systems). Systems of this type are well suited to applications where the function of the microcomputer can be divided into modular tasks, each one of them fixed assigned to a CPU in the system and where the communication traffic between CPUs is not very high. The disadvantage of course is that efficiency of resource usage is low since the resource of one CPU (slave or master) is not directly accessible to another CPU in the same system.

7.7.2 Multiple Master Type of Distributed Processing Systems

In the master-slave type of distributed system, a major disadvantage was the fact that resources (memory or I/O) of one CPU was not directly accessible to another CPU. This disadvantage is removed to a certain extent in Multiple-Master type of distributed systems such as the one shown in Fig. 7.7.2. Such a system has two or more master CPUs, anyone of which can drive the

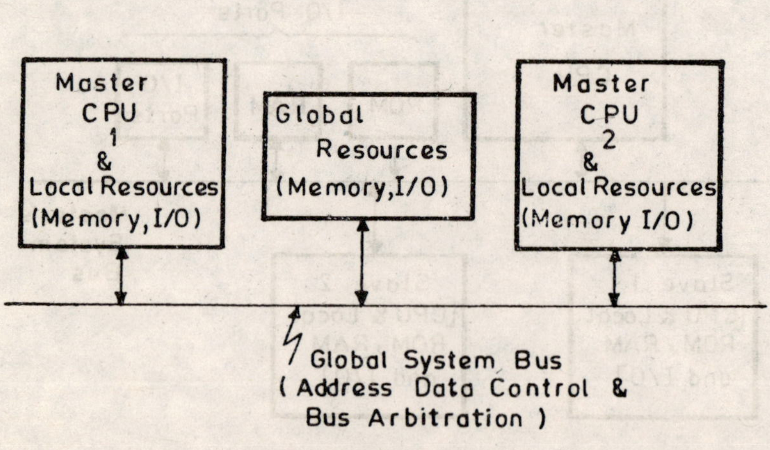

Fig. 7.7.2 Multiple Master Multiple Processor Architecture

global system bus used for inter-CPU communications. Each of these master CPUs also have their own local system bus which is used to control the CPU's own local resources, i.e. memory and I/O. Apart from the local resources of each master CPU, the system also has global resources which can be used by any one of the master CPUs. Communications between the master CPUs and the global resources are also carried by the global system bus.

Functionally, the operation of such a system is divided into tasks where these tasks are assigned to a particular master CPU at the design stage itself. A CPU executes its task when it is called upon to do so. In order to do this, it may require its own local resource, the global resources or even data from other CPUs in the system.

The designer of such a system needs to contend with two problems from the very initial stages—sharing of the global resources and sharing of the global system bus. The system should be designed such that a global resource being used by one CPU is not preempted by another CPU; unless, of course, the system designer is implementing priorities in such a manner as to allow that (e.g. the second CPU has a higher priority). For sharing the global bus, the system must ensure that at any given time only one of the master CPUs is actually the bus-master, i.e. controlling the bus. This requires coordination and cooperation between the various master CPUs. The information needed to do this must be present either on the system bus itself or should be provided via some other signal lines. MULTIBUS, GPIB, HPIB etc. are industry standards for protocols and signals which allow such coordination in Multiple Master configurations. A detailed description of such bus interface standards are beyond the scope of this book. However the reader should be aware of the fact that reasonably well defined standards do exist which cater to such applications.

PROBLEMS

1. Show a simple bussed architecture for a microcomputer having a CPU, ROM, RAM, two input ports and two output ports. Is it possible to implement a system using the same address for an input port and an output port? If not, give reasons why it cannot be done. If yes, show how one can do it.

2. For the microcomputer in (1), I would like to place
   (a) a (1024 × 8) ROM in address space $ØØØØ_H$ — $Ø3FF_H$
   (b) a (2048 × 8) RAM in address space $Ø8ØØ_H$ — $ØFFF_H$
   (c) Input Ports in addresses $1ØØØ_H$ and $2ØØØ_H$
   (d) Output Ports in addresses $3ØØØ_H$ and $4ØØØ_H$

   Assuming a 16-bit address bus, an $IO/\overline{M}$ signal, an 8-bit data bus and $\overline{RD}$ and $\overline{WR}$ signals, show the architecture of (1) in more detail. Show how the various chip selects will be generated.

3. In (2), suppose you are given the freedom to move the RAM addresses and I/O port addresses to whatever location you desire
   (a) What will be the simplest way of generating the proper chip selects?
   (b) What will be the most 'address space efficient' way of generating the chip selects?

   Repeat (a) and (b) for the case where the $IO/\overline{M}$ signal is not available.

4. What are the relative merits and demerits of the various memory addressing schemes described in Sec. 7.4?
5. Give a brief summary highlighting the advantages and disadvantages of the various I/O techniques described in Sec. 7.5. Give application examples where a particular I/O technique will be the logical choice over the others. Give reasons to support your claim.
6. Describe why a system using DMA is similar to a multiprocessor architecture. In some applications a DMA technique should not be used—can you think of situations where this will be true?

Appendix A7

A7.1 INTEL 8253 PROGRAMMABLE TIMER/COUNTER

Intel's 8253 Programmable Timer/Counter LSI chip is a very popular device specially designed for microcomputer applications which require timing and counting operations. Many real time applications require timing and counting functions which may be implemented through software. The software solution, however, typically places severe constraints on the overall system design. Using a device like the 8253, one can also have a hardware solution which does not suffer from these problems. The tradeoff in such situations is between the cost of the hardware (i.e. the cost of the 8253 or similar device) and the cost of the software solution in terms of memory and execution times.

In order to see the advantages of a hardware solution, as opposed to a software solution, consider the following example. A microcomputer system is required to execute N different tasks. Suppose it is required to do Task j exactly T seconds after executing Task i (i and j could be equal). The software solution would be to call a delay routine to count out a T seconds interval after Task i is completed and then do Task j. In order to maintain the precision of the delay, it will be difficult, if not impossible, for the microcomputer to execute any other task during this interval. From the viewpoint of the N tasks, the microcomputer is forced to stay 'idle' during this T seconds interval, which, obviously, will not be a very desirable thing to do. For the hardware solution, the microcomputer starts an external counter to count out T seconds (using an appropriate clock) after completing Task i. The counter interrupts the processor after T seconds. The microcomputer executes Task j once it gets this interrupt. The reader should note that, in this case, having started the counter, the microcomputer can now go and execute other tasks during this T second interval; unlike the previous case, it is not forced to effectively idle during this waiting period.

Apart from applications like the one described above, the 8253 device (and similar devices) may also be used for the following purposes:

- (i) Event Counter
- (ii) Real-time Clock
- (iii) Programmable one shot
- (iv) Delay generation and measurement

A7.1.1 Pin-out of the 8253

The pin-out of the 8253 device is shown in Fig. A7.1.1. A simplified symbolic representation of the device is shown in Fig. A7.1.2. As seen from the latter, the 8253 consists of three independent 16-bit counters which may be operated in various modes by proper programming. Before describing these modes and the programming aspects of this device, we present below a brief description of the various pins of the 8253.

V_{cc} +5V } (Power)
GND Ground }

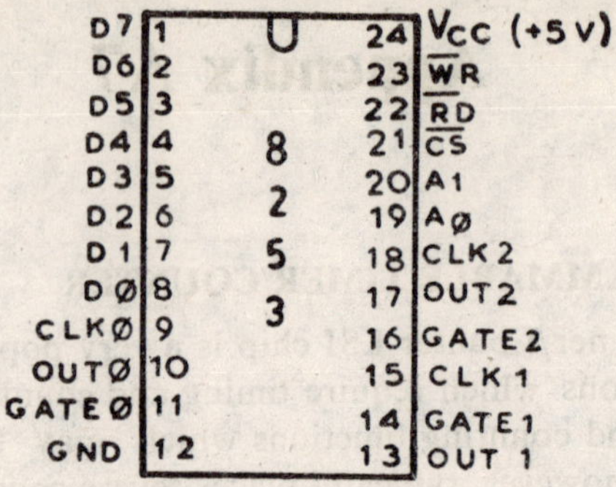

Fig. A7.1.1 Pin-out of Intel 8253

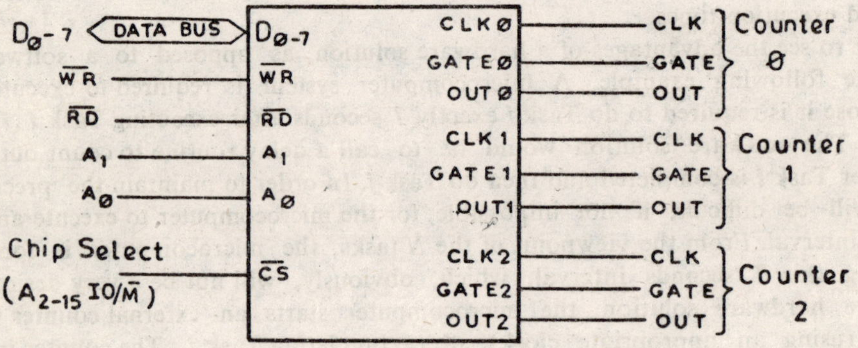

Fig. A7.1.2 Symbolic Representation of the Intel 8253

$D_7 - D_\emptyset$ Data bus (bidirectional)

These pins connect to the data bus of the microcomputer system. The microcomputer uses these pins to program the device, to preset each counter with appropriate values or to read the contents of these counters.

$\overline{RD}$ The microcomputer drives this input Lo in order to read data from the 8253. When $\overline{RD} = \emptyset$ (and $\overline{CS} = \emptyset$), the 8253 drives the data bus $D_7 - D_\emptyset$ with appropriate values.

$\overline{WR}$ The microcomputer drives this input Lo (with $\overline{CS} = \emptyset$ at the same time) in order to load data or commands into the 8253.

$\overline{CS}$ This input should be Lo (i.e. $\overline{CS} = \emptyset$) in order to do a proper read or write operation with the 8253. The microcomputer system views the 8253 as 4 locations either on its memory address space (memory mapped I/O) or on its I/O address space (I/O mapped I/O). Depending on the addressing technique being used and the actual address locations where the device is to be placed, the microcomputer system should provide an appropriate $\overline{CS}$ signal to the 8253.

$A_1, A_\emptyset$ These are generally connected to the lowest two bits of the microcomputer's address bus. The A_1, $A_\emptyset$ values determine, whether for a particular read or write operation, the microcomputer is accessing the Control Register of the 8253 (to load a command, i.e. program the device) or one of the three counters. This is done according to the following table.

$\overline{CS}$	$\overline{RD}$	$\overline{WR}$	A_1	$A_\emptyset$	
∅	1	∅	∅	∅	Load Counter # ∅
∅	1	∅	∅	1	Load Counter # 1
∅	1	∅	1	∅	Load Counter # 2
∅	1	∅	1	1	Write Mode Word in Control Register
∅	∅	1	∅	∅	Read Counter # ∅
∅	∅	1	∅	1	Read Counter # 1
∅	∅	1	1	∅	Read Counter # 2
∅	∅	1	1	1	No operation—Tristate data bus
1	X	X	X	X	No operation—Tristate data bus
∅	1	1	X	X	No operation—Tristate data bus
∅	∅	∅	X	X	Illegal operation

CLKn $n = 0, 1, 2$

This is the clock input to counter *n*. The 8253 requires this clock to have a frequency in the range d.c.—2 MHz. Note that the counters operate at the *negative-edge* (Hi to Lo) of this clock input.

GATEn $n = 0, 1, 2$

This is an external input to counter # *n* which, depending on the mode programmed, may be used to either enable or disable counter *n*. Further details of this input will be given later (Sec. A7.1.2); however, it should be noted that in general (with some exceptions)

(a) Lo or Hi → Lo at GATEn disables counting on counter # *n*
(b) Lo → Hi at GATEn initiates counting on counter # *n*
(c) Hi at GATEn enables counting on counter # *n*

OUTn $n = 0, 1, 2$

OUT∅, OUT1, and OUT2 are the respective outputs of counter # 0, counter # 1 and counter # 2.

A7.1.2 Modes of Operation of the 8253's Counters

Each of the three counters of an 8253 device can be programmed to operate in six modes of operation. These modes are described in this section. It should be noted that each of the 8253's counters is a *negative edge-triggered down-counter*. Using $\overline{CS} = \overline{WR} = \emptyset$ and a proper combination of A_1 and $A_\emptyset$, the user loads counter # *n* with a 16-bit (or 8-bit) number; once counter # *n* is properly enabled, it starts decrementing its contents by 1 at each Hi → Lo edge of CLKn. At any instant, the current count value remaining in the counter may be read by putting $\overline{CS} = \emptyset$, $\overline{RD} = \emptyset$ and choosing a proper combination of A_1 and $A_\emptyset$.

When the clock to one of the 8253's counters is present, the following events affect the counter's behaviour and its output:

(i) Setting the mode of the counter
(ii) Loading a 16-bit (or 8-bit) data on the counter
(iii) GATE = Ø or Hi → Lo transition of GATE input
(iv) GATE = 1
(v) GATE input going from Lo → Hi

The operation of the counter and the signal at its OUT pin reflect the response of the counter to the above events as per its present mode of operation. The six possible modes are described next.

Mode Ø: Interrupt on Terminal Count (Fig. A7.1.3)

The counter output OUT stays Lo once Mode Ø has been programmed. Now suppose a count value of N is loaded into the counter. With GATE = 1, the counter starts decrementing the count value. The output OUT stays Lo until the count value reaches Ø ($N + 1$ clock pulses after loading count value); when this occurs, OUT goes to 1 and stays at 1 until a new count is loaded. (The counter continues counting even after OUT = 1 occurs; this, however, does not change the OUT value.) When a new count is loaded, OUT returns to Ø and the above operation resumes.

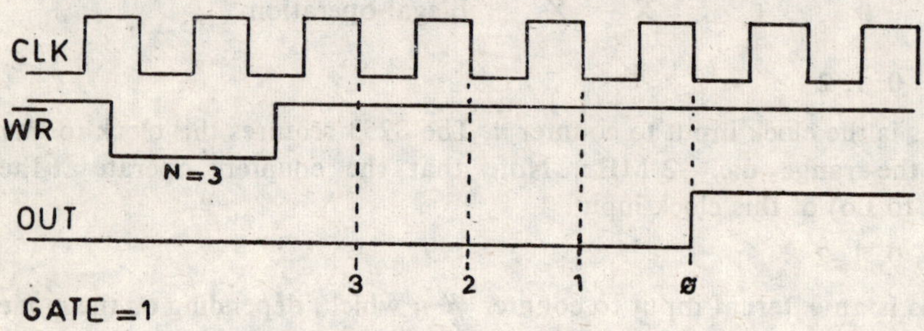

Fig. A7.1.3 Mode Ø, Interrupt on Terminal Count

Inputting GATE = Ø disables the counter. The count value stays 'frozen' if GATE goes to Ø while counting is underway (i.e. the counter essentially stops counting when GATE = Ø). Counting resumes (from the previously frozen value) when GATE returns to 1.

The count value may be changed at any instant even when the counting operation is going on. Loading a new count value essentially restarts the counter's operation with this new value. Note that if the count value to be loaded is 8 bits (one byte), the effect of loading the new count value is immediate. If the count value to be loaded has 16 bits, the counter stops counting after the first byte is loaded and resumes counting only after the second byte is loaded.

The OUT signal from the counter may be used as an interrupt signal to the microcomputer. This mode of operation is commonly used if the microcomputer wants to execute Task *j* precisely *T* seconds after another task (say Task *i*). Task *i*, then, on completion, starts the counter with the proper count value so that the microcomputer is interrupted by the OUT signal *T* seconds later. The interrupt service routine then executes Task *j*. (Of course for such applications, GATE should always be Hi and the count value should be properly calculated.)

Mode 1: Programmable One Shot (Retriggerable) (Fig. A7.1.4)

In this mode, the 8253's counter acts as a retriggerable, programmable one shot with the Lo → Hi edge of the GATE input acting as the trigger signal. Programming this mode sets OUT immediately Hi. Consider a situation where the microcomputer loads a count value of N into the counter. Now for every Lo → Hi transition of the GATE input, OUT goes to Lo for N clock pulses following the first negative edge of the CLK after the Lo → Hi transition of GATE. This is demonstrated in Fig. A7.1.4. This is exactly the behaviour one would expect from a monostable or one-shot. Since in this case, the width of the OUT pulse is programmable, this mode is referred to as the programmable one-shot mode.

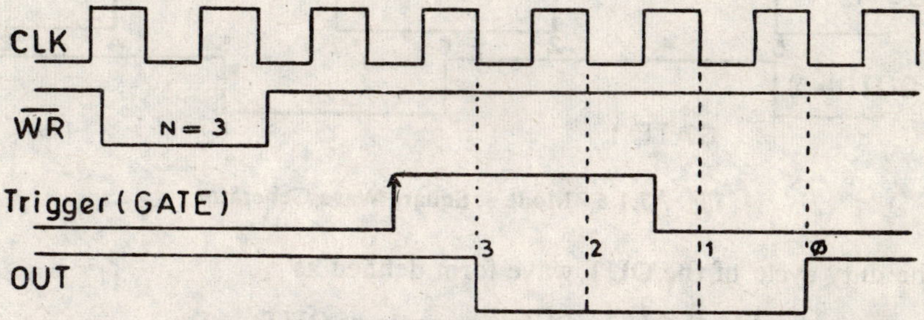

Fig. A7.1.4 Mode 1, Programmable One Shot (Retriggerable)

Consider a situation where a Lo → Hi edge of GATE has forced OUT = Ø or N clock pulses. Suppose that another Lo → Hi edge of the GATE signal occurs while OUT is Lo. In this case, OUT will stay Lo for N clock pulses after this second Lo → Hi edge, i.e. the one-shot gets retriggered. This is the reason why the one-shot in this mode is referred to as a retriggerable one.

Mode 2: Rate Generator (Fig. A7.1.5)

Programmed in this mode, the 8253's counter behaves as a simple divide-by-N counter if the microcomputer initially loads a count value of N into the counter. As shown in Fig. A7.1.5, in

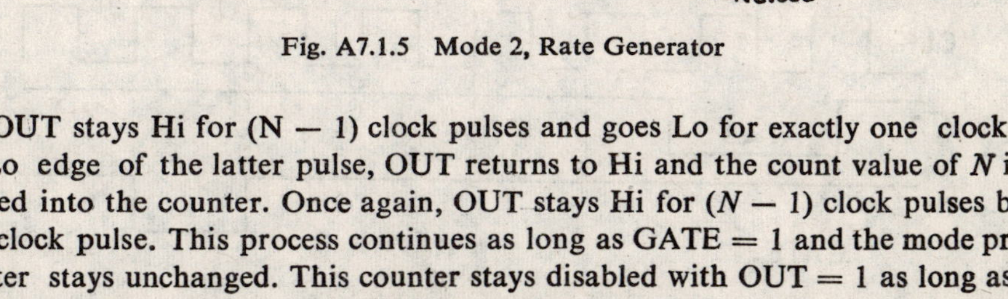

Fig. A7.1.5 Mode 2, Rate Generator

this mode, OUT stays Hi for (N — 1) clock pulses and goes Lo for exactly one clock pulse. At the Hi → Lo edge of the latter pulse, OUT returns to Hi and the count value of N is automatically reloaded into the counter. Once again, OUT stays Hi for (N — 1) clock pulses before going Lo for one clock pulse. This process continues as long as GATE = 1 and the mode programming of the counter stays unchanged. This counter stays disabled with OUT = 1 as long as GATE is Lo, resuming normal operation when GATE returns to Hi.

Mode 3: Square Wave Generator (Fig. A7.1.6)

A 8253-counter, in this mode, works as a divide-by-N counter (N = count value initially loaded into counter) just as in Mode 2. The only difference with the latter is in the duty cycle of the OUT output.

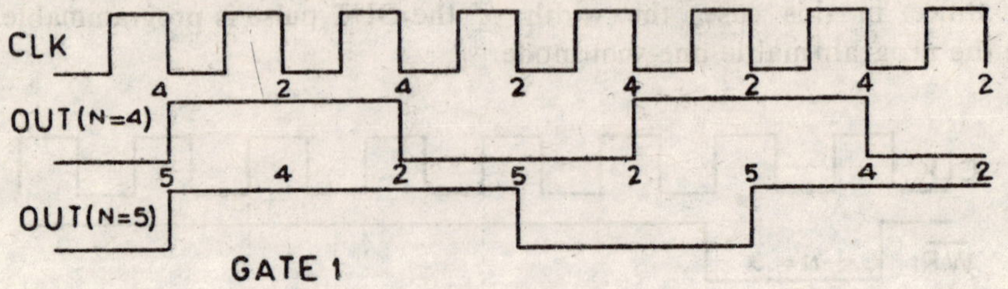

Fig. A7.1.6 Mode 3, Square Wave Generator

Let d be the duty cycle of the OUT wave form defined as

$$d = \frac{\text{duration for which OUT} = 1 \text{ in one cycle of OUT}}{\text{duration for one cycle of OUT}}$$

In Mode 3, d is given by

$$d = \frac{N/2}{N} \quad \text{for } N \text{ even}$$

$$= \frac{(N+1)/2}{N} \quad \text{for } N \text{ odd}$$

$\left(\text{Compare with Mode 2 for which } d = \frac{N-1}{N}.\right)$

The most common application of Mode 2 is as a baud rate generator for serial data transfer. By choosing N properly via software and initializing this as the count value, one can easily obtain a programmable baud rate generator.

Note that, as in Mode 2, the counter stays disabled with OUT = 1 as long as GATE is Lo. Normal operation resumes when GATE returns to Hi.

Mode 4: Software Triggered Strobe (Fig. A7.1.7)

The counter output OUT stays Hi once the counter has been programmed in this mode. The

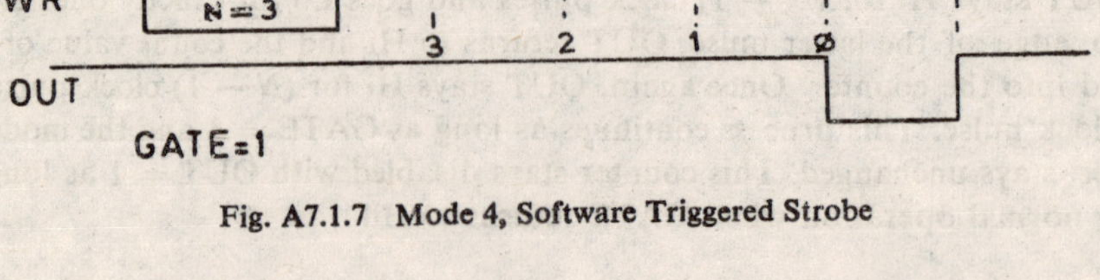

Fig. A7.1.7 Mode 4, Software Triggered Strobe

counter starts decrementing once the count value is loaded, provided GATE = 1. Once terminal count (counter decrements to Ø) is reached, OUT goes Lo for one clock pulse and the returns to Hi. This OUT pulse may be used as a strobe signal in a particular application. The counter is disabled when GATE = Ø; GATE = 1 is needed for operation in this mode. Note that in this mode, the generation of the strobe signal at OUT is triggered by the act of writing the count value into the counter. This is the reason why this mode is referred to be software triggered.

Mode 5: Hardware Triggered Strobe (Retriggerable) (Fig. A7.1.8)

The OUT pin of a counter programmed in this mode is initially set at 1. Consider a situation where a count value of *N* is loaded into the counter. Following a Lo → Hi edge of the GATE input (this acts as the trigger), the counter starts decrementing and generates a Lo pulse for one

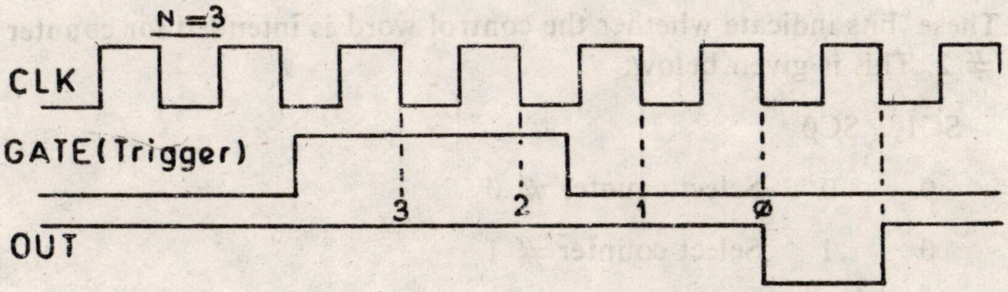

Fig. A7.1.8 Mode 5, Hardware Triggered Strobe (Retriggerable)

clock period once the counter reaches terminal count (i.e. counter content = Ø). Note that unlike Mode 4, the triggering event comes from the hardware, i.e. Lo → Hi edge of GATE, which is why this mode is referred to as a Hardware Triggered Strobe.

This mode is also retriggerable like Mode 1. Consider a situation where the counter has started to decrement from the count value *N* following a Lo → Hi edge of GATE. Suppose another Lo → Hi edge of GATE appears before a strobe pulse is generated. The effect of this second Lo → Hi GATE-edge is to reload the counter with the initial count value of *N*. Now the counter starts decrementing once again from *N*, generating a Lo pulse once terminal count is reached.

The effect of the GATE input on the various modes of the 8253's counters is summarised below:

Mode	GATE = Ø or GATE = (1 → Ø)	GATE = (Ø → 1)	GATE = 1
Ø, 4	Disables counting	—	Enables counting
1	—	(i) Initiates counting (ii) Resets OUT after next clock	—
2, 3	(i) Disables counting (ii) Sets OUT immediately Hi	Initiates counting	Enables counting
5	—	(i) Initiates counting (ii) Counter initialized to count value of *N*	—

A7.1.3 Programming the 8253

Programming the 8253 is achieved by writing an appropriate bit pattern (control word) into the Control Register ($A_1A_\emptyset = 11$) of the 8253. For each of the 8253's counters, the user can select the following features by choosing the proper control word:

(i) Mode of operation, i.e. Mode Ø – 5 (see Sec. A7.1.2)

(ii) 16-bit or 8-bit data to be read (or written)

(iii) 16-bit binary counter or 4 digit BCD counter operation

The control word format for choosing the above features is given below:

D7	D6	D5	D4	D3	D2	D1	DØ
SC1	SCØ	RL1	RLØ	M2	M1	MØ	BCD

SC1, SCØ: These bits indicate whether the control word is intended for counter # Ø, # 1 or # 2. This is given below:

SC1,	SCØ	
Ø	Ø	Select counter # Ø
Ø	1	Select counter # 1
1	Ø	Select counter # 2
1	1	Illegal

RL1, RLØ: Each of the 8253's counters has 16 bits. Since they are accessed via a 8-bit data bus, two separate read or write operations are required to completely access one counter. The 8253 allows the user a substantial degree of freedom in this regard. One can, for example, read (or load) only the most significant byte of the selected counter without worrying about the least significant byte or vice versa. The various choices are decided by the values of RL1 and RLØ as shown below:

RL1,	RLØ	
Ø	Ø	Counter Latching Operation (see Sec. A7.1.4)
Ø	1	Read/Load least significant byte only
1	Ø	Read/Load most significant byte only
1	1	Read/Load least significant byte first, then the most significant byte.

If RL1 = RLØ = Ø is programmed (i.e. counter latching) the four lower order bits (M2, M1, MØ, BCD) are 'don't cares'.

Note that the Read/Load operation specified through RL1, RLØ must be strictly followed by the software handling the 8253; otherwise erroneous counter operations will occur.

M2, M1, MØ: For the selected counter, the mode of operation is specified by these bits as per the following table.

M2	M1	MØ	
Ø	Ø	Ø	Mode Ø
Ø	Ø	1	Mode 1
X	1	Ø	Mode 2
X	1	1	Mode 3
1	Ø	Ø	Mode 4
1	Ø	1	Mode 5

BCD: A Ø in this bit commands the selected counter to operate as a 16-bit binary counter.

A 1 in this bit commands the selected counter to operate as a 4-digit BCD (Binary Coded Decimal) counter.

A7.1.4 Reading/Loading of the 8253's Counters

The microcomputer's software is required to initialize each counter's control register before using it. This initialization not only sets the mode of operation of the counter but also determines if it is going to operate as a 16-bit binary counter or a 4-digit BCD one. The RL1 and RLØ bits of the Mode Set command loaded into the control register also determines how the count value is to be loaded into the counter. The writing of this count value must follow strictly the choice made by the Mode Set Command.

The contents of a specific counter may be read at any time by the microcomputer outputting $\overline{RD} = \overline{CS} = \emptyset$ and by properly choosing A_1 and $A_\emptyset$. The byte output by the 8253 in this case depends once again on the RL1, RLØ bits of the mode set command. (Note that if RL1 = RLØ = 1, the microcomputer should issue two READ commands to the 8253 in order to read both the MSByte and the LSByte.)

Even though the counter contents may be read at any time, a problem of ambiguity remains. This is because the counter contents may change while they are being read. This problem will be especially severe if RL1 = RLØ = 1 option has been specified in the Mode Set Command (why?). The 8253 provides a Counter Latching Command to tackle this problem.

In order to issue a Counter Latching Command to a particular counter, the microcomputer writes a command to the 8253 with SC1, SCØ indicating the counter to be read and with RL1 = RLØ = Ø. (Other bits—D3, D2, D1, DØ—are 'don't cares' in this case.) Whenever the 8253 gets this command, it latches the contents of the selected counter into a storage register. By now issuing a normal read command, the microcomputer subsequently reads the contents of this storage register. Note that, once again, it is mandatory to do the entire read operation as specified by the RL1, RLØ bits of the original mode set command. Moreover, the Counter Latching Command must be issued every time the microcomputer wants to first latch the counter contents and then read it. Finally, it should be noted that the Counter Latching Command does not affect the counter's mode or its operation.

Chapter 8

Assembly Language Programming Using the 8085A

In this chapter, we will concentrate on Intel's 8085A as the microprocessor to be studied in detail. The 8085A is a 8-bit microprocessor which can address up to 64 K (65536) memory bytes using a 16-bit address. The memory space actually being used can be configured using RAMs or ROMs in any fashion. The only restriction in this case is that depending on usage certain locations must have program instructions. These are generally the locations to which the program will jump in response to inputs like RESET, TRAP or the various other interrupts. These will be discussed later.

8.1 ORGANIZATION OF THE 8085A AND THE NOTATION USED

Organizationally, the 8085 can be looked upon as a CPU having the following features

(i) Seven 8-bit working registers—Accumulator (A), B, C, D, E, H and L
(ii) 16-bit Program Counter
(iii) 16-bit Stack Pointer
(iv) Ability to address 256 I/O ports for input and output
(v) 8-bit Flag Register storing the following flags in the corresponding bits.

The other bits of the FLAG register are not relevant

Bit	FLAG	
7	S	Sign
6	Z	Zero
4	AC	Auxiliar Carry
2	P	Parity
0	CY	Carry

S is set if the MSB of the result of an operation is 1, i.e. if the result is a negative number in 2's complement representation

Z is set if the result of an operation is 'Ø'

P is set if the result has even parity (even number of 1s)

CY is set if as a result of an operation, a carry out is generated from the MSB (addition) or a borrow occurs from the MSB (subtraction)

AC is set if carry out occurs from the MSB of lower order nibble.

8.1.1 Notation Used

In order to describe the Instruction Set of the 8085A the following symbols and abbreviations will be used:

Accumulator	Reg. A with A_7 = MSB and $A_\emptyset$ = LSB
Addr.	16-bit data quantity; A_{15} = MSB, A_0 = LSB
Data	8-bit data; D_7 = MSB, D_0 = LSB
Data 16	16-bit data; D_{15} = MSB, D_0 = LSB
Byte 2	II byte of Instruction
Byte 3	III byte of Instruction
Port	8-bit address of an I/O device
r	One of the registers A, B, C, D, E, H or L as indicated
DST, SRC	Register bit patterns corresponding to the various registers A = 111 E = ∅11 B = ∅∅∅ H = 1∅∅ C = ∅∅1 L = 1∅1 D = ∅1∅ where DST = destination register and SRC = source register
r16	Register Pairs B = Concatenation of B-C D = Concatenation of D-E H = Concatenation of H-L SP = 16-bit Stack Pointer
RR	Bit pattern for the above register pairs B-C = ∅∅ D-E = ∅1 H-L = 1∅ SP = 11
rh	high order reg. of register pair (e.g. B of B-C)
rl	low order reg. of register pair (e.g. C of B-C)
(·)	Contents of register or memory location enclosed within parentheses
$D \leftarrow S$	transfer from S to D
$X \leftrightarrow Y$	interchange X and Y
$\wedge$	AND

V	OR
⊻	Ex-OR

We will write bytes as (MSB ...... LSB).

8.1.2 Addressing Modes of the 8085A

Addressing data with the 8085A can be done using Direct, Register, or Register Indirect modes. One can also provide Immediate Data where the instruction contains the data itself—this could be either a 8-bit quantity or a 16-bit quantity. A 16-bit quantity is stored with the low order byte in the lower memory location and the higher order byte next in the next higher memory location.

Subroutine linkage is done using the stack to store the return address. Branch instructions can provide the branch address in two ways, Direct or Register Indirect. Direct addressing instructions typically provide the lower order byte of the address in byte 2 of the instruction and higher order one in byte 3. (The RSTn instructions are exceptions to this since they are single byte instructions.)

8.1.3 Instruction Groups of the 8085A

The instructions available in the 8085A can be divided into five groups depending on their function.

1. *Data Transfer Group*: Moves data between registers, or between registers and memory locations (e.g. moves, loads, stores and exchanges).
2. *Arithmetic Group*: Add, subtract, increment or decrement contents of register or memory.
3. *Logic Group*: AND, OR, EX-OR, complement compare and rotate data in register or memory.
4. *Branch Group*: Conditional or unconditional jumps, calls, returns and restarts.
5. *Stack, I/O and Machine Control Group*: Manipulate stack, reading from ports, writing to ports, setting and reading interrupt masks and setting and clearing flags.

In the following (Sections 8.2–8.6) we have used the assembly language mnemonics for the various instructions. Typical assemblers also provide some instructions which facilitate program writing and development. Some possible Assembler Directives will be described in Section 8.8. Note that statements following ';' should be interpreted as comments.

8.2 DATA TRANSFER GROUP

These do not affect condition flags S, Z, AC, P and CY.

MOV X, Y ; Move

X, Y = r or M   MOV M, M is not allowed

where M = (H)(L)

Function: (X) ← (Y)

Op Code:	Ø1DST SRC	if X = r1, Y = r2
	Ø1DST 11Ø	if X = r, Y = M
	Ø111Ø SRC	if X = M, Y = r

Single Byte Instruction

(All mnemonics copyright Intel Corporation 1985)

MVI X, Data; Move Immediate
$X = r$ or M
where $M = (H)(L)$
Function: (X) ← Data
Op Code: ØØDST 11Ø if X = r
ØØ11Ø 11Ø if X = M
Single Byte Instruction

LXI rp, Data 16; Load register pair immediate, e.g. LXI H, 2ØØØH
Function: (rh) ← Byte 3
(rl) ← Byte 2
Op Code: ØØRRØØØ1
Format: ØØRRØØØ1, low order byte, high order byte

LDA Addr; Load Acc Direct
Function: (A) ← ((byte 3)(byte 2))
Op Code: $3A_H$
Format: $3A_H$, Low order Addr byte, High order Addr byte

STA Addr; Store Acc Direct
Function: ((byte 3)(byte 2)) ← (A)
Op Code: 32_H
Format: 32_H, Low order Addr byte, High order Addr byte

LHLD Addr; Load H and L direct
Function: (L) ← ((byte 3)(byte 2))
(H) ← ((byte 3)(byte 2) + 1)
Op Code: $2A_H$
Format: $2A_H$, Low order Addr byte, High order Addr byte

SHLD Addr; Store H and L direct
Function: ((byte 3)(byte 2)) ← (L)
((byte 3)(byte 2) + 1) ← (H)
Op Code: 22_H
Format: 22_H, Low order Addr byte, High order Addr byte

LDAX r16; Load Acc indirect with ((r16)) where r16 is either B-C or D-E, e.g. LDAX B
Function: (A) ← ((r16))
Op Code: ØØRR1Ø1Ø
Format: Single byte instruction

STAX r16; Store Acc indirect at ((r16)) where r16 can be either B-C or D-E., e.g. STAX D

Function: ((r16)) ← (A)

Op Code: ØØRRØØ1Ø

Format: Single byte instruction

XCHG; Exchange H-L with D-E

Function: (H) ↔ (D), (L) ↔ (E)

Op Code: EB_H

Format: Single byte instruction

8.3 ARITHMETIC GROUP

These instructions affect some or all of the FLAGs (with some exceptions). The particular FLAGs affected are indicated for each instruction. The subtraction instructions in this group give proper results with signed data if 2's complement representation is used. These set flag CY = 1 to indicate borrow into bit 7 (MSB); otherwise they set flag CY = Ø. The addition instructions similarly set CY = 1 to indicate carry out of bit 7.

ADD X; Add (X) to (A)

X = r or M
where M = (H)(L)

Function: (A) ← (A) + (X)

Op Code: 1ØØØØ SRC for X = r

86_H for X = M

Flags: Z, S, P, CY, AC

Format: Single byte instruction

ADI Data; Add Immediate data to (A)

Function: (A) ← (A) + Data

Op Code: $C6_H$

Flags: Z, S, P, CY, AC

Format: $C6_H$, Data

ADC X; Add (X) to (A) with carry

X = r or M, where M = (H)(L)

Function: (A) ← (A) + (X) + (CY)

Op Code: 1ØØØ1 SRC if X = r

$8E_H$ if X = M

Flags: Z, S, P, CY, AC

Format: Single byte instruction

ACI Data; Add Immediate Data to (A) with carry

Function: (A) ← (A) + Data + (CY)

Op Code: CE_H

Flags: Z, S, P, CY, AC

Format: CE_H, Data

SUB X; Subtract (X) from (A)

X = r or M where M = (H)(L)

Function: (A) ← (A) − (X)

Op Code: 1ØØ1Ø SRC for X = r

96_H for X = M

Flags: Z, S, P, CY, AC

Format: Single byte instruction

SUI Data; Subtract Immediate Data from (A)

Function: (A) ← (A) − Data

Op Code: $D6_H$

Flags: Z, S, P, CY, AC

Format: $D6_H$, Data

SBB X; Subtract (X) from (A) with borrow

X = r or M where M = (H)(L)

Function: (A) ← (A) − (X) − (CY)

Op Code: 1ØØ11 SRC for X = r

$9E_H$ for X = M

Flags: Z, S, P, CY, AC

Format: Single byte instruction

SBI Data; Subtract Immediate Data from (A) with borrow

Function: (A) ← (A) − Data − (CY)

Op Code: DE_H

Flags: Z, S, P, CY, AC

Format: DE_H, Data

Note that all the Arithmetic Instructions given above provide single byte arithmetic. By properly using the CY bit, the Subtract-with-Borrow and Add-with-Carry instructions can be used to provide multibyte Subtractions and Additions. Consider the double byte addition routine given below:

```
; Add B-C to D-E;  double byte addition
 MOV A, C;         get low order byte from C
 ADD E;            add C and E
 MOV C, A;         and store in C
 MOV A, B;         get high order byte from D
 ADC D;            and D and E with carry from previous addition
 MOV B, A;         and store in B
; Result in B-C
```

For addition type of instructions, CY = 1 indicates a result overflow out of the MSB, i.e. Result > 255 and AC = 1 indicates a carry out from bit-3 to 4. In subtraction instructions, CY = 1 indicates that the borrow out of bit-7 was 1, AC however is still set to 1 if there is a carry out from bit-3 to bit-4 in the 2's Complement Addition done internally for subtraction and not if there was a borrow out of bit-3. Consider the double byte subtraction routine given below.

```
; Subtract D-E from B-C,   double byte subtraction
 MOV A, C;                 get C to A
 SUB E;                    Subtract E
 MOV C, A;                 and store in C
 MOV A, B;                 get B
 SBB D;                    subtract D with borrow
 MOV B, A;                 store B
; Result in B-C
```

The CY and the AC flags are also useful for the manipulation of BCD data (see the DAA instruction later). The Z flag can be used to test if the result of an operation is Ø.

The 8085 does provide one instruction to perform double byte addition. This is described below:

DAD r16; Double add-add (r16) to (H)(L)

Function: (H)(L) ← (H)(L) + (r16)

Op Code: ØØRR1ØØ1

Flag: *CY only*; CY = 1 if there is carry out of the double precision add, otherwise CY = Ø

Format: Single byte instruction

Apart from the fact that DAD allows double precision addition via a single instruction, it has one other unique use. *It is the only instruction in the 8085's instruction set which will allow the user to read the contents of SP*. This can be done as follows:

```
; Get SP contents into B-C
  MVI H, ØØH;    clear H and L
  MVI L, ØØH;    (LXI H, ØØØØH could also have been done)
  DAD SP;        get SP into H-L
  MOV B, H;      transfer H-L
  MOV C, L;      to B-C
; SP contents in B-C
```

The 8085 provides one special instruction DAA to assist the user in doing BCD arithmetic. This is described below:

DAA; Decimal Adjust Accumulator

- Op Code: 27_H
- Flags: Z, S, P, CY, AC
- Format: Single byte instruction
- Function: DAA is used after an addition instruction for adding two 2-digit BCD numbers. It is used to change the result to a valid 2-digit BCD number. This is done as follows (result of addition in A, result of DAA also in A).

1. If the lower digit is greater than 9, or if AC = 1, the entire A register is incremented by 6.
2. If the higher digit is now greater than 9 or if CY = 1, increment this digit by 6.

CY = 1 if step 2 produces a carry out of bit 7 (MSB of A). (If Result > 99, CY = 1.)

Example

```
; Decimal Add two 4-digit BCD numbers stored in B-C and H-L
  MOV A, C
  ADD L;      add lower order 2 digits
  DAA;        decimal adjust
  MOV C, A;   store lower order 2 digits of result
  MOV A, B
  ADC H;      add higher 2 digits with CY from lower 2 digits
  DAA;        decimal adjust
  MOV B, A
; Result in B-C
```

The 8085A also provides some instructions which either increment or decrement the contents of single registers or the contents of register pairs. One important use of these instructions is in Register Indirect Addressing (especially when accessing a block of data). In this kind of usage, successive data bytes can be accessed using Register Indirect Addressing and by properly incrementing or decrementing the designated register pair (usually H-L). This class of instructions also provide some instructions which directly increment or decrement contents of designated memory locations. (These can be used for loop count storage in a looping program.) These instructions are given below:

INR X; increment (X) by 1

X = r or M where M = (H)(L)

Function: (X) ← (X) + 1

Op Code: ØØ DST 1ØØ for X = r

34_H for X = M

Flags: Z, S, P, AC

Format: Single byte instruction

DCR X; decrement (X) by 1

X = r or M where M = (H)(L)

Function: (X) ← (X) − 1

Op Code: ØØ DST 1Ø1 for X = r

35_H for X = M

Flags: Z, S, P, AC

Format: Single byte instruction

INX r16; Increment (r16) by 1

Function: (rh)(rl) ← (rh)(rl) + 1

Op Code: ØØ RRØØ11

Flags: None

Format: Single byte instruction

DCX r16; Decrement (r16) by 1

Function: (rh)(rl) ← (rh)(rl) − 1

Op Code: ØØRR1Ø11

Flags: None

Format: Single byte instruction

; Example of a looping program looping 7 times and accessing 7 successive memory locations for each execution of the loop.

```
START:      -----
                 :
            LXI H, ØØ1ØH;   ØØ1ØH—Start address of data table
            MVI B, Ø7H;     Load B = Loop count
LOOP:       MOV A, M;       get data from data table
            INX H;          increment data table pointer
                 :
            DCR B;          decrement loop count
            JNZ LOOP;       repeat if loop count ≠ 0 (JNZ discussed in Sec. 8.5)
LPOUT:      -----
                 :
```

Note that the INR and DCR instructions are typically used to increment or decrement loop counters whereas INX and DCX are typically used to increment or decrement address pointers.

8.4 LOGICAL GROUP

This group implements logical operations (AND, OR, EX-OR etc.) compare functions and shifting

ANA X; AND (A) with (X) in bit by bit fashion X = r or M where M = (H)(L)

Function: (A) ← (A) ∧ (X), Bit i of A = Bit i of A (AND) Bit i of X, i = 0, . . . , 7

Op Code: 1Ø1ØØ SRC for X = r

$A6_H$ for X = M

Flags: Z, S, P, CY, AC; CY = Ø; AC = 1

Format: Single byte instruction

ANI Data; AND immediate

Function: (A) ← (A) ∧ Data

Op Code: $E6_H$

Flags: Z, S, P, CY, AC; CY = Ø, AC = 1

Format: E6, Data

The AND functions can be used to set selected bits of the Accumulator to 'Ø'. The mask byte should have Øs in the bit positions to be set to 'Ø'.

Example

ANI $Ø6_H$; set bit 7, 6, 5, 4, 3, Ø to 'Ø' regardless of previous A contents

ORA X; OR (A) with (X), X = r or M where M = (H)(L)

Function: (A) ← (A) ∨ (X), bit 'i' of (A) = bit i of A (OR) bit 'i' of (X)
i = 0, . . . , 7

Op Code: 1Ø11Ø SRC for X = r

$B6_H$ for X = M

Flags: Z, S, P, CY, AC; CY = AC = Ø

Format: Single byte instruction

ORI Data; OR Immediate

Function: (A) ← (A) ∨ Data

Op Code: $F6_H$

Flags: Z, S, P, CY, AC; CY = AC = Ø

Format: F6, Data

The OR instructions can be used to set selected bits of the Accumulator to 1. This can be achieved by choosing a mask byte with 1s in the positions which we want to set to 1.

Example

ORI Ø2H; set bit 1 of accumulator to 1, rest unchanged

XRA X; Ex-OR (A) with (X)

X = r or M where M = (H)(L)

Function: (A) ← (A) ⊻ (X) bit i of (A) = bit i of (A) (EX-OR) bit i of (X);
i = Ø, . . . , 7

Op Code: 1Ø1Ø1 SRC for X = r

AE_H for X = M

Flags: Z, S, P, CY, AC; CY = AC = Ø

Format: Single byte instruction

XRI Data; Ex-OR immediate data

Function: (A) ← (A) ⊻ Data

Op Code: EE_H

Flags: Z, S, P, CY, AC; CY = AC = Ø

Format: EE_H, Data

The Ex-OR function $C = A \oplus B = A\bar{B} + \bar{A}B$ is used to selectively toggle Boolean variables—i.e. the selected bit is made a Ø if it was previously 1 and vice versa. The byte containing the variable to be complemented is Ex-ORed with a mask byte. The mask contains a 1 in the bit locations containing the Boolean variable to be toggled.

Example

XRI 82H; complement bit 2 and 7 of
; accumulator, other bits are
; unaffected.

CMP X; Compare (X) with (A)

X = r or M

where M = (H)(L)

Function: Calculates (A) − (X) without changing either A or X

$Z = 1, CY = \emptyset$ if $(A) = (X)$

$Z = \emptyset, CY = 1$ if $(A) < (X)$

$Z = \emptyset, CY = \emptyset$ if $(A) > (X)$

Op Code: 1∅111 SRC if X = r

BE_H if X = M

Flags: Z, S, P, CY, AC

Format: Single byte instruction

CPI Data; Compare immediate

Function: $Z = 1, CY = \emptyset$ if (A) = Data

$Z = \emptyset, CY = 1$ if (A) < Data

$Z = \emptyset, CY = \emptyset$ if (A) > Data

Op Code: FE_H

Flags: Z, S, P, CY, AC

Format: FE_H, Data

If the compare instructions are being used for unsigned or positive operands, the value of the CY flag is what was described earlier. Additional care must be taken in case the operands are signed quantities. To see this assume that $\tilde{A}$ = contents of A and $\tilde{B}$ is the contents of the other operand B. Then consider the following cases after executing a compare (CMP, CPI) instruction.

(i) $\tilde{A}$ and $\tilde{B}$ have same sign

$$CY = 1 \text{ if } \tilde{A} < \tilde{B}$$
$$= \emptyset \text{ if } \tilde{A} \geqslant \tilde{B}$$

(ii) A and B have different signs

$$CY = 1 \text{ if } \tilde{A} \geqslant \tilde{B}$$
$$= \emptyset \text{ if } \tilde{A} < \tilde{B}$$

Note that the sign bit in this case indicates the sign of the $(\tilde{A} - \tilde{B})$ result. These compare instructions usually precede a conditional instruction, where the execution of the latter depends on the status of the FLAG bits set by the compare instruction.

RLC; Rotate left

Function: $(A_{n+1}) \leftarrow (A_n)$, $(A_0) \leftarrow (A_7)$, $CY \leftarrow (A7)$

Op Code: $Ø7_H$

Flag: CY

Format: Single byte instruction

RRC; Rotate Right

Function: $(A_n) \leftarrow (A_{n+1})$, $(A_7) \leftarrow (A_0)$, $CY \leftarrow (A_0)$

Op Code: $ØF_H$

Flag: CY

Format: Single byte instruction

RAL; Rotate left through carry

Function: $(A_{n+1}) \leftarrow (A_n)$, $CY \leftarrow (A_7)$, $(A_0) \leftarrow CY$

Op Code: 17_H

Flags: CY

Format: Single byte instruction

RAR; Rotate right through carry

Function: $(A_n) \leftarrow (A_{n+1})$, $CY \leftarrow (A_0)$, $(A_7) \leftarrow CY$

Op Code: $1F_H$

Flags: CY

Format: Single byte instruction

Note the difference between the simple rotate and the rotate through carry instructions. In the rotate through carry instructions, we are actually operating with a 9-bit shift register—one CY and $A_7 - A_0$. In the other case, after the shift, the CY flag merely reflects the contents of A_7 (rotate left) or A_0 (rotate right).

CMA; Complement accumulator

Function: $(A_i) \leftarrow (\bar{A}_i) \quad i = 0, \ldots, 7$

Op Code: $2F_H$

Flags: None

Format: Single byte instruction

CMC; Complement carry

Function: $CY \leftarrow \overline{CY}$

Op Code: $3F_H$

Flags: CY

Format: Single byte instruction

STC; Set carry

Function: CY ← 1

Op Code: 37_H

Flags: CY

Format: Single byte instruction

Note that there are no specific instructions that reset CY (CY = Ø). This can be done using other instructions e.g.

STC } or ORA A if { Accumulator contents
CMC }   { cannot be changed

or by

XRA A if Accumulator contents can be changed (actually A = $ØØ_H$ after the operation)

8.5 BRANCH GROUP

This group of instructions is used to alter the normal sequential program flow and force the program (may be conditionally) to proceed from a different point. No FLAGS are affected.

Branch instructions can be of two types—conditional or unconditional. Unconditional branch instructions cause the program to branch to the indicated instruction whenever it is encountered. Conditional branch instructions, on the other hand, cause a branching to occur only if a tested condition (i.e. the status of a Condition FLAG) is found to be true. The 8 conditions that are tested (and their binary representation used here) are given below:

⟨cond⟩ = Condition		CND
NZ	(Z = Ø)	ØØØ
Z	(Z = 1)	ØØ1
NC	(CY = Ø)	Ø1Ø
C	(CY – 1)	Ø11
PO	Parity odd (P = Ø)	1ØØ
PE	Parity even (P = 1)	1Ø1
P	Positive (S = Ø)	11Ø
M	minus (S = 1)	111

It is important to keep in mind the interpretation of these Condition FLAGS and the instructions that do affect them. For example, consider the situation where somewhere in a program we have a compare instruction and that we want to branch to a different program sequence if the result of this test is Z = Ø. We do not have to introduce our 'Jump On Zero' instruction right after the compare instruction provided the instructions interposed between these two instructions do not affect the Zero FLAG. If such instructions do have to come in, special programming considerations will have to be taken into account (use PUSH PSW and POP PSW to save and restore flags respectively).

The 8085A supports four kinds of branching instructions. These are

(i) Jump type

(ii) Call type
(iii) Return type
(iv) RST type

The Jump type instructions are the simplest. All they do is change the PC contents to that of the new instruction. Call and Return type instructions are used to implement subroutines—Call to branch to a subroutine and Return to come back to the calling program from the subroutine. The RST instructions are basically subroutine call instructions where the address of the subroutine to which the program should branch is calculated in a special way.

JMP Addr; Jump Unconditional

Function: (PC) ← Addr (16-bit)
Transfer control to instruction pointed to by Addr

Op Code: C3

Format: C3, Low order byte of Addr, High order byte of Addr

J⟨cond⟩ Addr; Conditional Jump to Addr only if tested condition is true

Function: (PC) ← Addr if and only if tested condition (CND) is true (otherwise continue sequentially)

Op Code: 11 CND Ø1Ø

Format: 11 CND Ø1Ø, Low order byte of Addr, High order byte of Addr

CALL Addr; Call unconditional

Function: ((SP) − 1) ← (PCH)
((SP) − 2) ← (PCL)
(SP) ← (SP) − 2
(PC) ← Addr

Store return address at the top of the stack, decrement the Stack Pointer by 2 and jump to instruction at Addr.

Op Code: CD_H

Format: CD, Low order byte of Addr, High order byte of Addr

C⟨cond⟩ Addr; Call conditional

Function: If the specified condition (CND) is true—
((SP) − 1) ← (PCH)
((SP) − 2) ← (PCL)
(SP) ← (SP) − 2
(PC) ← Addr
Otherwise, continue sequentially

Op Code: 11 CND 1ØØ

Format: 11 CND 1ØØ, Low order byte of Addr, High order byte of Addr

RET; Return unconditional

Function: (PCL) ← ((SP))
(PCH) ← ((SP) + 1)
(SP) ← (SP) + 2
Retrieve the return address to the calling program from the top of the stack, increment the stack pointer and jump to that return address

Op Code: $C9_H$

Format: Single byte instruction

R<cond>; Return conditional

Function: If tested condition (CND) is true
(PCL) ← ((SP))
(PCH) ← ((SP) + 1)
(SP) ← (SP) + 2
Otherwise, continue sequentially
Essentially executes a RET instruction if tested condition is true

Op Code: 11 CND ØØØ

Format: Single byte instruction

RST n; Restart with parameter n, n = 0, 1, . . . , 7

Function: ((SP) − 1) ← (PCH)
((SP) − 2) ← (PCL)
(SP) ← (SP) − 2
(PC) ← 8 * (n)
Execute a Call instruction to location
8 * (n) e.g. RST 2 ≡ CALL ØØ1ØH
(Call to location ØØ1ØH)

Op Code: 11××× 111 ××× = 3-bit representation of n

Format: Single byte instruction

Note that RST n is really a short form (1 byte compared to 3 bytes) of a CALL instruction which provides a limited addressing capability. RST n is typically used in conjunction with Interrupts. Interrupts are actions taken by a device to get attention from the CPU. If the CPU acknowledges its Interrupt Request, the requesting device may provide an RST n instruction to the CPU. The location 8 * n in that case will typically have a Jump instruction to the program sequence which contains the service routine needed to attend to the device. (More about interrupts will be said later in Chapter 12). It can be easily seen that only a maximum of 8 devices can be directly controlled by the 8085A using the Interrupt feature and the appropriate RST n instruction. It can also be seen that if the service routine servicing the RST n instruction needs only 8 bytes, one does not have to do the kind of Jump instruction approach described earlier. However, that would be rather rare in practice. In most applications the RST n instruction actually points to a

'Jump Address Table' (at locations ØØØØH, ØØØ8H, ØØ1ØH, ØØ18H, ØØ2ØH, ØØ28H, ØØØ3H, ØØ38H)—the service routine is actually accessed via a Jump instruction appropriately placed in this 'Jump Address Table'. The 8085A also supports other kinds of Interrupts. These are described in greater detail in Chapter 12.

PCHL; Jump to (H)(L) location, Register Indirect Jump Instruction

Function: (PCH) ← (H)

(PCL) ← (L)

Jump to address contained in H-L register

Op Code: $E9_H$

Format: Single byte instruction

It is interesting to note that this is the only instruction available in the 8085A which allows the user to obtain a Jump address from a Register. All other branch instructions (except RET, Rcondition) use immediate addressing. (The Return instructions do use an Indirect mode to obtain the proper jump address). This paucity of instructions using a Register Addressing mode to obtain jump addresses is a serious shortcoming of the 8085's instruction set.

Another point worth noting is that (except for the rather special RST n instructions) all the branch instructions in the 8085's instruction set allow the user to make jumps from anywhere in the 65,536 byte address space to anywhere else. Some microprocessors also provide instructions which allow limited jumps (e.g. the Jump Address should have the same $A_{15} - A_8$ as the location of the current address so that the Jump Address is obtained as a ± 128 displacement from the current address). These limited jump instructions typically use less bytes than the 3 bytes required by the Jump and Call instructions of the 8085A (For our example above only 2-bytes are needed. The first byte is the Op Code whereas the second byte contains either $A_7 - A_\emptyset$ or the ± 128 displacement value.) It is easy to see that the addressing range of such an instruction (if available) would be limited.

8.6 STACK, I/O AND MACHINE CONTROL GROUP

These operations perform I/O, manipulate the stack as well as provide the user the ability to alter internal control flags. No condition FLAG is affected by any of these instructions (except the POP PSW instruction).

Regarding stack manipulation operations in this group (as well as Call and Return type instructions in the previous group), it should be noted that it is the user's responsibility to define some designated RAM spaces at the stack area and to appropriately initialize the stack pointer as the *highest* address in this RAM space (probably via a LXI SP, XXXX instruction placed in the initialization sequence). Unless this is done, the program should not use any instruction which needs the stack (erroneous results will, otherwise, occur).

PUSH r16; Push contents of r16 into stack

Function: ((SP) − 1) ← (rh)

((SP) − 2) ← (rl)

(SP) ← (SP) − 2

Higher order register byte is stored on the top of the stack, at (SP) − 1;

lower order register byte is stored at (SP) − 2 location. The stack pointer is decremented by 2 to point to the current top of the stack

r16 = SP cannot be specified

Op Code: 11 RR Ø1Ø1

Format: Single byte instruction

POP r16; Pop contents from the top of the stack into rp, r16 = SP cannot be specified

Function: (rl) ← ((SP))

(rh) ← (SP) + 1

(SP) ← (SP) + 2

This instruction does exactly the opposite of the POP r16 instruction

Op Code: 11 RR ØØØ1

Format: Single byte instruction

PUSH and POP instructions typically go in pairs. They can be used to store data in the stack (16 bits per instruction) and retrieve them as needed. Judiciously used, these instructions can also help in passing parameters between different portions of the program or between a calling program and a subroutine. They can also be used to store jump addresses and recover them as needed. There should usually be a POP associated with every PUSH or (as mentioned before) special care is needed to see that the stack stays justified during program execution.

PUSH PSW; save Accumulator contents and FLAG register contents (Program Status Word)

Function: ((SP) − 1) ← (A)

((SP) − 2) ← *S Z X AC X P X CY*

X = undefined

(SP) ← (SP) − 2

Op Code: $F5_H$

Format: Single byte instruction

POP PSW; pop processor status word from top of stack to Accumulator and FLAG register

Function: *S Z X AC X P X CY* ← ((SP))

X = undefined

(A) ← ((SP) + 1)

(SP) ← (SP) + 2

Op Code: $F1_H$

Flags: Z, S, P, CY, AC (obviously!)

Format: Single byte instruction

POP PSW is the only instruction in the instruction set which lets the user directly access the contents of the FLAG register. A combination of PUSH PSW and POP PSW can be used to set the FLAG register contents to whatever one desires.

The PUSH PSW and POP PSW instructions are also useful in handling subroutines. Suppose we want to ensure that when the program returns from a subroutine, the Processor Status Word should be the same as when the subroutine was called. This can be easily achieved by putting a PUSH PSW instruction in the beginning of the subroutine (or before the Call instruction) and a POP PSW just before the return instruction (or as the first instruction when one returns to the calling program). Note that the technique given in brackets will be acceptable for standard subroutines but not for Interrupts since Interrupts can occur at any point in a program's execution! Even when there are no subroutines or interrupts, an appropriate combination of the PUSH PSW and POP PSW instructions can be used to restore the processor status word to the value it had at an earlier stage in the program's execution.

NOP; no operations

Function: processor idle during this instruction, i.e. no operation performed, no registers affected

Op Code: $ØØ_H$

Format: Single byte instruction

NOP has two main uses. It is frequently used in delay loops to count out precise delay intervals. It is also used to let slower peripheral devices or slave processors catch up with the 8085A CPU.

XTHL; exchange the contents at the top of the stack with the contents of H-L

Function: $(L) \leftrightarrow ((SP))$

$(H) \leftrightarrow ((SP) + 1)$

(Note contents of SP is not changed)

Op Code: $E3_H$

Format: Single byte instruction

A common use of the XTHL instruction is to modify the Return address from a subroutine depending on the results of the subroutine. (It should be executed before the Return instruction and after the modified Return address is placed in H-L.) It may also be used to access data from the top of the stack without losing the contents of H-L.

SPHL; Move H-L to SP

Function: $(SP) \leftarrow (H)(L)$

Op Code: $F9_H$

Format: Single byte instruction

Like the "LXI SP, Addr" instruction SPHL may be used to initialize the stack pointer. The advantage here is that while the LXI SP, XXXX instruction will only initialize the SP to a predetermined constant, the SPHL instruction will allow the user to set SP to any value calculated in the H-L registers under program control. (It is also possible to use this instruction to implement multiple stacks in memory and choose the stack to be used under program control.)

IN Port; input data from the I/O device designated by Port

Function: (A) ← Data from Port

Port = $ØØ_H$ to FF_H

Op Code: DB_H

Format: DB, Port

OUT Port; Output data to the I/O device designated by Port

Function: (A) → Data to Port Port = $ØØ_H$ to FF_H

Op Code: $D3_H$

Format: D3, Port

The IN and OUT instructions are used to respectively read data from an I/O Port or write data into an I/O Port. The second byte of the instruction can be used to identify one of 256 I/O ports to be accessed. (Note that 'port' in all of the above may also imply an external hardware register having $\overline{RD}$ and $\overline{WR}$ controls and a $\overline{CS}$ Select input.)

As mentioned earlier, the data appearing on the system's data bus can be data, instruction op code or I/O port address. It is the responsibility of the CPU and the program to identify the data bus contents correctly. The 8085A CPU outputs a signal IO/$\overline{M}$ which is 1 when an I/O Port is being accessed and Ø when Memory is being accessed. This can be used to distinguish between I/O and Memory READ and WRITE operations. If this (distinguishing between I/O and Memory) is indeed done the system is said to employ *Standard I/O or I/O Mapped I/O*. If this distinction is not made (IO/$\overline{M}$ is not used), the system will not distinguish between I/O and Memory. An I/O port will then be treated as just another Memory location. This is referred to as *Memory Mapped I/O*. Microprocessor system designers often make a distinction between these two techniques. While designing a system it is generally worthwhile to examine the advantages and disadvantages of using one or both of these techniques (see also Chapter 10 and Sec. 7.3).

EI; Enable Interrupts

Function: After EI is executed, the 8085A will recognize external interrupt requests except those that are masked

Op Code: FB_H

Format: Single byte instruction

DI; Disable Interrupts

Function: After DI is executed, the 8085A will not recognize external interrupt request (except TRAP)

Op Code: $F3_H$

Format: Single byte instruction

Interrupts will also not be recognized while EI or DI is being executed. More on these instructions will be said when the interrupt structure of the 8085A is described in greater detail in Sec. 12.5.

HLT; Halt processor

Function: The processor halts without affecting either registers or flags. No further instructions are executed till an external hardware interrupt is applied (provided an EI has been executed) or if RESET or TRAP inputs are applied

Op Code: 76_H

Format: Single byte instruction

The two remaining instructions present in the instruction set of the 8085A (not available to the 8080) are the RIM and SIM instructions. Before describing these instructions, the Hardware Interrupts available on the 8085A and its Serial I/O feature needs to be briefly described.

The 8085A has three inputs, RST 5.5, RST 6.5 and RST 7.5, which are referred to as Vectored interrupts. A proper input on these 8085A inputs will cause the program being executed by the CPU to branch to some known locations i.e.

$ØØ2C_H$ for RST 5.5

$ØØ34_H$ for RST 6.5

$ØØ3C_H$ for RST 7.5

(These locations should have the first instruction of the corresponding service routine). The RST 5.5 and 6.5 inputs are level sensitive whereas RST 7.5 is edge sensitive. The RST 5.5 and 6.5 inputs will have to be held high till the 8085A acknowledges the interrupt. A $Ø \rightarrow 1$ transition of RST 7.5 sets an internal flip-flop, so this input can then be lowered (i.e. a $1 \rightarrow Ø$ transition) without losing the interrupt request. The 8085A also has an INTR interrupt.

The 8085A also provides one *non maskable* vectored interrupt called TRAP. A 1 at the TRAP input after a $Ø \rightarrow 1$ edge causes the program to branch to ØØ24H.

The 8085A also provides some primitive on-chip facility for Serial I/O via the SID and SOD lines. The SID line can be used to input serial data to the 8085A while the SOD line outputs serial data from the 8085A. Inputting and outputting these bits are also achieved using the RIM and SIM instructions.

SIM; Set Interrupt Mask

Function: The SIM instruction uses the contents of the accumulator (A) to perform the following functions.

(i) Program the interrupt mask for RST 7.5, 6.5 and 5.5. A '1' in $A_Ø$, A_1 and A_2 disables RST 5.5, 6.5 and 7.5 provided the 'Mast Set Enable'—MSE—bit A_3 is also 1. If A_3 is 'Ø' when the SIM instruction is executed, the interrupt mask is not changed.

(ii) Recall that RST 7.5 is edge sensitive (LO → HI). A pulse at the RST 7.5 sets an internal RST 7.5 flip-flop. This is the reason why the RST 7.5 can be requested by a pulse at the corresponding input. If interrupts are disabled at the time the RST 7.5 pulse occurs, this input will still be recognized later (when interrupts are enabled) since the RST 7.5 flip-flop has already been set by the pulse. This flip-flop can be reset by keeping $A_4 = 1$ when SIM is executed. (We may want to do this if we do not want to service an earlier RST 7-5 interrupt.)

Note that the *RST 7.5* flip-flop is also cleared by a RESET IN input or when the CPU acknowledges a RST 7.5 interrupt. All three mask bits (A_2, A_1, A_0) are also set by $\overline{\text{RESET IN}}$, i.e. all vectored interrupts (except TRAP) are masked and therefore disabled.

(iii) Load the SOD (Serial Output Data) latch.

The SOD output pin of the 8085A shows the status of a 1-bit output port. Execution of the SIM instruction sets the output port contents to that of A_7 provided the Serial Output Enable (SOE) bit, A_6, is also 1. If SOE = A_6 = Ø, the contents of the output port is unaffected. SOD is reset by the $\overline{\text{RESET IN}}$ input.

Using SOE (A_6) and MSE (A_3) properly one can use the SIM instruction in any of the following ways:

(a) Change Interrupt Mask and SOD

(SOE = 1, MSE = 1)

(b) Change SOD but not Interrupt Mask

(SOE = 1, MSE = Ø)

(c) Change Interrupt Mask but not SOD

(SOE = Ø, MSE = 1)

(d) Do not change either Interrupt Mask or SOD

(SOE = Ø, MSE = Ø)

This, for example, can be used if we just want to reset RST 7.5 flip-flop without affecting Interrupt Mask or SOD. Before SIM is executed, content of A should be—

A_7 = SOD	A_3 = MSE (enable = 1)
A_6 = SOE (enable = 1)	A_2 = Mask (disable) RST 7.5 if 1
A_5 = Undefined	A_1 = Mask (disable) RST 6.5 if 1
A_4 = Reset RST 7.5 flip-flop if 1	$A_Ø$ = Mask (disable) RST 5.5 if 1

Op Code: $3Ø_H$

Format: Single byte instruction (operates on contents of A before SIM is executed)

RIM: Reset Interrupt Mask

Function: The RIM instruction loads A with information on interrupts and the data at the Serial Input Data (SID) pin. After RIM is executed the contents of A are as follows:

(i) If RST 7.5, 6.5, or 5.5 is currently disabled, A_2, A_1, or $A_Ø$ are respectively 1 ('Ø' if they are enabled).

(ii) If interrupts are currently enabled IE = A_3 = 1, except following a TRAP interrupt.

Following a TRAP interrupt, IE = A_3 reports the status of interrupts ($\overline{\text{enabled/disabled}}$) prior to the TRAP interrupt.

This is useful to retrieve current interrupt status following TRAP

(This is important since TRAP is a nonmaskable interrupt which can happen at any time).

(iii) Hardware interrupts pending (received but not serviced) on RST 5.5, 6.5, and 7.5. A 1 at A_6, A_5, and A_4 respectively indicate that RST 7.5, 6.5 or 5.5 interrupts are pending.

(iv) Transfer the bit present at 8085A's SID input to A_7.

After executing RIM, A contains

$A_7 = $ SID

$A_6 = $ I 7.5
$A_5 = $ I 6.5 } 1 indicates corresponding interrupt is pending
$A_4 = $ I 5.5

$A_3 = $ IE: 1 indicates interrupts enabled at present (note exception for TRAP interrupt)

$A_2 = $ M 7.5
$A_1 = $ M 6.5 } 1 indicates corresponding interrupt is currently disabled (masked)
$A_\emptyset = $ M 5.5

Apart from letting the CPU get the SID input, the RIM instruction is primarily used to monitor interrupt status, e.g.

(a) Monitor whether or not an interrupt is pending without actually servicing it.

(b) Check, using IE properly, if CPU is currently servicing an interrupt.

(c) The 8085A gives RST 7.5 the highest priority (among RST 7.5, 6.5 and 5.5) and RST 5.5 the lowest. By properly using RIM and SIM one can design any other priority structure under program control. (See Sec. 12.4)

Op Code: $2\emptyset_H$

Format: Single byte instruction (outputs data in A after RIM is executed).

8.7 Unannounced OP Codes of the 8085A

If one examines the instruction set of the 8085A, one finds that out of the 256 possible op codes with 8 bits, Intel announces only 246; there are no announced instructions corresponding to the 10 'missing op codes'. Users have since reported that the 8085A does have instructions corresponding to these missing op codes. These are given below.

(*Note*: The user should be careful when using these instructions. Intel does not guarantee in any way that these instructions will indeed function!)

The FLAGs specified by Intel are S, Z, AC, P and CY located in the FLAG register as

$$F_7 = S, F_6 = Z, F_4 = AC, F_2 = P, F_0 = CY$$

Users have reported that 2 (out of the 3) missing FLAG bits F_5 and F_1 actually have some meaning.

F_1 is 1 if Overflow occurs in 2's complement addition/subtraction/DCX/INX for 8 and 16-bit arithmetic operation

$F_5 = S1 \cdot S2 + S1 \cdot R + S2 \cdot R$

where S1 = Sign of operand 1

S2 = Sign of operand 2

R = Sign of result

(for subtraction and comparison replace S2 with $\overline{S2}$ where operand 2 is the subtrehand, i.e. Result = Opd. 1 − Opd. 2)

The new instructions are (arbitrary mnemonics used)

DSB; double subtraction

Op Code $\varnothing 8_H$

Function $(H)(L) \leftarrow (H)(L) - (B)(C)$

Flags Z, S, P, CY, AC, F1, F5

Format Single byte instruction

ASHL; Arithmetic shift of H-L to the right

Op Code $1\varnothing_H$

Function $(H_7) \leftarrow (H_7), (H_{n-1}) \leftarrow (H_n)$

$(L_7) \leftarrow (H_0), (L_{n-1}) \leftarrow (L_n)$

$CY \leftarrow (L_0)$

Flags CY

Format Single byte instruction

RDLC; Rotate D-E left through carry

Op Code 18_H

Function $(D_{n+1}) \leftarrow (D_n), (D_0) \leftarrow (E_7)$

$CY \leftarrow (D_7), (E_{n+1}) \leftarrow (E_n), (E_0) \leftarrow CY$

Flags CY, F1

Format Single byte instruction

DADI Data; Load D-E with H-L plus immediate data

Op Code 28_H

Function $(D)(E) \leftarrow (H)(L) + \text{Data}$

Flags None

Format 28, Data

DSPI Data; Load D-E with SP plus Data

Op Code 38_H

Function $(D)(E) \leftarrow (SP) + \text{Data}$

	Flags	None
	Format	38_H, Data
RSTOV;	restart on overflow	
	Op Code	CB_H
	Function	If F1 = 0, continue sequentially
		If F1 = 1
		((SP) − 1) ← (PCH)
		((SP) − 2) ← (PCL)
		(SP) ← (SP) − 2
		(PC) ← $ØØ4Ø_H$
		(If F1 = 1 push PC on stack, increment SP and JUMP to $4Ø_H$; otherwise continue sequentially)
	Flags	None
	Format	Single byte instruction
SHIX;	store (H) and (L) indirectly through D-E	
	Op Code	$D9_H$
	Function	((D)(E)) ← (L)
		((D)(E) + 1) ← (H)
	Flags	None
	Format	Single byte instruction
LHIX;	Load H and L indirectly through D and E	
	Op Code	ED_H
	Function	(L) ← ((D)(E)), (H) ← ((D)(E) + 1)
	Flags	None
	Format	Single byte instruction
JNF5;	Addr;	Jump to Addr if F5 = 0
	Op Code	$DØ_H$
	Function	If F5 = 1 proceed sequentially
		If F5 = Ø, (PC) ← Addr
	Flags	None
	Format	DØ, Lo-order byte of Addr, Hi-order byte of Addr
JF5	Addr;	Jump to Addr if F5 = 1
	Op Code	FD_H

Function	If F5 = Ø, proceed sequentially
	If F5 = 1, jump to Addr
Flags	None
Format	FD, Lo-order Addr byte, Hi-order Addr byte

Once again the reader is warned to use these instructions with due caution since they are not guaranteed by Intel. Also they should only be used at the machine level since typical Assemblers, Cross-assemblers, Simulators etc. will not support these instructions.

8.8 Basic Assembler Directives (Pseudo Instructions)

In the previous sections, we have presented the instruction set of the 8085A. For convenience, we have given the Assembly Language mnemonics of the various instructions. In order to execute a program, the processor would require the program to be in machine code, i.e. a sequence of bytes each containing 1s and Øs.

It is difficult for the user to write his program directly in machine code. It is more common to find the user writing his program in Assembly Language and then using an Assembler program to translate the Assembly Language program into machine code. This Assembler program also allows certain other conveniences to the user. In this section, we will present the highlights of a typical Assembler to show some of the basic features an Assembler provides.

The Assembler typically requires the user to write his program in a particular way, e.g.

Label	Op Code Mnemonic	Operand	Comment
MOVE:	MVI	A, ØØH;	Clear A

Label: The optional Label field serves to identify the lines in an Assembly Language program. After Assembly, Label contains the value (2 bytes) of the memory location of the first byte (i.e. the Op Code) of the assembled code corresponding to that line. If that Label appears in the Operand field elsewhere (as an Address or a double byte constant), this will be the value the Operand will have. If the Label is present, one must use an appropriate delimiter after it. We have used ':' as the delimiter.

Op Code Mnemonic: This is the Assembly Language Mnemonic corresponding to the 8085A instruction, e.g. MOV, JMP, MVI etc. Delimiter between Op Code Mnemonic and Operand is shown to be a blank (or space).

Operand Field: This contains the information to be used in conjunction with the Op Code field in order to define precisely the operation to be done by that Op Code. Depending on the instruction, the operand field may be absent, may contain one item or two items separated by a comma (,). It may also contain Labels defined elsewhere. The operands may also contain numbers where one needs to specify whether the number is hexadecimal (H), decimal (D) or binary (B), e.g.

1010 0011 B

A3H

96D

etc.

Most Assemblers require that the first character of a hexadecimal number should not be A, B, C, D, E or F, e.g. A3H should be written as ØA3H. If the operand field is followed by a comment field, one must use a delimiter between the two. We have used ';' as this delimiter.

Comment: Anything following ';' on a line is considered to be a comment. This optional field usually contains explanatory remarks on what that Assembly Program line does.

One advantage of using an Assembler and writing in Assembly Language is that memory locations can be referred to by names representing their addresses rather than by cumbersome hexadecimal codes. Consider the following program where START is assumed to denote ØØ4ØH.

```
ØØ4Ø   START:   MOV   A, B      START = ØØ4ØH
ØØ41            ADD   C         STAR2 = ØØ45H
ØØ42            JC    LOOP      LOOP  = Ø1ØØH
ØØ45   STAR2:   ADD   D
ØØ46            JC    START
 :                :
 :                :
Ø1ØØ   LOOP:    JMP   STAR2
```

After Assembly this would look like

```
ØØ4Ø   78   MOV A, B
ØØ41   81   ADD C
ØØ42   DA   JC Ø1ØØ H
ØØ43   ØØ
ØØ44   Ø1
ØØ45   82   ADD D
ØØ46   DA   JC ØØ4Ø H
ØØ47   4Ø
ØØ48   ØØ
 :
 :
Ø1ØØ   C3   JMP ØØ45 H
Ø1Ø1   45
Ø1Ø2   ØØ
```

The Assembler here has been shown to substitute the correct addresses for the labels it encountered in the program.

There should be some facility available to the user so that he can tell the Assembler what should be the location of the first memory address or the memory addresses of locations not in

---

numerical order. (The latter case can arise in many situations, e.g. we want to locate different program segments in nonconsecutive memory locations or we want to define our RAM area in a location not consecutive after our ROM area). The ORGN pseudo instruction allows us to do this.

There is no machine code for the ORGN instruction; it is actually a command for the assembler. It tells the assembler that the first byte of machine code generated for the program segment following the ORGN statement should be placed at the location which is the operand of the ORGN instruction, e.g. in the earlier example, we can ensure that START = ØØ4ØH and LOOP = Ø1ØØH by doing the following:

```
        ORGN    ØØ4ØH
START:  MOV     A, B
          :
          .
        ORGN    Ø1ØØH
LOOP:   JMP     STAR2
```

(An ORGN may also precede a DFB, DFW or DFS statement described later. In these cases, the location of the first byte of storage or variable 8-bit quantity or 8-bit constant is the same as the operand of the ORGN pseudo instruction. These instructions are discussed later).

Some other useful Assembler pseudo-instructions typically available are given below. Note that they are all commands to the assembler and are not translated (nor, for that matter, are they translatable) to machine code. They are basically conveniences meant to make writing the program in Assembly Language easier for the user.

EQ

```
NAME    EQ    EXPRESSION
```

The assembler evaluates EXPRESSION and substitutes that value every time NAME appears in the program. The EQ instructions are usually written at the beginning of the program. The advantage of using the EQ statement is evident when one wants to change the value of NAME. Without EQ, one will have to look for every occurrence of NAME in the program and change each appropriately. Using EQ, one merely has to change the EQ expression, the assembler takes care of the rest. EQ can also be used to define the value of a constant.

END

```
END;    Program End
```

This tells the assembler that the end of the program has been reached and that there are no more instructions to be assembled.

DFB (Define byte)

```
OPER1:    DFB    LIST
```

LIST is a list of either

(i) Arithmetic or Logical expressions evaluating to one or more 8-bit data quantities

(ii) Strings of ASCII characters enclosed appropriately, e.g. in Quotes such as 'AX'

The 8-bit value of each expression or the 8-bit ASCII representation for each character is stored in the next available memory byte starting with the byte address of OPER1. Note that OPLAB should be in ROM area; in other words, DFB will not load a constant into RAM as that is not a job which can be done at assembly time. The usefulness of DFB is to define constants or reserve and label RAM locations. (The latter is better done via the DFS instruction).

For examples

```
HERE:     DFB   ØA3H          ; ØA3
WORD1:    DFB   5*2, 2FH-ØAH  ; ØA, 25
```

DFW (Define word)

```
OPER2:    DFW   LIST
```

DFW is identical to DFB except that it operates on 16-bit data quantities. LIST has the same meaning as above. The significant 8-bits of the first word of LIST are stored at OPER2 and the most significant 8-bits are stored at OPER2 + 1; successive words are stored in successive address locations.

Typically, DFB is used to specify data (8-bit) constants whereas DFW is used to specify addresses (16-bits), e.g.,

```
ADD3:     DFW   3CØ1H, 3CAEH, Ø13CH, ØAE3CH
```

DFS (Define Storage)

```
OPER2:    DFS   EXP
```

'EXP' is a single arithmetic or logical expression which can be evaluated at assembly time to generate a value in the range ØØØØH to ØFFFFH. The DFS instructions tells the assembler to reserve memory locations OPER2 to OPER2 + EXP -- 1 for data storage. (These should be RAM locations). Note that the assembler does not initialize the contents of these locations in any way.

The reader should note that the assembler features described are what is generally provided by such a routine. A specific assembler may require a different syntax and may or may not provide all these features. The reader is advised to examine carefully the appropriate manuals before actually using a specific assembler.

PROBLEMS

1. A table of data is stored in memory starting at location 2ØØ3H. The length of the table (number of bytes) is stored in location 2ØØØH. Write a subroutine SERCH which finds the maximum and minimum data bytes stored in this table. On return from SERCH location 2ØØ1H should contain the maximum value and 2ØØ2H the minimum. The original contents of the table should not change.
2. A logical shift operation shifts one bit of data out of a register while feeding a Ø as the bit shifted in. How would you implement a 1-bit logical left or right shift using the 8085A? Assume that the Accumulator contains the byte to be shifted.
3. An arithmetic left shift is exactly similar to a logical left shift except that the Sign Bit (the

MSB) does not change. An arithmetic right shift is like a logical right shift except that the Sign Bit stays unchanged and the Sign Bit value is the bit fed into the bit position before the MSB. How would you implement a 1-bit arithmetic left or right shift using the 8085A? Assume that the Accumulator contains the byte to be shifted.

4. For some reason, a designer wants to implement a subroutine JNAC which will cause a jump to (HL) if AC = Ø. Write JNAC such that register contents are not affected by its execution.

5. Without using any of the subtraction instructions of the 8085A write a subroutine DSUB which does

$$(HL) \leftarrow (HL) - (DE)$$

Assume that both operands are signed quantities using 2's Complement Representation. The result should also be in the same representation.

6. (a) Write a subroutine MAXTAB which will search a table of data and locate the maximum data value in the table. When calling MAXTAB, the starting address of the table must be specified in RAM locations TABLE, TABLE+1 and its length (=no. of bytes $\leqslant$FF) in RAM location LENGTH. Without affecting any of the registers, MAXTAB returns the address of the maximum data value (in the table) in RAM locations MAXVAL, MAXVAL + 1. (If multiple entities in the table have the maximum value, MAXTAB returns with the address of only one of them.)
   (b) Using MAXTAB, write a subroutine SORT to sort a table of data in locations 4ØØØH to 4ØFØH (i.e. length = F1H). SORT should rearrange the data in descending order in locations 4ØØØH to 4ØFØH with the maximum data value in 4ØØØH and the minimum data value in 4ØFØH.
   (c) Redo (b) for the case where one wants the data table rearranged in an ascending order.
   (d) Can you suggest a better (faster) sorting algorithm than the one outlined in (b)? If so, write SORT using this algorithm. (*Hint*: Use Bubble-Sort.)

7. (a) A n-bit binary number $b_{n-1} \ldots b_{\emptyset}$ is converted to a n-bit Gray Code number $g_{n-1} \ldots g_{\emptyset}$ using the following

$$g_{n-1} = b_{n-1} \quad \text{for MSB}$$

$$g_i = b_{i+1} \oplus b_i \quad (\emptyset \leqslant i \leqslant n-2) \text{ for non-MSB}$$

   Write a program which converts the 8-bit contents of A-reg to a Gray Code representation with the result in the A-reg.
   (b) A n-bit Gray Code number $g_{n-1} \ldots g_{\emptyset}$ is converted to a n-bit binary number $b_{n-1} \ldots b_{\emptyset}$ using the following algorithm.

$$b_{n-1} = g_{n-1} \quad \text{for MSB}$$

$$b_i = b_{i+1} \oplus g_i \quad (\emptyset \leqslant i \leqslant n-2) \text{ for non-MSB}$$

   Write a program which converts the 8-bit contents of A-reg from Gray Code to binary with the result in the A-reg.
   (c) Check that if the programs in (a) and (b) are executed one after another, the A contents remain unchanged.

8. Consider the following program:

```
                 :
                 :
           LXI   D, DISP;  (DE) = displacement
LABEL:     CALL  PCRJP
           NOP
                 :
                 :
```

Write a subroutine PCRJP such that the RET instruction in that routine forces (PC) to LABEL + DISP + 3. Keep in mind the fact that the CALL PCRJP instruction may be used in a number of places in the program.

(Note that PCRJP allows a 'program relative jump'—an operation not provided otherwise in the 8085A's instruction set.)

9. Write a subroutine SERIAL which when called, outputs serially (LSB first, MSB last) the 8-bit contents of A via SOD. The serial transmission should be preceded by a START bit (=Ø) and should end with a STOP bit (=1). Execution of SERIAL should not affect the register contents, the interrupt masking, the interrupt enable/disable status or the state of the RST 7.5 flip-flop. The interrupts (other than TRAP) should, however, be disabled during the serial transmission. One should also ensure that the SOD output for each bit should be valid for at least 50 cycles of CLK OUT (i.e. 50 T_{CLK} = 50 T-states). SOD = 1 before the transmission starts and after the transmission is over.

10. (a) Write a subroutine TEN, to do the following

$$(HL) \leftarrow (HL) * 10_D$$

without affecting any of the registers.

(b) Using the subroutine TEN, write a subroutine BCDTB to convert a 4-digit BCD number to its binary equivalent. Assume that the decimal digits are stored in RAM locations (DE), . . . , (DE) + 3 with the least significant digit in location (DE). (Note: one digit/location; BCDTB should return with the result in HL.

11. Suppose the DAA instruction were not available on the 8085A. Using the other instructions of the 8085A write a subroutine DCADJ which can be used just like DAA for multidigit decimal additions. DCADJ should (at least) affect the CY and Z flags in the same way as DAA for such additions.

12. How can one do multidigit decimal subtraction with the 8085A? Write a program to illustrate your technique.

13.
```
DELAY:  NOP           ;  enter with (HL) ≠ Ø
DLY1 :  MVI A, ØFFH;  initialize A to FF
DLY2 :  DCR A         ;  decrement A
        NOP           ;  and loop till
        JNZ DLY2      ;  (A) = ØØ
        DCX H         ;  decrement HL
```

(All mnemonics copyright Intel Corporation 1985).

```
JNZ DLY1    ; and loop
RET         ; till (HL) = Ø, then return
```

(a) Correct the above program.

(b) Modify it such that DELAY can be entered with any value of (HL), not necessarily with (HL) $\neq$ Ø.

14. Write a program to multiply two BCD digits using a look-up table. Generate the look-up table. How can one use this to multiply two multi-digit numbers?

15. Give the status of the various FLAGs and contents of A when each one of the following instructions is executed separately. The following are specified prior to the execution of these instructions

(A) = 89H	(B) = 7FH	(C) = 9BH	(H) = 3ØH
(L) = ØØH	(F) = ØØH	(3ØØØ) = 77H	

(i) MOV A, B (ii) MOV A, M (iii) ADD B (iv) ADD M (v) SUB C
(vi) SUB M (vii) ADD M followed by DAA (ix) XRA M

16. A decimal integer in the range 0 to 199 is present in register B in binary form. Write a program CONV to convert this to BCD format. The result should be in A and CY with the MSD (Ø or 1) in CY.

17. In computer applications one often uses a stack to pass parameters between a main program and a subroutine. Can you suggest a way of doing this with the 8085A?

18. A device generates a train of equally spaced pulses which are fed to the RST 7.5 input of an 8085A.

(a) When the pulse repetition rate is high, an approximate estimate of the pulse repetition rate can be formed by counting the number of pulses that appear in a given (fixed) time interval. How would you do this via software alone? Using $f_{crystal}$ = 2 MHz, what will be the maximum pulse rate that you can measure?

(b) When the pulse repetition rate is sufficiently low, it is more useful to make an approximate measurement of the time interval between two pulses. In this case, the measurement is done in units of T sec. where the choice of T is left to the designer. How would you do this via software alone? If $f_{crystal}$ = 2MHz, what is the value of T for your realization?

(c) Repeat (a) using an 8253.

(d) Repeat (b) using an 8253.

For parts (c) and (d), you need not use the RST 7.5 interrupt or you may use it in some other fashion of your choice.

19. Simplify the software of (9) by using an 8253 (and possibly some of the interrupts) to generate the bit timing.

Appendix A8

A8.1 DECIMAL FLOATING-POINT ARITHMETIC WITH THE 8085A, A PROGRAMMING EXAMPLE

In this appendix, we present an assembly language program package for the 8085A which can be used to implement arithmetic operations like addition, subtraction, multiplication and division. Though the programs have been extensively tested, the reader is urged to regard them only as examples and test them prior to use.* Extensive comments have been provided for each program in the package. However, the reader may find the following summary helpful in understanding the package.

A8.1.1 Representation of Numbers

The package assumes that the floating-point operands are stored in 11 consecutive memory locations. The mantissa is represented in 10's complement unpacked BCD format with 8 locations for the magnitude part and one additional location for the sign digit (0 or 9). The exponent occupies one location and is represented in binary excess —128 (e.g. an exponent of —23 is represented as Ø11Ø 1ØØ1 or 69H)

Registers HL and DE are used to represent the operands. They essentially act as pointers to the memory locations containing the floating point number. If POINTER represents the contents of these register-pairs, then

(POINTER) = Exponent in excess —128

(POINTER + 1) to (POINTER + 8) = Magnitude part of Mantissa in unpacked BCD with lowest digit in (POINTER + 1)

(POINTER + 9) = Sign Digit, 0 for positive and 9 for negative, Mantissa is in 10's Complement Representation

(POINTER + 10) = extra location which should be left empty by the user for internal usage of the program.

The mantissa is always assumed to be less than 1.00 . . . 0 in magnitude. The symbols HL^ and DE^ have sometimes been used here to represent the floating point numbers pointed at by HL and DE respectively.

A8.1.2 Floating-Point Addition

To add two numbers HL^ and DE^, the subroutine NORML is first used to make their exponents equal (to the larger of the two values) while doing a sign-extended shift of their mantissae. ADFLP is then called to add the numbers storing the result in the locations pointed at by BC.

BC^ = HL^ + DE^

The subroutine FLADD may be directly called to do the above signed floating-point addition.

*No responsibility is taken or implied for errors or misprints in the program package.

A8.1.3 Floating-Point Subtraction

To do BC^ = HL^ − DE^ with signed numbers, one can directly call FLSUB. This calculates the 10's complement of the mantissa of DE^ before doing essentially the same operations as for addition.

A8.1.4 Floating-Point Multiplication

MLDGT provides a subroutine for multiplying two BCD digits using look-up tables DGMT1 (for the lower digit of the result) and DGMT2 (for the higher digit of the result). This is used by the subroutine NUMLT to multiply a N-digit unsigned BCD number by a single digit and is used in turn by NUMST to multiply a 8 digit number (in the form 0.× × × × × × × ×) by a single digit, giving the result in an exponential floating-point format. These subroutines are used by MPYFL to calculate

BC^ = HL^ * DE^

where all the numbers are assumed to be positive. Finally, signed multiplication BC^ = HL^ * DE^ is done by the subroutine FLMUL which adjusts sign appropriately before and after MPYFL is called to give proper signed results.

A8.1.5 Floating-Point Division

Inversion (i.e. HL^ = 1.0/HL^) is done by the subroutine INVERT. Division is achieved by inverting the divisor and then multiplying it to the dividend. The subroutine FLDIV does this to calculate

BC^ = HL^/DE^

where the numbers are signed quantities.

The subroutine INVERT calculates y = 1/x by means of the recursion formula

$$y_0 = 1.0$$

$$y_{n+1} = y_n * [2 - x * y_n]$$

The iterations are terminated when the error $y_{n+1} - y_n$ becomes sufficiently small.

A8.1.6 The Floating Point Package

```
; * * * * * * * * * * * * * * * * * * * * * * * *
;
; BCD FLOATING POINT ARITHMETIC
;           ROUTINES ENTERED WITH DE & HL
;           (DE) ≫ 8-BIT EXPONENT EXCESS 128 REP.
;                   LSD(1)  TO  MSD(8)
;                   AND SIGN  + =00   − =09
;                           TOTAL = 10 LOCATIONS
;
;           (HL) ≫ SIMILAR TO DE
;
;
; 8085A     NORMALIZE EXPONENTS OF TWO NUMBERS
```

```
NORML:   LDAX D
         CMP M          ; COMPARE EXPONENTS
         RZ             ; RETURN IF EXPONENTS EQUAL
         PUSH B
         PUSH D
         PUSH H         ; SAVE DE AND HL
         JNC NORM1      ; NORM1 IF ((DE)) > ((HL))
         XCHG
         LDAX D
NORM1:   SUB M          ; HERE ((DE)) > ((HL))
         CPI 08H        ; IF ((DE)) > ((HL)) + 8
         JNC NORMZ      ;        GO TO NORMZ
         MOV C, A       ; (C) = ((DE)) - ((HL))
         MVI A, 09H
         SUB C
         MOV B, A       ; (B) = 9 - (C)
         PUSH D
         PUSH H
         INX H
         XCHG
         MVI H, 00H     ; (DE) = OLD(HL) + 1
         MOV L, C       ; (HL) = OLD(HL) + 1 + DIFF IN EXPO
         DAD D
NLOOP:   MOV A, M
         STAX D         ; SHIFT ONE POSITION
         INX D
         INX H
         DCR B
         JNZ NLOOP      ; EXECUTE (B) TIMES
SLOOP:   STAX D         ; (A) = SIGN DIGIT
         INX D
         DCR C          ; ARITHMETIC SHIFT
         JNZ SLOOP      ; EXECUTE (C) TIMES
```

```
         POP H
         POP D
         LDAX D
         MOV M, A        ; FIX EXPONENTS
         POP H
         POP D           ; RESTORE HL AND DE
         POP B
         RET
NORMZ:   LDAX D
         MOV M, A        ; FIX EXPONENTS
         MVI B, 09H      ; FIX EXPONENTS
ZLOOP:   INX H           ; SET DIGITS = 00
         MVI M, 00H
         DCR B
         JNZ ZLOOP
         POP H
         POP D
         POP B
         RET
;
;
; FLOATING POINT ADDITION
;                  SIGNED 10'S COMPLEMENT   8 DIGITS + SIGN
; DE, HL POINT TO OPERAND STORAGE AREA
; BC POINTS TO RESULT STORAGE AREA
; ***11 LOCATIONS IN EACH CASE WITH ONE EXTRA BYTE IN LOCATIONS
;                        (DE) + 0A, (HL) + 0A, (BC) + 0A
; ****ACCURACY = 8 SIG DIGITS GUARANTEED******
ADFLP:   CALL NORML      ; EQUATE OPERAND EXPONENTS
         PUSH B
         PUSH H
         PUSH D          ; HL = X, DE = Y, BC = R     SAVE ON STACK
;                                      STACK = Y]   X, R
```

```
          CALL ADJST     ; INCREASE ONE MSD IN <X>
                           E.G. 0180 TO 00180, 9240 TO 99240
          POP D
          POP H          ; RESTORE DE AND HL REG
          PUSH H
          PUSH D         ; STACK = Y] X, R
          XCHG           ; H = Y   D = X
          CALL ADJST     ; INCREASE ONE MSD IN <Y>
          POP D
          POP H
          PUSH H
          PUSH D
          LDAX D         ; TENTATIVE EXPO OF RESULT = EXPO OF
          STAX B         ;             NORMALIZED OPERAND
          ANA A          ; CY = 0
          MVI A, 0AH     ; LOOP COUNTER = 0A = (LC)
          PUSH PSW       ; STACK = FL], LC, Y, X, R
ADDLP:    INX D          ; DE = Y + 1      POINT TO PROPER
          INX H          ; HL = X + 1      (LEAST) SIG. DIGITS
          INX B
          LDAX D
          ADC M          ; ADD DIGITS AND
          DAA            ;            DECIMAL ADJUST
          PUSH PSW       ; STACK = FL'], A', FL, LC, Y, X, R
          ANI 10H
          JZ LESS0
          POP PSW        ; STACK = FL], LC, Y, X, R
          STC            ; SET CY = 1 TO INDICATE CARRY OUT OF ADDITION
          JMP LESS1
LESS0:    POP PSW        ; STACK = FL], LC, Y, X, R
LESS1:    PUSH PSW       ; STACK = FL'], A', FL, LC, Y, X, R
          ANI 0FH
```

```
        STAX B          ; STORE ADDITION RESULT DIGIT
        POP PSW         ; STACK = FL], LC, Y, X, R
        XTHL            ; L = FL   H = LC      STACK = L], H, Y, X, R
        DCR H           ; LC = LC - 1
        JZ ADOUT        : LC = 0 IF ADDITION OVER
        XTHL            ; RESTORE HL AND STACK      STACK = FL] LC, Y, X, R
        JMP ADDLP       ; REPEAT TILL LC = 0
ADOUT:  XTHL
        POP PSW         ; STACK = Y], X, R
; AT THIS POINT  HL = X + 0A,  DE = Y + 0A,  BC = R + 0A  STACK = Y], X, R
        DCX B           ; BC = R + 09
        LDAX B
        CPI 09H
        JZ ADFIN        ; JUMP TO ADFIN IF
        CPI 00H         ;         THE EXTRA DIGIT INTRODUCED
        JZ ADFIN        ;              WAS NOT NEEDED
        POP D
        POP H
        POP B           ; RESTORE POINTERS
        PUSH B
        PUSH H
        PUSH D          ;         AND STORE BACK IN STACK
; DE=Y HL=X BC=R         STACK = Y] X, R
        MVI D, 09H      ; D = 09 FOR COUNTER
        MOV L, C
        MOV H, B        ; H = R
        INR M           ; INCREMENT EXPO OF RESULT BY 1
        INX H           ; H = R + 1
ANLOP:  INX H
        INX B
        MOV A, M
        STAX B
        DCR D
```

```
          JNZ ANLOP     ; EXECUTE SHIFT TO GET 8-DIGIT ACCURACY
ADFIN:    POP D
          POP H
          POP B
          RET
ADJST:    LXI D, 0009H
          DAD D
          MOV A, M
          INX H
          MOV M, A
          RET
;
;
; TENSC—10'S COMPLEMENT OF N {9} DIGIT BCD NUMBER IN EXPO FORMAT
; ***** IMPORTANT ****** (C) = N {9}
; HL POINTS TO LSD OF N-DIGIT {9} BCD NUMBER IN EXPO FORMAT
; DE POINTS TO LSD OF RESULT BCD NUMBER IN EXPO FORMAT
TENSC:    PUSH PSW
          PUSH B
          PUSH D
          PUSH H        ; SAVE REGISTERS
          MOV A, M
          STAX D        ; SET EXPO OF RESULT
          INX H
          INX D
          MVI A, 0AH    ; SET FIRST MINUEND DIGIT TO 0AH
TENLP:    SUB M         ; SUBTRACT
          ADI 00H       ; DECIMAL ADJUST DIFFERENCE
          DAA
          CALL DCADJ    ; SET CARRY AND ACC PROPERLY
          STAX D        ; SAVE 10'S COMPLEMENTED DIGIT
          DCR C
          JZ TNOUT      ; EXIT IF DONE
```

```
        INX H
        INX D
        MVI A, 09H      ; SET MINUEND DIGIT TO 09H OR
        ACI 00H         ;   0AH IF PREVIOUS DIGIT WAS 00
        JMP TENLP       ; REPEAT TILL DONE
TNOUT:  POP H
        POP D
        POP B
        POP PSW
        RET
DCADJ:  PUSH B
        MOV B, A
        ANI 10H
        JZ DCO          ; NO CARRY
        MOV A, B
        ANI 0FH
        STC
        POP B
        RET
DCO:    MOV A, B
        POP B
        RET
;
; MLDGT  MULTIPLIES TWO BCD DIGITS USING LOOK-UP TABLES DGMT 1
;                AND DGMT 2
;        ONE OPERAND IN B, OTHER OPERAND IN C
;        RETURNS WITH MSD OF PRODUCT IN B AND LSD IN C
;        OTHER THAN REG B-C NO OTHET REG AFFECTED
;                NO RAM REQUIRED
MLDGT:  PUSH PSW
        PUSH H
        MOV A, B
        RLC
```

(All mnemonics copyright Intel Corporation 1985).

```
        RLC
        RLC
        RLC
        ANI 0F0H
        ORA C
        MOV C, A        ; PACK BOTH DIGITS IN C
        MVI B, 00H      ; SET BC TO OFFSET ADDRESS
        LXI H, DGMT2
        DAD B           ; HL = LOOK UP TABLE 2 FOR MSD
        MOV A, M
        LXI H, DGMT1
        DAD B           ; HL = LOOK UP TABLE 1 FOR LSD
        MOV C, M
        MOV B, A
        POP H.
        POP PSW         ; RESTORE PSW AND HL
        RET
;
; * * * * * * * * * * * * * * * * * * * * * * * *
;               NUMLT
;    THIS MULTIPLIES AN N-DIGIT BCD NUMBER WITH A SINGLE DIGIT
; HL = SOURCE ADDRESS       N DIGITS HL POINTING TO LSD
; B  = SINGLE DIGIT NUMBER
; C  = NO. OF DIGITS = N
; DE = DEST ADDRESS         N + 1 DIGITS DE POINTING TO LSD
; RAM REQUIRED IN LOCATIONS WRAM-WRAM + 3
;       REGISTERS UNAFFECTED EXCEPT REG C, SOURCE DATA UNAFFECTED
; ALL OPERANDS AND REG RESULT ASSUMED POSITIVE
;   ON RETURN C = 00 FOR N-BIT RESULT, C = FF FOR (N + 1) BIT RESULT
NMSRC   EQ    WRAM     ; EQ ASSEMBLER PSEUDO-OP OF
NMDST   EQ    WRAM + 2; HYPOTHETICAL ASSEMBLER OF SEC. 8.8
;
NUMLT:  PUSH PSW
        PUSH H
```

```
         PUSH D
         PUSH B          ; SAVE REGISTERS ON STACK
         XCHG
         SHLD NMDST      ; SAVE DE IN NMDST
         XCHG
         MOV C, M        ; GET DIGIT FROM NUM
         INX H
         SHLD NMSRC      ; SAVE HL IN NMSRC
         CALL MLDGT      ; ON CALLING B & C HAVE DIGITS TO BE MULTIPLIED
                           ON RETURN B & C HAVE RESULT
         LHLD NMDST
         MOV M, C        ; SAVE RESULT
         INX H
         MOV M, B
         SHLD NMDST
         POP B
         DCR C
         PUSH B
         JZ NMOUT        ; TEST IF DONE
NMLP:    LHLD NMSRC
         MOV C, M        ; GET NEXT DIGIT
         INX H
         SHLD NMSRC      ; UPDATE NMSRC
         CALL MLDGT      ; MULTIPLY DIGITS
         LHLD NMDST
         MOV A, C        ; ADD LOWER CURRENT DIGIT TO HIGHER
         ADD M           :      PREVIOUS DIGIT
         DAA
         CALL DCADJ      ; ADJUST THE SUM
         MOV M, A        ; STORE RESULT IN MEMORY
         MOV A, B
         ACI 00H         ; ADJUST NEXT DIGIT
         INX H
```

```
         MOV M, A        ; AND STORE IN MEMORY
         SHLD NMDST
         POP B
         DCR C
         PUSH B
         JNZ NMLP
NMOUT:   POP B
         POP D
         LHLD NMDST
         MOV A, M
         CPI 00H
         JZ CSET
         MVI C, OFFH
CSET:    POP H
         POP PSW
         RET
;  * * * * * * * * * * * * * * * * * * * * * * * * *
;   NUMST    MULTIPLY 8 DIGIT NO. WITH ONE DIGIT AND GIVE THE
:     RESULT IN EXPONENT FORMAT    EXPO, 8 DIGITS, SIGN
;       ×××××××× * (B)
;     HL POINTS TO LSD OF ××××××××   POSITIVE NUMBERS ASSUMED
;     DE POINTS TO RESULT IN EXPO FORMAT
:   REGISTERS NOT AFFECTED
NUMST:   PUSH PSW
         PUSH B
         PUSH H
         PUSH D
         MVI C, 08H
         INX D
         CALL NUMLT
         XCHG
         DCX H
         MVI M, 00H
```

```
         MOV A, C
         CPI 00H
         JZ NSOUT
         MVI M, 01H
         INX H
         INX H
         POP D
         PUSH D
         INX D
         MVI C, 08H
STOOP:   MOV A, M
         STAX D
         INX H
         INX D
         DCR C
         JNZ STOOP
NSOUT:   POP H
         PUSH H
         LXI D, 0009H
         DAD D
         MVI M, 00H
         POP D
         POP H
         POP B
         POP PSW
         RET
;
; * * * * * * * * * * * * * * * * * * * * * * * * *
;
; MPYFL     FLOATING POINT MULTIPLICATION ROUTINE
;          MULTIPLYING FLOATING POINT NUMBERS MULT1 AND MULT2
;          HL POINTS TO MULT1 STORED IN MEMORY IN EXPO FORMAT
;                10 LOCATIONS
```

```
;            DE POINTS TO MULT2 STORED IN MEMORY IN FXPO FORMAT
;                 10 LOCATIONS
;            BC POINTS TO RESULT STORED IN MEMORY IN EXPO FORMAT
;                 11 LOCATIONS      (ONE EXTRA LOC FOR WORKSPACE)
; DOES NOT MANIPULATE SIGN OF OPERANDS
; DOES NOT SET SIGN OF RESULT
; REGISTERS NOT AFFECTED
; **** ASSUMES ALL OPERANDS AND RESULT TO BE POSITIVE ****
HLSTOR   EQ WRAM+4; EQ ASSEMBLER PSEUDO-OP OF
DESTOR   EQ WRAM+6; HYPOTHETICAL ASSEMBLER OF SEC. 8.8
BCSTOR   EQ WRAM+8
WRK1     EQ WRAM+10   ; 11 LOCATIONS REQUIRED
WRK2     EQ WRAM+21   ; 11 LOCATIONS REQUIRED
WRK3     EQ WRAM+32
;   TOTAL RAM REQUIREMENT FOR WORKSPACE=43 BYTES, WRAM TO WRAM+42
MPYFL:   PUSH PSW     ; SAVE PSW AND BC
         PUSH B
         XTHL
         SHLD BCSTOR  ; LOAD BCSTOR = (BC)
         XTHL
         PUSH D
         PUSH H       ; SAVE HL AND DE
         INX H
         SHLD HLSTOR  ; LOAD HLSTOR
         XCHG
         INX H
         MOV B, M     ; GET MULT2 DIGIT
         INX H
         SHLD DESTOR  ; LOAD DESTOR
         MVI C, 00H   ; SET COUNTER VALUE IN C
         LHLD BCSTOR
         XCHG         ; HL POINTING TO MULT1 + 1, DE TO RESULT
         CALL NUMST
         LXI D, WRK1
```

```
; *** ON CALL       HL = MULT1 + 1, B = DIGIT, C = COUNT NO., DE = RESULT
; *** ON RETURN                 SAME AS ABOVE
MUPLP:   INR C          ; INCREMENT COUNT
         LHLD DESTOR
         MOV B, M       ; GET NEXT DIGIT
         INX H
         SHLD DESTOR    ; RESTORE DESTOR CONTENTS
         LHLD HLSTOR
         CALL NUMST     ; MULTIPLY NEXT DIGIT, RESULT IN WRK1 AREA
         PUSH B         ; SAVE BC FOR LATER USE
         LDAX D
         ADD C          ; FIX EXPONENT OF RESULT
         STAX D
         LHLD BCSTOR    ; SET FLOATING POINT ADDITION OF PARTIAL RESULT
         MOV C, L
         MOV B, H       ; RESULT IN RESULT AREA
; ONE OPERAND HL^ IN RESULT AREA, OTHER DE^ IN WRK1 AREA
         CALL ADFLP
         POP B          ; RESTORE BC
         MOV A, C
         CPI 07H        ; OVER IF (C) = 07H
         JNZ MUPLP
         POP H
         POP D
         POP B
         LDAX B
         ADD M
         SUI 48H
         XCHG
         ADD M
         SUI 40H
         XCHG           ; SET EXPO OF RESULT
         STAX B
```

```
        POP PSW
        RET
;
; FLOATING POINT ADDITION ROUTINE ****** FLADD
; EXPO FORMAT
;       BC^ = HL^ + DE^  WHERE HL DE BC ARE POINTERS TO THE (EXPONENT)
; REGISTERS NOT AFFECTED
FLADD:  PUSH PSW
        CALL ADFLP
        POP PSW
        RET
; FLOATING POINT SUBTRACTION ROUTINE*****FLSUB
; EXPO FORMAT
;       BC^ = HL^ - DE^ WHERE HL DE BC ARE POINTERS (TO THE EXPONENT)
; REGISTERS NOT AFFECTED
; RAM REQUIRED AS WORKSPACE IN WRK2
FLSUB:  PUSH PSW
        PUSH B
        PUSH H
        PUSH D
        MVI C, 09H
        XCHG
        LXI D, WRK2
        CALL TENSC
        POP D
        POP H
        POP B
        PUSH D
        LXI D, WRK2
        CALL ADFLP
        POP D
        POP PSW
        RET
```

```
; ************ TRANSFER FROM HL^ TO DE^ NUMBERS IN EXPO FORMAT*******
; REGISTERS NOT AFFECTED
TRANS:   PUSH PSW
         PUSH H
         PUSH D
         PUSH B
         MVI C, 0AH
TROOP:   MOV A, M
         STAX D
         INX H
         INX D
         DCR C
         JNZ TROOP
         POP B
         POP D
         POP H
         POP PSW
         RET
;
;
; FLOATING POINT SIGNED MULTIPLICATION ****** FLMUL
; EXPO FORMAT
;        BC^ = HL^ * DE^ WHERE HL DE BC ARE POINTERS (TO THE EXPONENT)
; REGISTERS NOT AFFECTED
; RAM NEEDED FOR WORKSPACE AT WRK2 AND WRK3 (OF 11 BYTES EACH)
FLMUL:   PUSH PSW         ; SAVE PSW ON STACK
         PUSH B
         XTHL
         SHLD BCSTOR
         XTHL
         SHLD HLSTOR
         XCHG
         SHLD DESTOR
```

```
        XCHG
        PUSH H
        PUSH D          ; SAVE BC DE HL ON STACK AS WELL AS IN RAM
        MVI A, 00H      ; INITIALIZE A
        PUSH PSW        ;     AND THEN ONCE AGAIN SAVE ON STACK
        LXI B, 0009H
        DAD B
        MOV A, M        ; CHECK SIGN OF HL^
        CPI 09H
        JNZ FL1         ; JMP TO FL1 IF +IVE
        LHLD HLSTOR     ; CHANGE SIGN OF NUMBER AND STORE IN WRK2
        LXI D, WRK2
        CALL TENSC
        POP PSW
        INR A
        JMP FL2
FL1:    LHLD HLSTOR     ; DONT CHANGE SIGN BUT TRANSFER TO WRK2
        LXI D, WRK2
        CALL TRANS
        POP PSW
FL2:    LHLD DESTOR
        DAD B
        PUSH PSW
        MOV A, M        ; CHECK SIGN OF DE^
        CPI 09H
        JNZ FL3         ; JUMP TO FL3 IF +IVE
        LHLD DESTOR     ; CHANGE SIGN OF DE^ AND STORE IN WRK3
        LXI D, WRK3
        CALL TENSC
        POP PSW
        INR A
        JMP FL4
FL3:    LHLD DESTOR     ; DONT CHANGE SIGN BUT TRANSFER TO WRK3
```

```
        LXI D, WRK3
        CALL TRANS
        POP PSW
FL4:    LHLD BCSTOR
        MOV B, H
        MOV C, L
        LXI H, WRK2
        CALL MPYFL      ; ACTUAL MULT OF +IVE NOS.
        ANI 01H         ; DOES SIGN OF RESULT HAVE TO BE CHANGED?
        JZ FLOUT        ; IF YES, CONTINUE ELSE GO TO FLOUT
        PUSH B
        POP H
        PUSH B
        POP D
        MVI C, 09H
        CALL TENSC
FLOUT:  POP D
        POP H
        POP B
        POP PSW         ; RESTORE SAVED REGISTERS
        RET
;* * * * * * * * * * * * * * * * * * * * * * * * *
; ROUTINES FOR FLOATING POINT INVERSION
; WORKSPACE NEEDED IN WRK4, WRK5, WRK6, WRK7, WRK8, IN EXPO FORMAT
; DIVISION CONDITION FLAGS IN DIVFLG
;       DIVFLG  BIT0 = 1    FOR SIGN, INTERNAL USE
;               BIT6 = 1    FOR DIVISION ERROR
;               BIT7 = 1    TRYING TO NORMALIZE AN ALL 0 QUANTITY
DIVFLG  EQ      WRAM + 43 ; EQ ASSEMBLER PSEUD-OP DESCRIBED
WRK4    EQ      WRAM + 44 ; EARLIER FOR A HYPOTHETICAL ASSEMBLER
WRK5    EQ      WRAM + 55 ; OF SEC. 8.8
WRK6    EQ      WRAM + 66
WRK7    EQ      WRAM + 77
WRK8    EQ      WRAM + 88
```

```
;* * * * * * * * * * * * * * * * * * * * * * * * *
;
; NORMLZ IS AROUTINE TO ELIMINATE THE LEADING ZEROES IN THE MANTISSA
;          OF A FLOATING POINT NUMBER PROPERLY UPDATING THE EXPONENT
; NUMBER IN EXPO FORMAT IN HL^, RESULT RETURNED IN HL^
;          EXPO, LSD ------ MSD, SIGN
; NUMBER MUST BE POSITIVE **** IMPORTANT *****
; REGISTERS NOT AFFECTED
; WORKSPACE NOT NEEDED
;
NORMLZ:  PUSH PSW
         PUSH B
         PUSH D
         PUSH H            ; SAVE REGISTERS ON STACK
         LDA DIVFLG
         ANI 7FH
         STA DIVFLG        ; RESET BIT 7 OF DIVFLG, FLAG FOR NORMLZ
         LXI D, 0008H      ;     TO EXAMINE MSD
         DAD D
         MVI C, 00H        ; (C) = NO. OF LEADING ZEROES
CRCT1:   MOV A, M
         ORA A
         JNZ CRCTOV        ; NO MORE LEADING 0, JUMP TO CRCTOV
         INR C
         DCX H
         MOV A, C
         CPI 08H           ; IF (C) = 08, ERROR ALL DIGITS ZERO
         JNZ CRCT1         ; REPEAT FOR MORE LEADING ZEROES
CRCTOV:  MOV A, C
         ORA A
         JZ CRCTOT         ; NORMALIZATION OVER IF (C) = 0
         CPI 08H
         JNZ CRCT2
         LDA DIVFLG
```

```
         ORI 80H         ; IF (C) = 8 SET BIT 7 OF DIVFLG TO SHOW ERROR
         STA DIVFLG      ;    ALL DIGITS ZERO
         JMP CRCTOT      ;         AND EXIT
CRCT2:   POP H
         PUSH H          ; GET HL
         MVI A, 08H
         SUB C
         MOV B, A        ; (B) = 08 - (C)
         PUSH B          ; SAVE BC IN STACK
         MOV E, L
         MOV D, H        ; HL = DE
         LXI B, 0008H
         DAD B
         XCHG            ; HL = OLD HL    DE = OLD HL + 08
         POP B
         PUSH B          ; RESTORE BC
         MOV C, B
         MVI B, 00H
         DAD B           ; HL = OLD HL + OLD B SOURCE, DE DESTINATION
         POP B           ; RESTORE BC
CRCT3:   MOV A, M
         STAX D          ; SHIFT ONE DIGIT TO MSD
         DCX H
         DCX D
         DCR B           ; REPEAT (B) TIMES
         JNZ CRCT3
         POP H
         PUSH H          ; RESTORE HL
         MOV A, M
         SUB C
         MOV M, A
CRCT4:   INX H
         MVI M, 00H      ; INSERT ZEROES AT THE TAIL END
```

```
          DCR C
          JNZ CRCT4
CRCTOT:   POP H
          POP D
          POP B
          POP PSW        ; RESTORE ALL REGISTERS
          RET            ;     AND RETURN
;
;* * * * * * * * * * * * * * * * * * * * * * * * * * *
; INVERT  CALCULATES 1/X
;         ORIGINAL NO. STORED IN HL^ IN EXPO FORMAT
;         RESULT STORED IN HL^ IN EXPO FORMAT
; REGISTERS NOT AFFECTED
; WORKSPACE NEEDED IN WRK1-WRK8
;   IF Y = 1/X, RECURSION USED IS Y(N + 1) = Y(N) * [2.0 - X * Y(N)]
;         WITH INITIAL GUESS Y(0) = 1.0
;   SIGNED NUMBERS ALLOWED
;   EXPO SHOULD BE BETWEEN -64 TO +64 IN EXCESS 128 REPRESENTATION
INVERT:   PUSH PSW
          PUSH B
          PUSH D
          PUSH H         ; SAVE REGISTERS ON STACK
          MVI A, 00H
          STA DIVFLG     ; CLEAR DIVFLG CONTENTS TO 00
          LXI D, 0009H
          DAD D          ; CHECK SIGN OF NUMBER
          MOV A, M
          ORA A
          JZ SGNFIN      ; IF SIGN + , GO TO SGNFIN TO SET BIT0 OF DIVFLG
;                                                            TO 1
          CPI 09H
          JNZ DIVERR     ; ERROR IF SIGN DIGIT IS NEITHER 0 NOR 9
          POP H
```

```
        PUSH H          ; RESTORE HL
        MOV E, L
        MOV D, H        ; DE=HL
        MVI C, 09H
        CALL TENSC
        MVI A, 01H
        STA DIVFLG      ; SET DIVFLG BIT 0 TO 0 AND COMPLEMENT NUMBER
        JMP SGNOV
SGNFIN: MVI A, 00H
        STA DIVFLG
SGNOV:  POP H
        PUSH H          ; RESTORE HL
        CALL NORMLZ     ; NORMALIZE NUMBER
        LXI D, WRK 4
        CALL TRANS      ; TRANSFER NUMBER TO WRK4 WITH EXPO=128
                                                        (NORMAL 0)
        XCHG
        MVI M, 80H
        XCHG
        MOV B, M        ; ADJUST FINAL EXPONENT OF ANSWER TENTATIVELY
        MVI A, 0FFH
        SUB B
        INR A
        MOV M, A
        MVI B, 08H
        LXI H, WRK5
        MVI M, 80H
CRCT5:  INX H           ; WRK5=+0.999999999 EXP 128
        MVI M, 09H
        DCR B
        JNZ CRCT5
        INX H
        MVI M, 00H
```

```
DIVRPT:   MVI B, 07H
          LXI H, WRK6      ; WRK6=+0.20000000 EXP 129
          MVI M, 81H
CRCT6:    INX H
          MVI M, 00H
          DCR B
          JNZ CRCT6
          INX H
          MVI M, 02H
          INX H
          MVI M, 00H
          LXI H, WRK4
          LXI D, WRK5
          LXI B, WRK7
          CALL FLMUL       ; WRK7=WRK4*WRK5=X*Y(N)
          LXI H, WRK7
          CALL NORMLZ
          LXI H, WRK6
          LXI D, WRK7
          LXI B, WRK8
          CALL FLSUB       ; WRK8=WRK6-WRK7=2-X*Y(N)
          LXI H, WRK8
          CALL NORMLZ
          LXI D, WRK5
          LXI B, WRK7
          CALL FLMUL       ; WRK7=WRK5*WRK8=Y(N)*[2-X*Y(N)]=Y(N+1)
          LXI H, WRK7
          CALL NORMLZ
          LXI D, WRK5
          LXI B, WRK8
          CALL FLSUB
          LXI H, WRK8
          LXI D, 0009H
```

```
         DAD D
         MOV A, M
         ORA A
         JZ DIVCHK
         CPI 09H
         JNZ DIVERR
         MVI C, 09H
         LXI H, WRK8
         MOV E, L
         MOV D, H
         CALL TENSC
DIVCHK:  LXI H, WRK8     ; WRK8=ABS(ERR) ERR=Y(N+1)-Y(N)
         CALL NORMLZ
         LDA DIVFLG
         ANI 80H
         JNZ DIVOUT
         MOV A, M
         CPI 7AH
         JNC DIVCK1
         JMP DIVOUT
DIVCK1:  LDA DIVFLG
         ANI 7FH
         STA DIVFLG
         LXI H, WRK7
         LXI D, WRK5
         CALL TRANS
         JMP DIVRPT      ; REPEAT WITH WRK5=Y(N+1)=WRK7
DIVOUT:  POP H
         NOP             ; RESTORE HL
         LXI D, WRK7
         LDAX D
         SUI 40H
         ADD M
```

```
          SUI 40H
          STAX D
          XCHG
          CALL TRANS
          XCHG
          LDA DIVFLG
          ANI 01H        ; CHECK SIGN TO DO COMPLEMENTATION OR NOT
          JZ DIVFIN      ; GO TO DIVFIN IF +
          MOV E, L       ; ELSE INVERT SIGN
          MOV D, H
          MVI C, 09H
          CALL TENSC
DIVFIN:   POP D
          POP B
          POP PSW
          RET            ; RESTORE REG AND RETURN
DIVERR:   LDA DIVFLG
          ORI 40H        ; SET BIT 6 OF DIVFLG TO SHOW ERROR IN INVERSION
          STA DIVFLG
          POP H
          JMP DIVFIN
;
;* * * * * * * * * * * * * * * * * * * * * * * * *
; FLOATING POINT SIGNED DIVISION          FLDIV
;
;         BC^=HL^/DE^
;
; ALL NUMBERS IN EXPO FORMAT
;
FLDIV:    XCHG
          CALL INVERT
          XCHG
```

```
         CALL FLMUL
         RET                ; MULTIPLICATION TABLES
DGMT1:   DFB     00, 00, 00, 00, 00, 00; DFB ASSEMBLER PSEUD-OP DESCRIBED
         DFB     00, 00, 00, 00        ; EARLIER FOR A HYPOTHETICAL
         DFB     99, 99, 99, 99, 99, 99; ASSEMBLER OF SEC. 8.8
         DFB     00, 01, 02, 03, 04, 05
         DFB     06, 07, 08, 09
         DFB     99, 99, 99, 99, 99, 99
         DFB     00, 02, 04, 06, 08, 00
         DFB     02, 04, 06, 08
         DFB     99, 99, 99, 99, 99, 99
         DFB     00, 03, 06, 09, 02, 05
         DFB     08, 01, 04, 07
         DFB     99, 99, 99, 99, 99, 99
         DFB     00, 04, 08, 02, 06, 00
         DFB     04, 08, 02, 06
         DFB     99, 99, 99, 99, 99, 99
         DFB     00, 05, 00, 05, 00, 05
         DFB     00, 05, 00, 05
         DFB     99, 99, 99, 99, 99, 99
         DFB     00, 06, 02, 08, 04, 00
         DFB     06, 02, 08, 04
         DFB     99, 99, 99, 99, 99, 99
         DFB     00, 07, 04, 01, 08, 05
         DFB     02, 09, 06, 03
         DFB     99, 99, 99, 99, 99, 99
         DFB     00, 08, 06, 04, 02, 00
         DFB     08, 06, 04, 02
         DFB     99, 99, 99, 99, 99, 99
         DFB     00, 09, 08, 07, 06, 05
         DFB     04, 03, 02, 01
         DFB     99, 99, 99, 99, 99, 99
DGMT2:   DFB     00, 00, 00, 00, 00, 00
```

```
        DFB     00, 00, 00, 00
        DFB     99, 99, 99, 99, 99, 99
        DFB     00, 00, 00, 00, 00, 00
        DFB     00, 00, 00, 00
        DFB     99, 99, 99, 99, 99, 99
        DFB     00, 00, 00, 00, 00, 01
        DFB     01, 01, 01, 01
        DFB     99, 99, 99, 99, 99, 99
        DFB     00, 00, 00, 00, 01, 01
        DFB     01, 02, 02, 02
        DFB     99, 99, 99, 99, 99, 99
        DFB     00, 00, 00, 01, 01, 02
        DFB     02, 02, 03, 03
        DFB     99, 99, 99, 99, 99, 99
        DFB     00, 00, 01, 01, 02, 02
        DFB     03, 03, 04, 04
        DFB     99, 99, 99, 99, 99, 99
        DFB     00, 00, 01, 01, 02, 03
        DFB     03, 04, 04, 05
        DFB     99, 99, 99, 99, 99, 99
        DFB     00, 00, 01, 02, 02, 03
        DFB     04, 04, 05, 06
        DFB     99, 99, 99, 99, 99, 99
        DFB     00, 00, 01, 02, 03, 04
        DFB     04, 05, 06, 07
        DFB     99, 99, 99, 99, 99, 99
        DFB     00, 00, 01, 02, 03, 04
        DFB     05, 06, 07, 08
        DFB     99, 99, 99, 99, 99, 99
; WRAM to WRAM + 64H        RAM required as workspace
        END
```

Chapter 9

Pin-out and Timing of the 8085A

The Intel 8085A is one of the most popular 8-bit microprocessors available in the market today. In this chapter we present the specific hardware details of this device—namely, its pin-out and timing.

9.1 PIN-OUT OF THE 8085A

Referring to Fig. 9.1.1, the pin-out of the 8085A CPU is given below. The number in brackets

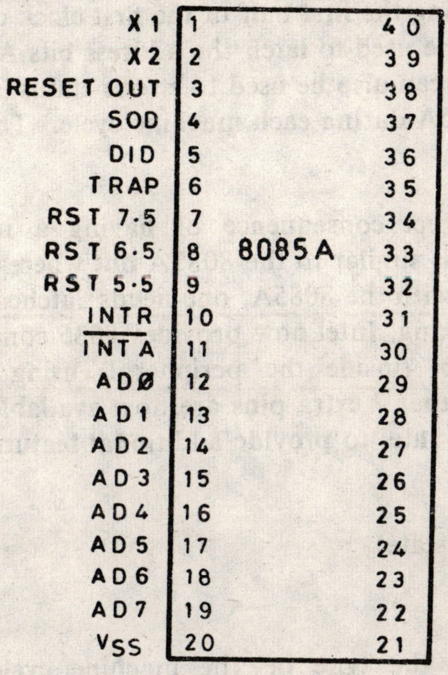

Fig. 9.11 Pin-out of the 8085A Microprocessor

indicates the pin number on the 40 pin device package. The reader is advised to re-read this section after reading about machine cycles in Sec. 9.2.

A8–A15 Output, Tri State
(21–28)

Higher order byte of ADDRESS BUS.

These pins output the most significant 8-bits of the memory address or the 8-bits of the I/O address. Output is tristate during HOLD and HALT modes and during RESET (see Sec. 9.2).

AD∅–AD7 (12–19) Input/Output, Tri State

Lower order byte of ADDRESS BUS and Data Bus (Multiplexed).

Lower order address byte is output during the first clock cycle of a machine cycle (see Sec. 9.2). It then becomes the data bus during the second and third clock cycles of that machine cycle. During the latter, the bus is floated if the 8085 does not output data on the data bus and no other device drives it.

ALE (30) Output

ALE or Address Latch Enable is used to distinguish whether the $AD_{\emptyset-7}$ bus contains address bits $A_{\emptyset-7}$ or data $D_{\emptyset-7}$. ALE can itself be used to latch the address bits $A_{\emptyset-7}$ which appear on $AD_{\emptyset-7}$ during the first clock cycle of a machine cycle.

ALE is output High during the first half of the first clock cycle of a machine cycle and its falling edge can be used to latch the address bits $A_{\emptyset-7}$ in a latch (i.e. 8212). The falling edge of ALE can also be used to strobe status information (S∅, S1, IO/$\overline{\text{M}}$) being output by the 8085A during each machine cycle. The ALE output is never tristated in the 8085A.

The fact that ALE is required is a direct consequence of having a multiplexed Data/Address Bus. This is unlike the 8080 which is similar to the 8085A but where these buses are not multiplexed. Using 8080 type peripherals with the 8085A, one needs latches (e.g. 8212s) strobed by ALE in order to do the demultiplexing. Intel now provides 8085 compatible peripherals which can demultiplex these busses 'on chip' (inside the peripheral) using $AD_{\emptyset-7}$ and ALE. The advantage of using multiplexing is that 7 extra pins are now available on the same 40 pin DIP as the 8080; these have been used by Intel to provide additional features like vectored interrupts and serial I/O lines.

S∅ (29), S1 (33) and IO/$\overline{\text{M}}$ (34) Output (IO/$\overline{\text{M}}$ can be tristate)

These outputs identify the type of the machine cycle being executed by the 8085A. Seven types of machine cycles are possible. These will be described later (Sec. 9.2).

S1 can be used as an advanced RD/$\overline{\text{WR}}$ status. It becomes valid at the beginning of the machine cycle and stays valid throughout the cycle. S∅ and IO/$\overline{\text{M}}$ also become valid at the beginning of the machine cycle and stay valid throughout the cycle.

As mentioned earlier, one can also latch these signals, if necessary, using the falling edge of ALE.

Machine Cycle	Status			Control		
	IO/$\overline{M}$	S1	SØ	$\overline{RD}$	$\overline{WR}$	$\overline{INTA}$
1. Op Code Fetch (OF)	Ø	1	1	Ø	1	1
2. Memory Read (MR)	Ø	1	Ø	Ø	1	1
3. Memory Read (MW)	Ø	Ø	1	1	Ø	1
4. I/O Read (IOR)	1	1	Ø	Ø	1	1
5. I/O Write (IOW)	1	Ø	1	1	Ø	1
6. INTR Ack (INA)	1	1	1	1	1	Ø
7. Bus Idle (BI)						
(a) DAD	Ø	1	Ø	1	1	1
(b) RST/TRAP	1	1	1	1	1	1
(c) HALT	TS	Ø	Ø	TS	TS	1

TS = Tri State

During HOLD and RESET Conditions, IO/$\overline{M}$ goes tristate and the outputs on SØ and S1 are 'Don't Care's.

$\overline{RD}$ (32) Output, Active Low, Tri State only during HOLD, HALT and RESET.

The 8085A pulls $\overline{RD}$ output Low to indicate that the contents of the selected memory address or I/O port is to be read from the Data Bus into the 8085A. The selected memory or I/O port also uses this signal to present its data on the DATA BUS.

$\overline{WR}$ (31) Output, Active Low, Tri State during HOLD, HALT and RESET.

The 8085A pulls $\overline{WR}$ low to indicate that the contents of the Data Bus are to be written into the selected memory address or I/O port. The data to be written is put on the Data Bus via the $AD_{Ø-7}$ lines at the falling edge of $\overline{WR}$.

READY (35) Input

A memory or peripheral device indicates to the 8085A that it is ready to send or receive data by keeping READY high. If READY is pulled low, the 8085A will wait for an integral number of clock cycles till READY goes high once again.

This feature is useful when interfacing the 8085A with slow memory or I/O peripherals. (Those which require more time than the normal $\overline{RD}$ or $\overline{WR}$ outputs of the 8085A.) The 8085A samples the READY line in the middle of the second clock cycle of (almost) every machine cycle. It proceeds to the third clock cycle in a normal fashion if READY is 1. On the other hand, if READY is Ø at this time, the 8085A inserts waiting clock cycles till it sees READY go to 1; only then does it proceed to the third clock cycle of that machine cycle. The 8085A outputs remain unchanged during the waiting clock cycles, i.e. they remain what they were at the end of the second clock cycle. Its internal status also remains unchanged during this time.

This feature essentially provides a slow peripheral more time to respond than what it would normally have.

HOLD (39) Input

The HOLD input is used to make the 8085A relinquish the system bus—data, address and control ($\overline{RD}$, $\overline{WR}$, $IO/\overline{M}$)—when asked to do so by some other device.

Inputting HOLD = 1 indicates a HOLD request. Upon receiving this request, the 8085A acknowledge the request by outputting HLDA = 1 and relinquishes its use of the Address, Data and Control Bus (after the completion of the current bus transfer, if any). Internal processing, not requiring the use of the system busses may continue. (Note that every instruction requires an Op Code Fetch operation which is essentially a Memory Read operation and will consequently need the system bus. Therefore, the HOLD mode will at least stop the CPU at the next instruction). This HOLD state can be excited (i.e. the 8085A regains control of the bus) only after HOLD returns to 'Ø'. The 8085A floats its Address and Data bus as well as the $\overline{RD}$, $\overline{WR}$ and $IO/\overline{M}$ outputs while in the HOLD state.

This input is useful in systems where there are more than one CPU like devices (e.g. a DMA controller) sharing the same system bus. If CPU_i wants to gain control of the bus, the HOLD inputs of the other devices are used to make them enter the HOLD state.

HLDA (38) Output

HLDA or Hold Acknowledge is output high by the 8085A after it gets the HOLD request. This indicates that the 8085A has received the HOLD request properly and will relinquish the system bus after the next clock cycle (following Ø → 1 edge of HLDA). HLDA goes low after HOLD goes low, the 8085A regains control of the bus half a clock cycle after HLDA returns to 'Ø'.

INTR (10) Input

This is the lowest priority Interrupt Request in the 8085A processor and is used as a general purpose interrupt. An input of INTR = 1 implies some external device has put up an interrupt and wants the CPU to execute an appropriate service routine. The 8085A monitors the status of the INTR line by sampling it in the last-but-one clock cycle of each instruction and during HOLD and HALT states.

If the Interrupt Structure of the 8085A is enabled when INTR is sampled High, the Program Counter (PC) will not be incremented. Instead an $\overline{INTA} = 0$ is output. It is now the responsibility of the Interrupting Device (e.g. 8259 Interrupt Controller) to issue a RSTn or CALL ADDR instruction so that the 8085A can jump to the proper interrupt service routine. (This is really done as a subroutine call!)

The Interrupt Structure of the 8085A is disabled (INTR will not be acknowledged) following Reset, Acknowledgement of another interrupt or execution of a DI instruction. It is enabled by executing an EI instruction.

$\overline{\text{INTA}}$ (11) Output

The 8085A outputs $\overline{\text{INTA}}$ (Interrupt Acknowledge) Low when it acknowledges an Interrupt Request. It can be used to activate the 8259 Interrupt Controller or any other device which will provide the 8085A with the required RSTn or CALL ADDR instruction.

RST 5.5(9), 6.5(8) and 7.5(7) Inputs

These are the 8085A's maskable vectored interrupts. They operate exactly like INTR except for the following:

(i) The Restart instruction is automatically inserted by internal logic. It does not have to be provided from outside. These instructions are

RST 5.5	CALL	ØØ2CH
RST 6.5	CALL	ØØ34H
RST 7.5	CALL	ØØ3CH

(ii) RST 7.5 is edge (Ø → 1) sensitive unlike RST 5.5, 6.5 or INTR which are level (1) sensitive.

(iii) These are individually maskable (enable/disable) using SIM instruction.

(iv) They have higher priority than INTR. Among them, RST 7.5 has the highest priority and RST 5.5 the lowest.

Like the INTR, recognizing any of these interrupts also disables interrupts. Therefore EI must be appropriately placed in the program to enable interrupts once again. (See Sec. 12.6.)

TRAP (6) Input

Trap is a non-maskable (always enabled) vectored interrupt which is similar to the RST X.5 (X = 5, 6, 7) and INTR interrupts. On being recognized, it forces the 8085A to perform a 'CALL ØØ24H' instruction.

TRAP has the highest priority of any interrupt. It is both edge and level sensitive i.e. it becomes active at the Lo → Hi edge but must stay High until it is sampled. (See Sec. 12.6.)

$\overline{\text{RESET IN}}$ (36) Input

If held low for 600 ns (3 clock periods), this input forces the 8085A to do the following:

(i) (PC) = ØØØØH

(ii) Instruction Register cleared (on-going instruction discontinued)

(iii) All Interrupts (except TRAP) disabled, with RST 5.5, 6.5 and 7.5 also masked

(iv) SOD is forced to Ø

(v) During $\overline{\text{RESET IN}}$ Low, data, address and control bus are floated.
$\overline{\text{RESET IN}}$ is a Schmidt-triggered input allowing connection to an R·C network

as in Fig. 9.1.2 for power-on or manual RESET. The 8085A stays RESET as long as $\overline{\text{RESET IN}}$ is low.

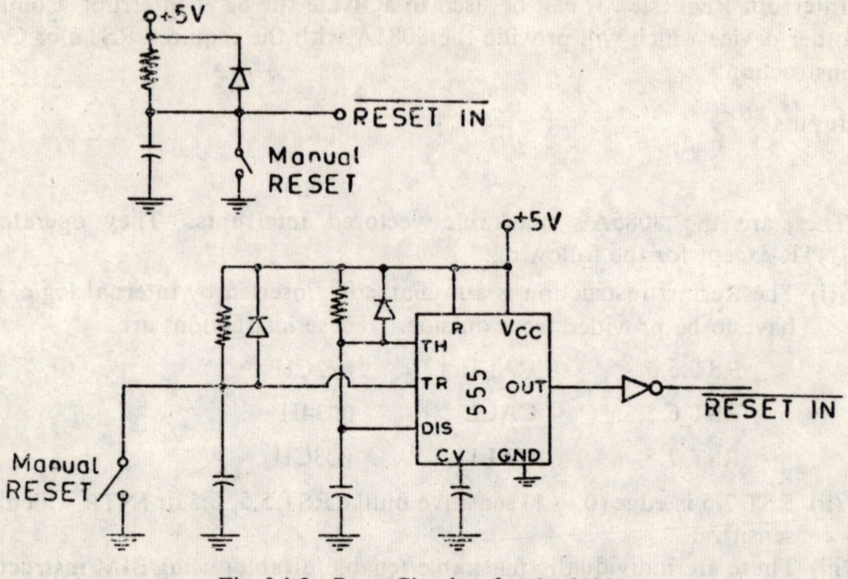

Fig. 9.1.2 Reset Circuitry for the 8085A

RESET OUT (3) — Output

Indicates 8085A is being reset. This output stays HIGH as long as $\overline{\text{RESET IN}}$ is Low and for one integral clock cycle after that. (If $\overline{\text{RESET IN}}$ goes Lo → Hi in the middle of a clock cycle, RESET will fall to Low only after that clock cycle is over.)

This output is used to reset other devices used in the microprocessor system.

XI (1) & X2 (2) — Input

X1, X2 can be connected to a crystal or R-C (or L-C) network to generate the clock internally as shown in Fig. 9.1.3.

X1 can also be a clock signal (TTL) externally provided as shown in Fig. 9.1.4.

The driving frequency must be at least 1 MHz and must be twice the desired internal clock frequency. The 8085A can use up to a 6 MHz crystal (for a 3 MHz clock) whereas the 8085 AH can use up to a 10 MHz crystal (for a 5 MHz clock).

CLK (37) — Output

The 8085A outputs this signal for use as a system clock. This CLK is used internally to mark clock cycles for machine cycles. Its period is twice the period of the clock input to the device at X1, X2.

SID (5) — Input

Serial Input Data (SID) essentially provides a 1-bit input port in the 8085A. A

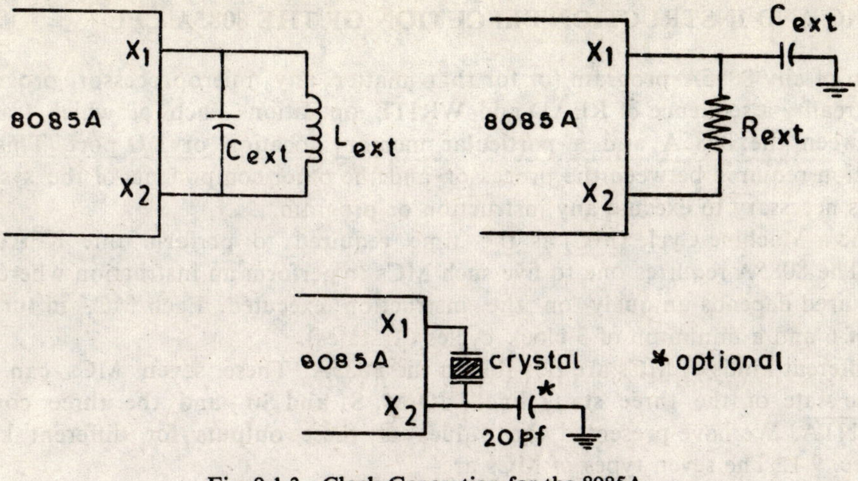

Fig. 9.1.3 Clock Generation for the 8085A

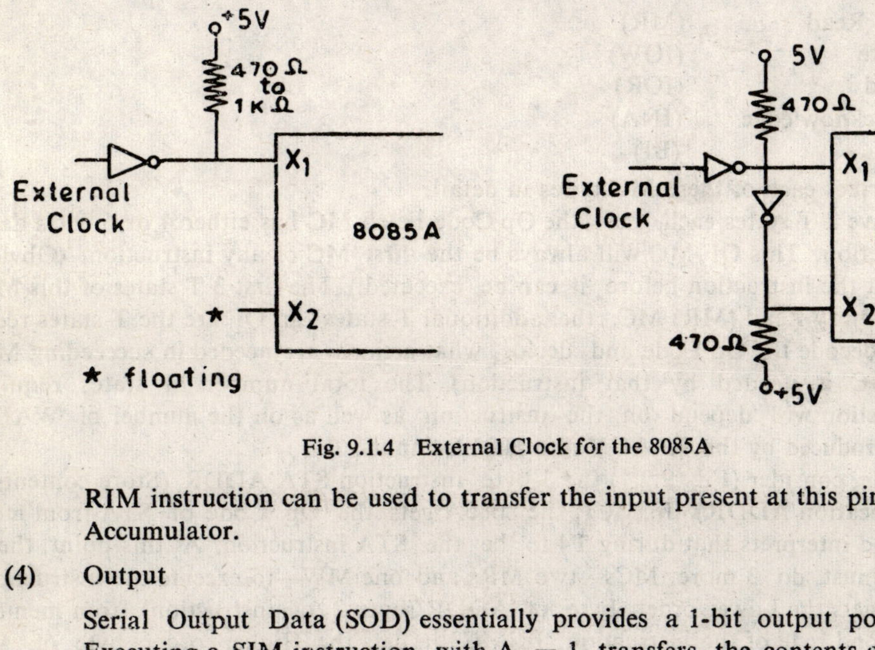

Fig. 9.1.4 External Clock for the 8085A

RIM instruction can be used to transfer the input present at this pin to bit 7 of the Accumulator.

SOD (4) Output

Serial Output Data (SOD) essentially provides a 1-bit output port on the 8085A. Executing a SIM instruction with $A_6 = 1$ transfers the contents of A_7 to the SOD output where it stays latched (until another SIM is executed with $A_6 = 1$).

The combination of SID and SOD (along with the RIM and SIM instruction) can be used to provide a very primitive Serial I/O capability on the 8085A chip itself. A more useful application, however, would be to use SID and RIM to read *single* signal status and SOD and SIM to output single signal controls.

V_{cc}(40) + 5 volt supply (170 ma current)

V_{ss}(20) GND

9.2 TIMING AND INSTRUCTION EXECUTION OF THE 8085A CPU

Examination of any 8085A program (or for that matter any microprocessor program), reveals that it is in reality a sequence of READ and WRITE operations each of which transfers a byte of data between the 8085A and a particular memory location or I/O port. This is the only communication required between the processor and the other components of the system and will be all that is necessary to execute any instruction or program.

We define a Machine Cycle (MC) as the time required to perform one READ or WRITE operation. The 8085A requires one to five such MCs to perform an instruction where the number of MCs required depends uniquely on the instruction executed. Each MC, in turn, contains a maximum of 6 and a minimum of 3 clock cycles (T states).

Seven different kinds of MCs are possible in the 8085A. These seven MCs can be differentiated by the state of the three status lines—$IO/\overline{M}$, S1 and SØ—and the three control signals $\overline{RD}$, $\overline{WR}$, $\overline{INTA}$. We have presented the values of these outputs for different kinds of MCs earlier in Sec. 9.1. The seven types of MCs are—

(i)	Op Code Fetch	(OF)
(ii)	Memory Write	(MW)
(iii)	Memory Read	(MR)
(iv)	I/O Write	(IOW)
(v)	I/O Read	(IOR)
(vi)	INTR Acknowledge	(INA)
(vii)	Bus Idle	(BI)

This section describes each of these MC types in detail.

Most MCs have 3 T states each. Only the Op Code Fetch MC has either 4 or 6 MCs depending on the instruction. This OF-MC will always be the first MC of any instruction. (Obviously, the CPU must get the instruction before it can be executed!) The first 3 T states of this MC are identical to a Memory Read (MR) MC; the additional T states in OF are the T states required by the 8085A to decode the Op Code and decide what actions are needed in succeeding MCs (if more than one MC is needed by that instruction). The total number of states required to execute an instruction will depend on the instruction as well as on the number of WAIT and HOLD states introduced by the READY and HOLD inputs.

As an example, consider (Fig. 9.2.1) the 3-byte instruction STA ADDR (Store contents of A in the memory location ADDR). In MC1, the 8085A gets the Op Code of STA from memory during T1–T3 and interprets that during T4 to be the STA instruction. At this point, the CPU knows that it must do 3 more MCs—two MRs and one MW—to execute the instruction. In MC2 the 8085A gets the Lower order byte of ADDR (byte 2 of instruction) from memory by accessing the second byte of the instruction. In MC3 it gets the Higher order byte of ADDR from memory by accessing the third byte of the instruction. At this point all three byte of the STA instruction is available to the 8085A. It must now execute this instruction. It does this by placing the data obtained in MC2 and MC3 on the Address bus and the contents of accumulator, A, on the data bus and executing a MW-MC; this is MC4. After MC_4 is over, the 8085A will get the next instruction (op code first) in the next MC1 and continue in a similar fashion.

The first T state (CLK Cycle) of every MC is distinguishable by the fact that an ALE pulse (=1) will be present during the first half of the CLK cycle. The first MC (i.e. OF) of every instruction can be distinguished by the fact that SØ = S1 = 1. These outputs can be monitored by external logic if one wants to know the current status of the 8085A.

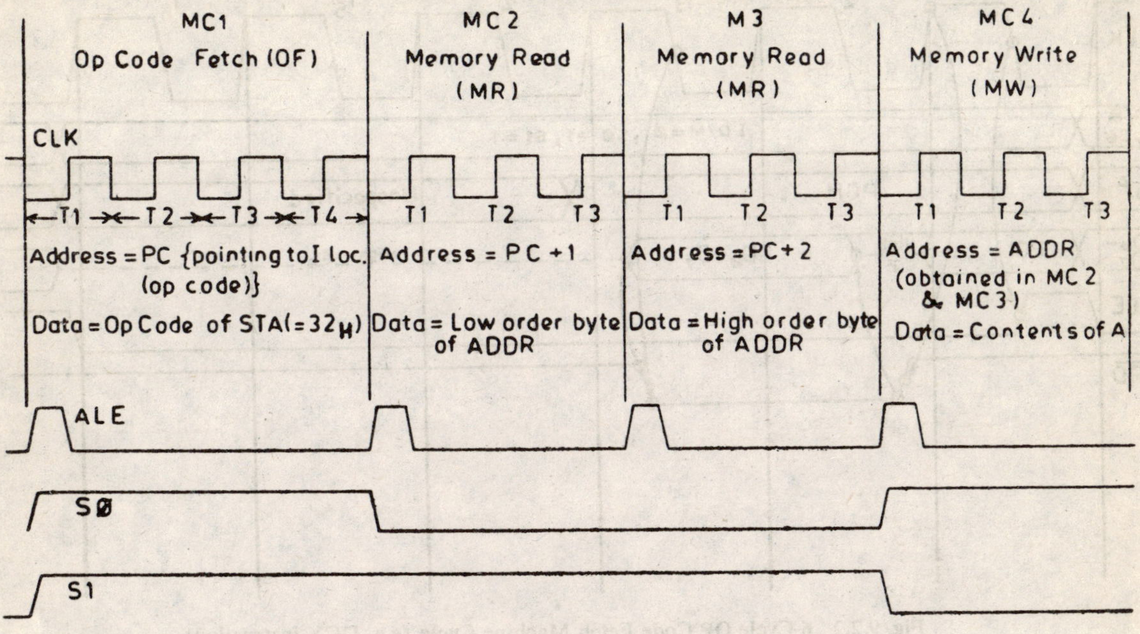

Fig. 9.2.1 CPU Timing for the STA Addr Instruction

We now present the timing details for each of the seven types of MCs. As we shall see, the timings for each of the MCs are almost identical.

9.2.1 OP Code Fetch (OF) Machine Cycle

Unlike other MCs, the OF-MC requires more than 3 T states. The first three T states essentially perform a Memory Read operation from the location pointed at by PC. The fourth T state decodes the OP Code. At this point, depending on the Op Code, the 8085A decides whether to go to the next MC or go for two more T states—T5 and T6. It has been observed that if a particular instruction requires subsequent memory accesses (RD or WR) after the OF-MC, OF for that instruction will only require 4T states; otherwise, the OF *may* have 6T states. An OF-MC is identified by $S1 = 1$ and $SØ = 1$. Since a Memory Read is required, $IO/\overline{M} = Ø$. It should be noted that these status lines SØ, S1 and $IO/\overline{M}$ stay true throughout the cycle.

As an example of an OF-MC, we shall consider the DCX instruction which has 6T-states. (The advantage of doing this is that it also shows what an OF with 4-T states would look like; for this, one can just omit the last two states, T5 and T6.) The timing diagram for this instruction is shown in Fig. 9.2.2.

In the first T state of this machine cycle, the 8085A outputs the high order byte of PC on A_8–A_{15} and the low order byte of PC on $AD_Ø$–AD_7. (PC points to the location of the instruction.) Since $AD_Ø$–AD_7 is a multiplexed Address/Data bus which will be turned around and used as the data bus later, external hardware (e.g. using a 8212 buffer) must latch the lower order address byte appearing during T1. This is done by using the ALE signal. A circuit for doing this is shown in Fig. 9.2.3. Some support devices for the 8085A do this latching internally.

At the end of T1, the output drivers of $AD_Ø$–AD_7 are turned off in preparation for the selected

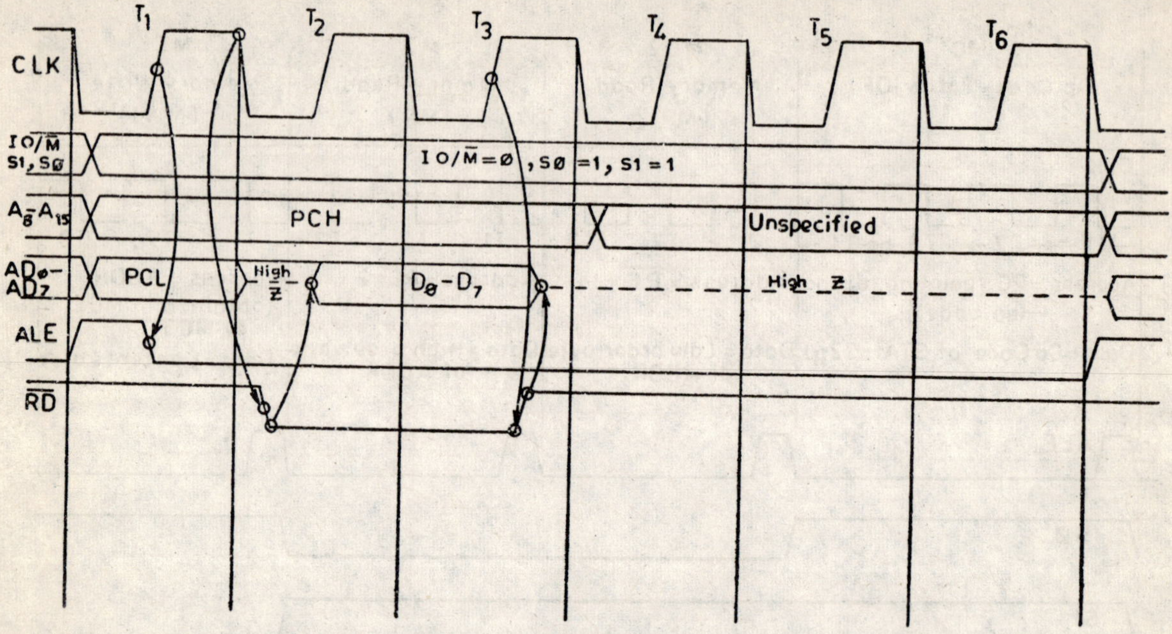

Fig. 9.2.2 6-Cycle OP Code Fetch Machine Cycle (e.g. DCX instruction)

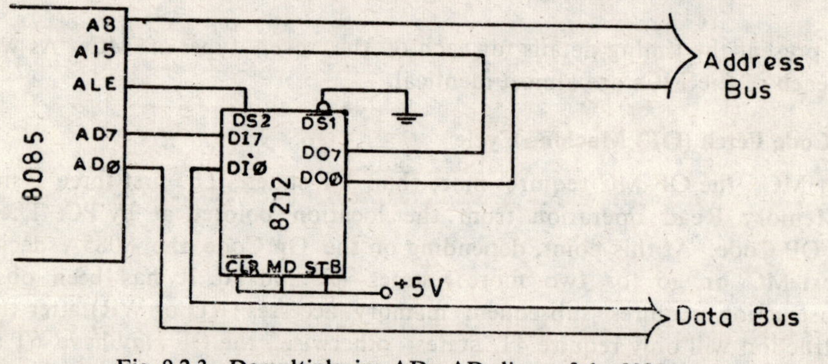

Fig. 9.2.3 Demultiplexing ADØ-AD$_7$ lines of the 8085A

memory driving this bus with data. For some time during T2—at the beginning—the ADØ-AD$_7$ bus essentially floats (till the memory device drives it). The Address Bus becomes valid at the beginning of T1 (Fig. 9.2.3 circuit used) and therefore the proper memory chip is selected well before T1 ends and T2 begins. The actual $\overline{RD}$ pulse (asking selected memory to output on the data bus) goes Low at the beginning of T2 and stays Low till the middle of T3. The accessed memory should be fast enough to output its data before $\overline{RD}$ goes high. Slower memories can gain more time by pulling the READY input of 8085A LO. This will introduce an integral number of T_{WAIT} states between T2 and T3 as long as READY is LO. Once $\overline{RD}$ goes HI, the memory device which was driving the data bus previously, stops doing so and the DØ-D$_7$ bus returns to the High-Z states (floats) and stays that way till the next MC. At the end of T3, the 8085A has got the OP Code (DCX in this case) inside and starts decoding it. Actions indicated by DCX are taken inside the CPU during T4, T5 and T6. Note that during T4, T5 and T6 the address bus may take on unspecified values. This can lead to multiple devices being selected in the following ways:

(i) IO Mapped I/O Port and a memory module both being selected if $IO/\overline{M}$ is not used in the selection logic.

(ii) If linear memory selection is being used, two memory modules might get simultaneously selected.

This can be potentially damaging and should be avoided. One way of doing so is to qualify Memory and I/O Selects with $\overline{RD}$, $\overline{WR}$ and $IO/\overline{M}$. This is normally done by deriving Chip Selects ($\overline{CS}$s) based on $IO/\overline{M}$ and the address value and connecting $\overline{RD}$ and $\overline{WR}$ directly to the device. It can also be done by decoding $IO/\overline{M}$, $\overline{RD}$ and $\overline{WR}$ into $\overline{MEMR} = IO/\overline{M} + \overline{RD}$, $\overline{MEMW} = IO/\overline{M} + \overline{WR}$, $\overline{IOR} = (\overline{IO/\overline{M}}) + \overline{RD}$, $\overline{IOW} = (\overline{IO/\overline{M}}) + \overline{WR}$. Chip Selects may now be derived directly from the address bus with the $\overline{MEMR}$, $\overline{MEMW}$, $\overline{IOR}$ and $\overline{IOW}$ connected properly to the $\overline{READ}$ and $\overline{WRITE}$ inputs of the memory or I/O device.

9.2.2 Memory Read (MR) Machine Cycle

This is a 3 T-state machine cycle. T1, T2 and T3 of MR are identical to T1, T2 and T3 of OF except that $S1 = 1$ and $S\emptyset = \emptyset$. In all other aspects, these three states of the two types of MCs are identical ($IO/\overline{M} = \emptyset$). As for the OF-MC, address latching is done using the ALE signal.

9.2.3 I/O Read (IOR) Machine Cycle

This is identical to MR in all aspects except that $IO/\overline{M} = 1$ ($S1 = 1$, $S\emptyset = \emptyset$).

9.2.4 Memory Write (MW) Machine Cycle (Fig. 9.2.4)

This is identical to MR except that $S1 = \emptyset$, $S\emptyset = 1$ and that $AD_{\emptyset}$–AD_7 is driven with the data from the CPU during T2 and T3. In all other aspects, the two MCs are identical ($IO/\overline{M} = \emptyset$).

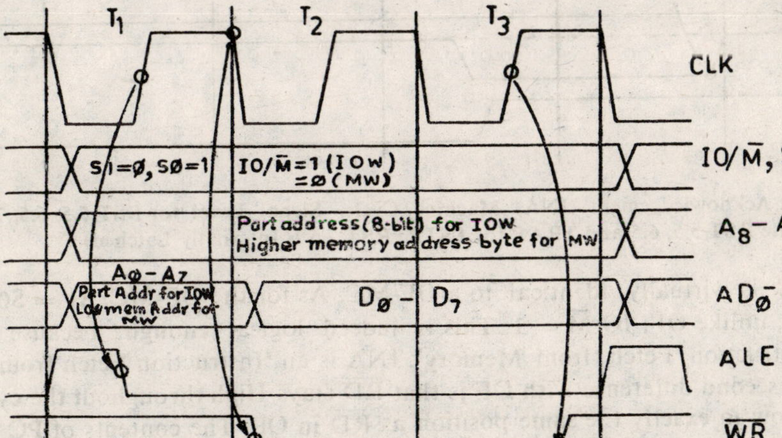

Fig. 9.2.4 Memory Write or I/O Write Machine Cycle

In this case we replace $\overline{RD}$ with $\overline{WR}$ in the timing diagram and the A-Bus and the AD-Bus are held constant till the next T1 (till after T3).

9.2.5 I/O Write (IOW) Machine Cycle (Fig. 9.2.4)

This is identical to MW in all aspects except that $IO/\overline{M} = 1$ ($S1 = \emptyset$, $S\emptyset = 1$).

One important difference between IO Machine Cycles (IOW, IOR) and Memory Machine Cycles (MR, MW) is that the demultiplexed address bus output in the former is such that the higher and lower address bytes are identical and equal to the 8-bit address of the I/O port; in the latter case the address bus output is the true 16-bit address.

9.2.6 Interrupt Acknowledge (INA) Machine Cycle (Fig. 9.2.5)

This MC is executed by the 8085A if interrupts are enabled and the INTR input goes High. The status of TRAP, RST 5.5–7.5 and INTR is sampled between the last and last-but-one T-state of the last MC of every instruction. An INA-MC is initiated if interrupts are enabled and if INTR is the *only* valid interrupt at this time.

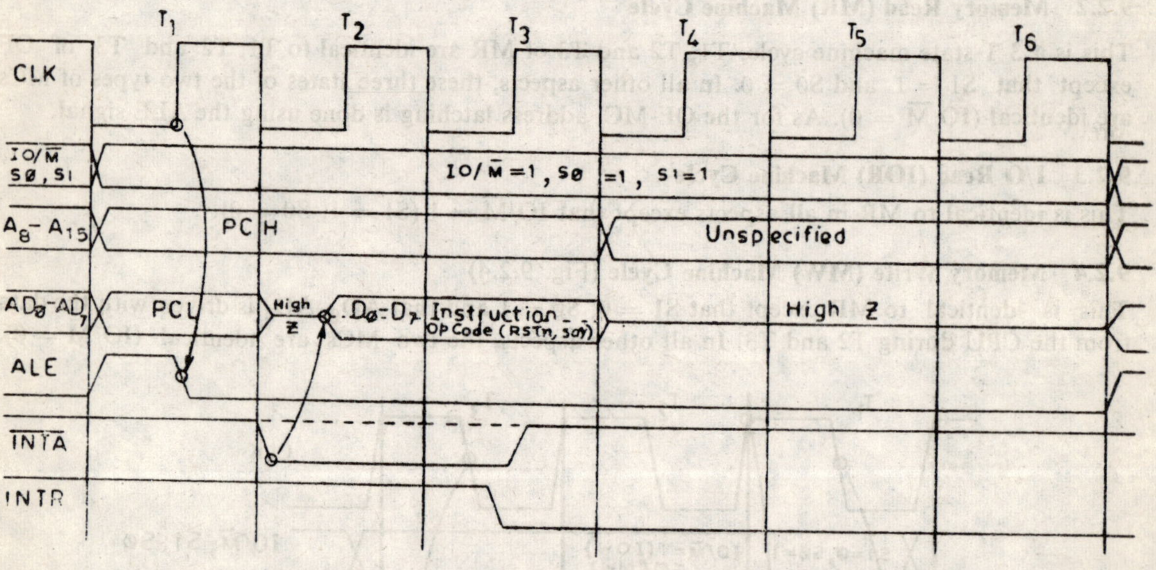

Fig. 9.2.5 Interrupt Acknowledgement (INA) Machine Cycle—Signal Level for RST 5.5, 6.5, 7.5 and TRAP (Substitute RST 5.5, 6.5 and TRAP for INTR; RST 7.5 is Internally Latched)

This INA-MC is virtually identical to a OF-MC. As for the OF-MC, $S1 = S\emptyset = 1$ during this MC, however, unlike OF, $IO/\overline{M} = 1$. This is indeed logical enough, because unlike OF, which is an 'Instruction Fetch from Memory', INA is an 'Instruction Fetch from I/O' (as we shall see later). A second difference with OF is that $\overline{RD}$ stays High throughout the cycle; instead, $\overline{INTA}$ is pulsed low in exactly the same position as $\overline{RD}$ in OF. The contents of PC are sent out on A_8–A_{15} (T1, T2, T3) and $AD_\emptyset$–AD_7 (T1 only) and ALE is pulsed during T1 as in OF. However, for the INA-MC, address bus and ALE are useless and their contents can be ignored. As in the OF-MC, A_8–A_{15} are undefined after T3 and memory (or IO) select logic should be disabled during that time. Moreover, an IO device other than the one outputting INTR = 1 may also get selected during T1, T2 and T3. This does not cause a problem if the system has been carefully designed, i.e. if $\overline{RD}$ and $\overline{WR}$ of the I/O ports are properly qualified. Note that PC contents are not incremented during INA-MC.

As mentioned above, PCL is output on $AD_{\emptyset}$–AD_7 during T1 and PCH is output on A_8–A_{15} during T1, T2 and T3. An ALE pulse is also output during T1 causing the Address Bus to have PC during T1, T2 and T3 if properly latched. The 8085A outputs $\overline{INTA} = \emptyset$ from the beginning T2 to the middle of T3. This should be used to strobe the interrupting device and make it put on the data bus the *Op Code* of whatever instruction it wants executed following the interrupt. This can be any instruction (except EI or DI) but the two instructions commonly used in practice are either RST n or CALL ADDR. This must be done while INTR is low and wait states may have to be used (by pulsing READY low appropriately) if the interrupting device is slow at doing this. If the instruction to be executed is an RST n (Single byte instruction) additional INA–MCs are not required. For CALL ADDR, two more bytes are needed by 8085A. For this, the 8085A issues two more 3 T-state INA-MCs. These are similar to MR-MCs except that $\overline{INTA}$ is pulsed and not $\overline{RD}$. It is important to note that PC does not change in these MCs; this is necessary to ensure that the correct PC value is pushed into the stack.

As mentioned in the discussion of the OF-MC, the memory (or I/O) select signals should be properly qualified with $\overline{RD}$ and $\overline{WR}$ in order to prevent multiple device selection during T4, T5 and T6 of OF. The procedure outlined there to avoid this should be sufficient to prevent I/O or memory devices from being selected during the INA-MCs.

One final problem of selection still remains. From our description it seems that the $\overline{INTA}$ low pulse not only selects the interrupting device but also gives it a Read Strobe, so that it can place the proper Op Code on the data bus. This does not pose a problem if the interrupt requesting device is fast enough. Otherwise, it may be beneficial to select the interrupting device with an interrupt acknowledgement generated using $\overline{(IO/\overline{M})\cdot S1\cdot S0}$ (active low) and strobe it for reading with the $\overline{INTA}$ pulse. It should also be noted that unlike other MCs described earlier, PC contents are not incremented during the INA machine cycle. This is done so that the proper value of PC can be stored as the return address (to return to after the interrupt has been serviced).

9.2.7 Bus Idle (BI) Machine Cycles

The 8085A response to a RST 5.5, RST 6.5, RST 7.5 (internally latched request) or TRAP interrupt is very similar to its response to the INTR interrupt. The following differences should however be noted

(i) $\overline{INTA}$ is not pulsed low in acknowledgement.

(ii) Restart vector is internally generated and is not provided by the interrupting device. (Hence the $AD_{\emptyset}$–AD_7 bus floats during T2, T3, T4, T5 and T6.)

(iii) ALE is pulsed but $\overline{RD}$ is not generated. In other respects it is similar to an OF machine cycle.

This is an example of a Bus Idle (BI) Instruction Fetch machine cycle where the instruction is fetched internally but the 8085A needs the Bus Idle time to create the appropriate instruction object code internally.

During a Bus Idle (BI) MC control signals being output by the 8085A do not change state. Three types of BI–MCs are possible.

(i) An instruction fetch BI executed after a RST 5.5–7.5 or TRAP interrupt. This has been described earlier. It requires 6 T-states for execution. For this, $IO/\overline{M} = S\emptyset = S1 = \overline{RD} = \overline{WR} = \overline{INTA} = 1$.

(ii) An instruction execute BI following the DAD instruction (DAD requires two of these MCs following the OF-MC). This MC requires 3 T-states. During T1, PCH is output on A_8–A_{15} and PCL on $AD_{\emptyset}$–AD_7 just as in a MR machine cycle but ALE and $\overline{RD}$ are not generated by the 8085A. $AD_{\emptyset}$-AD_7 are floated for the rest of the MC, i.e. T2 and T3. For this MC, $IO/\overline{M} = \emptyset$, $S\emptyset = \emptyset$ and $S1 = 1$ during the complete MC and $\overline{RD} = \overline{WR} = \overline{INTA} = 1$.

(iii) HALT Bus Idle is generated following a HALT instruction. The HALT instruction needs 4 T-states to be executed. It is then followed by HALT states. In the first HALT state PCH, PCL are output respectively on A_8–A_{15} and $AD_{\emptyset}$–AD_7 and an ALE pulse is provided. In the succeeding HALT states, A_8–A_{15}, $AD_{\emptyset}$–AD_7, $IO/\overline{M}$, $\overline{RD}$ and $\overline{WR}$ are floated while $\overline{INTA} = 1$, $S\emptyset = 1$, $S1 = \emptyset$ and ALE is not pulsed. An indeterminate number of HALT BIs can be executed before the 8085A leaves this state (via an interrupt or RESET). Interrupts are sampled in each HALT state. If interrupts are enabled and an interrupt is sensed in the n-th HALT state, the 8085A executes one more HALT state (i.e. $n + 1$) before responding to the interrupt with an appropriate machine cycle.

9.2.8 WAIT, HOLD and HALT States

The WAIT state is entered by inputting $READY = \emptyset$ to the 8085A. The effect is to insert extra clock cycles—T_{WS}—between T2 and T3. The WAIT state is exited by returning READY to 1. The purpose of the WAIT state is to stretch MC times so that slow peripherals can respond properly (exception: WAIT state cannot be introduced in the instruction execute BI machine cycle, e.g. MC_2 and MC_3 of DAD, but then WAIT states here would be useless).

The HOLD state is entered by raising HOLD HI. During this state, system busses are floated and ALE stays low allowing other CPUs to capture the system bus. It is exited by returning HOLD to LO. Interrupts are not recognized during HOLD.

The HALT state follows a HALT instruction when HALT BI-MCs are created. It can be exited via interrupts or RESET.

It is possible to enter and exit a HOLD state while being in the HALT state. If you are in a HOLD state within a HALT state, you cannot exit the HALT state using interrupts; the HALT state can be exited only after exiting from the HOLD state. Resetting the 8085A will of course allow you to exit both the HALT and the HOLD states.

9.2.9 Initiation and Termination of the WAIT, HOLD and HALT States

As mentioned earlier, the 8085A initiates WAIT states if it finds the READY input LOW. This is done by sampling the status of the READY line in the middle of T2. If READY is HI, it proceeds normally to T3; if it is LOW, it inserts WAIT states TW, between T2 and T3. It also samples READY during the middle of each WAIT state (WAIT state duration = 1 clock cycle). It continues to state T3, if it samples READY to be HI at this time; else, it introduces WAIT states until READY goes HI and is sampled to be HI in the middle of a WAIT state.

A HALT state is initiated by software. The 8085A enters a HALT state only if it executes the HLT instruction. It can exit a HALT state only via an interrupt, a RESET or by going into a HOLD state. If the 8085A enters the HOLD state from a HALT state, it returns to the HALT state once the HOLD state terminates.

A HOLD state is initiated by providing a HI level at the 8085A's HOLD input. In every machine cycle the HOLD input is sampled during T2 of every machine cycle. If $HOLD = 1$ at this time, HLDA is output HI during T3 and the 8085A initiates the HOLD state from the next clock cycle. For a six-cycle machine cycle, if HOLD is sampled LO during T2 then HOLD will be sampled again during T4. If it is sampled HI at this time, then HLDA is output HI during T5 of this machine cycle and the 8085A enters the HOLD state from T6 of this machine cycle. The HOLD input is also sampled during every clock period of a HALT state. In this case, if HOLD is sampled HI, HLDA is output HI in the next clock cycle and the actual HOLD state is initiated from the following clock cycle. The HOLD state terminates two clock periods after the HOLD signal goes LO.

9.3 COMPARING THE 8085A WITH OTHER 8-BIT MICROPROCESSORS

If we survey the field of microprocessor applications today, we will find that a number of 8-bit CPU's are currently in use. The most popular 8-bit CPU's now are probably the Intel 8085A, Zilog Z-80, Motorola 6800 and the MOS Technology 6502 along with a number of somewhat less popular devices. Examining these processors critically, it is very difficult to select one which is outstandingly better than the others. As a matter of fact, it is fair to say that comparing any two CPU's, while one may outperform the other in some respects, in other respects it may actually be inferior. It is also worthwhile to note that the reason why the 8085A has been dealt with in detail in this text is not because it is a better processor but rather because of its popularity and the excellent hardware and software support available for it.

The 8085A is certainly better than all the other CPU's mentioned above in one respect—its ability to support a number of vectored interrupts, e.g. TRAP, RST 5.5, RST 6.5 and RST 7.5, apart from the usual general purpose interrupt. These vectored interrupts make it possible for the user to build low-cost, interrupt controlled systems without investing in an interrupt controller. Apart from the excellent hardware and software support available for the 8085A, the presence of these vectored interrupts was probably one of the main reasons for this CPU's popularity.

Another good feature of the 8085A, compared to the 6800 or the 6502, is the presence of a large number of data registers in the 8085A. By using these registers for storing intermediate data, memory access operations can be effectively minimized. The paucity of general purpose registers in the 6800 and the 6502 imply that memory access is done more often. The Z-80 provides more registers than the 8085A. In this CPU, one finds the register set of the 8085A (A, PSW, B, C, D, E, H, L) to be duplicated so that one can change from one set of registers (register bank) to the other, under software control. Apart from these duplicated registers, the Z-80 also has some other additional registers not present in the 8085A (i.e. two Index Registers, an Interrupt Vector Register and a Memory Refresh Counter).

It is fairly common to find different microprocessors being compared on the basis of their instruction sets. Strictly speaking, it is not always possible to do so because an instruction set may be better than another, only in some respects. Moreover, the architecture of the CPU usually has a strong influence on what is available in its instruction set. If we examine the four CPU's mentioned above (i.e., the 8085A, Z-80, 6800 and 6502), we find that though they are roughly equivalent in terms of their arithmetic and logical operations, they differ substantially in their modes of data addressing.

With its large number of registers, the 8085A provides a very large set of instructions to move data between them. However, its capability of addressing memory is rather poor; it can only cater to immediate data, direct addressing and register-indirect addressing. With very few internal registers, the 6800 and the 6502 need to provide more addressing modes to make memory accessing easier. The 6800 provides one 16-bit Index Register to support indexed addressing apart from the usual direct addressing mode. The 6502 provides two 8-bit Index Registers

Table 9.1 A Comparison of the 8085A, Z-80, 6800 and 6502 CPU's

Criteria for Comparison	8085A	Z-80	6800	6502
Address Data Length	16-bit address, 8-bit data, ADØ-AD7 multiplexes DØ–D7 with AØ–A7	16-bit address 8-bit data No multiplexing	16-bit address 8-bit data No multiplexing	16-bit address 8-bit data No multiplexing
I/O Mapped I/O	Yes 8-bit I/O Address	Yes 8-bit I/O Address	No	No
Read/Write Control	Separate $\overline{RD}$ and $\overline{WR}$ pulses	Separate $\overline{RD}$ and $\overline{WR}$ pulses	Single RD/$\overline{WR}$ (R/$\overline{W}$)	Single RD/$\overline{WR}$ (R/$\overline{W}$)
DMA Control Lines	HOLD, HLDA	$\overline{BUSRQ}$, $\overline{BUSAK}$	$\overline{HALT}$, TSC BA, DBE	Extra logic
WAIT States for slow peripherals	READY	$\overline{WAIT}$	TSC, DBE, $\overline{HALT}$ Stretch clock cycles	RDY
Dynamic Memory Control	None	Yes using $\overline{RFSH}$	None	None
Reset	$\overline{RESET\ IN}$ RESET OUT	RESET	RESET	$\overline{RESET}$
Serial I/O	SID, SOD	None	None	None
Non-maskable Interrupt	TRAP	$\overline{NMI}$	$\overline{NMI}$	$\overline{NMI}$

General Purpose Interrupt	INTR Forced Instruction	$\overline{INT}$ Modes 0, 1 and 2 Mode 0: like 8085A RST n response Mode 1: branch to ØØ38H Mode 2: through interrupt vector table	$\overline{IRQ}$ Execute instruction at ((FFF8), (FFF9))	$\overline{IRQ}$ Same as 6800
Other Interrupts	RST 5.5, RST 6.5, RST 7.5	None	None	None
Status Signals	S_1, $S_\emptyset$, IO/$\overline{M}$	$\overline{MREQ}$, $\overline{IORQ}$ $\overline{M1}$, $\overline{HALT}$	VMA, BA	SYNC
Software Interrupts	None	None	SWI	BRK
Data Addressing	Immediate, Direct, Register Indirect	Immediate, Direct, Register Indirect, Indexed Direct (or Base Relative)	Immediate, Direct, Direct Indexed	Immediate, Direct, Direct Indexed, Indirect Indexed (pre and post indexing)
Internal Registers	16-bit PC and SP 8-bit A, B, C, D, E, H, L, PSW	16-bit PC, SP, IX and IY (IX, IY are Index Registers) 8-bit A, B, C, D, E, H, L, PSW (double) 8-bit Interrupt Vector Register 8-bit Memory Refresh Counter	16-bit PC, SP, and X (X: Index Register) 8-bit Accumulators A and B 8-bit Status Register	16-bit PC 8-bit SP, X, Y (X, Y: Index Registers) 8-bit Accumulator 8-bit Status Register
Flags	S, Z, P, CY, AC	S, Z, P/O, CY, AC, N O: Overflow N: Flag for BCD Subtraction	S, Z, AC, O, CY	S, Z, O, CY, D, B D: Decimal Mode B: Break

(*contd*)

Table 9.1 *(contd)*

Criteria for Comparison	8085A	Z-80	6800	6502
Branching	Direct, Register Indirect	Direct, Register Indirect, Short Program Relative Jump, Decrement and Jump on Not Zero	Short and Long Direct, Short Jump on Condition, Program Relative Branch, Branch through Index Register	Direct, Indirect Short Jump on Condition
Arithmetic Operations	Byte Add and Subtract, Double Byte Add, BCD Add	Byte Add and Subtract, Double Byte Add and Subtract, BCD Add and Subtract	Byte Add and Subtract, BCD Add	Byte Add and Subtract, BCD Add and Subtract
Logical Operations	AND, OR, EX-OR, Complement	AND, OR, EX-OR, Complement	AND, OR, EX-OR Complement	AND, OR, EX-OR
Block Move Operations	None	Memory to Memory Memory to I/O I/O to Memory	None	None
Block Search Instructions	None	Yes	None	None
Bit Testing	None directly	Possible to directly test or alter bits of registers or memory	None directly	None directly

for indexed addressing. Apart from direct indexed addressing and base page, indexed addressing, the 6502 also supports indirect, indexed addressing with either pre-indexing or post-indexing. The Z-80 also provides two 16-bit Index Registers for direct indexed addressing.

The Z-80 CPU represents an upward compatible enhancement of the 8085A's predecessor, the 8080A, in terms of its instruction set. Comparing the Z-80 with the 8085A, the Z-80 supports all the instructions of the 8085A (except RIM and SIM) and more. Therefore, the instruction set of the Z-80 is certainly as powerful as that of the 8085A. In addition, it supports a variety of indexed addressing modes, and a number of other powerful instructions. To the beginner, the complexity of the Z-80's instructions may be somewhat worrying, which is probably one reason why it is perhaps not as popular as it should have been. One should also remember that whereas all instruction op-codes of the 8085A are only one-byte long, the Z-80's op-codes may have either one or two bytes; because of these longer instructions, the Z-80 does not give as much of an advantage over the 8085A as might otherwise be expected. Unlike the 8085A, the Z-80 provides some on-chip logic for dynamic RAM refresh, as well as instructions to move blocks of data between two memory areas or between memory and I/O, test or alter the condition of individual register or memory bits. It also has an internal Interrupt Vector register for somewhat simpler interrupt vectoring (in some respects) than the 8085A. The fact that the 8085A is still competing with the Z-80 in the market place today, testifies to the usefulness of the former's vectored interrupts and Intel's prominence in providing hardware and software support. [These are the reasons leading to the choice of the 8085A as the CPU of interest in this text.] In Table 9.1, we have highlighted some of the comparative features of the 8085A, Z-80, 6800, and the 6502.

PROBLEMS

1. Give a circuit which will generate one WAIT state in every Op Code Fetch Machine Cycle (OF-MC).
2. A 8085A can be made to single step machine cycle by machine cycle using appropriate circuitry to manipulate READY. This circuitry should be such that whenever there is a Hi $\rightarrow$ Lo edge in an input SS, the 8085A should execute one MC. It should then WAIT in the next MC until the next Hi $\rightarrow$ Lo edge of SS.
3. Repeat (2) for the case where one wants to step the 8085A from one OF-MC to another.
4. A 8085A microcomputer system uses a particular RAM chip which needs $\overline{RD}$ or $\overline{WR}$ pulses of at least 2.5 μs for proper operation. The system designer has placed this RAM chip in locations Ø8ØØ-ØFFFH.

   Can this chip be used directly in a system where $f_{crystal} = 2$ MHz? If not, what additional circuitry is needed?
5. Give a circuit which will introduce two WAIT states in each MC (except BI-MC). Modify your circuit (or give a new one) such that two WAIT states are introduced only in the OF-MC. In each case, show appropriate timing diagrams which should explain the operation of your circuits.
6. For the 8085A, give a circuit which will introduce two WAIT states in the MR, MW and OF machine cycles and three WAIT states in the IOR, IOW and INA machine cycles.
7. Assuming $f_{crystal} = 2$ MHz, write a software delay loop which should take approximately 1 ms to execute.

8. Design a circuit to introduce two WAIT states, whenever a memory write operation is attempted on memory locations 8ØØØH-83FFH.

9. Design a circuit which will detect whenever the 8085A tries to write data into a ROM area. This circuit should force a TRAP interrupt in case such an event occurs. Assume ROM area to be from ØØØØH to Ø7FFH.

10. How can one modify the approach of (9) to guard against stack overflowing into memory areas not designated for the stack? Design the software and hardware to do this.

11.
```
DELAY:  LXI B, N
LOOP:   DCX B
        MOV A, B
        ORA C
        JNZ LOOP
```

(a) For the delay loop shown what is the total delay as a function of N given that $f_{crystal} = 2$ MHz (i.e. $T_{CLK} = 1$ μsec).

(b) Suggest a modification to the program so that the delay can be increased by 2N T_{CLK} sec.

Note the programming trick employed to check (BC) = ØØØØH.

12. A 8085A system is configured using a (2K × 8) ROM chip, one 2K × 8 RAM chip, three input ports (IPOR1, IPOR2, IPOR3) and three output ports (OPOR1, OPOR2, OPOR3). Subject to the requirement that the ROM must be in locations ØØØØH-Ø7FFH give a circuit to generate the proper chip selects. Describe the addressing of the various components of this system using the chip select circuitry designed.

13. Consider Prob. 8.13(a) with (HL) = N on entering DELAY. Compute the total number of T-states required to execute DELAY as a function of N.

14. State how many MCs each of the following instructions needs and identify their type:

(i) LDA Addr, (ii) MOV A, M, (iii) STAX B, (iv) LHLD Addr, (v) IN port, (vi) DAD D, (vii) ADD M, (viii) CALL Addr, (ix) RET, (x) PUSH PSW, (xi) POP C, (xii) JZ Addr, (both possibilities).

15. During a MR-MC the READY line goes Low and the HOLD line Hi in the middle of T1. They revert back to their original states after $3\frac{1}{2}$ and 5 clock states respectively. Draw the complete timing diagram for this case.

16. A RST 5.5 interrupt occurs at the end of the instruction SUB B stored in location Ø82ØH. Assuming (SP) = ØB1ØH, what are the contents of A_8–A_{15} and $AD_Ø$–AD_7 during the MCs following the SUB B instruction? Draw the timing diagram.

Appendix A9

A9.1 MACHINE CYCLES OF THE 8085A's INSTRUCTIONS

Instructions		Bytes	T States	Machine Cycles
(1) Data Transfer Group				
MOV	r, r/XCHG	1	4	1
MOV	r, M/LDAX rp	1	7	1, 3
MOV	M, r/STAX rp	1	7	1, 5
MVI	r, Data	2	7	1, 3
MVI	M, Data	2	10	1, 3, 5
LXI	rp, Data 16	3	10	1, 3, 3
LDA	Addr	3	13	1, 3, 3, 3
STA	Addr	3	13	1, 3, 3, 5
LHLD	Addr	3	16	1, 3, 3, 3, 3
SHLD	Addr	3	16	1, 3, 3, 5, 5
(2) Arithmetic Group				
ADD	r/ADC r/SUB r/SBB r/INR r/DCR r/DAA	1	4	1
ADD	M/ADC M/SUB M/SBB M	1	7	1, 3
ADI	Data/ACI Data/SUI Data/SBI Data	2	7	1, 3
DCR	M/INR M	1	10	1, 3, 5
DCX	rp/INX rp	1	6	2
DAD	rp	1	10	1, 7, 7
(3) Logical Group				
ANA	r/ORA r/XRA r/CMP r/RLC/RRC/RAL/RAR/ CMA/CMC/STC	1	4	1
ANA	M/ORA M/XRA M/CMP M	1	7	1, 3
ANI	Data/ORI Data/XRI Data/CPI Data	2	7	1, 3
(4) Branch Group				
PCHL		1	6	2

Instructions	Bytes	T States	Machine Cycles
JMP Addr/J<cond> Addr if TRUE	3	10	1, 3, 3
J<cond> Addr if FALSE	3	7	1, 3
CALL Addr/C<cond> Addr if TRUE	3	18	2, 3, 3, 5, 5
C<cond> Addr if FALSE	3	9	2, 3
RET/R<cond> if TRUE	1	10	1, 3, 3
R<cond> if FALSE	1	7	1, 3
RSTn	1	12	2, 5, 5
(5) Stack, I/O and Machine Control			
RIM/SIM/EI/DI/NOP	1	4	1
HLT	1	5	1*
SPHL	1	6	2
IN Port	2	10	1, 3, 4
OUT Port	2	10	1, 3, 6
XTHL	1	16	1, 3, 3, 5, 5
PUSH rp/PUSH PSW	1	12	2, 5, 5
POP rp/POP PSW	1	10	1, 3, 3

Types of Machine Cycles (symbols used above)

1	4 T-State	Op Code Fetch (OF)
2	6 T-State	Op Code Fetch (OF)
3	3 T-State	Memory Read (MR)
4	3 T-State	I/O Read (IOR)
5	3 T-State	Memory Write (MW)
6	3 T-State	I/O Write (IOW)
7	3 T-State	Bus Idle (BI)
*	Additional T-States	

A9.2 THE INTEL 8282 AND 8283 OCTAL LATCHES

The Intel 8282 is an Octal Latch with tristate control for the outputs. This device is very useful in microprocessor systems and may be used to implement the following:

(i) Simple (unlatched) input ports,

(ii) Latched input ports,

(iii) Latched output ports with outputs always enabled,
(iv) Latched output ports with tristate control on the outputs,
(v) Generate 8-bit interrupt vectors (e.g. RSTn for the 8085A),
(vi) Demultiplex $AD_{\emptyset}$–AD_7 lines of the 8085A.

The dependency notation symbol of the Intel 8282 is shown in Fig. A9.2.1 with the pin numbers of the various inputs and outputs. From this symbol it is evident that data into the 8 D-latches is latched from DIØ-DI7 by a High input at STB. Note that the data latch contents

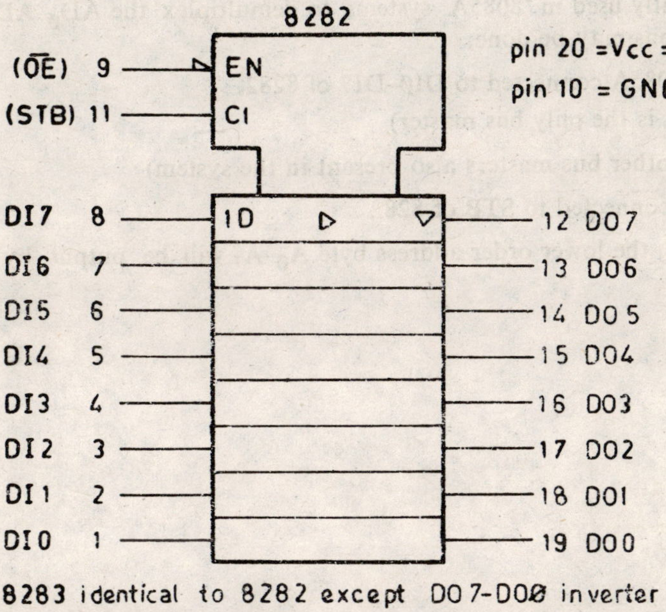

Fig. A9.2.1 Intel 8282 (8283) Noninverting (Inverting) Octal Latch

follow the DI inputs as long as STB is held High. The latch contents are output on the DOØ–DO7 lines as long as the $\overline{OE}$ input is Low; with $\overline{OE} = 1$, the data output lines stay tristate.

The 8283 is an Inverting Octal Latch which is very similar to the 8282. The only difference between the two is the 8283's outputs are the complements of what the 8282 will output in the same situation. Except for this difference, the two devices are identical (including the pin-out) as indicated in Fig. A9.2.1.

Normal TTL (LS Series) parts also exist which perform essentially the same functions as the 8282 and the 8283. These are the 74LS363 (non-inverting) and the 74LS364 (inverting) octal latches, respectively. It is important to note that the pin-out of the 74LS363 and the 74LS364 devices are not the same as those of the 8282 and the 8283.

To use the 8282 as an input port, DIØ–DI7 should be connected to the external device and DOØ–DO7 to the microcomputer's data bus. The mirocomputer should then generate

$$\overline{OE} = \overline{RD} + \overline{CS}$$

such that $\overline{OE} = \emptyset$ whenever the port is to be read. The external device may latch data into this port using STB or keep STB = 1, permanently, for an unlatched operation.

To use the 8282 as an output port, DIØ–DI7 is connected to the microcomputer's data bus with DOØ–DO7 being the output lines to the external world. The microcomputer should then generate

$$STB = \overline{\overline{WR} + \overline{CS}})$$

such that STB = 1 whenever it wants to write to this port. The port outputs DOØ–DO7 may be kept continuously active by keeping $\overline{OE}$ = Ø permanently; $\overline{OE}$ should be properly controlled if DOØ–DO7 are required to be tristate lines (e.g. when connecting to a bus).

The 8282 is frequently used in 8085A systems to demultiplex the $AD_Ø$–AD_7 lines. For this, the following connections may be done:

(a) $AD_Ø$–AD7 of 8085A connected to DIØ–DI7 of 8282,

(b) $\overline{OE}$ = Ø (8085A is the only bus master)

or $\overline{OE}$ = $\overline{HLDA}$ (other bus masters also present in the system)

(c) ALE of 8085A connected to STB of 8282

With these connections, the lower order address byte $A_Ø$–A_7 will be output by the 8282 on its DOØ–DO7 outputs.

Chapter 10

Input/Output (I/O) Techniques

10.1 I/O COMMUNICATING WITH DEVICES OUTSIDE THE MICROCOMPUTER

Functionally, a microcomputer may be viewed as an instrument which accepts data from devices outside the system, manipulates them according to its resident software and outputs the results to other external devices. Data may be input to a microcomputer from such diverse devices as switches, keyboard, data monitoring and acquisition devices, sensors and transducers or even other microcomputers or digital peripheral devices. Similarly, the microcomputer may output data to display elements, storage devices, controllers and actuators or other microcomputers and digital peripheral devices. I/O ports are devices which actually allow such communications between devices external to the microcomputer system and the microcomputer itself.

In order to appreciate the need for I/O ports, one should realize that the microcomputer does not (and, indeed, cannot) exercise any appreciable control on devices outside its own system. However, it must have full control over its own bus (Bussed Architecture, see Sec. 7.2) in order to function properly. An external device inputting data to the microcomputer should put its data on the system's data bus only when allowed to do so by the microcomputer. Otherwise, if the external device puts data on the data bus when it wanted to, the result, in terms of the microcomputer's operation, could very well be catastrophic in nature. Input Ports are devices used in this case by the microcomputer to ensure that the data from the external device appears on its data bus only when the microcomputer wants it to. (In essence, an input port is a gateway for data coming from an external device which can be opened or closed under the microcomputer's control.) Similarly, an Output Port is a device used to control the transfer of data from the microcomputer's system bus to an external device. This is necessary because all the data appearing on the data bus is never intended for the external device; data intended for the external device appears on the bus at very specific time instants. These time instants are known to the microcomputer but not to the external device. An output port is needed to capture the data intended for an external device and pass it on to that device in an appropriate fashion. (In essence, an output port is a 'filter' which filters the data appearing on the data bus for the particular data intended for the external device it is serving).

On occasion, an external device may be such that it both supplies data to the microcomputer and expects data from it. Such a device may be connected to the microcomputer's data bus using an appropriate combination of input and output ports. An alternative approach would be to use Bidirectional Ports for this situation. In contrast to input and output ports where the direction of data transfer is fixed during design, bidirectional ports can be used by the CPU to operate in either direction (to the CPU or from it), as needed.

A discussion of I/O addressing has been presented earlier in Sec. 7.3.3. Chapter 7 also presented a brief description of the various I/O techniques commonly used. A more detailed description of these I/O techniques will be presented in this chapter along with a discussion of other considerations which arise in I/O operations. Brief descriptions of some LSI devices commonly used as I/O ports will also be presented in this chapter (as well as in Appendix A10.1–A10.4). Specific examples, wherever presented, will be in the context of the 8085A CPU.

10.2 I/O HANDSHAKING AND CONTROL

The simplest I/O operations we encounter in microcomputers are ones where the CPU executes the I/O operation, input or output, without paying any regard to the status of the external device—i.e. whether it is ready for the data transaction or not. This simple I/O technique, referred to as Unconditional Program Controlled I/O in Sec. 7.5, is initiated by the microcomputer and done completely under its control. In effect, the microcomputer assumes, in this case, that the device on the other side of the I/O port is always ready for the microcomputer-initiated data transaction. As mentioned earlier, an I/O transaction of this type is most commonly used to read switch positions or drive displays.

Unfortunately, most of the time, an external device interfacing to the microcomputer system will also have its own requirements regarding when it can participate in a data transaction with the microcomputer. In such situations, the simple Unconditional Program Controlled I/O mentioned above will not be applicable and more sophisticated I/O techniques are needed. A brief description of these (more sophisticated) I/O techniques—Conditional Program Controlled, Interrupt Controlled and Hardware Controlled I/O—have been given earlier in Sec. 7.5 and a more detailed discussion is also presented later in this chapter. Common features of these I/O techniques are given below.

(a) An external device (or an I/O port) which needs to carry out a data transaction with the microcomputer somehow signals this.

(b) The microcomputer carries out the I/O operation when both the microcomputer and the external device (and its I/O port) are ready for it. Only then is the 'request for service' actually attended to by the CPU.

(c) Carrying out the requested data transaction with the requesting device's I/O port actually amounts to servicing the 'request for service' generated by that device. The 'request for service' should then be removed.

As (a), (b) and (c) show, the more sophisticated I/O techniques require close coordination between the microcomputer and the external device (and its interfacing I/O port). From the point of view of the external device, such coordination is necessary so that it can get service from the CPU when it is needed. From the microcomputer's view point, such coordination is needed for the microcomputer to find out when service is being requested by an external device (and its I/O port) and when an I/O transaction (input data or output data) can indeed be carried on with this device. 'Handshaking' between the microcomputer and the external device using appropriate control lines provides this coordination. It also allows reliable data transfers between a microcomputer and the external devices (using the I/O ports as the interfacing devices between them). The handshaking procedure needed for I/O ports—input, output and bidirectional—are described separately next, along with a discussion of the appropriate control signals.

10.2.1 Handshaking for Data Inputs from External Device

For the simple Unconditional Program Controlled Input (Sec. 7.5) case, it is assumed that the external device always has the data available at the inputs to the input port and that the microcomputer can obtain this data at any time by issuing a READ command to this input port. (Note that, in this case, the input port need not latch the data. In that case, the data that the microcomputer gets on reading such an input port, is the actual data currently being provided by the external device). The only control lines needed for this I/O port, from the microcomputer side, will be a $\overline{RD}$ (Read) signal and a $\overline{CS}$ (Chip Select) signal; the I/O port drives the data bus only when both $\overline{CS}$ and $\overline{RD}$ are Ø. If desired, these two signals may be combined into one, $(\overline{RD} + \overline{CS})$; they are derived from the $\overline{RD}$, $IO/\overline{M}$ and the address lines $A_Ø$–A_{15} for the 8085A CPU. The exact method will depend on whether I/O Mapped or Memory Mapped I/O is being used. For Unconditional Program Controlled I/O, no further control lines or handshaking is needed.

In most applications, one very frequently comes across situations where it is no longer possible for the microcomputer to expect that data from the external device will always be there for it to read. In such situations, the external device will typically load data into an input port whenever it has such data ready and will expect the microprocessor to read this data before the next data word is made available. The following considerations should be taken into account.

(a) The microprocessor should somehow be made aware of the fact that there is data in the input port for it to read. It should also know if the data currently in the input port is an old one, which it has read before, or a new one it has not read yet. This information is passed between the input port and the microprocessor by means of handshaking signals.

(b) From the point of view of the external device, the microprocessor should read the data loaded into the input port as soon as possible, at least before it loads or tries to load a new data word into the port. In other words, the time interval between the 'request for service' from an external device and its associated input port, and the 'granting of such service' from the microprocessor should be as brief as possible. The timing considerations involved here are discussed in greater detail in the next section, Sec. 10.3.

Fig. 10.2.1 shows how an external device is interfaced to a microcomputer via an input port in a situation where handshaking is needed to transfer valid data to the microcomputer. Specifi-

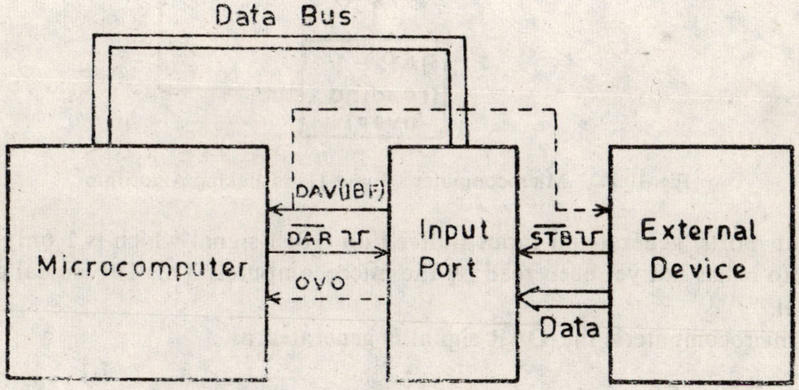

Fig. 10.2.1 Handshaking Interface for Input

cally, the handshake signals between the microcomputer and the input port are DAV and $\overline{DAR}$ with directions as shown. These signals do the following:

DAV = 1, Data Available at the input port for the microcomputer to read; this is new data which the microcomputer has not read before

= Ø, otherwise

$\overline{DAR} = \emptyset$, (Data Read Signal) asks the input port to place the data it wants the microcomputer to read on the data bus

$= 1$, input port should not drive the data bus

In order to function properly, the microcomputer is expected to monitor the DAV line periodically (e.g. using interrupts or another input port). When it senses DAV = 1, it issues a read command to the input port by outputting $\overline{DAR} = \emptyset$. After the reading is over, $\overline{DAR}$ stays at 1, until the external device loads another data word into it. At that time the reading process is repeated. This algorithm is illustrated by the block diagram of Fig. 10.2.2.

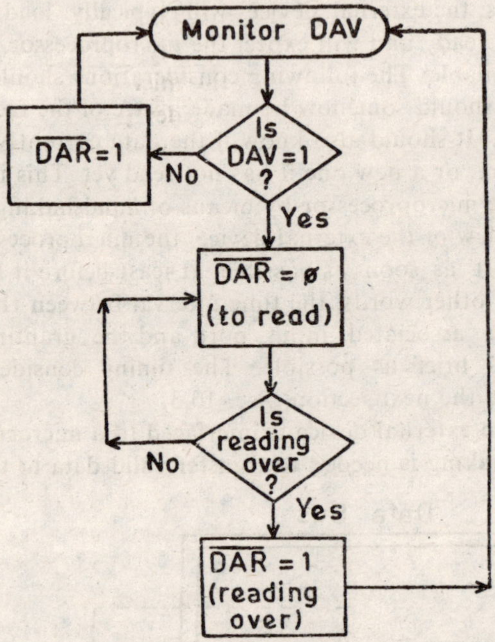

Fig. 10.2.2 Microcomputer's Input Handshaking Algorithm

Some input ports generate an Input Buffer Full (IBF) signal which is 1 only if the previous data loaded into it has not yet been read by the microcomputer. This IBF signal can be used as the DAV signal.

In typical microcomputers, the $\overline{DAR}$ signal is generated as

$$\overline{DAR} = \overline{RD} + \overline{CS}$$

It is also possible that the input port accepts separate $\overline{RD}$ and $\overline{CS}$ inputs rather than the single $\overline{DAR}$ input. In most cases, $\overline{RD}$ and $\overline{CS}$ for the input port are derived using the address outputs of the microprocessor and its IO/$\overline{M}$ (if present) and $\overline{RD}$ outputs; the exact decoding used will depend on the I/O addressing mode and the address of the input port.

The external device has been shown (Fig. 10.2.1) to load data into the port using a $\overline{STB}$ (Strobe) pulse. It is possible to prevent data from being overwritten into the port (i.e. a new data word loaded before the previous data could be read) if the external device also monitors

DAV and loads a new data word only if DAV = Ø. Sometimes the port also provides an output OVO (overflow) to the microcomputer to signal the condition when a data word has been lost because the external device has overwritten into the input port (i.e. OVO = 1, if data was overwritten; else OVO = Ø).

It is also possible that the DAV and OVO information is presented to the microcomputer as the contents of a register, read as a specifically designated input port (different from the input port containing the data). This is read in order to find the values of these signals. In this case, the signal lines themselves may or may not be present. Finally, one may come across situations where the input ports available do not support the handshaking shown in Fig. 10.2.1. In that case, external hardware, as shown in Fig. 10.2.3, may be used to provide this handshaking.

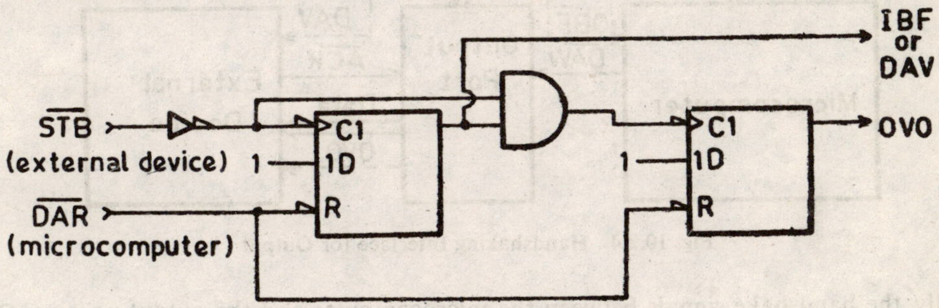

Fig. 10.2.3 Handshaking Signals for an Input Port

10.2.2 Handshaking for Data Outputs to External Device

Consider a situation where the microcomputer is to be interfaced to an external device which is always ready to accept data from the microcomputer. In that case, the output port interfacing this device to the microcomputer will be one whose outputs are always active and which latches whatever data is written into it by the microcomputer. The microcomputer can use Unconditional Program Controlled I/O (Sec. 7.5) to write data into this output port whenever it has data for the external device. (The latching of data at the output port is necessary to maintain valid port outputs, in between the microcomputer's write operations). From the microprocessor's side, the only control lines needed by this output port are $\overline{WR}$ (write) and $\overline{CS}$ (Chip Select). The port latches the data appearing on the data bus at the time instant when $\overline{CS}$ and $\overline{WR}$ are both Ø; this data is then output to the external device. If so desired, these two control signals may be combined into one, $\overline{WR} + \overline{CS}$; for the 8Ø85A CPU, they are derived from the $\overline{WR}$, IO/$\overline{M}$ and $A_{Ø}$-A_{15} address outputs of the microprocessor. As for the input case, the exact derivation will depend on the address of the I/O port and the I/O addressing mode being used.

In most applications, one frequently comes across situations where the external device is not always ready to accept data from the microcomputer. This, for example, could be the case when the external device is another microcomputer system which is able to read the port contents only at very specific instants in its own program execution. In such a situation, the following considerations will arise.

(a) Having loaded one data word into the output port, the microcomputer should avoid loading another data into it until the previous data is read by the external device. This is ensured by the handshaking described here.

(b) From the point of view of the microprocessor, the external device should read the contents of the output port as soon as it is loaded with data by the microprocessor; at least before it loads, or tries to load, new data into the port. These timing conditions are described in detail in Sec. 10.3.

Figure 10.2.4 shows how an external device is interfaced to a microcomputer via an output port in a situation where handshaking is needed to transfer valid data to the external device.

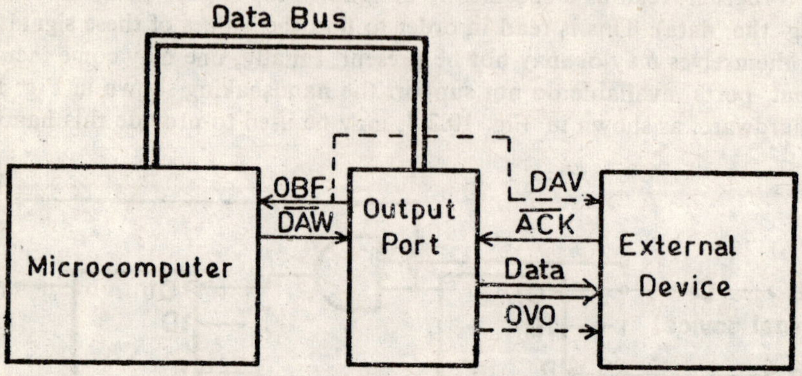

Fig. 10.2.4 Handshaking Interface for Output

Specifically, the handshake signals between the microcomputer and the output port are OBF and $\overline{\text{DAW}}$ which indicate the following:

OBF = 1; Output Buffer Full (OBF) since last data loaded into the port by the microcomputer has not yet been read by the external device

= ∅; otherwise

$\overline{\text{DAW}}$ = ∅; (Data Write Signal) latches the data on the data bus of the microcomputer into the output port

= 1; otherwise (output port's latch contents do not change)

In this case, the microcomputer should monitor the status of OBF before outputting data to the output port. If data needs to be sent to the external device and if OBF = ∅, then the microcomputer loads data into the port by outputting $\overline{\text{DAW}}$ = ∅. When $\overline{\text{DAW}}$ = ∅, the contents of the data bus are latched into the port. Once this latching operation is over, $\overline{\text{DAW}}$ returns to 1 and OBF also becomes 1 inhibiting further data transfers to the port. The OBF = 1. signal can serve as an indication to the external device telling it that there is data in the port for it to read. The external device reads this data by presenting $\overline{\text{ACK}}$ = ∅ and returning $\overline{\text{ACK}}$ to 1 once the data reading is over. This pulsing Hi-Lo-Hi of $\overline{\text{ACK}}$ resets OBF = ∅ signalling the microcomputer that it can now load fresh data into the port if more data transfers are required. Note that the OBF output of the port may be used as the DAV to the external device and that $\overline{\text{ACK}}$ of the port is actually the $\overline{\text{DAR}}$ output of the external device (see earlier Sec. 10.2.1). Similarly, the STB signal of Sec. 7.2.1 is actually the $\overline{\text{DAW}}$ input to the port. It is easy to see that the input port shown in Fig. 10.2.1 may be used as the output port of Fig. 10.2.4 by appropriately reversing the connections between it and the external device and the microcomputer. The algorithm used for this output handshaking is illustrated by the block diagram of Fig. 10.2.5.

In Sec. 10.2.1, we had shown that the DAV signal from the input port is the one actually

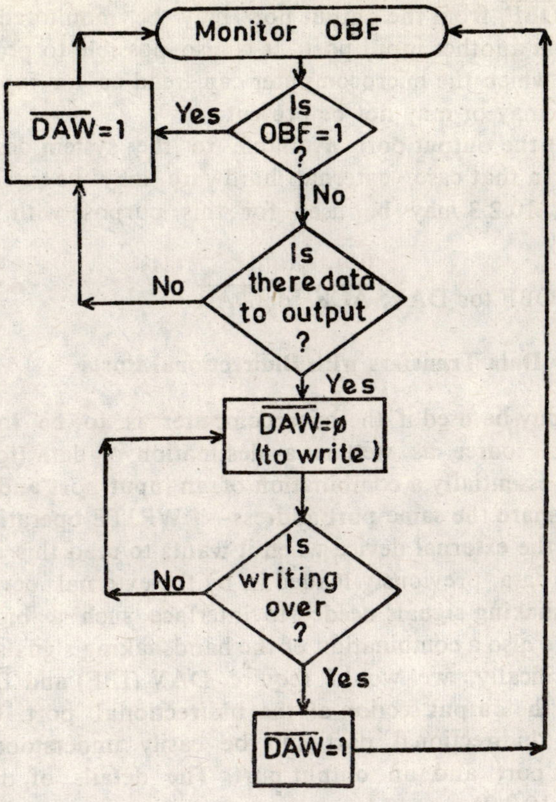

Fig. 10.2.5 Microcomputer's Output Handshaking Algorithm

placing a 'Request for Service' (when DAV = 1) to the microcomputer. For the output port, a similar service request can be placed using $\overline{OBF}$ (i.e. request for service when OBF = Ø or $\overline{OBF}$ = 1). In this regard, it should be noted that the microcomputer may decide on its own, when to pay attention to a service request and when to disregard it; it need not always pay attention to such a request.

In typical microcomputers, the $\overline{DAW}$ output is generated as

$$\overline{DAW} = \overline{WR} + \overline{CS}$$

It is also possible that the output port accepts separate $\overline{WR}$ and $\overline{CS}$ inputs rather than the single $\overline{DAW}$ signal. In most cases, the $\overline{WR}$ and $\overline{CS}$ inputs to the output port may be derived using the $\overline{WR}$, IO/$\overline{M}$ (if present) and address outputs of the microprocessor. The exact decoding used will depend on the I/O addressing mode and the address of the output port.

The microcomputer will usually be able to avoid overwriting data into the output port by monitoring the OBF line and taking appropriate action. However, situations may arise where it is forced to load fresh data into a port before the previous data can be read by the external device. In such a situation, this overwriting can be signalled to the external device if an OVO output is provided by the output port.

OVO = 1; overwriting has occurred
= Ø: otherwise

If OVO = 1, it is reset whenever the contents of the port are read.

The service request, $\overline{\text{OBF}}$, from the output port may be monitored by the microcomputer either via interrupts or via another input port. It is also possible to provide this information as the content of a register which the microcomputer can read as an input port. In this case, the physical signal line $\overline{\text{OBF}}$ may or may not be present.

It is also possible that the output port available to the system designer does not provide the above handshaking. In that case, external hardware may be used to provide this facility. The circuit shown in Fig. 10.2.3 may be used for this purpose with the following notational changes:

$\overline{\text{DAW}}$ for $\overline{\text{STB}}$, OBF for DAV, $\overline{\text{ACK}}$ for $\overline{\text{DAR}}$

10.2.3 Handshaking for Data Transfers with Bidirectional Ports

Bidirectional I/O ports may be used if the microcomputer is to be interfaced to an external device which is both a source as well as a destination of data (for the microcomputer). A bidirectional I/O port is essentially a combination of an input port and an output port where both the ports typically share the same port address—a WRITE operation on this port address loads data for output to the external device when it wants to read this data, a READ operation from this address inputs data, previously loaded in by the external device, into the microcomputer's CPU. The handshaking signals needed to interface such a bidirectional port with the microcomputer system are also a combination of the handshaking signals needed to interface input and output ports. Specifically, we would require DAV (IBF) and $\overline{\text{DAR}}$ for the input section and OBF and $\overline{\text{DAW}}$ for the output section of the bidirectional port. Using these handshaking lines, the operation of a bidirectional port can be easily understood by considering it as a combination of an input port and an output port. The details of this are left to the reader (refer to Sec. 10.2.1 and 10.2.2).

10.3 I/O TIMING

The handshaking signals, described in the previous section, were needed to ensure reliable data transfer between the microcomputer and an I/O port when such a transfer is deemed necessary. For the microcomputer, these handshaking signals essentially provide information about when a data transaction should be done with the corresponding I/O port as well as the signals (i.e. $\overline{\text{DAR}}$, $\overline{\text{DAW}}$) needed in order to effect this data transfer. Two kinds of timing considerations arise when one examines the I/O transactions in a microcomputer system—those that arise in the execution of a single (isolated) I/O transaction and those involved when one or more I/O devices want to engage the microcomputer in a series of I/O operations.

Every I/O port in a real system will have certain physical constraints which must be satisfied in order to successfully carry out an I/O operation between it and the microcomputer. Here we are assuming that the microcomputer somehow knows that such and such I/O port needs an I/O transaction. The timing considerations involved in this transaction are the 'physical constraints' mentioned earlier. These constitute the first kind of timing considerations one has to pay attention to while doing I/O operations. These timing considerations are further illustrated by means of the timing diagrams of Fig. 10.3.1a and Fig. 10.3.1b. These figures illustrate the timing diagrams of the $\overline{\text{CS}}$ and $\overline{\text{RD}}$ or $\overline{\text{WR}}$ signals for an input and an output port, respectively. In both cases, several timing parameters have been indicated. The manufacturer of an I/O port (input or output) would typically specify min-max limits on the values of these parameters. A microcomputer system using this I/O port should generate control signals satisfying these limits in order to carry out an I/O operation successfully. (Fig. 10.3.1a and Fig. 10.3.1b show the most

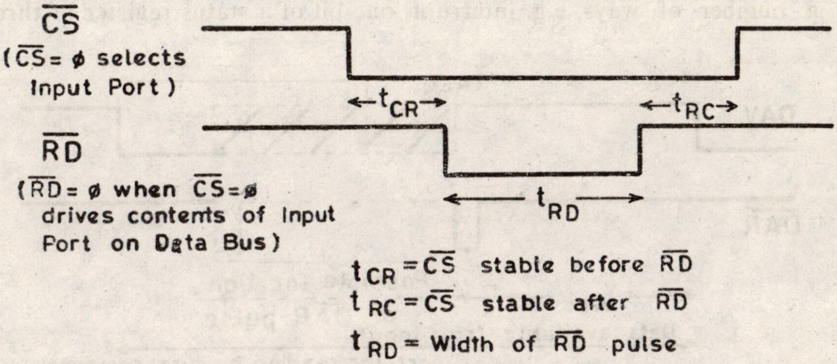

Fig. 10.3.1 (a) Timing for Reading from an Input Port

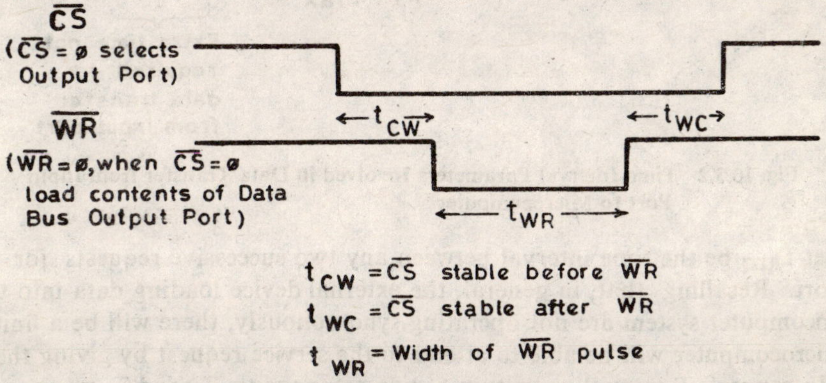

Fig. 10.3.1(b) Timing for Writing to an Output Port

important timing parameters. For a particular I/O port, the manufacturer may need to specify the limits of more parameters than are shown in these figures, e.g. when a port has more than one $\overline{CS}$ inputs). In order to simplify these considerations, a microprocessor manufacturer typically provides compatible I/O ports such that these timing considerations are automatically satisfied by its own microprocessor.

We have seen earlier that I/O devices request service from the microcomputer by making DAV = 1 (input port) or OBF = Ø (output port). The microcomputer should respond to this request by inputting data (from an input port) or outputting data (to an output port). This is done by appropriately providing a $\overline{DAR}$ (input port) or a $\overline{DAW}$ (output port) pulse to the port. If the number of I/O ports in a system is small and if the frequency with which a particular I/O port requests service is low, the probability of having potential timing problems is low. This, however, is not true if either the number of I/O ports is large or if a port tends to request service too often. Taking care of the potential timing problems that might be caused in this case, gives rise to a second kind of timing considerations.

In order to examine this more closely, consider Fig. 10.3.2 which illustrates a situation where an input port wants to transfer a sequence of data bytes to the microcomputer. The request for service is signalled by DAV = 1, where this request may be monitored by the microcomputer in a number of ways, e.g. interrupt, one bit of a status register or through another

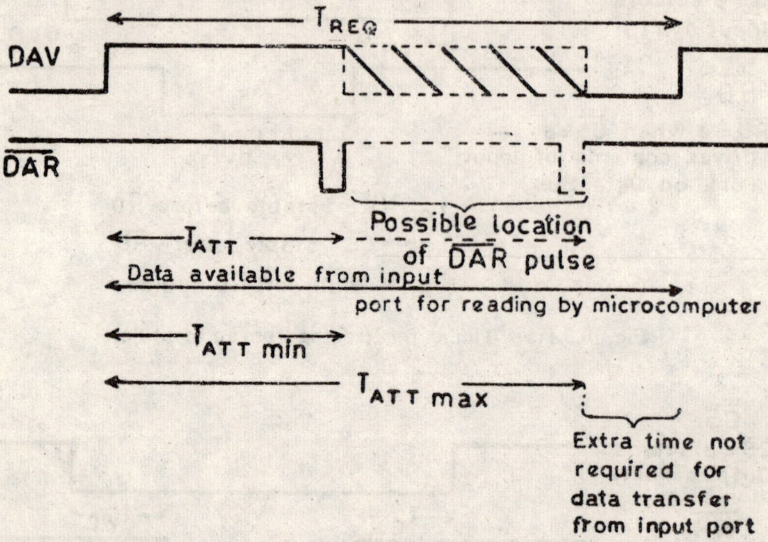

Fig. 10.3.2 Time Interval Parameters Involved in Data Transfer from Input Port to Microcomputer

input port. Let T_{REQ} be the time interval between any two successive requests for service from the input port. Recalling that, in general, the external device loading data into the input port and the microcomputer system are not operating synchronously, there will be a finite time T_{ATT} before the microcomputer will be able to attend to the service request by giving the input port a $\overline{DAR}$ pulse. In general, T_{ATT} will constitute of two components T_1 and T_2 where

T_1 = time taken by microcomputer to branch from whatever it is doing to the routine that services the requesting I/O port.

and T_2 = time taken by that particular service routine which services the I/O port.

Note that typically T_{ATT} will vary between T_2—assuming $T_1 \simeq 0$—and $(T_1 + T_2)_{max}$ and that one should satisfy the inequality

$$T_{ATT} < T_{REQ} \tag{1}$$

for proper operation. In case the inequality is violated for an input port, the phenomenon is referred to as Overwriting. The corresponding timing diagram for an output port is shown in Fig. 10.3.3. The comments made above are equally applicable for this case provided one substitutes $\overline{OBF}$ for DAV as the 'service request signal'. In case this inequality is not satisfied for an output port, the phenomenon is referred to as 'Underwriting'.

Referring to Fig. 10.3.2, and Fig. 10.3.3, one immediately realizes that the potential timing problems arising in this case are caused by the need to satisfy inequality (1) for all possible values of T_{REQ} and T_{ATT}. This is further complicated by the fact that both T_{REQ} and T_{ATT}

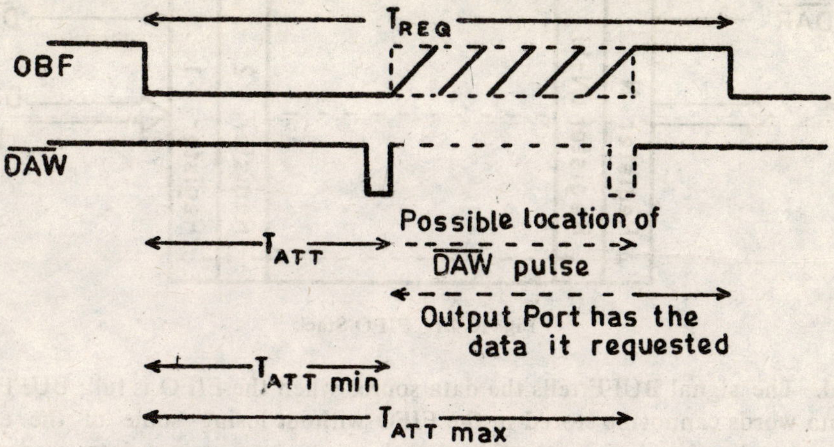

Fig. 10.3.3 Time Interval Parameters Involved in Data Transfer from Microcomputer to Output Port

might be random and that, where there are more than one I/O ports, inequality (1) will have to be satisfied for each one of them, individually. Two approaches may be taken to alleviate timing problems of this kind.

(a) Provide buffering (may be using FIFOs) to reduce the random nature of T_{REQ}. This is discussed in the next section (Sec. 10.4).

(b) Provide techniques for improving the speed with which the microcomputer responds to a service request. (Sec. 10.5, 10.6 and 10.7).

10.4 I/O BUFFERING-LATCHES AND FIFOs

The inequality relationship between T_{ATT} and T_{REQ} mentioned in the previous section can be translated into physical terms. What that inequality implies is that the microcomputer must pay attention to the request for service initiated by a particular I/O port before that I/O port has reasons to demand service once again. In order to do this, one has to ensure that the microcomputer responds to a service request as fast as it can. Various techniques to do this will be described in later sections. However, in case these techniques still do not provide proper operation, buffering can be employed to make sure that underwriting or overwriting does not occur.

In order to appreciate the need for buffering, consider a situation where an external device wants to transfer data to a microcomputer system through an input port. From the external device's point of view, the data transfer may become extremely awkward if it were required to hold the data inputs to the port until the microprocessor reads it. This problem has been alleviated, to some extent, in our earlier discussions by assuming that the input port latches the data it receives from the external device. (This leaves the external device free to do other things, if needed.) This provided one level of buffering. Multiple-level buffering using First In First Out (FIFO) stacks may be used to reduce the timing problems even further.

A FIFO stack is a block of N registers connected as shown in Fig. 10.4.1. Four handshake signals (or variations thereof) are provided for the FIFO—DAV, BUFF, $\overline{DAR}$, $\overline{DAW}$. The data source can load upon N data words in the FIFO by pulsing $\overline{DAW} = \emptyset$ for each data word

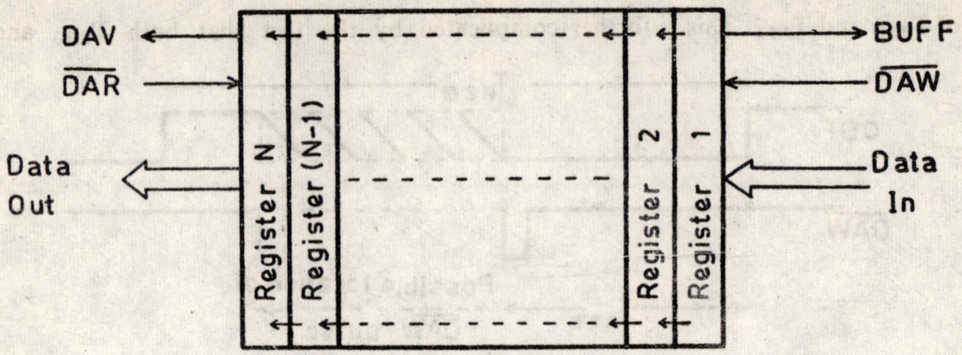

Fig. 10.4.1 FIFO Stack

to be stored. The signal BUFF tells the data source when the FIFO is full; BUFF = 1 implies that more data words cannot be stored in the FIFO without losing some of the earlier stored data. DAV = 1 tells the data destination that there are one or more data words stored in the FIFO for it to read These data words can be read by pulsing $\overline{DAR}$ = Ø, once for each data word, until DAV = Ø is detected.

Consider a FIFO which is initially empty, i.e. DAV = Ø, BUFF = Ø. Now let the data source write one word into the FIFO. This data 'ripples' down the FIFO from Register 1 to Register N where it is stored (now DAV = 1, BUFF = Ø). If the source writes one more data word before the previous data is read, the new data is stored in Register (N − 1).

The timing required to output from the FIFO is completely independent of the input. A data word input to the FIFO is available to be read as long as it is not lost due to FIFO-overflow. In a typical situation, the external device will load data into the FIFO as long as BUFF = Ø. From the moment the first data word is loaded, DAV becomes 1 signalling the microprocessor that the FIFO needs service. When the microprocessor is finally able to provide such service, it does not read merely one data word but rather all the data words currently in the FIFO; it empties the FIFO, thereby making DAV = Ø. It should be obvious that, with a large enough value of N, the FIFO would tend to smooth out the input-output process such that there will be no loss of data as long as the average data arrival rate from the source is less than or equal to the average data extraction rate of the destination.

A FIFO which is N-word deep will provide N levels of buffering compared to the one-level buffering provided by a simple latched port. Such a FIFO can be used either to input data to a microcomputer or to output data from it by providing appropriate handshaking.

10.5 PROGRAM CONTROLLED I/O—CONDITIONAL TRANSFERS

In the earlier sections of this chapter, we have described how an I/O port—input or output—places a service request to the microcomputer when it wants a data transfer operation to be done. This and later sections (10.6 and 10.7) describe how the microcomputer actually handles such requests. There are two aspects to this issue—(a) Sensing the request, (b) Initiating and executing the program to service this request.

Referring to the earlier sections, we notice that the request for service is actually signalled by the status of a particular signal output by the I/O port (DAV = 1 for input, OBF = Ø for output). The microcomputer can sense the status of these lines (for each I/O port) through some other input ports or directly through 'Interrupt' lines, if the interrupt feature is available. Alternatively, I/O port may internally provide a status register (read as an input port), particular bits of which indicate the required signal status. This second approach is commonly encountered in LSI I/O ports.

Consider a situation where the microcomputer is required to sense a request for service either by inputting data from an input port monitoring these signal lines or by reading the contents of an appropriate status register. As an example, we consider the circuit of Fig. 10.2.3 where DAV and OVO are input to the microcomputer as bits Ø and 1 of PORTØ. Assuming PORT1 to be the actual data input port, a routine servicing PORT1 is given below:

```
; 8085A
; PORT1 = Data Input
: PORTØ — bitØ = IBF (DAV), bit 1 = OVO
; D = Data Counter, HL = Data Address Reg (DAR)
CHECK:  IN PORTØ      ; stay in CHECK loop if IBF = Ø
        ANI Ø1H       ; since data not available
        JZ CHECK      ; exit if IBF = 1
SERVE:  IN PORT1      ; input data
        MOV M, A      ; and move to memory
        INX H         ; increment DAR
        DCR D         ; decrement data counter and
        JNZ CHECK     ; loop until Ø,
        RET           ; else exit.
```

A microcomputer executing the program CHECK given above repeatedly checks PORTØ to see if input port PORT1 needs service. If service is required, it executes the program SERVE to actually provide the desired service, i.e. input data from PORT1. A similar program could be written if PORT1 was an output rather than an input port. Note that in the above example, OVO is not checked for overflow.

In the example given above, the microcomputer checks, *under program control*, whether the particular I/O port needs service or not. This technique is referred to as Conditional Program Controlled I/O. Note that, in this case, the I/O port cannot actually demand service. It merely reports to the microcomputer, when asked, whether service is required. It is finally the microcomputer itself which decides (may be conditioned on other factors, e.g. priority, busy/idle status etc.), whether the I/O port is to be granted service or not. In an example such as the one given above, this is not a problem since the microcomputer does very little else other than monitor the service request. In a real application with probably more than one I/O port this could potentially pose a serious problem. This, for example, will be the case where the microcomputer has a number of other tasks to perform and can afford to check the service request status of the I/O ports only at very infrequent intervals.

If more than one device needs to be controlled using Conditional Program Controlled I/O, a Polling Strategy is usually recommended. Using this technique, the microcomputer's program tests the status of each I/O port in sequence at periodic intervals to see if service is required. The software designer may prioritize this checking sequence so that the I/O ports are serviced according to their priorities. Moreover, this priority structure may be modified, as desired, under program control. If a particular set of I/O ports in the system are more active than others, it can be taken into account in the software, i.e. by checking their service request status more often than those of the others.

Example: Consider a system with eight Input Ports—PORT1 to PORT8—and one status port, PORTØ, where Bit i of PORTØ reflects the condition of the DAV signal for PORT i. A polling sequence with PORT1 having the highest priority and PORT8 the lowest, is given below:

```
POLL:  IN PORTØ    ; (A) = (PORTØ)
       ANI Ø1H     ; check DAV1
       CNZ INP1    ; INP1 services PORT1 if DAV1 = 1
       IN PORTØ
       ANI Ø2H
       CNZ INP2
       :
       :
       IN PORTØ
       ANI 8ØH
       CNZ INP8
       RET
```

From the preceding description, the problems associated with Conditional Program Controlled I/O should be quite evident. The handshaking required for proper operation, i.e. testing of service request status, needs additional software and hardware as well as time to execute. This overhead is inherent in this I/O technique. A more serious problem is that of timing. As mentioned earlier, the microcomputer should service each I/O port at at least the same speed as the one at which the external device uses this port. In other words, the program testing the status bits should be executed in a loop 'tight' enough so that any request from an I/O port is serviced before the next service request from that port arrives. In time critical applications, it may not be possible to satisfy this condition, especially where the microcomputer is also required to perform a number of other equally time critical tasks. Moreover, using this technique, the I/O overhead increases with the number of I/O ports in the system even when the I/O ports require service only very intermittently. Some of these problems are avoided or reduced in the Interrupt Controlled I/O technique presented next.

10.6 INTERRUPT CONTROLLED I/O

A typical microprocessor usually provides one or more special inputs referred to as Interrupt inputs. Though the specific nature of these inputs vary considerably, they share one common feature—an appropriate signal (e.g. HI, LO, HI → LO, LO → HI etc.) at an interrupt input causes the microcomputer to execute immediately a CALL Addr type of subroutine call. The address Addr is either generated internally or supplied by external hardware and is commonly referred to as the Interrupt Vector. As a result of this interrupt signal, the microcomputer branches out from whatever program it was currently executing and, instead, begins executing the Interrupt Service Routine starting at address Addr. Since the interrupt acts as a kind of subroutine call, on terminating the interrupt service routine with a RETURN type of instruction, the microcomputer will resume execution of the previously interrupted program from the point of interruption. Finally, it should be noted that the interrupt input is a completely asynchronous input; at least to the extent that the microcomputer has no direct control over when it occurs, i.e. the interrupt input can interrupt the microprocessor's normal program execution at any point. In reality, most microprocessors can enable/disable their interrupt inputs under software control, e.g. executing the EI or DI instructions in the 8085A. These are typically used (a) to allow the microcomputer time to do proper initialization at start-up time without fear of interruption and (b) to permit the software designer to write critical sections of programs which he does not want interrupted. More will be said about Interrupts and how they are handled in Chapter 12.

Using the interrupt inputs of the microcomputer, an I/O technique more efficient than Conditional Program Controlled I/O can be devised. Such an I/O technique is referred to as Interrupt Controlled I/O. Using this technique, the I/O port requiring service drives the interrupt input of the microcomputer with the appropriate interrupt request signal. The interrupt Service Routine executed in response to this interrupt is also the service routine for that I/O port. Therefore from the viewpoint of the I/O port, it can get the desired service from the microcomputer merely by providing the proper interrupt signal. Note that unlike Conditional Program Controlled I/O, Interrupt Controlled I/O does not waste time and software checking if the I/O port needs service. (The interrupt is placed and the corresponding service routine executed only when service is actually needed.) The absence of the I/O overhead present in Conditional Program Controlled I/O is what makes this technique more efficient.

In typical microcomputer systems with a large number of I/O ports, one usually finds that only the time critical I/O requirements are serviced using Interrupt Controlled I/O techniques. There are two reasons for this approach. Firstly, the number of interrupt inputs available on a microprocessor are very limited (usually 1 to 5). If more I/O ports are to be served using interrupts than the number of interrupts available, extra hardware is needed to build an Interrupt Expanders, e.g. INTEL 8214 or 8259, Programmable Interrupt Control Unit (PICU). Moreover, extra hardware is also needed if the microprocessor's interrupt input requires that the interrupt vector be provided by external hardware. (The PICUs mentioned above do have this hardware for providing interrupt vectors.) This extra hardware implies greater cost which is another reason why the use of Interrupt Controlled I/O is generally minimized.

Interrupt Controlled I/O allows the I/O port to actually initiate I/O transfer operations using hardware external to the microprocessor. Even though it is faster and more efficient than Conditional Program Controlled I/O, it still has the following shortcomings.

(a) Even though the process is initiated by the I/O port, the actual data transfer operation still involves the CPU; the CPU has to execute the proper data transfer instructions. Ideally we would like to make the data transfer process completely transparent to the CPU's normal program execution.

(b) If a block of data is to be transferred between I/O port and memory, Interrupt Controlled I/O will do so by moving each data byte into the microprocessor and then transferring it to the destination (I/O or memory). Obviously, a more efficient method will be to allow such transfers to occur directly without the intermediate involvement of the CPU.

(c) As mentioned in (a), interrupts interfere with the CPU's normal program execution. Since these interrupt inputs are completely asynchronous in nature, debugging and testing a system using these is difficult, if not impossible. Generally, the software designer has to do very conservative programming (e.g. enable interrupts only at selected places in the program) in order to ensure that the interrupts do not cause the system to malfunction.

Some of these shortcomings of Interrupt Controlled I/O can be removed, or at least reduced, by using Hardware Controlled I/O or Direct Memory Access (DMA).

10.7 HARDWARE CONTROLLED I/O OR DIRECT MEMORY ACCESS (DMA)

Consider a situation where an I/O operation is required to transfer N data words (bytes for 8-bit systems) between the microcomputer's memory and an I/O port. This is done most easily if we somehow provide external hardware in the system which will take over the system bus from the CPU, and directly effect the transfer. This is what is done in Hardware Controlled I/O or DMA. The external hardware needed to implement this I/O technique is very often referred to as a DMA Controller. Note that since the entire set of N I/O operations (input or output) are effected directly, without active CPU intervention, this technique will have less timing and software overhead than either Interrupt Controlled I/O or Conditional Program Controlled I/O.

In order to provide the DMA I/O in a microcomputer system, the microprocessor used should have one important feature—it should be able to vacate the system bus (tristate it from the CPU side) on request from external hardware. The 8085A provides this feature using the HOLD input and the HLDA output where the HOLD-HLDA pair provides the handshaking required for DMA operations. Inputting HOLD = 1 forces the 8085A to float its address and data bus (ADØ–AD7, A8–A15) as well as the control signals $\overline{RD}$, $\overline{WR}$ and IO/$\overline{M}$. The 8085A also outputs HLDA = 1 signalling to the other system components that the CPU is currently in the HOLD state. It should be noted that in the HOLD state, 8085A is essentially 'marking time' and continues to do so while HOLD = 1. During this time, the microprocessor can continue operations which does not need the system bus. Its operation freezes once it reaches a point where memory or I/O references must be made. The HOLD state is terminated when the HOLD input returns to Ø. The HLDA output then also returns to Ø and normal CPU operation is resumed. Lines similar to HOLD and HLDA are usually present for all microprocessors which support DMA operations. The reader is advised to read the manufacturer's data sheets for more details regarding the use of these lines for microprocessors other than the 8085A.

Once the HOLD state is on, the DMA Controller takes over the system bus in order to actually execute the data transfer between I/O port and memory. Referring to Fig. 10.7.1, which shows a simplified schematic of a system using DMA, the DMA Controller needs to do the following in order to transfer one byte between memory and I/O port.

(i) Initiate the HOLD state of the microprocessor by placing a Hold Request (HRQ = 1). Proceed to (ii) only when HLDA = 1 signalling that the HOLD state is on.

(ii) DMA controller takes control of the system bus. It ensures that the appropriate memory location is selected by putting out the proper address on the Address Bus. It also selects the concerned I/O port by driving $\overline{DACK}$ = Ø ($\overline{DACK}$ connected to $\overline{CS}$ of I/O port).

(iii) (a) To input data from I/O port to memory, it generates HI → LO → HI pulses on $\overline{MEMW}$ and $\overline{IOR}$.

(b) To output data from memory to I/O port, it generates HI → LO → HI pulses on $\overline{MEMR}$ and $\overline{IOW}$.

Note: DMA operations configured as in Fig. 10.7.1 require decoded $\overline{MEMR}$, $\overline{MEMW}$, $\overline{IOR}$ and $\overline{IOW}$ rather than the undecoded signals $\overline{RD}$, $\overline{WR}$ and IO/$\overline{M}$. This is a direct consequence of the fact that a DMA operation requires the simultaneous presence of a WRITE pulse and a READ pulse where one of them is directed towards memory and the other towards I/O.

If more than one byte is to be transferred, the DMA Controller repeats steps (i) to (iii) with a different memory address (typically incremented or decremented by one). This, for example, could be done by properly programming the DMA Controller, if one wants to transfer blocks of data. The microcomputer programs the DMA Controller by writing the proper parameters into its internal registers. Typical parameters needed are the number of data words to be transferred, the memory addresses involved in the transfer and the direction of the transfer. A popular DMA Controller, INTEL 8257, is described in detail in Appendix A10.1.

Referring once again to Fig. 10.7.1, it should be noted that once the DMA Controller has been properly programmed, the I/O port can itself initiate the DMA operation by placing a DMA

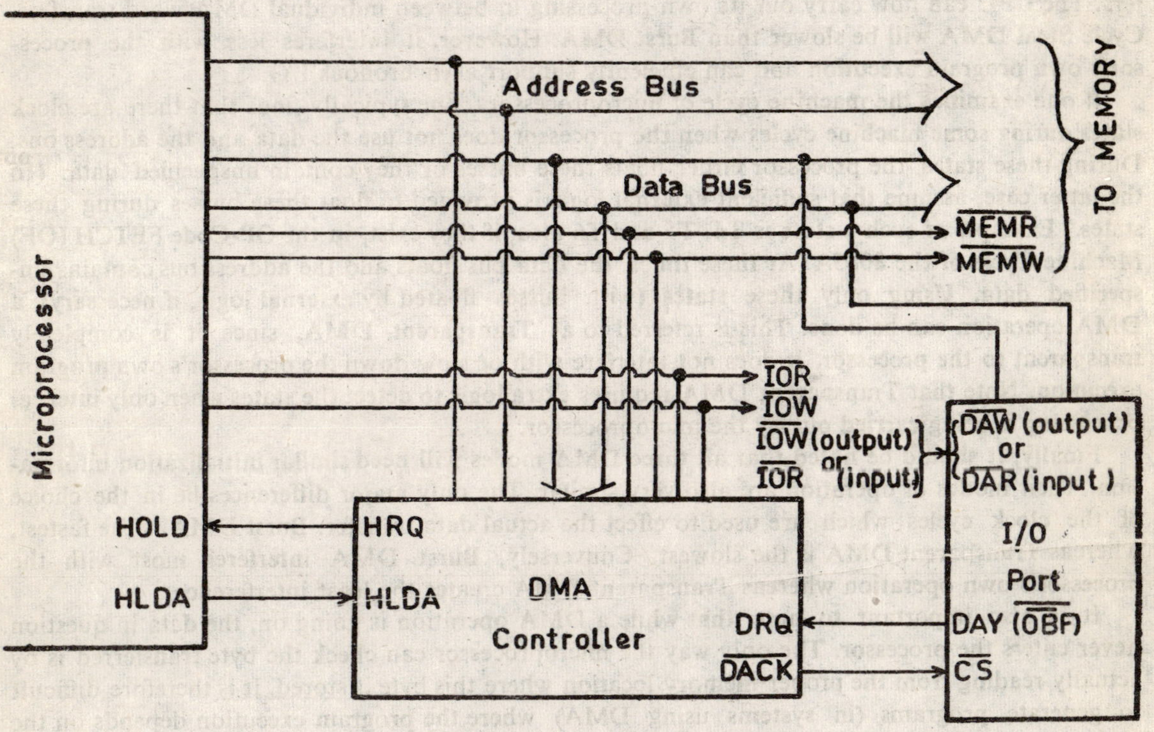

Fig. 10.7.1 DMA I/O

Request (DRQ = 1) to the controller. From our earlier description of handshaking, this request may be provided by the DAV output of an input port or by $\overline{OBF}$ output of an output port.

Depending on the way a DMA operation of N data words is actually executed, three types of DMA modes can be defined. These are

(i) Burst DMA,
(ii) Cycle Steal DMA,
(iii) Transparent DMA.

In a Burst DMA operation, the DMA Controller keeps the CPU in the HOLD state until all N data words have been transferred. It is the fastest DMA mode, where the speed of the DMA transfer is limited only by the speed with which the I/O port and memory can read or write data and the speed of the DMA Controller itself. Note that while a Burst DMA operation is going on, the CPU is actually sitting idle since it cannot execute its own program during this time. If the DMA operation is requested too often or if the number of data words, N, to be transferred per Burst DMA operation is large, this by itself would cause timing problems with the microcomputer's program execution. These problems tend to get worse if one uses Burst DMA for asynchronous I/O, i.e. random intervals between transfer of individual data words in a N-word transfer; this is the major reason why one normally does not use Burst DMA for such applications.

Cycle Steal DMA is very similar to Burst DMA operation with one major difference. The DMA Controller removes its HOLD REQUEST (HRQ = 1 to HRQ = Ø) after each word has been transferred, and reasserts HRQ = 1, only when the next word actually needs to be transferred. Functionally, this tends to give control back to the CPU in between successive word transfers. The CPU can now carry out its own processing in between individual DMA word transfers. Cycle Steal DMA will be slower than Burst DMA. However, it interferes less with the processor's own program execution and can efficiently support asynchronous I/O.

If one examines the machine cycle of microprocessors, one typically finds that there are clock states during some machine cycles when the processor does not use the data and the address bus. During these states, the processor either floats these busses or they contain unspecified data. (In the latter case, assume that sufficient external logic is provided to float these busses during these states.) Example of such a state is T4, T5 and T6 also, if they exist, in the OP-Code FETCH (OF) Machine Cycle of the 8085A. At these times, the data bus floats and the address bus contains unspecified data. Using only these states (with busses floated by external logic, if necessary), a DMA operation can be done. This is referred to as Transparent DMA, since it is completely transparent to the processor. It does not interfere with or slow down the processor's own program execution. Note that Transparent DMA requires extra logic to detect the states when only internal processing is being carried out by the microprocessor.

Finally, it should be noted that all three DMA modes will need similar initialization information. Their modes of operation are also very similar. The only major differences lie in the choice of the clock cycles which are used to effect the actual data transfer. Burst DMA is the fastest, whereas Transparent DMA is the slowest. Conversely, Burst DMA interferes most with the processor's own operation whereas Transparent DMA creates the least interference.

It is also important to note that while a DMA operation is going on, the data in question never enters the processor. The only way the microprocessor can check the byte transferred is by actually reading from the proper memory location where this byte is stored. It is therefore difficult to generate programs (in systems using DMA) where the program execution depends on the value of the last byte transferred. This is one disadvantage of DMA.

The microprocessor can usually check how far a particular DMA operation has progressed by reading contents of some registers in the DMA Controller. It can then modify its program logic accordingly and may even terminate the DMA operation, if necessary. This is sometimes referred to as 'DMA-on-the-fly'.

10.8 LSI I/O PORTS

In its simplest form, an I/O port can be configured using parallel input–parallel output shift registers and tristate drivers; whether it would serve as an input or output port is decided by the direction of data transfer. Fig. 10.8.1 illustrates an I/O port created in this fashion. Note that the figure does not show the handshaking lines (DAV, OBF) that will be required to interface this I/O port to the microcomputer system; it is assumed that a circuit similar to that of Fig. 10.2.3 will be used to generate the handshaking lines. Using two such I/O ports and appropriate external logic, one can also configure a Bidirectional I/O port, if needed.

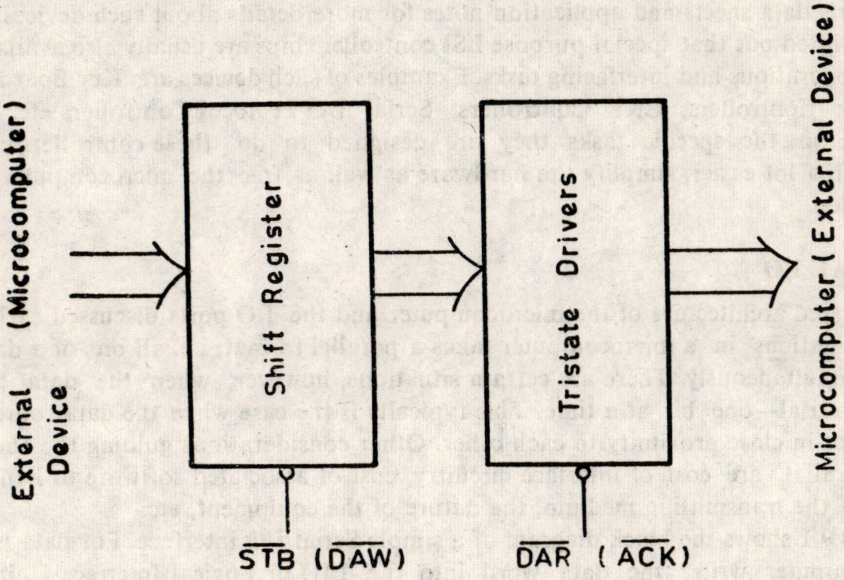

Fig. 10.8.1 I/O Port [Input (Output)] Using Shift Registers and Tristate Drivers

The circuit shown in Fig. 10.8.1, along with that needed for handshaking, is very often provided on one LSI chip. A good example of such a chip is INTEL's 8212 device, a detailed description of which may be found in Appendix A10.2. Properly configured, the 8212 may be used as either an input or output port, with or without handshaking. In microcomputer systems, this extremely versatile chip also finds many other uses, some of which are outlined in the Appendix.

When an 8212 or similar device is used as an I/O port in a system, whether it is going to be used for input or output depends on the way it is wired into the circuit; once the decision (input or output) is made in the circuit layout, it cannot be changed. It is sometimes desirable to leave this choice to the software being executed by the microcomputer so that appropriate changes, if needed, can be easily made. This facility is provided by programmable I/O ports, such as INTEL's 8255A-Programmable Peripheral Interface. This LSI device described in Appendix A10.3, provides three I/O ports which may be configured as either input or output ports. Moreover, it allows three modes of operation—as simple I/O ports, I/O ports with handshaking and bidirectional I/O. All these choices are available to the designer through software programming, i.e. by writing an appropriate data word into a Control Register internal to the 8255A chip. The reader is referred to the Appendix for greater details regarding this LSI device.

On occasions, especially when the microcomputer is busy and has to execute a number of time critical tasks, it would be desirable to delegate some of the system's intelligence to the I/O device itself. In such situations, it is desirable to use one of the several single chip microcomputers (e.g. INTEL 8748, 8741 etc.) as the I/O device interfacing between the I/O port and the microcomputer system. Such an interface is not difficult to design, especially since ROM (or EPROM), RAM and I/O ports (and sometimes a Timer/Counter) are usually available on that one single chip itself. This concept has been carried even further in the 16-bit systems put out by INTEL; there the user has the option of using a completely separate I/O processor (8089) to handle the bulk of I/O operations and its associated problems. The reader is referred to the manufacturer's data sheets and application notes for more details about such devices. Finally, it should be pointed out that special purpose LSI controller chips are usually also available for some specific I/O operations and interfacing tasks. Examples of such devices are: Key Board Controllers, Floppy Disk Controllers, CRT Controllers, Serial I/O Protocol Controllers etc. Used appropriately and in the specific tasks they are designed to do, these controllers can make the designer's job a lot easier, simplify the hardware as well as free the microcomputer to do other tasks, if needed.

10.9 SERIAL I/O

Using the bussed architecture of the microcomputer and the I/O ports discussed earlier, most of the communications in a microcomputer takes a parallel format, i.e. all bits of a data word are transferred simultaneously. There are certain situations, however, when the data transfer will need to be serial—one bit at a time. This typically is the case when the data source and destination are not in close proximity to each other. Other considerations guiding the choice of link, serial or parallel, are cost of interface circuitry, cost of associated software to handle the link, the nature of the transmitting medium, the nature of the equipment, etc.

Figure 10.9.1 shows the block diagram of a simple Serial I/O interface. For data transmission, the microcomputer writes the data word into the LIU or Logical Interface Unit (typically a UART or a USART, discussed later). This converts the data word into a serial format appending extra bits, if needed, by the serial link protocol. The serially formatted bits are passed on, one by one, to the Electrical Interface Unit. This converts the logical bit stream into the appropriate electrical signals for transmission over the medium; for RF links, this would require a modem for generating carrier modulation, whereas, for base band data links, simple line drivers would usually suffice. For data reception, exactly the opposite strategy is followed. The EIU monitors the medium, and converts the electrical signals received into a logical bit stream. The LIU extracts the data words from this stream and does a serial to parallel conversion, stripping away any extra bits introduced by the serial link protocol. These data words are then presented to the microcomputer.

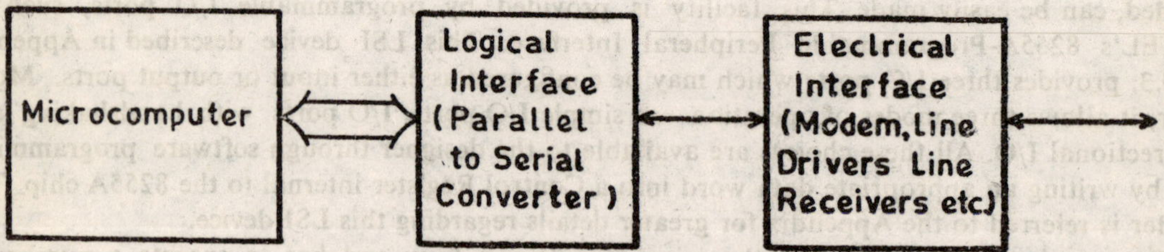

Fig. 10.9.1 A Simple Serial I/O Interface

Serial data transfer links are either Simplex, Half Duplex or Full Duplex. In a simplex link, data is sent in only one direction. In a half duplex link, data may be transferred in both directions but only in one direction at a time. In a full duplex link, data transfer may occur in both directions simultaneously. The Baud Rate of a serial link is the speed with which data can be transferred over the link. In links transmitting binary data (1s and Øs), the Baud Rate is equal to the number of bits per second.

The two data formats most commonly encountered in serial data links are NRZ (non-return to zero) and NRZI (non-return to zero inverted). NRZ treats low level signals as 'Ø' bits and high level signals as '1' bits. NRZI represents signal level changes rather than the signals themselves. A signal level change represents a 1 bit, no signal level change represents a Ø. An illustration of NRZ and NRZI signalling is shown in Fig. 10.9.2.

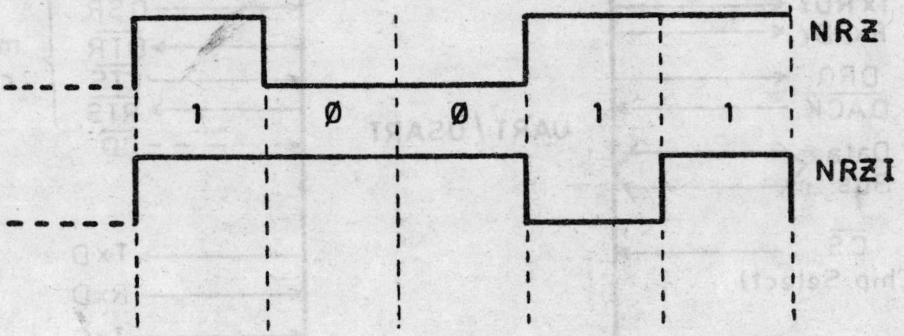

Fig. 10.9.2 NRZ and NRZI Data

Serial Data Link Protocols provide simple rules governing the transfer of serial data over a link. A protocol, for example, needs to specify if the data transmission is synchronous or asynchronous. In a synchronous link, the data transmission is continuous. A synchronous link cannot stay idle even if there is nothing to transmit; it typically sends a special character, e.g. SYNC, if it runs out of data to transmit. In an asynchronous link, a character is sent whenever one is available for transmission; the link idles when no characters are available. In an asynchronous link, the time interval between characters is variable; however, the time interval between the bits constituting the character is fixed. In contrast, the time interval between every bit being sent on a synchronous link is fixed. Synchronous links are costlier and more complex to implement but can support much higher rates of information transfer than asynchronous links. The data link protocols also specify (typically for synchronous links) whether the basic unit in the serial data stream is a byte (8 bits) or a bit—i.e. Byte Oriented and Bit Oriented Protocols.

10.10 UARTs/USARTs AND THEIR INTERFACING TO MICROPROCESSOR SYSTEMS

A number of devices are now available which greatly simplify the job of setting up serial data links in microprocessor systems. Devices which support asynchronous data transfers are referred to as UARTs (Universal Asynchronous Receiver Transmitter); devices for supporting both synchronous and asynchronous data transfers are called USARTs (Universal Synchronous/Asynchronous Receiver Transmitter). These interface directly to the microprocessor data bus on one side and to the serial data channel on the other side, possibly through a modem. In the transmitting mode, they accept parallel words (bytes) from the microprocessor system and transmit a serial data stream according to the rules of the particular protocol being used. In the receiving mode, they accept serial data from the link and following the same protocol rules, convert them to parallel words (bytes) and then pass them on to the microprocessor system via its data bus. These devices not only implement some of the rules and conditions of the protocol being used (without involving the microprocessor) but also detect and report various error conditions that might occur during transmission or reception. A typical UART/USART is illustrated in Fig. 10.10.1.

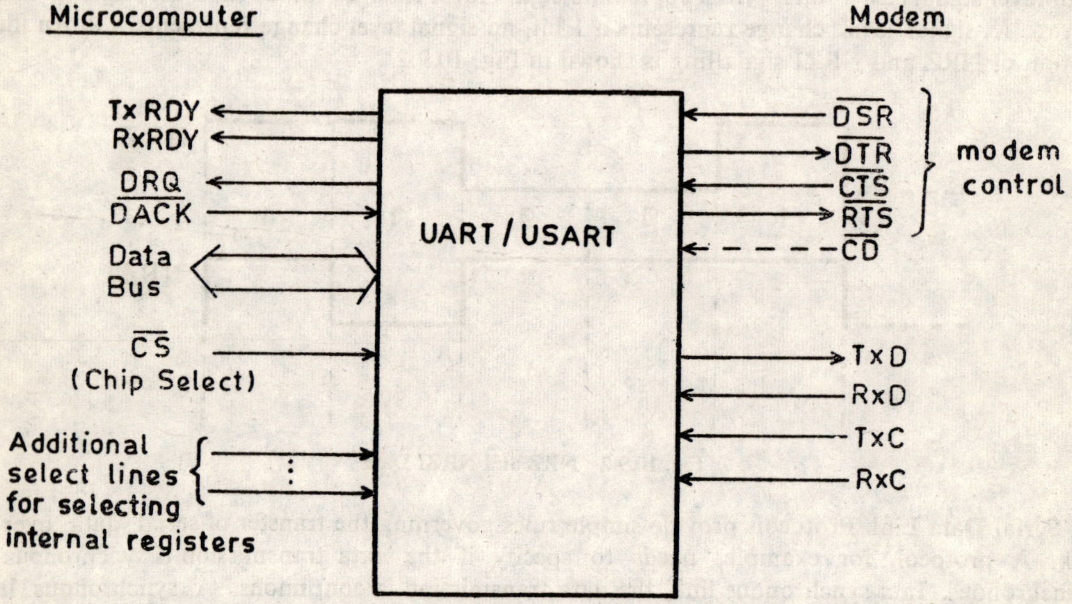

Fig. 10.10.1 UART/USART

Depending on the particular device being used, these UART/USARTs can be accessed by the microprocessor system using the I/O techniques described earlier, i.e. Program Controlled, Interrupt Controlled or DMA I/O. Note that this interface is not standard. It is not unusual to find UARTs/USARTs which offer the user a choice of I/O techniques that can be used. If Interrupt Controlled I/O is being used, the UART/USART typically provides two outputs, TxRDY and RxRDY which should be used as interrupt inputs to the microprocessor. In this case, TxRDY = 1 indicates that the UART/USART wants a new byte from the microprocessor for serial transmission; similarly, RxRDY = 1 indicates that the UART/USART has a parallel byte assembled from the serial data which it would like the microprocessor to read. Instead of providing output pins for TxRDY and RxRDY, the UART/USART may also report their condition via bits of a status register which the microprocessor can read. In that case, Program Controlled or polled I/O can be used. Of course the UART/USART can provide both the options and leave the system designer the choice of which I/O technique will be used. For DMA I/O technique to be used, the UART/USART should have provisions for a DMA Request (DRQ) output as well as be able to accept a DMA Acknowledge ($\overline{DACK}$) input. Both DRQ and $\overline{DACK}$ will be needed for transmission and reception. For the two cases, these pins will either have to be provided separately or should be suitably multiplexed. In addition to these pins, a UART/USART providing DMA I/O should preferably have one or more interrupt lines to signal the beginning and end of transmission or reception. (Remember that in DMA I/O, the microprocessor is no longer directly involved. So unlike the other types of I/O techniques, the microprocessor has no way of knowing without interrupts whether a message has been completely transmitted or received.)

On the serial data link or modem side, the pins of the UART/USRT are fairly standard and are given below.

$\overline{DSR}$ (Data Set Ready) INPUT
The modem puts $\overline{DSR} = \emptyset$ when it is ready to transfer data; otherwise, this signal is 1.

$\overline{DTR}$ (Data Terminal Ready) OUTPUT
The UART/USART puts $\overline{DTR} = \emptyset$ when it is ready to communicate; otherwise this signal is 1.

$\overline{RTS}$ (Request To Send) OUTPUT
When both the modem and the UART/USART are ready to communicate, both $\overline{DSR}$ and $\overline{DTR}$ will be low. At this point, the UART/USART outputs $\overline{RTS} = \emptyset$ to tell the modem that it is ready to send serial data; otherwise this signal is 1.

$\overline{CTS}$ (Clear to Send) INPUT
When $\overline{DSR}$, $\overline{DTR}$ and $\overline{RTS}$ are all $\emptyset$, the modem outputs $\overline{CTS} = \emptyset$ to indicate that it is also able and willing to transmit the serial data being provided by the UART/USART.

$\overline{DSR}$, $\overline{DTR}$, $\overline{RTS}$ and $\overline{CTS}$ are the standard modem control signals. In addition, at times a $\overline{CD}$ (Carrier Detect) input is also provided to the UART/USART. In RF data links, when serial data is being received, the modem outputs $\overline{CD} = \emptyset$ if the received carrier level is found to be above a predetermined threshold, i.e. when carrier is present.

In addition to the above, a UART/USART will also have the following pins for data and clock

RxD Receive Data Input

TxD Transmit Data Output

RxC Receive Clock Input

TxC Transmit Clock Input

Note that TxC and RxC clocks may either be the same (X1) as the baud rate clock or some multiple thereof, e.g. X4, X16, X32, X64 etc. A typical UART/USART may either specify what should be chosen, or leave it as a user selectable option. A typical UART/USART—INTEL's 8251A device—is described in detail in Appendix A10.4.

PROBLEMS

1. A 8085A system uses an input device having a RUN input and a $\overline{DAV}$ output. When RUN = 1, the device outputs data asynchronously to the μP system, where the $1 \rightarrow \emptyset$ transition of $\overline{DAV}$ indicates that a data byte is available for reading. Reading this new data

byte returns $\overline{DAV}$ to 1 and the device cannot output a new data byte until the old byte has been read. The device is turned OFF by resetting RUN to Ø. RUN, $\overline{DAV}$, $\overline{RD}$ and $D_\emptyset$–D_7 are the only connections to the device.

(a) Design the hardware interfacing this device to the μP system with the following port assignments.

RUN	Bit 7 of Output Port Ø
$\overline{DAV}$	Bit 7 of Input Port Ø
Data	8-bits of Input Port 1

(b) Write a subroutine that starts the device, inputs 128 bytes of data from it, stores them in memory locations BUF to BUF + $7F_H$ and then turns the device OFF.

(c) Repeat (a) and (b) for the situation where RUN is output via SOD, but where $\overline{DAV}$ is monitored via SID.

(d) Repeat (a) and (b) for the situation where RUN is output via SOD, but where $\overline{DAV}$, or some signal appropriately derived from $\overline{DAV}$, drives an interrupt line of the 8085A.

2. Determine the control words needed to program an 8255 as follows (refer to Appendix A10.3):
   (a) Port A input, Port B output, Port C input—all in Mode 0.
   (b) Port A and C lower output, Port B and C upper input—all in Mode 0.
   (c) Port A and C upper input, Port B and C lower output—all in Mode 0.
   (d) Port A Mode 2, Port B output Mode 1.
3. An I/O device which can function both as an input or output device is interfaced to Channel Ø of a 8257 DMA Controller; the other DMA channels are not being used. The DRQ and $\overline{DACK}$ lines of the device are connected to the DRQØ and $\overline{DACK\emptyset}$ pins of the 8257. The I/O device transfers data in blocks of length FF_H using three outputs ININT, OUTINT and OVER to signal the type of DMA operation, input or output; that is desired. This is done as follows (refer to Appendix A10.1):

INPUT: ININT = 1 to indicate request to transfer FF_H bytes to memory. ININT returns to Ø after the first byte is transferred. A pulse (Lo → Hi → Lo) appears at OVER when the entire transfer is over.

OUTPUT: OUTINT = 1 to indicate request for FF_H bytes from memory. OUTINT returns to Ø after the first byte has been transferred. A pulse (Lo → Hi → Lo) appears at OVER when the entire transfer is over.

The I/O device has been designed such that OUTINT will not go High between the rising edge of ININT and the falling edge of the corresponding OVER pulse. Similarly ININT will not go High between the rising edge of OUTINT and the falling edge of the corresponding OVER pulse.

The system designer has interfaced the 8257 as an I/O Mapped device with its internal registers occupying I/O addresses $4\emptyset_H$ to 48_H. Data input from the I/O device is to be stored in locations $2\emptyset\emptyset\emptyset_H$ – $2\emptyset FE_H$ and data to be output to the device is taken from $21\emptyset\emptyset_H$ – $21FE_H$.

(a) The system designer wants to connect the ININT, OUTINT and OVER lines directly to the Maskable Hardware Interrupts of the 8085A. Give reasons for your choice of connection.

(b) Write the appropriate interrupt service routines including the instructions to properly program the 8257 and enable/disable DMA channel Ø.

(c) Suppose the OVER output is not available. What modifications would be required in (a) and (b)?

4. A High Speed Data Acquisition Device has its data lines connected to the data bus of an 8085A system. In addition, it has a RUN input, a $\overline{RD}$ input and a DAV (data available) output. The device is turned ON by giving it RUN = 1 and is turned OFF by RUN = Ø. The output DAV = 1 implies that the device has data available to be read and that on getting a $\overline{RD}$ pulse it will put this data on the data bus. DAV returns to Ø on the falling edge of $\overline{RD}$ if the next data byte is not available to be read for some reason. DAV stays at 1 if the next byte is indeed available for reading (as will normally be the case). In the former case (i.e. next byte not available) the DAV output of the device goes to 1 when a data byte is available to be read. In this way, the device provides a sequence of data bytes for reading as long as RUN is held at 1; when RUN = Ø, the device is turned off and DAV stays at Ø.

(a) Draw a rough circuit diagram of how you can interface this device to a 8085A microcomputer system using the 8257 DMA Controller to handle the actual transfer.

(b) Write a subroutine TRANS which is used to transfer data from the input device to memory by giving it a proper RUN input and appropriately programming the 8257 as per the hardware of (a). TRANS should assume that the acquired bytes are to be stored from memory locations (HL) to (HL) + (DE) − 1.

(c) What software other than TRANS is needed? Write them giving brief descriptions of their purpose.

5. Making appropriate assumptions, write subroutines to implement the handshaking algorithms of Figs. 10.2.2 and 10.2.5 using a 8085A based microcomputer.

6. (a) Port A of a 8255 is to be used to connect a 8085A to an input device in Mode 1. Show the interface and draw the appropriate timing diagram.

(b) In what respects does the 8255 have better handshaking features compared to the 8212.

(c) What additional circuitry will be needed with an 8212 to provide a 8255 kind of handshaking? Discuss the input port and output port cases separately.

7. 16 keys are arranged in a 4 × 4 matrix. Design a scheme to scan the rows and read the columns to decode the keyboard using 8212s. Write the associated software. When a key is pressed, the accumulator should get a word. The word has one 1 in the upper 4 bits indicating the row number and one 1 in the lower 4 bits indicating the column number.

8. Two microprocessor systems, each having an 8255, communicate with each other through these 8255s. Show the interconnection between the 8255s using whatever other hardware is needed. Draw a timing diagram for the situation where system 1 sends a byte to system 2 and system 2 receives it.

9. An 8255 is to be used to interface a multiplier chip to an 8085 processor. The multiplier has 2 four-bit inputs and an eight-bit output. It also has a START input and an EOC output. Multiplication is initiated by a Lo → Hi edge of START which also makes EOC Low. EOC return to High once the multiplication is over.

(a) Design the interface described above.

(b) Redo (a) using 8212s instead of the 8255.

Appendix A10

A10.1 PROGRAMMABLE DMA CONTROLLERS

A10.1.1 The INTEL 8257 DMA Controller

DMA controllers may be as simple as one built using various SSIs and MSIs or as complex as special LSI devices. The INTEL 8257 is a good example of the latter. This controller may be programmed to allow special options of various kinds and can control the operation of up to four I/O devices—one DMA channel for each device. Figure A10.1.1 shows the pin-out of this device. A brief description of these pins is given below.

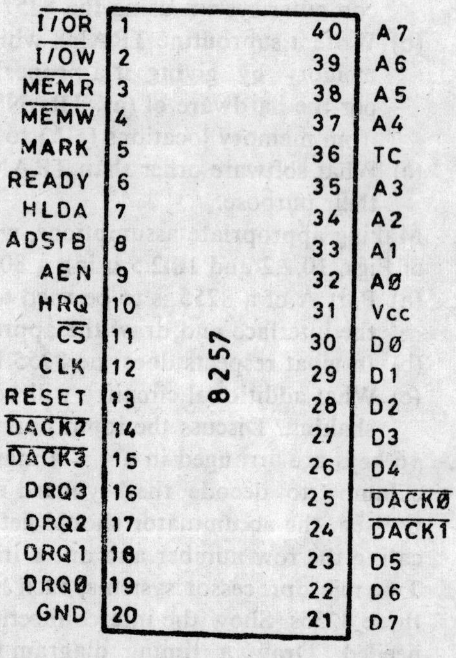

Fig. A10.1.1 Intel 8257 DMA Controller

V_{CC}, GND	Power, $V_{CC} = +5$ V
DØ–D7	Data Bus when CPU is writing data into the 8257's internal registers or reading from them. During DMA, 8257 outputs Address A8–A15 on this bus while ADSTB is Hi. When ADSTB goes low these pins go tristate so that other devices can output on the data bus
AØ–A7	Address lower, AØ–A3 are bidirectional, A4–A7 are tristate outputs.
$\overline{I/OR}$	I/O Read; $\overline{I/OW}$ I/O Write
$\overline{MEMR}$	Memory Read; $\overline{MEMW}$ Memory Write
$\overline{CS}$	Chip Select Input to write to 8257 or read from it
DRQØ–3	DMA Request from I/O ports for channels Ø–3
$\overline{DACKØ–3}$	DMA Acknowledge to I/O ports for channels Ø–3
ADSTB	Output high when DØ–D7 has address (upper)
HRQ	Hold Request to microprocessor
HLDA	Hold Acknowledge from microprocessor
AEN	Address Enable, output high during DMA
TC	Terminal Count signal, when any one of DMA channels Ø–3 reaches terminal count

MARK Output signalling 128 bytes have been transferred on one channel

CLK System Clock; RESET System Reset

READY Ready Low is used to stretch memory read or write cycles.

The 8257 multiplexes the upper address byte A_{15}–A_8 and the data byte D_7–$D_\emptyset$ on the same pins. As shown in Fig. A10.1.2, it therefore needs a latch (e.g. a 8212) to demultiplex the bus using ADSTB as the Strobe signal. It needs 4 clock cycles to transfer one data byte and is restricted to using a clock with a minimum period of 320 ns. Hence, it will need at least 1280 ns to transfer one byte of data (Rate of 781.25 Kbytes/sec).

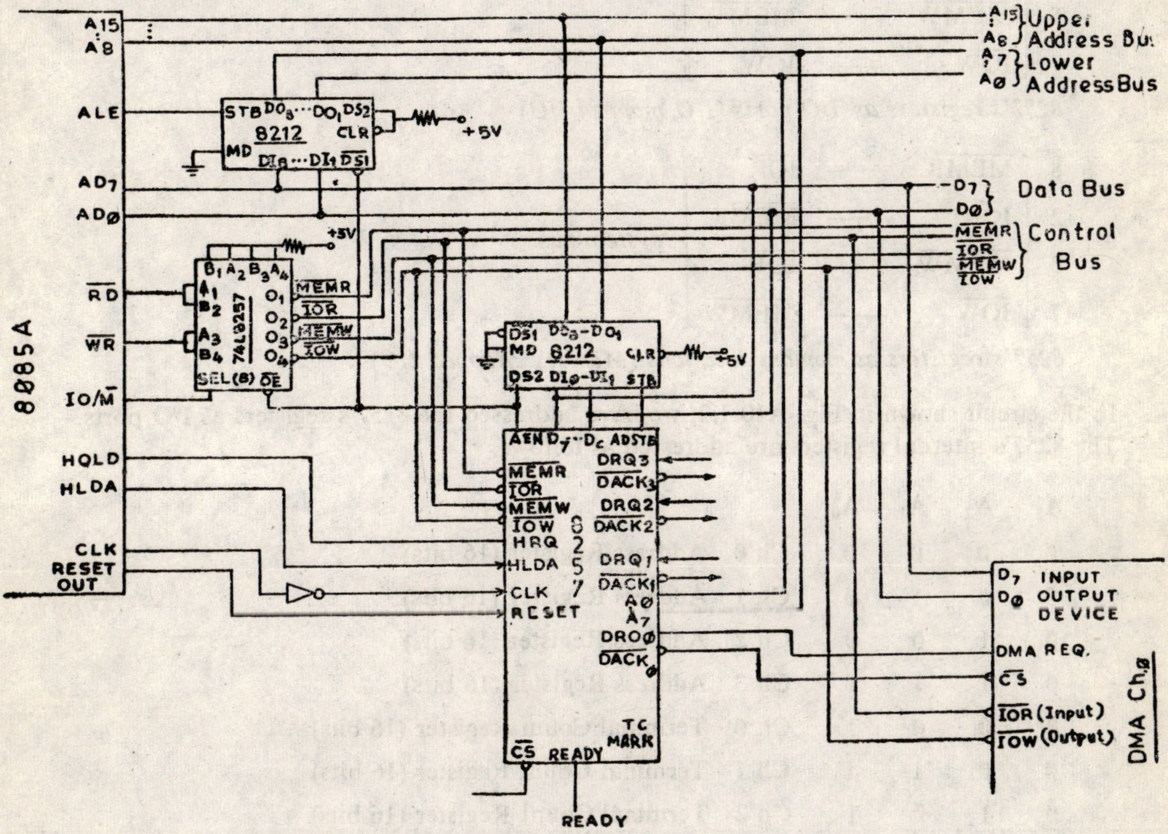

Fig. A10.1.2 Using the 8257 DMA Controller with the 8Ø85A CPU

The 8257 works in two modes—slave and master. The slave mode is used when the microprocessor is writing data into the 8257's internal registers during initialization or when it is reading the contents of these internal registers. The 8257 operates in the master mode when it is actually executing a DMA transfer and has control of the system bus. The two modes can be distinguished by the AEN output of the 8257. When AEN = 1, the microprocessor should relinquish control of the system bus since the DMA controller is then the bus master and wants control of the bus. This can be seen in the application circuit shown in Fig. A10.1.2. Note that the circuit shown in Fig. A10.1.2 uses DMA Channel Ø.

The 8257 DMA Controller has a number of internal registers which have to be properly initialized before a DMA operation can be started. These registers are accessed by setting $\overline{CS} = \emptyset$ and by properly choosing the lower order address bits A_3, A_2, A_1 and $A_\emptyset$. For writing into these registers, $\overline{IOW}$ should go Lo; whereas, for reading, $\overline{IOR}$ should go low. By means of the connections given below, the internal registers of the 8257 can be accessed as either memory locations or I/O ports.

	8257		system bus
8	$\overline{MEMR}$	—	$\overline{MEMR}$
2	$\overline{IOR}$	—	$\overline{IOR}$
5	$\overline{MEMW}$	—	$\overline{MEMW}$
7	$\overline{IOW}$	—	$\overline{IOW}$

8257's registers as I/O ports (I/O mapped I/O)

	8257		system bus
8	$\overline{MEMR}$	—	$\overline{IOR}$
2	$\overline{IOR}$	—	$\overline{MEMR}$
5	$\overline{MEMW}$	—	$\overline{IOW}$
7	$\overline{IOW}$	—	$\overline{MEMW}$

8257's registers as memory locations (Memory Mapped I/O)

In the circuit shown in Fig. A10.1.2, we have addressed the 8257's registers as I/O ports. The 8257's internal registers are addressed as follows.

A_3	A_2	A_1	A_0			
Ø	Ø	Ø	Ø	Ch Ø Address Register (16 bits)		
Ø	Ø	1	Ø	Ch 1 Address Register (16 bits)		
Ø	1	Ø	Ø	Ch 2 Address Register (16 bits)		
Ø	1	1	Ø	Ch 3 Address Register (16 bits)		
Ø	Ø	Ø	1	Ch Ø Terminal Count Register (16 bits)		
Ø	Ø	1	1	Ch 1 Terminal Count Register (16 bits)		
Ø	1	Ø	1	Ch 2 Terminal Count Register (16 bits)		
Ø	1	1	1	Ch 3 Terminal Count Register (16 bits)		
1	Ø	Ø	Ø	Mode Set Register (Control Register)	*WRITE ONLY*	(8 bits)
1	Ø	Ø	Ø	Status Register	*READ ONLY*	(8 bits)

Each of the Address and Terminal Count Registers have 16 bits (2 data bytes). However, each of them is accessed using one external address (one value of A_3–$A_\emptyset$). The 8257 device allows reading or writing 2 bytes with one address by providing an internal first/last (F/L) flip-flop. On RESET, this flip-flop is reset to Ø. Every time an Address or Terminal Count

Register is accessed (for Read or Write), the F/L flip-flop is toggled. This, in essence, provides two internal addresses (one for F/L = Ø, another for F/L = 1) for one external address. In order to load a 16-bit quantity into one of these 16-bit internal registers, the low order byte is written first, followed by the high order byte. When reading the contents of any of these registers, the low order byte is outputted on the first read with the high order byte outputted on the next read. It is very important that the program reading these registers or writing into them should perform every read and write operation in pairs. Otherwise, erroneous results will arise.

The microprocessor initializes the Address Register of each DMA channel with the lowest address of the memory locations to be used by the DMA channel. This is because the 8257 increments the Address Register of the DMA channel by one when one byte is transferred using that channel. Note that the address is 16 bits long; therefore the 8257 can address any of the 64 K bytes in the address space of a 8080 or 8085 type microprocessor. If the Address Register is read in the middle of a DMA operation, it outputs the address of the next memory location to be accessed for a byte transfer.

The Terminal Count Register has two fields. The actual Terminal Count is 14 bits long (Ø to 16 K bytes) and is contained in TC_{13}–$TC_{Ø}$. The remaining bits—TC_{14}, TC_{15}—specify the direction of data transfer. The meanings attached to these bits are different depending on whether the registers of the 8257 device are being accessed as I/O ports or Memory Locations. These two cases are given below:

• TC_{15}	TC_{14}	I/O port	Memory location
Ø	Ø	Verify DMA Cycle	Verify DMA Cycle
Ø	1	DMA Write Cycle	DMA Read Cycle
1	Ø	DMA Read Cycle	DMA Write Cycle
1	1	ILLEGAL	ILLEGAL

DMA Read Cycles transfer data from memory to I/O whereas DMA Write Cycles transfer data from I/O to memory.

The DMA Verify Cycle is a special kind of DMA Cycle where the 8257 takes control of the system just as it would in DMA READ or WRITE but no data is actually transferred during this cycle.

The Terminal Count Register is decremented by one after one byte has been transferred. If this register is read in the middle of a DMA operation, the contents of $TC_{Ø}$–TC_{13} specify the number of data bytes still to be transferred and bits TC_{14} and TC_{15} the type of DMA Cycle needed for this channel, i.e. the direction of data transfer. Note that the counts in the Terminal Count Register should be interpreted as 'Zero-inclusive' both for reading these registers as well as writing into them. For example, (a) if you want n-bytes transferred, load $TC_{Ø}$–TC_{13} with (n — 1), (b) if you read $TC_{Ø}$–TC_{13} contents as K, it implies that (K + 1) bytes remain to be transferred.

The Mode Set or Control Register (MSØ–MS7) of the 8257 has to be written into with the proper data byte during the initialization process. (This Register is a WRITE only type, its contents cannot be read.) The bits, MSØ–MS7, of this register will be interpreted by the 8257 as follows:

MSØ	Channel Ø Enable/$\overline{\text{Disable}}$ (Enable = 1, Disable = Ø)
MS1	Channel 1 Enable/$\overline{\text{Disable}}$ (Enable = 1, Disable = Ø)
MS2	Channel 2 Enable/$\overline{\text{Disable}}$ (Enable = 1, Disable = Ø)
MS3	Channel 3 Enable/$\overline{\text{Disable}}$ (Enable = 1, Disable = Ø)
MS4	Ø = Sequential Priority, 1 = Rotating Priority
MS5	Ø = Normal write, 1 = Extended write
MS6	1 = TC Stop option
MS7	1 = Auto Load option (on DMA Channel 2 only using Channel 3 registers for storage)

MSØ–MS3 selectively enable or disable the DMA Channels Ø, 1, 2 or 3.

MS4 indicates the priority arbitration scheme which will be used if the DMA request is active (DRQ = 1) for more than one channel at the same time. The DMA request of only the higher priority channel is acknowledged with the appropriate $\overline{\text{DACK}}$ ($\overline{\text{DACK}}$ = Ø). If sequential priority is specified, Channel Ø always has the highest priority and Channel 3 has the lowest priority. If rotating priority is specified, the priority of the channels has a circular sequence depending on the last channel which was served. If Channel N was the last channel to be served, it will now have the lowest priority.

If MS5 = Ø, normal $\overline{\text{WR}}$ pulses are generated for $\overline{\text{MEMW}}$ and $\overline{\text{IOW}}$ outputs. If MS5 = 1, both $\overline{\text{MEMW}}$ and $\overline{\text{IOW}}$ will go low one clock cycle earlier. This not only gives external logic and memory more time to do the indicated write operation, but also allows more time to external logic for deciding if additional WAIT states are needed. (WAIT states are introduced by pulling the READY input of the 8257 low.)

If MS6=1, the DMA channels which have reached their terminal count are automatically disabled. Such a disabled channel must be explicitly reenabled by the microprocessor by writing a 1 in the proper bit location, MSØ to MS3. If MS6 = Ø, it is the responsibility of the microcomputer to stop the DMA Requests in order to terminate a DMA operation—the contents of $TC_Ø$ to TC_{13} do not have any importance and can be regarded as Don't Cares for both reading and initializing these registers. (Note that you cannot specify a TC Stop option for some of the channels but not for the others. The same option has to be specified for all the channels.)

If MS7 = 1, a special AUTO LOAD option is specified for DMA Channel 2. In this option, if terminal count on Channel 2 has been reached, the Address Register and the Terminal Count Register of Channel 2 are *reinitialized* with the values put in them during the original initialization and the channel (Channel 2) *stays enabled.* If this option is specified, Channel 3 can no longer be used. This is because if Channel 2 is in the Auto Load mode, its initialization values are also stored in the corresponding Channel 3 registers (which are then used to reinitialize Channel 2 after terminal count is reached).

The 8257 DMA Controller also has a STATUS Register which can be read by the microprocessor to monitor the status of the ongoing DMA operations. Note that this is a READ ONLY register which cannot be written into. The individual bits, SØ–S7, have the following meaning:

SØ	= 1 if DMA Channel Ø has reached Terminal Count; Ø otherwise	
S1	= 1 if DMA Channel 1 has reached Terminal Count; Ø otherwise	
S2	= 1 if DMA Channel 2 has reached Terminal Count; Ø otherwise	
S3	= 1 if DMA Channel 3 has reached Terminal Count; Ø otherwise	
S4	Update Status	
S5–S7	Unassigned	

The 8257 has one TC output which is shared by all four DMA channels. (A pulse is output on TC whenever any of these channels reaches terminal count and if the TC Stop option has been chosen). When a TC pulse is detected, the microprocessor should read the contents of SØ–S3 in order to determine which channel had actually reached the terminal count.

If the Auto Load option (mentioned earlier) is being used, the 8257 'steals' some extra clock cycles after the terminal count on Channel 2 is reached. This is used to reinitialize the Channel 2 registers. This is referred to as an 'Update Cycle'. If the Update Cycle is currently in progress, S4 will be 1; otherwise it will be Ø. Knowing this is necessary so that the CPU does not inadvertently write into Channel 3 registers while the Update Cycle is going on. This may give erroneous operation as, during this time, Channel 3 register contents are in the process of being transferred to the Channel 2 registers.

Suppose (Fig. A10.1.2) we want to use Channel Ø in DMA READ mode (Output to I/O port). We want to store 256 bytes starting from 1ØØØH to 1ØFFH. Moreover, assume that the 8257's registers are being used as I/O ports starting from 4ØH to 48H. The initialization program can be as follows:

```
DI              ;   disable interrupts
MVI  A, ØØH     ;   load 1ØØØH
OUT  4ØH        ;   into Channel 1's
MVI  A, 1ØH     ;   Address
OUT  4ØH        ;   Register
MVI  A, ØFFH;       Set TC count
OUT  41H            to FFH = 256
MVI  A, 8ØH     ;   and TC15, TC14 = 1Ø for
OUT  41H        ;   DMA Read
MVI  A, 41H     ;   enable Channel Ø only, with TC stop
OUT  48H        ;   option
```

When the output device wants the first byte of data, it outputs DRQØ HIGH to the 8257. The 8257 then outputs a Hold Request (HRQ) HIGH to the 8085A which returns a Hold Acknowledge (HLDA) high after completing its current machine cycle. It also relinquishes control of the system bus. The 8257 now takes control of the system bus. It outputs AEN High disabling the 8212 latch used for demultiplexing the 8085A ADØ–AD7 bus. AEN = 1 also ensures that (because of the 74LS257) the source for $\overline{IOR}$, $\overline{MEMR}$, $\overline{IOW}$ and $\overline{MEMW}$ will be the 8257 rather than the 8Ø85A. (Recall that, in the HOLD state, the 8Ø85A floats A8–A15 and ADØ–AD7).

The 8257 now outputs $\overline{\text{DACKØ}}$ Low acknowledging the DMA request placed by the output device. The output device uses this as a chip select signal to select itself. The 8257 also outputs the higher order address on DØ–D7 during the first clock cycle following HLDA = 1. This is latched by the 8212 device using ADSTB as the strobe pulse. After ADSTB goes Low, DØ–D7 are floated by the 8257. After ADSTB returns to Ø, the contents of the Address Bus will be identical to the contents of Channel Ø's Address Register. This points to the memory location that will be read in the DMA Read operation. The 8257 now outputs overlapping $\overline{\text{MEMR}}$ and $\overline{\text{IOW}}$ pulses. The selected memory location now drives the data bus and the data bus contents are sent to the output device being served by Channel Ø.

After the byte has been transferred, $\overline{\text{DACKØ}}$ returns to 1. Whether the DMA operation is going to be 'burst' or 'cycle steal' will depend on the output device. If Burst DMA is needed, the output device should keep DRQØ High and accept the data bytes sent to it as fast as they are presented. Otherwise, it should return DRQØ to Ø (which returns HOLD, HLDA and AEN to Ø) and raise it High only when it wants another data byte.

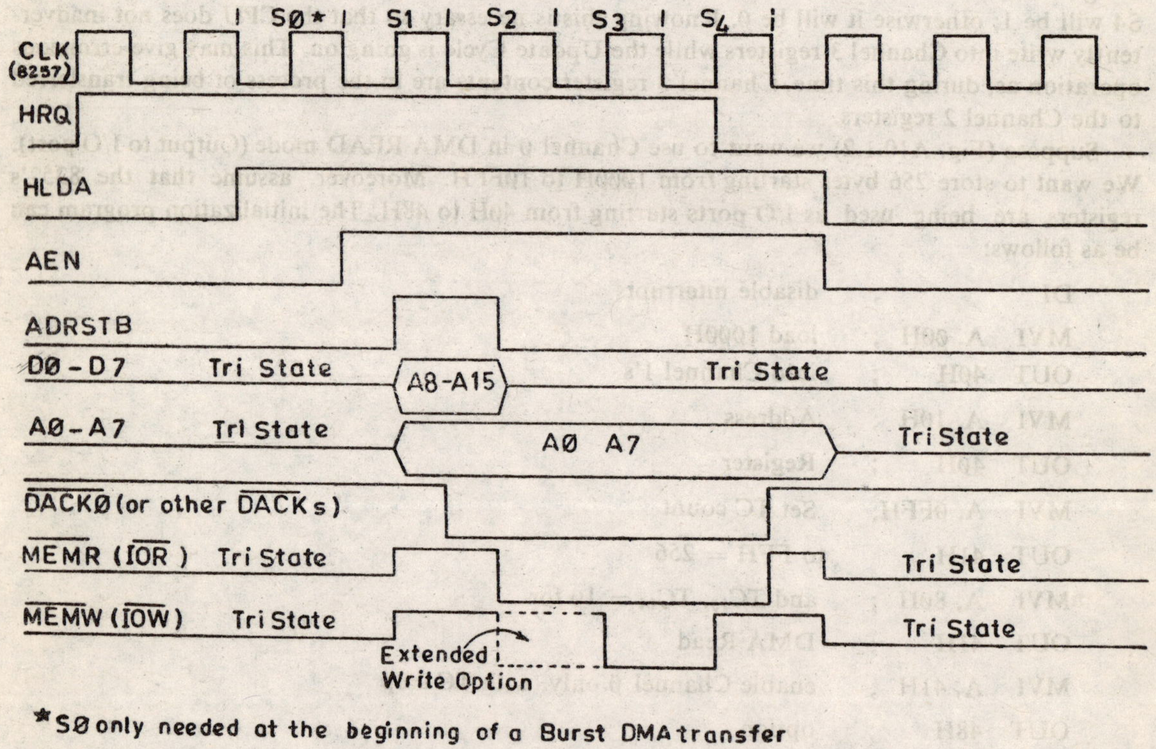

Fig. A10.1.3 DMA Timing Diagram for the Intel 8257 (showing a one byte transfer)

The DMA Read operation described above will terminate either when Terminal Count is reached or if the microprocessor disables Channel Ø by resetting MSØ. The timing diagram for this DMA operation is shown in Fig. A10.1.3. A DMA Write operation is essentially similar to a DMA Read operation except that the I/O port (selected by $\overline{\text{DACKØ}}$) drives the data to be written into the memory location selected by the Address Register's output for that channel.

A10.1.2 THE INTEL 8237A DMA CONTROLLER

The Intel 8237A DMA Controller is essentially an enhancement of the 8257 device described in Appendix A10.1.1. It offers more features than the latter and has a greater number of programmable options. In this appendix, we present a brief description of this device relying heavily on a comparison with the 8257 wherever possible. The reader should also note that Intel's 8237A DMA Controller is virtually identical to the AMD 9517 device produced by Advanced Micro Devices.

Compared to the 8257, the 8237A incorporates a number of desirable features. Some of these are summarized below:

(i) All four independent DMA channels may be given the Autoload option separately or together. (The 8257 provided this option only for Channel 2 in which case Channel 3 becomes virtually unusable.)

(ii) The 8237A devices may be cascaded to any desired depth to provide as many DMA channels as required.

(iii) Memory-to-Memory transfers may be done using Channels 0 and 1. Address for Channel 0 may be held constant to allow initialization of a memory block (addressed by Channel 1) with a constant value. Note that for such transfers Channel 0 addresses the source and Channel 1 the destination.

(iv) Memory addresses may be programmed to increment or decrement. (Increment only in the 8257).

(v) DMA requests may be placed through software.

(vi) The 8237A allows three types of DMA transfer modes. These are Single Transfer, Block Transfer and Demand Transfer. (The Demand Transfer mode is effectively the equivalent of the only transfer mode allowed by the 8257).

(vii) It allows faster operation than the 8257 by requiring less clock cycles for transferring a block. An even faster operation may be obtained by programming it in the Compressed Timing Mode.

(viii) An external $\overline{EOP}$ input may be input to the 8237A to terminate DMA transfers and auto initialize (if auto load option is specified).

(ix) The sense of the DRQO–3 inputs and the DACKO–3 outputs may be programmed to be either High or Low, as desired. Note that for the 8257, DRQ is active High and $\overline{DACK}$ is active Low.

(x) For each DMA channel, apart from showing whether TC stop has been reached or not, the Status register of the 8237A also reports whether a DMA request is currently pending on that channel.

(xi) Specific commands may be issued to the 8237A to
—clear the internal First/Last flip-flop (see 8257) so that it is in a known state,
—reset the device to the equivalent of a hardware reset, and
—enable all DMA channels simultaneously.

In the following, we give a brief description of the 8237A device. The reader, however, is urged to examine the manufacturer's data sheets (Intel 8237A or AMD 9517) of the relevant device prior to actual use. For the sake of brevity, we will not describe (but only indicate) the particular aspects in which the 8237A resembles closely the 8257 device.

Pin-Out of the 8237A Device

The pin-out of the Inter 8237A DMA Controller is given in Fig. A10.1.4. Comparing this with the 8257 device shown is Fig. A10.1.1, one finds the two pin-outs to be identical except for the differences described below:

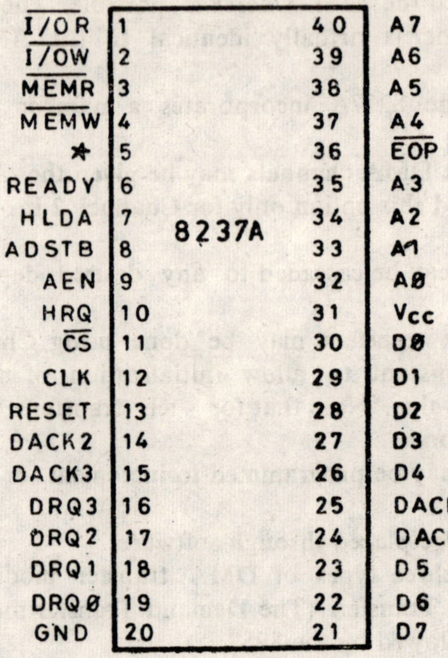

* Tie to V_{CC}=+5V

Fig. A10.1.4 Intel 8237A DMA Controller

(i) In Fig. A10.1.4 we have indicated the DMA Request inputs as DRQ 0–3. It should be noted that the manufacturer labels these as DREQ 0–3. We have not used these labels so as to emphasize the similarity between the pin-outs of the two devices. As mentioned earlier, the sense of these inputs (coming from the I/O ports requesting a DMA transfer) may be programmed to be either active High or active Low.

(ii) The DACK 0–3 outputs of the 8237A device are used to select the I/O port which is being allowed to transfer data using DMA. As for the DRQ lines, the DACK outputs may also be programmed to be active High or active Low. [Note that, as described later, the sense of the DRQs and the DACKs may be separately chosen. However, all DRQs will have the same sense; similarly, all DACKs will also have the same sense.]

(iii) Pin 5 of the 8237A device no longer outputs a MARK signal (unlike the 8257). Instead it is internally pulled up to V_{cc} if left floating. The manufacturer also specifies that it may be directly tied to $V_{cc} = +5$ V.

(iv) Pin 36 of the 8237A device is labelled $\overline{EOP}$. (This provided the TC pulse output for the 8257 device.) The $\overline{EOP}$ pin of the 8237A is bidirectional in nature. It outputs a pulse (similar to TC) whenever a channel reaches Terminal Count. As an input, a Low pulse at $\overline{EOP}$ terminates DMA transfer and auto initializes the channel if the Autoload option has been specified. In memory-to-memory transfers, the $\overline{EOP}$ pin outputs a pulse when Channel 1 reaches

Terminal Count. If it is not required as an input, the $\overline{EOP}$ pin should be tied to V_{cc} with a pull-up resistor.

The other pins of the 8237A device are functionally identical to those of the 8257 device even though there may be some minor differences, e.g. in timing. The similarity is so great that with minimal hardware modification (e.g. pull-up resistors on pins 5 and 36) one can replace the 8257 in a circuit with an 8237. Note that this will however entail some more software to tackle appropriately the greater number of programming options available on the 8237.

DMA Timing of the 8237A

Referring to Fig. A10.1.3, note that the 8257 DMA Controller requires four states S1, S2, S3 and S4 to complete one transfer. Successive transfers are achieved by a repetition of this four-state sequence. Of these four states, the only function of state S1 is to allow the upper order address bits, A8–A15, to be latched by a demultiplexing latch. This is inefficient (i.e. S1 extra wasted time) as in most cases A8–A15 will change on rather rare occasions. Note that A8–A15 changes only when there is a carry-out of A7 on a long transfer over one channel or when the 8257 switches from one DMA channel to another.

The 8237A removes this timing inefficiency by dropping state S1 for a DMA transfer except when A8–A15 needs to be changed. In a long sequence of transfers, the device usually manages to transfer bytes using only the states S2, S3 and S4. Otherwise, its timing for a DMA transfer between memory and I/O is essentially similar to that of the 8257. For memory–I/O transfers with the 8237A, one can also specify a Compressed Timing feature not available on the 8257. With Compressed Timing the 8237A also deletes state S3 (reducing the read pulses, $\overline{IOR}$ or $\overline{MEMR}$ allowing for even faster transfers, if such timing is acceptable to the memory and the I/O devices. (This option is made for all channels. It cannot be made if memory-to-memory transfers are desired.)

For memory-to-memory transfers, Channel 0 provides the source address and Channel 1 the destination address. Eight states, S11–S14 and S21–S24, are required to transfer one byte. Of these, S11–S14 are used to read data from the source memory location (Channel 0) into the Temporary Register of the 8237A (described later); the states S21–S24 then transfer the contents of this register to the destination memory location (Channel 1).

Initiation and Termination of DMA Transfers

In the 8237A, DMA transfers can be initiated in two ways, provided the particular channel is enabled. As in the 8257, the DMA transfer may be initiated by raising the DRQ of that channel appropriately. It may also be initiated, by software, i.e. by setting to 1 appropriate bits of the 8237A's Request Register (described later).

A DMA transfer may be terminated in a number of ways. It may be terminated by disabling all channels by a hardware or software reset. (A hardware reset is applied by an active High pulse at the RESET input. Software reset is applied by an appropriate software command to the 8237A). It may also be terminated for all channels by disabling the 8237A through software. DMA transfer on a selected channel may be disabled by setting the mask bit for that channel in the Mask Register. Moreover, as in the 8257, a Terminal Count may be specified for each channel, disabling the channel when Terminal Count is reached. (Unlike the 8257, one can specify a 16-bit Terminal Count in the 8237A.) Reaching Terminal Count is referred to as the internal $\overline{EOP}$ (End of Process). A channel may also be disabled by an external $\overline{EOP}$, i.e. by applying a pulse at the $\overline{EOP}$ pin from external hardware.

Auto-initialization

The 8257 provides auto-initialization through the auto-load option only for Channel 2 using Channel 3's address and terminal count registers to store the initializing values. In the 8237A, enough internal storage has been provided using Base Address and Word Count Registers (for each channel), to extend the auto-load option to all the channels. Auto-initialization through autoload may be done in the 8237A provided this option is chosen while programming a channel. If chosen, the channel's Address and Terminal Count (also referred to as Word Count) registers are reinitialized to their original values following an internal or external $\overline{EOP}$ (i.e. Terminal Count reached or external $\overline{EOP}$ pulse).

DMA Transfer Modes Allowed by the 8237A

The 8237A allows three types of DMA transfer modes for transferring a sequence of data bytes between memory and I/O. These are referred to as Demand Transfer, Block Transfer and Single Transfer modes. The Demand Transfer mode is essentially the mode of operation in which the 8257 operates, i.e. a sequence of bytes transferred with the CPU in HOLD by keeping DRQ active. The other two modes are not available for the 8257. In the Block Transfer mode, a sequence of bytes will be transferred until an external or internal $\overline{EOP}$ (i.e. TC stop reached) is encountered. Unlike Demand Transfer, however, an active DRQ is only needed to initiate the transfer, i.e. till the first DACK is received. Following this, the transfer continues until $\overline{EOP}$ occurs, irrespective of the value of DRQ; of course, autoinitialization is possible following $\overline{EOP}$, if the auto-load option has been specified. The Single Transfer mode is interesting because it comes closest to the Cycle Steel DMA mode described in Sec. 10.7. In this mode, even if DRQ is active, the 8237 relinquishes the system bus to the CPU after one byte has been transferred. This is done by lowering HRQ to Ø for a length of time sufficient to allow a 8085A-type CPU to execute one machine cycle on the bus. Following this, HRQ is reasserted to 1 (if DRQ is still active) so that the next byte can be transferred through DMA. This is repeated for all the bytes in the sequence.

Properly programmed, the 8237A may also be used for doing a memory-to-memory transfer in the DMA mode. This is equivalent to a block memory transfer. Channel 0 points to the source from which data is transferred to the Temporary Register; this data is then transferred to the destination memory address using Channel 1. Channel 1 addresses will be either incremented or decremented (following each byte transferred) as programmed. Channel 0 addresses may also be similarly incremented or decremented as per the programming of the 8237A. In addition, one may also program the 8237A to hold the Channel 0 address constant during the entire transfer sequence. This option is useful in initializing a memory block with a constant value (i.e. the data contained in the memory location pointed at by the fixed Channel 0 address). The transfer sequence is initiated by programming a memory-to-memory transfer via the 8237A's Command Register (described later) and Mode Register and enabling Channels 0 and 1 of the 8237A. Actual transfer is however initiated by putting a software DMA Request (DRQ) for Channel 0 via the 8237A's Request Register. The transfer is terminated by an internal or external $\overline{EOP}$; the internal $\overline{EOP}$ is provided by the TC stop option of Channel 1, whereas the external $\overline{EOP}$ is provided by a low pulse on $\overline{EOP}$. Note that the 8257 cannot do a memory-to-memory transfer like the 8237A.

Cascading 8237As for More DMA Channels

A number of 8237As may be cascaded as shown in Fig. A10.1.5 to provide more DMA channels than the four provided by one 8237A. The HRQ and HLDA lines of the higher level 8237As are connected to the DRQ and DACK lines of the next lower 8237A. For this latter 8237A, the

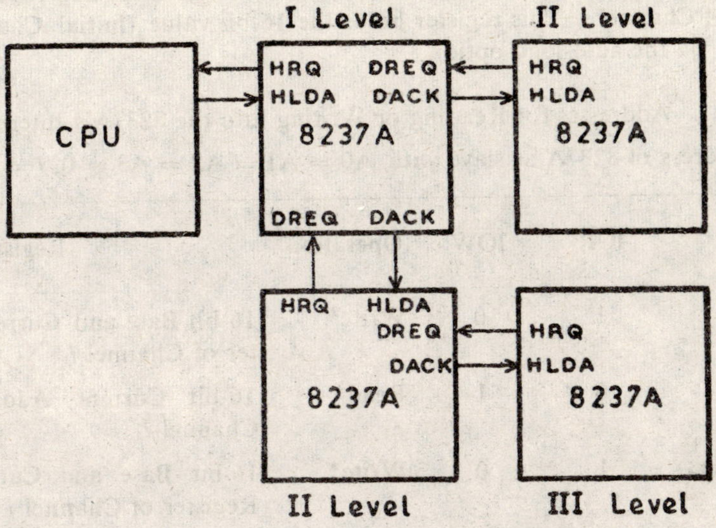

Fig. A10.1.5 Cascading Multiple 8237A Devices at Different Levels

DMA channel corresponding to this DRQ and DACK should be programmed in the Cascade Mode. This is all that is required to operate the 8237As in a cascade mode so as to provide more DMA channels in the system. For the arrangement shown in Fig. A10.1.5, a total of thirteen DMA channels are available. This useful cascading option is not present in the 8257 device.

Slave Mode of Operation of the 8237A

The 8237A operates in the slave mode for the purposes of programming it or reading its internal registers. Apart from the fact that it has a larger number of internal registers than the 8257, this mode of operation is identical for the two devices. Specifically, $\overline{CS}$ input of the 8237A should be Ø when reading from or writing into the 8237A is desired by the microcomputer and A0–A3 should indicate the internal register to be read or written; now applying $\overline{IOW}$ = Ø, $\overline{IOR}$ = 1 one gets a write operation and vice-versa for the read. Note that as for the 8257, the 8237A would then act as 16 I/O ports of the system and could be either memory mapped or I/O mapped. Specific addressing details are given in Table A10.1.

Note that in Table A10.1, we have chosen ADDR to be either a 16-bit (memory mapped) or 8-bit (I/O mapped) quantity whose lowest 4 bits, A0–A3, are Øs. The higher bits of address are used to derive the $\overline{CS}$ of the 8237A in the slave mode.

As in the 8257, the 8237A has an Internal First/Last (F/L) flip-flop for reading or writing 16-bit registers treating them as one location of the microcomputer's address space. The reader is urged to review this aspect of the 8257's operation which is identical to that of the 8237A. The 8237A further provides a software command to clear this flip-flop. This can be achieved by writing any data byte to the address ADDR + C_H as shown in Table A10.1. This is the *Clear First/Last flip-flop* command to the 8237A.

The initial address for Channel *i* of the 8237A is loaded into Channel *i*'s Current *Address*

Register at ADDR + $2i_H$ just as one would do it for the 8257. (Note that whether the address for Channel i will be incremented or decremented is decided by the programming of the Mode Register.) In the 8237A, writing into ADDR + $2i_H$ also writes these address bytes into the *Base Address* register of Channel i. This register holds the 16-bit value (initial Channel i address) for autoinitialization via the auto-load option.

Table A10.1 Addresses for Reading or Writing into the 8237A's Internal Registers
ADDR = Address of 8237A as slave with A0 = A1 = A2 = A3 = 0, i = 0, 1, 2, 3

Address	$\overline{IOR}$	$\overline{IOW}$	Operation	Register
ADDR + $2i_H$	1	Ø	Write*	16-bit Base and Current Address Register of Channel i
"	Ø	1	Read*	16-bit Current Address Register of Channel i
ADDR + $(2i + 1)_H$	1	Ø	Write*	16-bit Base and Current Word Count Register of Channel i
"	Ø	1	Read*	16-bit Current Word Count Register of Channel i
ADDR + 8_H	1	Ø	Write	Command Register (8-bit), Write Only
"	Ø	1	Read	Status Register (8-bit), Read Only
ADDR + 9_H	1	Ø	Write	Request Register (8-bit), Write Only
"	Ø	1	ILLEGAL	—
ADDR + A_H	1	Ø	Write	Set/Reset Single Mask Register bit
"	Ø	1	ILLEGAL	—
ADDR + B_H	1	Ø	Write	Mode Register (8-bit)
"	Ø	1	ILLEGAL	—
ADDR + C_H	1	Ø	Write	Clear Internal First/Last Flip-Flop
"	Ø	1	ILLEGAL	—
ADDR + D_H	1	Ø	Write	Master Clear (Software Reset)
"	Ø	1		Temporary, Register (8-bit), Read Only
ADDR + E_H	1	Ø	Write	Clear Mask Register (enable all channels)
"	Ø	1	Read	—
ADDR + F_H	1	Ø	Write	Mask Register (8-bits), Write Only
"	Ø	1	ILLEGAL	—

*16-bit register read or written into with 8-bit data using Internal First/Last Flip-Flop (procedure identical to that described for the 8257 in Appendix A10.1.1).

The *Current Word Count* register of Channel i at address ADDR + $(2i + 1)_H$ of the 8237A corresponds to the Terminal Count register of the 8257. However, unlike the 8257, one can specify 16-bit values of the Terminal Count for the 8237. Otherwise, the functions and operation of these two registers are identical for the two devices. In the 8237A, writing into ADDR + $(2i + 1)_H$ also writes the 16-bit Terminal Count value into the *Base Word Count* register of Channel i. This register holds the 16-bit value (initial Terminal Count) for autoinitialization through the auto load option.

Note that the two 16-bit Base registers mentioned above are present in the 8237A for each of the four channels, i.e. Channels 0, 1, 2 and 3. This is why the auto-load option can be exercised in the 8237A for all the channels. The reader should compare this with the rather limited auto-load facility of the 8257, i.e. auto-load only on Channel 2 reducing Channel 3 to a virtually-unusable state.

The *Command Register* (address = ADDR + 8_H) is a write-only register of the 8237A used to specify some basic command options for all the DMA channels. Its 8 bits, CR0–CR7, (LSB = CR0, MSB = CR7) are interpreted as follows:

CR0 = ∅	No memory-to-memory transfer
= 1	Memory-to-memory transfer on Channels 0 and 1
CR1 = ∅	Channel 0 address increments or decrements
= 1	Hold Channel 0 address constant
= ×	if CR0 = ∅ (i.e. Channel 0 address cannot be held constant for memory–I/O transfers)
CR2 = ∅	Enable 8237A DMA Controller (subject to Mask register)
= 1	Disable master mode of 8237A (all channels disabled, this should be done until programming the 8237A is complete)
CR3 = ∅	normal timing
= 1	compressed timing (no state S3)
= ×	if CR0 = 1 (i.e. compressed timing not possible if memory-to-memory transfer is enabled)
CR4 = ∅	fixed priority
= 1	rotating priority (i.e. as in 8257)
CR5 = ∅	normal Write pulses ($\overline{\text{IOW}}$, $\overline{\text{MEMW}}$)
= 1	extended Write pulses (i.e. as in 8257)
CR6 = ∅	DRQ sensed active High (i.e. as in 8257)
= 1	DRQ sensed active Low

CR7 = ∅ DACK output active Low (i.e. $\overline{DACK}$ as in 8257)

= 1 DACK output active High

Specific operational details for individual channels of the 8237A is programmed by writing data bytes (separately for each channel) in the *Mode Register* (Address = ADDR + B_H) of the device. This is a write-only register whose bits MR0–MR7 (LSB = MR0, MSB = MR7) are to be interpreted as follows:

MR1, MR0	∅∅	select Channel 0 for mode programming
	∅1	select Channel 1 for mode programming
	1∅	select Channel 2 for mode programming
	11	select Channel 3 for mode programming
MR3, MR2	∅∅	verify mode
	∅1	DMA Write mode (I/O to memory)
	1∅	DMA Read mode (memory to I/O)
	11	ILLEGAL Choice
	××	if MR7, MR6 = 11 (selected channel used to cascade other 8237As)
		(see Sec. A10.1.1 for description of DMA Read/Write modes)
MR4		Auto-load bit for selected channel
	∅	no autoinitialization
	1	autoinitialize (Address and Word Count registers) on internal or external $\overline{EOP}$
MR5	∅	increment this channel's address (i.e. Current Address register) following each byte transferred
	1	decrement address
MR7, MR6	∅∅	Demand Transfer mode selected
	∅1	Single Transfer mode selected
	1∅	Block Transfer mode selected
	11	Cascade mode selected (selected channel used to cascade other 8237As)

Note that bits MR2–MR7 select various operation features for the 8237A DMA channel indicated by the bits MR1 and MR0.

For each of the four DMA channels, the 8237A has a *Request Bit* which is not directly accessible, but may be Set/Reset by writing appropriate bytes into the *Request Register* (Address = ADDR + 9_H). The individual bits RR0–RR7 do the following:

RR1, RR0	∅∅	Select Channel 0
	∅1	Select Channel 1

	1Ø	Select Channel 2	
	11	Select Channel 3	
	(Selection implies setting/resetting the corresponding Request Bit as indicated by RR2)		
RR2	Ø	reset	Request Bit of selected channel
	1	set	Request Bit of selected channel
RR3–RR7	×××××		Don't Cares

The Request Bit for a channel gets cleared on internal or external $\overline{EOP}$. Moreover, the Request Bits for all four channels get simultaneously cleared on a software or hardware RESET. If a channel is programmed in the Block Transfer mode, a DMA transfer sequence may be initiated under software control by setting the corresponding Request Bit. (The transfer continues until an internal or external $\overline{EOP}$ is encountered.) Setting this bit is then functionally equivalent to making the DRQ of that channel active. A software controlled request for DMA through the Request Bit, will go through the priority structure specified by CR4 but cannot be masked (disabled) by the Mask Register. Memory-to-memory transfers are initiated setting the Request Bit for Channel 0, i.e. write ××××× 1ØØ to ADDR + 9_H.

Apart from the global (all channels) enabling/disabling of DMA channels possible through a software or hardware RESET or by set/reset of CR2, one can also selectively enable/disable channels by a reset/set of their corresponding *Mask Bits*. Even if CR2 was set in the Command Register, the Mask Bit for a channel has to be reset before that channel can respond to a hardware DMA Request from the DRQ input (software DMA requests through the Request Bits not disabled by masking). The Mask Bit for a particular channel can be set/reset (set for channel disable, reset for channel enable) under software control. It gets set whenever the channel encounters an internal or external $\overline{EOP}$ but has not been programmed in the auto-load mode. All the mask bits also get set by a software or hardware RESET; they can all be cleared (i.e., all channels enabled) by the Clear Mask Register command.

The *Clear Mask Register command* to the 8237A may be given by writing any 8-bit pattern to the address ADDR + E_H. This has the effect of enabling all the channels (if CR2 had been set previously) by clearing all the Mask Bits.

The four Mask Bits of the 8237A may be separately set or reset by writing a proper byte, MR0–MR7, (LSB = MR0) into its Mask Register (Address = ADDR + F_H). This is a write-only register where the bits MR0–MR7 do the following:

MRi ($i = 0, 1, 2, 3$)	Ø	clear mask bit of Channel i (enable subject to CR2)
	1	set mask bit of Channel i (disable)
MR4–MR7	××××	Don't Cares

The Mask Bit of a selected channel may also be individually set or reset by writing an appropriate byte, SM0–SM7 (SM0 = LSB) at address ADDR + A_H as shown in Table A10.1. The individual bits, SM0–SM7, should then be interpreted as follows:

SM1, SM0	ØØ	select Mask Bit of Channel 0
	Ø1	select Mask Bit of Channel 1
	1Ø	select Mask Bit of Channel 2

	11	select Mask Bit of Channel 3
SM2	Ø	clear selected Mask Bit
	1	set selected Mask Bit

Reading location ADDR + 8_H of the 8237A allows the controlling microcomputer to read the device's *Status Register*. This 8-bit read-only register reports the TC status of the DMA channels and whether any of them are currently requesting service. Its bits SR0–SR7 (SR0 = LSB), should be interpreted as follows after a read operation on location ADDR + 8_H:

SR_i	$i = 0, 1, 2, 3$	1	Channel i has reached Terminal Count (TC)
		Ø	otherwise
SR_{j+4}	$j = 0, 1, 2, 3$	1	Channel j currently requesting service
		Ø	otherwise

The 8237A contains another read-only register, referred to as the *Temporary Register* (address = ADDR + D_H). For memory-to-memory transfers, this register acts as the intermediate storage for the data byte actually being transferred. (This has been described earlier.) The microcomputer can, therefore, read this register to find the value of the last data byte transferred in this fashion.

The 8237A also accepts three software commands which may be issued to it by writing any data byte into locations ADDR + C_H, ADDR + D_H and ADDR + E_H. Of these, the commands 'Clear Internal First/Last Flip-Flop' (ADDR + C_H) and 'Clear Mask Register' (ADDR + E_H) have been described earlier. Writing any data byte into location ADDR + D_H is equivalent to giving a software RESET to the 8237A. This is referred to as the *Master Clear Command*. Like applying a hardware RESET (by a high pulse at the 8237A's Reset input), issuing this command sets the Mask Bits of all the channels, and clears the Command Register, Status Register, Request Register, Temporary Register and the Internal First/Last Flip-Flop.

Programming and Using the 8237A

Even though the 8237A device is considerably more powerful than the simpler 8257 allowing a much greater choice of options, the ways of using them do not differ much. Basically, the device has to be first programmed (in the slave mode) by writing into its various registers as described earlier. One can then monitor the progress of the DMA operation on a particular channel by reading the Status Register or the corresponding Current Address and Word Count registers. The $\overline{\text{EOP}}$ output (TC Stop pulse) may also be used to find out when Terminal Count is reached on a particular channel (i.e. by using interrupts). One can also do a hardware comparison of the byte being transferred with a pre-set value to generate an external $\overline{\text{EOP}}$ pulse (applied to the $\overline{\text{EOP}}$ pin) to terminate, and may be autoinitialize, the DMA transfers on a channel when the two bytes match.

It is considered advisable to disable all DMA channels via CR2 of the Command Register before programming the 8237A. This is done to ensure that the 8237A stays in the slave mode while the programming is being done. As the final step, one can then appropriately write into the Command Register with CR2 = Ø. This leaves the 8237A properly programmed to tackle active DMA request on the channels which have been unmasked.

Finally, note that the hardware similarity between the 8257 and the 8237A is so much that, in the circuit of Fig. A10.1.2, one can replace the 8257 with the 8237A almost directly. Properly programmed, the 8237A can do the DMA transfer required of the 8257 in that circuit.

A10.2 INTEL 8212, 8-BIT INPUT/OUTPUT PORT

We have used the 8212 earlier as a latch strobed by ALE to demultiplex the address lines from the ADØ–AD7 output of the 8085A. Here we describe the chip in detail along with simple configurations illustrating its use as an input or output port. (It cannot be used as a bidirectional port.) Its pin-out is given in Fig. A10.2.1. A brief description of its pins is given below.

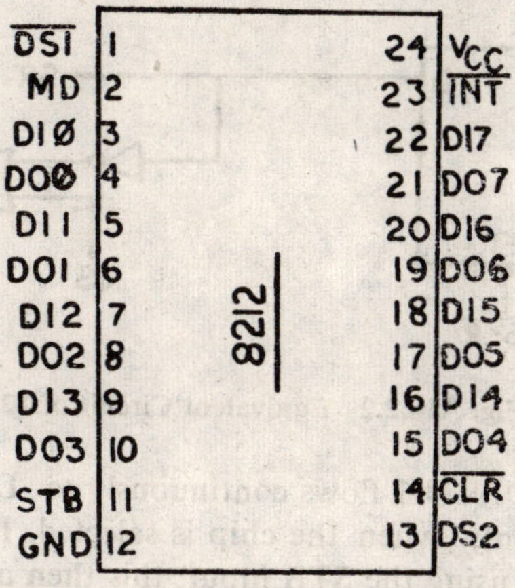

Fig. A10.2.1 Pin-out of the 8212

DIØ–DI7	: Input	; Input Data Bus
DOØ–DO7	: Tristate Output	; Output Data Bus
$\overline{DS1}$, DS2	: Input	Device Select = $(\overline{\overline{DS1}})\cdot DS2$
MD	: Input	Mode Select
STB	: Input	Data Strobe
$\overline{CLR}$	: Input	Device Clear
$\overline{INT}$	: Interrupt Request Output	
V_{cc}, GND	: Power, $V_{cc} = 5$ V	

The 8212 contains 8 D-type data latches and one Service Request (SR) FF (Fig. A10.2.2). DIØ–DI7 are the inputs to the data latches which are clocked by a combination of $\overline{STB}$, $\overline{DS1}$, DS2 and MD inputs. The $\overline{CLR}$ input resets the output of these latches to Ø.

$\overline{DS1}$ and DS2 are the chip select inputs. When $\overline{DS1}$ is low and DS2 is high, the 8212 is selected, its tristate outputs are enabled and the contents of the data latches appear at DOØ–DO7. This also sets the SR output.

The MD input decides the operating mode of the 8212. It decides the state of the output buffer and the source of the clock signal clocking the data latches. When MD = 1, the output buffers are always enabled and the clock to the data latches is equivalent to $(\overline{\overline{DS1}})\cdot DS2$. If this

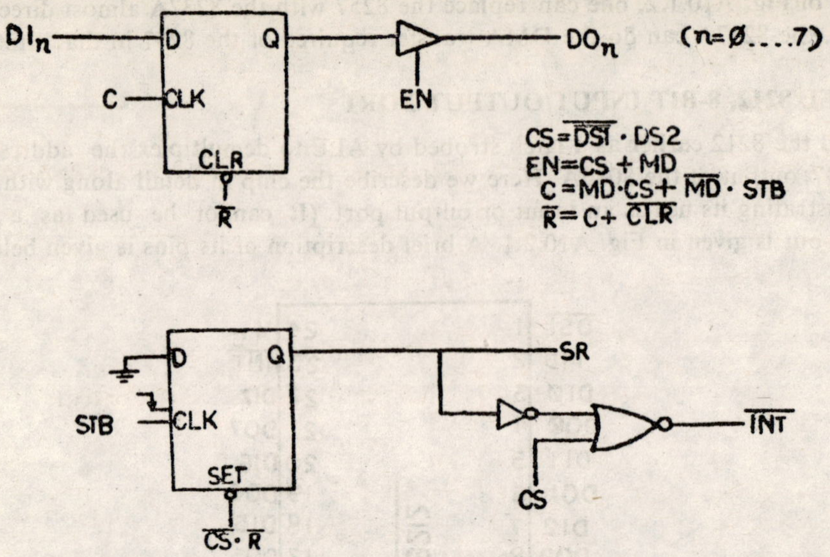

Fig. A10.2.2 Equivalent Circuit of 8212

clock is held high, the data at DIØ–DI7 flows continuously to DOØ–DO7. When MD = Ø, the output buffers are enabled only when the chip is selected. In this mode, data can be latched on the data latches at any time using the STB input; this then acts as the clock to these latches.

The STB input latches data in the data latches when MD = Ø; it also resets the SR flip-flop. It latches data at the rising edge of STB (positive level triggered data latches) and resets SR at the falling edge of STB (negative edge triggered SR FF).

If the 8212 is being used as an input port (with MD = Ø), then STB will latch data into the latches. This data will appear on DOØ–DO7 when the 8212 is selected using $\overline{DS1}$ and DS2. If the 8212 is being used as an output port we can either use STB to store the data into latches with MD = Ø and let external logic read the data using $\overline{DS1}$ and DS2 or we can tie MD and STB to 1 and use $\overline{DS1}$ and DS2 to transfer DIØ–DI7 contents to DOØ and DO7. In the former case, the data bus (DØ–D7) to the external device is active only when ($\overline{DS1}$) is Ø, DS2 is 1; otherwise it is in High-Z state. In the latter case, this is always active (since MD = 1). Let

$$(\overline{DS1})\cdot DS2 = CS \quad (CS = 1 \text{ if } \overline{DS1} \text{ is } \emptyset \text{ and DS2 is } 1)$$

Then examining the circuit for the 8212 given in Fig. A10.2.2, we see the following:

EN = Enable Output Buffers

= CS + MD

C = Clock to data latch (*level sensitive*; high level Data Latch Out = Data Latch In or DI_n)

= MD·CS + $\overline{MD}$·STB

if MD = 1, C = CS

if MD = Ø, C = STB

$\overline{R}$ = Reset data latch (active Low)

$= C + (\overline{CLR})$ (Reset if C is Ø and $\overline{CLR}$ is Ø)

or $R = \overline{C + \overline{CLR}}$

The behaviour of the 8212 can also be summarised by the three tables shown next.

STB	MD	CS	DOØ–DO7	
Ø	Ø	Ø	tristate	Output not selected
1	Ø	Ø	tristate	
Ø	1	Ø	= Data latch	Output selected Output = Data Latch Contents
1	1	Ø	= Data latch	
Ø	Ø	1	= Data latch	
Ø	1	1	$= DI_Ø–DI_7$	Output selected Output ≡ Input
1	Ø	1	$= DI_Ø–DI_7$	
1	1	1	$= DI_Ø–DI_7$	

With $\overline{CLR}$ = 1, the data stored in the (internal) data latches are given by the following table.

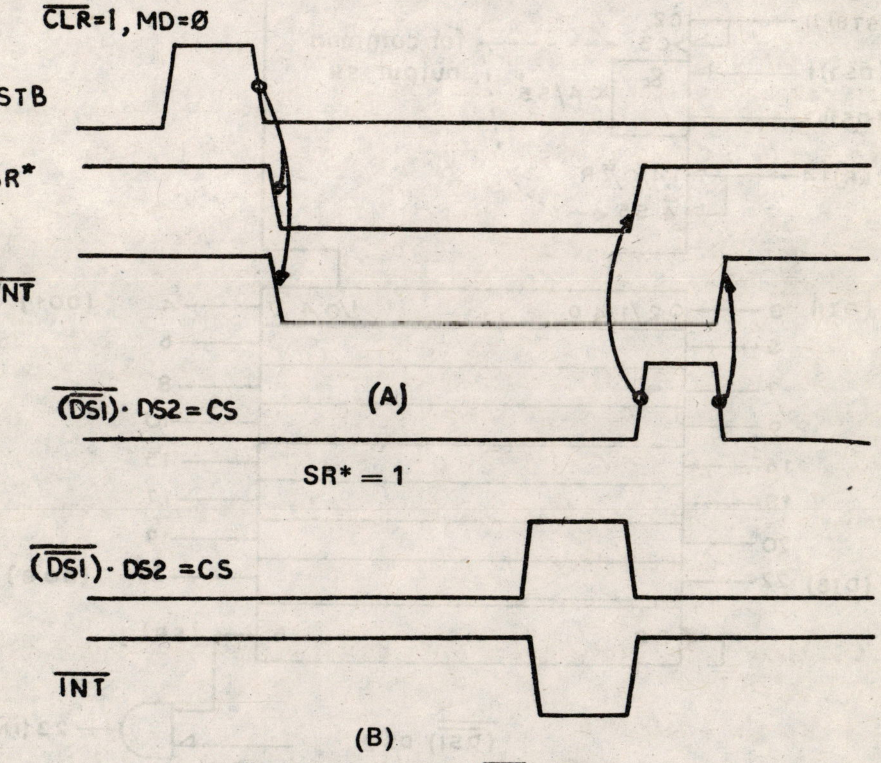

Fig. A10.2.3 Timing of $\overline{INT}$ Output

*Internal to the 8212.

(Note that $\overline{CLR} \equiv \emptyset$ clears all these latches.)

STB	MD	CS	Data Latch Contents
Ø	Ø	X	No change from earlier contents
X	1	Ø	
1	Ø	X	Latch Contents = DIØ–DI7
X	1	1	

The $\overline{INT}$ output of the 8212 essentially indicates a buffer full condition (i.e. $\overline{INT} = \emptyset$) when the device is being operated with MD = Ø. In this case, consider a situation where $\overline{INT} = 1$ to begin with. This, for example, can be achieved by applying a Hi → Lo → Hi pulse at the $\overline{CLR}$ input. Now suppose data is latched into the 8212 by a Lo → Hi → Lo pulse at the STB input. $\overline{INT}$ goes Lo at the Hi → Lo edge of STB indicating a buffer full condition. This buffer can be 'emptied' by applying a Ø → 1 → Ø pulse for CS by properly manipulating the $\overline{DS1}$ and DS2 inputs; the 1 → Ø transition of CS returns $\overline{INT}$ to 1 indicating the buffer empty condition. This timing relationship is shown in Fig. A10.2.3(a). Even with SR = 1, $\overline{INT} = \emptyset$ will also arise if CS = 1. This is illustrated in Fig. A10.2.3(b). The reader should be aware of this condition when using the $\overline{INT}$ output (inverted or non-inverted) as an interrupt to the microprocessor.

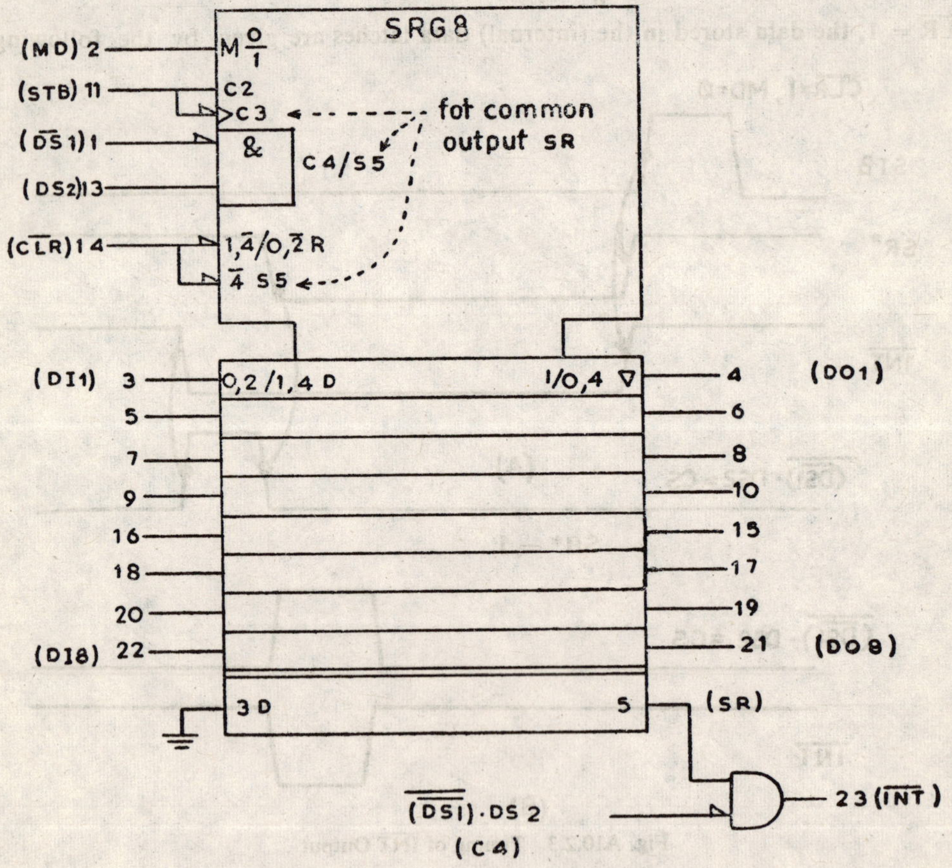

Fig. A10.2.4 Dependency Notation Symbol for the Intel 8212

The dependency notation symbol for the 8212 is given in Fig. A10.2.4. Note that this symbol, properly interpreted, summarises the complete behaviour of the 8212 described in detail earlier.

The 8212 is a very versatile device which may be used to serve a variety of functions in a microprocessor system. A very common use of this device is as an address/data demultiplexer for the 8085A. An example of such a usage may be seen in Fig. A10.1.2. Apart from this, a number of other usages of the 8212 is possible. Some examples of these are described briefly next.

A very simple use of the 8212 is as a Gated Buffer with Tristate outputs as shown in

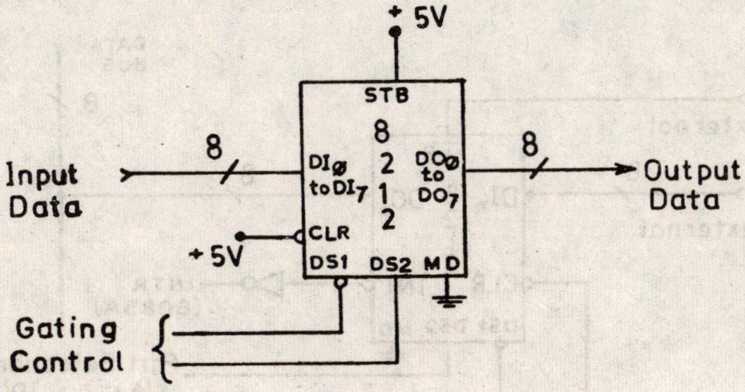

Fig. A10.2.5 Gated Buffer (tristate) Using 8212

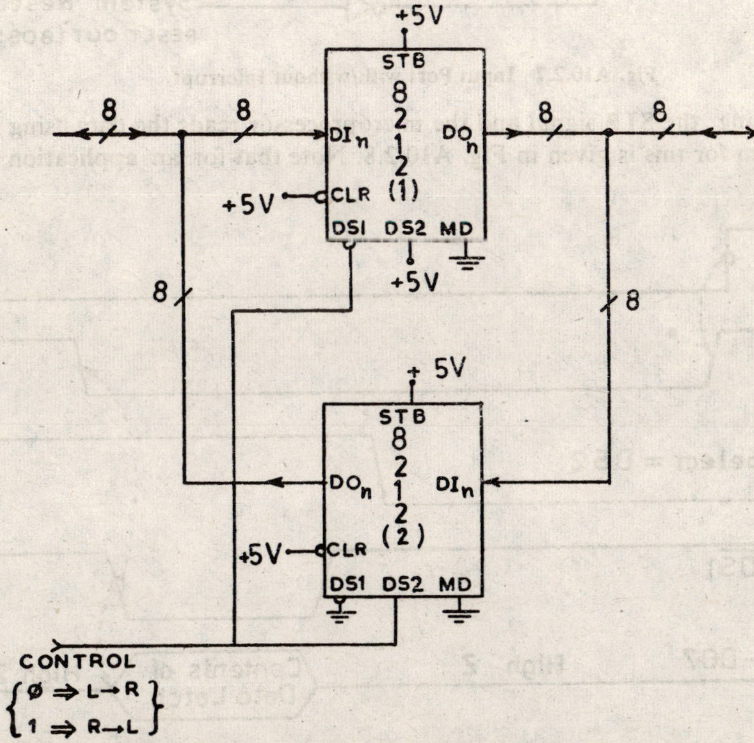

Fig. A10.2.6 Bidirectional Bus Driver

Fig. A10.2.5. Here Output Data = Input Data as long as $\overline{DS1} = \emptyset$ and DS2 = 1; otherwise, the outputs are tristate. This circuit is very often used when one needs a higher drive capability than that available from normal MOS devices.

One can also use 8212s to implement a Bidirectional Bus Driver as shown in Fig. A10.2.6. The mode of operation of the two 8212s here is very similar to that in Fig. A10.2.5. Note that the direction of data transfer is controlled by the binary value of the CONTROL input.

The 8212 is commonly used to implement an input port (with or without interrupt) in a microcomputer system. This is illustrated in Fig. A10.2.7 where the external device loads data

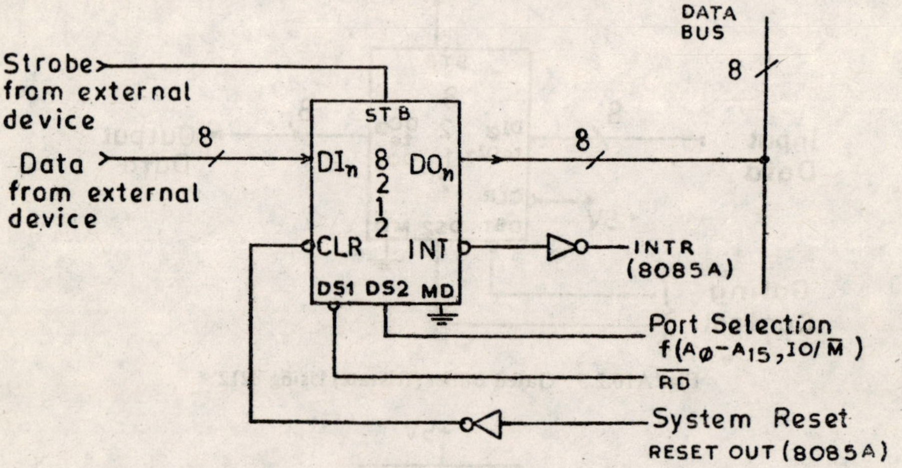

Fig. A10.2.7 Input Port with/without Interrupt

into the 8212 using the STB signal and the microprocessor reads the data using $\overline{DS1}$ and DS2. The timing diagram for this is given in Fig. A10.2.8. Note that for an application requiring an

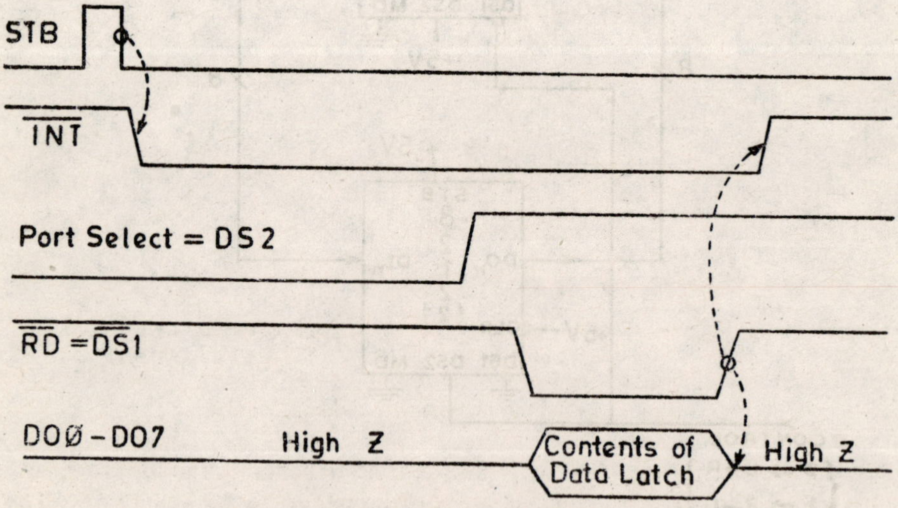

Fig. A10.2·8 Input Port Timing

interrupt controlled input port the $\overline{INT}$ output of the 8212 should be appropriately used.

The 8212 device may also be used as an output port. Two circuits for this are given in Fig. A10.2.9 and Fig. A10.2.10. In Fig. A10.2.9, the output of the 8212 is continuously on; this

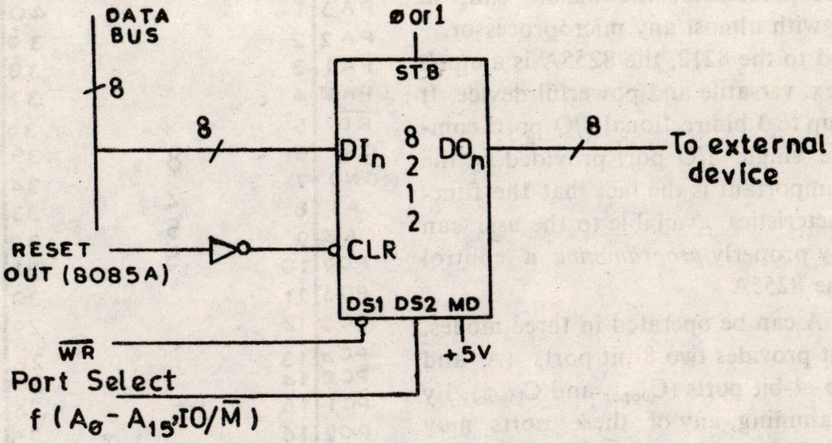

Fig. A10.2.9 Output with Output Continuously ON

may be useful when one wants the output port to drive a display. The circuit of Fig. A10.2.10 is useful when one wants the output port to act as the input port of another microprocessor. The $\overline{INT}$ output of the 8212 can then also be used appropriately to generate an interrupt.

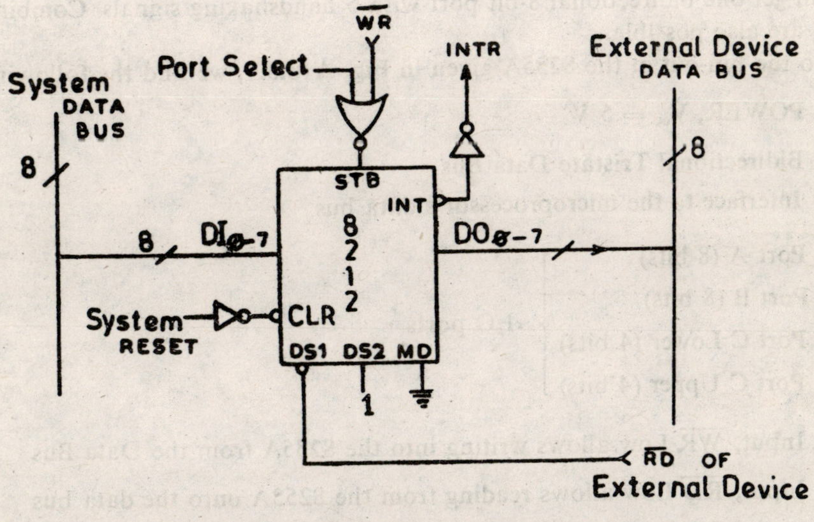

Fig. A10.2.10 Output Port with Output Control

For 8085A based systems, the 8212 may also be used to generate the interrupt vector for the INTR interrupt. Two circuits for doing this are shown in Figs. 12.5.2 and 12.5.3. The reader is referred to Sec. 12.5 for a description of these circuits.

A10.3 INTEL 8255A PROGRAMMABLE PERIPHERAL INTERFACE (PPI)

In this appendix, we describe the 8255A PPI which can be used as a general purpose I/O device. Though primarily intended for use with Intel's family of microprocessors, the 8255A can, in fact, be used with almost any microprocessor.

Compared to the 8212, the 8255A is a much more complex, versatile and powerful device. It can provide up to 3 bidirectional I/O ports compared to the single I/O port provided by the 8212. More important is the fact that the functional characteristics available to the user can be selected by properly *programming* a control register in the 8255A

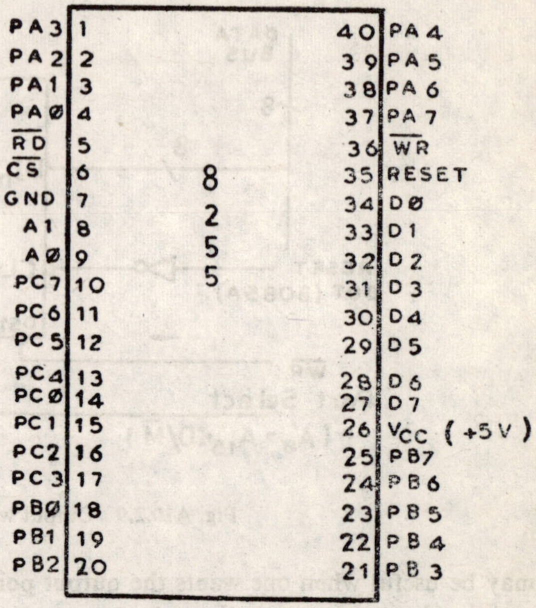

Fig. A10.3.1 Pin-out of the 8255A PPI

The 8255A can be operated in three modes. In Mode Ø, it provides two 8-bit ports (A and B) and two 4-bit ports (C_{upper} and C_{lower}). By proper programming, any of these ports may be used as either input or output ports (*but not bidirectional*). In Mode 1, it provides two 8-bit ports (A and B), each of which can be programmed as either input or output (*but not bidirectional*); each 8-bit port also has three handshaking signals associated with it for input or output with handshake. Finally, in Mode 2, we can get one bidirectional 8-bit port with 5 handshaking signals. Combinations of the various modes are also possible.

Referring to the pin-out of the 8255A given in Fig. A10.3.1, we find the following:

V_{cc}, GND	POWER, $V_{cc} = 5$ V
$D_Ø$–D_7	Bidirectional Tristate Data Bus Interface to the microprocessor's data bus
PAØ–PA7	Port A (8 bits) } I/O ports
PBØ-PB7	Port B (8 bits) } I/O ports
PCØ–PC3	Port C Lower (4 bits) } I/O ports
PC4–PC7	Port C Upper (4 bits) } I/O ports
$\overline{WR}$	Input, $\overline{WR}$ Low allows writing into the 8255A from the Data Bus
$\overline{RD}$	Input, $\overline{RD}$ Low allows reading from the 8255A onto the data bus
$\overline{CS}$	Input, $\overline{CS}$ Low selects the 8255A for read or write operations (derived from Address Bus and IO/$\overline{M}$)
A1, AØ	Input, selects one of four registers or ports inside the 8255A according to the following table

$\overline{CS}$	A1	AØ	
Ø	Ø	Ø	PORT A
Ø	Ø	1	PORT B
Ø	1	Ø	PORT C
Ø	1	1	Control Register

Depending on whether PORTs A, B and C are configured as input or output, one can read from or write into the corresponding register by using the $\overline{RD}$ or $\overline{WR}$ signals from the CPU with $\overline{CS}$ low and A1, AØ properly chosen. Note that one can *only write into the Control Register* and that reading from that register is not allowed.

RESET Input, RESET = 1 clears the Control Register and all ports (A, B, C) are set to the input mode.

As mentioned earlier, the 8255A operates its ports in three modes—Modes Ø, 1 and 2. Before discussing the programming of the 8255A, we present the operating features of a port operated in these modes.

Mode Ø

A port in Mode Ø provides simple input or output (*but not both*) depending on how it is programmed. No handshaking is required since the processor simply outputs data to a Mode Ø output port or inputs data from a Mode Ø input port. The inputs to this port are not latched but the outputs are latched. Functionally, this implies that an external device inputting data to a Mode Ø input port will have to hold the data at the port input till the processor reads the port contents. Conversely, if the processor outputs data to a Mode Ø output port, the port's output reflects this data until new data is outputted to the port. A typical example of a Mode Ø output port is one driving a display; an example of a Mode Ø input port is one that a processor uses to read external switch positions. Finally, it should be noted that since no handshaking is required, a Mode Ø port does not need any extra pins (i.e. more than those required by the port for actual input/output).

Mode 1

A port in Mode 1 provides means for transferring I/O data to or from the port along with strobes or 'handshake' signals.

A Mode 1 Input Port of the 8255A uses the following control signals. Their timing relationship is shown in Fig. A10.3.2.

Mode 1 Input Port Control Signals (Fig. A10.3.2)

$\overline{STB}$ A low on the $\overline{STB}$ loads data from the external device into the input latch of the input port

IBF (Input Buffer Full F/F)
A high on the IBF output indicates that the last byte loaded in by the external device has not yet been read by the CPU. It is set (=1) by a Low $\overline{STB}$ and reset (=Ø) when the CPU reads the data byte previously loaded. Normally the external device should not output more data to the port until IBF return to Ø. IBF also goes to Ø when the 8255A is RESET.

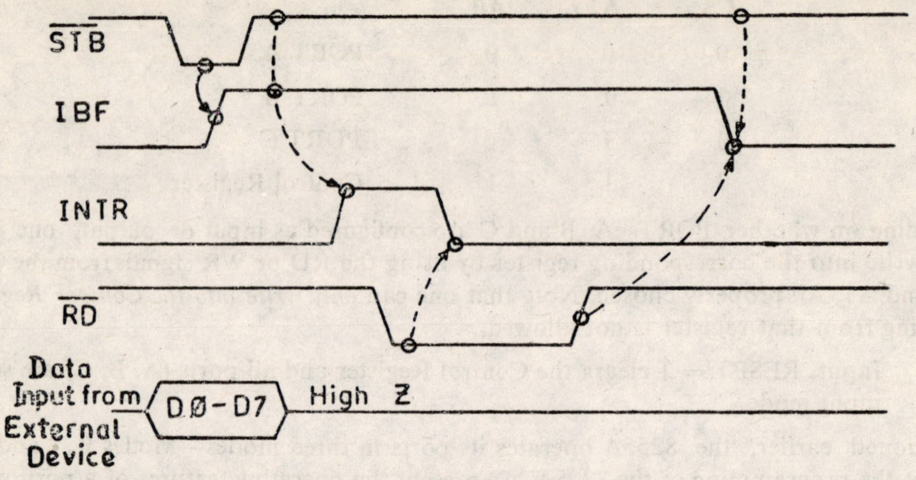

Fig. A10.3.2 Mode 1 Input Port Timing

INTR (Interrupt Request)

INTR = 1 indicates to the CPU that the port has got fresh data and wants the CPU to service it. INTR can be disabled by resetting an internal flip-flop INTE to Ø. INTR is set by $\overline{\overline{STB}}$·IBF·INTE (i.e. when $\overline{STB}$ is 1 and IBF and INTE are 1) and reset by the Lo → Hi transition of $\overline{RD}$ when the CPU reads the contents of this port.

The control signals needed for a Mode 1 Output Port in the 8255A are given below. The timing relationships of these control signals are shown in Fig. A10.3.3.

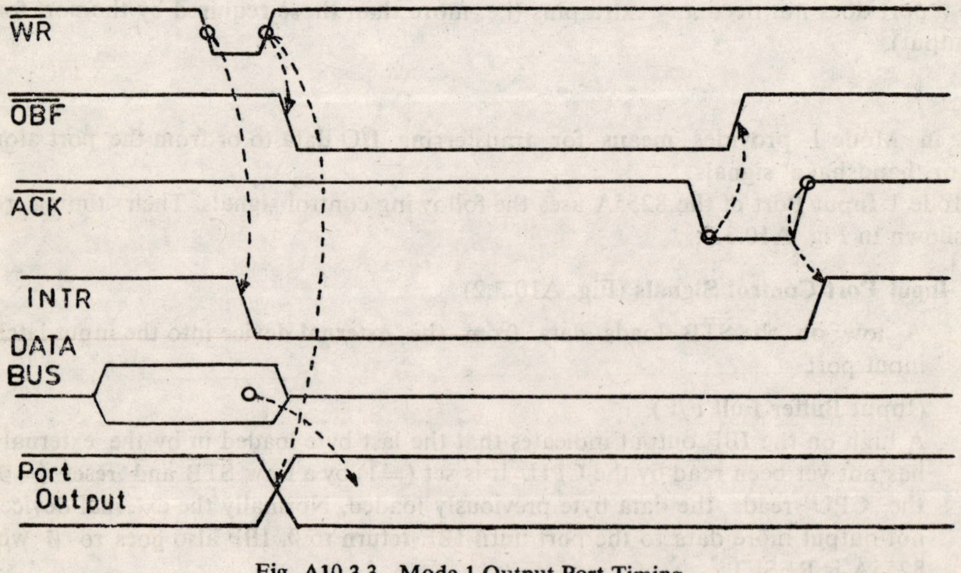

Fig. A10.3.3 Mode 1 Output Port Timing

Mode 1 Output Port Control Signals (Fig. A10.3.3)

$\overline{\text{OBF}}$ (Output Buffer Full F/F)

$\overline{\text{OBF}}$ output goes Low when the CPU writes data to the port (specifically at the Lo → Hi transition of $\overline{\text{WR}}$ when the port has been selected). It is set (=1) when the $\overline{\text{ACK}}$ input is Low. It is also '1' after RESET. $\overline{\text{OBF}} = \emptyset$ indicates to the CPU that the previous data written by it into the port has not yet been read by the external device. Normally, the CPU will not write in a fresh data byte to the port until $\overline{\text{OBF}}$ becomes High.

$\overline{\text{ACK}}$ (Acknowledge)

External device pulses $\overline{\text{ACK}}$ Low to indicate that it has accepted the new data byte.

INTR (Interrupt Request)

INTR = 1 may be used as an interrupt input to the CPU indicating that the external device has accepted the last data output by the CPU to this port and that the CPU should now output a new byte to the port.

INTR can be disabled by resetting an internal flip-flop INTE to Ø.

INTR is set when $\overline{\text{ACK}}$ and $\overline{\text{OBF}}$ are 1 and INTE is 1. It is reset by the Hi → Lo edge of $\overline{\text{WR}}$ if this port has been selected.

Note that in both Mode Ø and Mode 1 Output Ports the output of the ports are always active (drivers are ON). This in turn implies that these outputs cannot be directly connected to a bus. If interfacing to a bus is required we should interpose a latch or buffer with tristate output control between the port outputs and the bus. This tristate control is internally provided in Mode 2 for which the control signals are given next.

Mode 2 Bidirectional Input/Output Port

A Mode 2 port allows the microprocessor to communicate with a peripheral device or system for both data transmission and reception. Proper enable signals are provided so that the port can interface to a data bus on both sides. In this case, the following five handshake signals are needed.

$\overline{\text{OBF}}$: As in Mode 1, $\overline{\text{OBF}} = \emptyset$ indicates that the CPU has written data into the port which has not yet been read by the external device.

$\overline{\text{ACK}}$: The $\overline{\text{ACK}}$ input is provided by the external device. Putting $\overline{\text{ACK}} = \emptyset$ enables the tristate output buffer of the port. The external device can now read the port contents.

$\overline{\text{STB}}$: An external device uses $\overline{\text{STB}}$ to latch data (for the CPU) into the port.

IBF : As in Mode 1, IBF = 1 indicates that an external device has latched data into the port which has not yet been read by the CPU.

INTR : This is a common interrupt (active high) for the CPU used by both input and output operations in the same manner as in Mode 1.

A Mode 2 port provides two interrupt enable flip-flops INTEI and INTEO. Resetting INTEI to Ø disables interrupts coming from input operations and resetting INTEO to Ø disables interrupts arising from output operations. Note that since the interrupt INTR is shared by both input and output, the interrupt service routine executed by the CPU should test IBF and $\overline{\text{OBF}}$ in order to decide the reason for the Interrupt (i.e. input or output). Alternatively, INTEI and INTEO may be used to selectively mask out input or output interrupts.

Mode 2 control signals are basically a combination of the control signals for Mode 1 input and output ports. The timings of these signals also follow the same form. Summarising the timing we can say the following:

(a) When an external device pulses $\overline{\text{STB}}$, IBF goes high and INTR goes high (if INTEI = 1). Reading the port from the CPU returns IBF to Ø and INTR to Ø (provided an output interrupt is not pending).

(b) When the CPU writes data into the port, $\overline{\text{OBF}}$ and INTR (if INTEO = 1) both go low provided an input interrupt is not pending. The external device reads the port by pulsing $\overline{\text{ACK}}$. This returns $\overline{\text{OBF}}$ to 1 and also raises INTR high.

Inputs and outputs are latched in both Mode 1 and 2 operation of a port.

Programming the 8255A

Programming the 8255A for operation in various modes is achieved by writing a suitable Control Word into its Control Register. (As mentioned earlier the Control Register can be written into by setting $\overline{\text{CS}}$ = Ø, AØ = 1, A1 = 1 and pulsing $\overline{\text{WR}}$ low. Reading from this register is an illegal operation.)

For mode selection purposes, the 8255A's port pins are divided into two groups, A and B.

Group A

Port A	(PAØ–PA7)
Port C upper	(PC4–PC7)

Group B

Port B	(PBØ–PB7)
Port C lower	(PCØ–PC3)

The following mode configurations can be chosen

Mode Ø : Port A (8), Port B (8), Port C upper (4), Port C lower (4) — Any of these can be input or output

Mode 1 : Port A (8); Handshake signals provided by three pins of Port C as follows

Input $PC_4 = \overline{STB_A}$, $PC_5 = IBF_A$, $PC_3 = INTR_A$ — $INTE_A$ flip-flop = PC_4

Output $\quad PC7 = \overline{OBF_A}$, $PC6 = \overline{ACK_A}$, $PC3 = INTR_A$ } $INTE_A$ flip-flop = PC_6

Port B (8); Handshake signals provided by three pins of Port C

Input $\quad PC2 = \overline{STB_B}$, $PC1 = IBF_B$, $PC\emptyset = INTR_B$ } $INTE_B$ flip-flop = PC2

Output $\quad PC1 = \overline{OBF_B}$, $PC2 = \overline{ACK_B}$, $PC\emptyset = INTR_B$ } $INTE_B$ flip-flop = PC2

Note that, depending on whether PORT A has been programmed as input or output, either PC4, PC5 or PC6, PC7 will be unused. These can be separately programmed as a Mode Ø 2-bit input or output port.

Mode 2 : Port A (8); Handshake provided by Port C pins

$PC3 = INTR_A$, $PC4 = \overline{STB_A}$, $PC5 = IBF_A$, $PC6 = \overline{ACK_A}$, $PC7 = \overline{OBF_A}$ } INTEI flip-flop = PC4; INTEO flip-flop = PC6

If Port A is used in Mode 2, one can still do the following with other port pins:

(i) Use Port B in Mode 1 with PCØ, PC1 and PC2 for handshaking.

(ii) Use Port B in Mode Ø and PCØ–PC2 as another 3-bit Mode Ø port.

Apart from the mode combinations described above, several other mode combinations are possible. We will not describe them in detail besides noting that port pins not used by Modes 1 or 2 can be used for the corresponding lower modes provided operating those port pins in this mode is allowed.

While programming the 8255A, the Control Register CRØ–CR7 have the following meaning.

CR7 : = 1, Mode Set Active (Ø implies that the intent of this byte is to Set/Reset Port C bits)

CR6, CR5 Mode Selection for Group A

CR6	CR5	
Ø	Ø	Mode Ø (Port A, Port C upper)
Ø	1	Mode 1 (Port A)
1	×	Mode 2 (Port A)

CR4	:	Port A Input	1
		Port A Output	Ø
CR3	:	Port C upper Input	1
		Port C upper Output	Ø
CR2	:	Mode Selection for Group B	
		Ø = Mode Ø (Port B, Port C lower)	
		1 = Mode 1 (Port B)	
CR1	:	Port B Input	1
		Port B Output	Ø
CRØ	:	Port C lower Input	1
		Port C lower Output	Ø

Examples (of port programming) :

(i) Port A Mode Ø Input
Port B Mode Ø Output
Port C upper Mode Ø Output
Port C lower Mode Ø Input
Control word = 1ØØ1ØØØ1
$= 91_H$

(ii) Port A Mode 1 Output (PC_3, PC_6, PC_7 for handshaking)
Port B Mode 1 Input ($PC_Ø$-PC_2 for handshaking)
PC_4, PC_5 Mode Ø Output
Control word = 1Ø1ØØ11×
$= A6_H$ (choose × = Ø)

(iii) Port A Mode 2 Input/Output (PC_3–PC_7 needed for handshaking)
Port B Mode 1 Input ($PC_Ø$–PC_2 needed for handshaking)
Control Word = 11× × ×11×
Since Port A is in Mode 2 it can be either input or output

Bit Set/Reset Feature

By writing a control word with CR7 = Ø into the Control Register any of the eight bits of Port C can be set or reset. When Port C is being used for status/control of Ports A and B, these bits can be set or reset in this way just as if they were data output ports. The control word format for this is given below.

CR7 Ø

CR6, CR5, CR4	× (don't cares)
CR3, CR2, CR1	Select port C bit indicated by (CR3, CR2, CR1)
CRØ	1 = set selected bit
	Ø = reset selected bit

Note that separate control words have to be written in order to set/reset each bit of Port C. (One 'control word write' operation will not affect more than one bit.)

Interrupt Masking

In our discussion of Mode 1 and 2 control signals, we had mentioned that the interrupts provided in these modes can be disabled by resetting the corresponding interrupt enable (INTE) flip-flops to Ø. These flip-flops are $INTE_A$ and $INTE_B$ for Mode 1 Port A and B respectively (both input and output), INTEI for Mode 2 Port A input interrupt and INTEO for Mode 2 Port A output interrupt. These flip-flops are also set/reset using the Bit Set/Reset feature given above and the following table.

Flip-Flop	Port	Mode	Input/Output port	Port C bit to be set/reset
$INTE_A$	Port A	1	Input	PC_4
$INTE_B$	Port B	1	Input	PC_2
$INTE_A$	Port A	1	Output	PC_6
$INTE_B$	Port B	1	Output	PC_2
INTEI	Port A	2	Input Interrupt	PC_4
INTEO	Port A	2	Output Interrupt	PC_6

All INTE F/Fs are reset (—Ø) after RESET (all interrupts are therefore disabled).

Example

We want to operate Port A in Mode 2 and Port B as a Mode 1 input port. However, we want to disable INTEI of Port A. As mentioned before, the Mode is programmed by writing

11×××11×

in the control register.

When we want to disable INTEI of Port A, we need to reset PC_6 (from table given above). For this we need to write

Ø×××11ØØ

in the control register.

When we want to reenable the interrupt we need to write

Ø×××11Ø1

in the control register.

Status of the 8255A

When either (or both) Port A or Port B is programmed in Mode 1 or Port A is programmed in Mode 2, various Port C pins perform the required 'Handshaking' functions. In this case, some of the handshaking signals (specifically those being output by the 8255A, e.g. IBF, $\overline{OBF}$ etc.) as well as the status of the various Interrupt Enable, INTE, flip-flops can be read by reading Port C. This information is treated as the status of the 8255A. Note that since a Status READ is equivalent to a Port C READ, the following inputs will be required for this reading operation.

$$\overline{CS} = \emptyset, \quad A_1 = 1, \quad A_\emptyset = \emptyset, \quad \overline{RD} = \emptyset, \quad \overline{WR} = 1$$

Depending on the mode programming of the 8255A, the various bits $D_\emptyset$–D_7 obtained by reading Port C should be properly interpreted. The way these bits should be interpreted is given in Fig. A10.3.4, where the significance of the bits (i.e. handshake signals or Port C inputs) depend on the mode programming.

Data Bits on Status Read / Mode	D_7	D_6	D_5	D_4	D_3	D_2	D_1	$D_\emptyset$
Mode 1 Input	I/O	I/O	IBF_A	$INTE_A$	$INTR_A$	$INTE_B$	IBF_B	$INTR_B$
Mode 1 Output	$\overline{OBF_A}$	$INTE_A$	I/O	I/O	$INTR_A$	$INTE_B$	$\overline{OBF_B}$	$INTR_B$
Mode 2	$\overline{OBF_A}$	$INTE_1$	IBF_A	$INTE_1$	$INTR_A$	—	—	—
	GROUP A					GROUP B		

Fig. A10.3.4 Status Word Format of the 8255A (on reading the contents of Port C through the data bus)

It was mentioned earlier that, if all the ports have not been programmed as Mode Ø, some of the pins of Port C will be used for handshaking. In this case, one or more Port C pins may still be available for input or output—specifically, those that are not being used for handshaking. If these unused Port C pins have been programmed as input, then the 'Status (Port C) Read' will also report the input present at these pins. For example, if Port C pin PCi is not being used for handshaking of Port A and/or Port B, and has been programmed as input, the Status Read operation will report PCi on Di.

Examples

(1) Port A Mode 1 Input, Port B Mode Ø (Input or output)

Port C upper and lower as Input

On Status Read—

$D_5 = IBF_A, \quad D_4 = INTE_A, \quad D_3 = INTR_A$

$D_\emptyset, D_1, D_2, D_6, D_7 = PC_\emptyset, PC_1, PC_2, PC_6, PC_7$ (respectively)

(2) Port A Mode 2, Port B Mode 1 Input

On Status Read—

$D_7 = \overline{OBF_A}$, $D_6 = INTEO$, $D_5 = IBF_A$, $D_4 = INTEI$

$D_3 = INTR_A$, $D_2 = INTE_B$, $D_1 = IBF_B$, $D_\emptyset = INTR_B$

(No Port C pins left for input or output)

(3) Port A Mode Ø, Port B Mode 1 Output, Port C input

On Status Read—

$D_2 = INTE_B$

$D_1 = \overline{OBF_B}$

$D_\emptyset = INTR_B$

$D_7, D_6, D_5, D_4, D_3 = PC_7, PC_6, PC_5, PC_4, PC_3$ (respectively).

A10.4 INTEL 8251A—PROGRAMMABLE COMMUNICATION INTERFACE

The Intel 8251A is a LSI device which functions as a USART (Universal Synchronous/Asynchronous Receiver Transmitter) in microprocessor system. Originally designed for serial I/O in 8080A systems, this device is powerful enough to have become very commonly used device for asynchronous as well as BISYNC type synchronous byte oriented protocols.

The 8251A USART provides a number of useful features. A list of these features is given below.

Mode = Feature	Synchronous Mode	Asynchronous Mode
(a) Bits/characters	5–8	5–8
(b) Baud rate	0–64 Kbps	0–19.2 Kbps
(c) Synchronization	Internal or External	—
(d) Clock rate	Baud rate	x1, x16, x64 (times the baud rate)
(e) Stop bits	—	1, 1.5, 2
(f) Parity	even/odd/none	even/odd/none
(g) SYNC characters	1 or 2	—
(h) Errors detected	Parity, Overrun	Parity, Overrun, Framing

With a large set of software programmable options and a slightly cumbersome structure for writing commands to it, the 8251A is a somewhat difficult chip to use. The objective of this appendix is to familiarise the reader with the use of this chip; it should not, however, be considered as a substitute for the actual manufacturer's data sheets. In this appendix, we will describe the pin out of the 8251A and how it is interfaced both to the microprocessor system and the serial channel. This will be followed by a description of its command codes and how they should be issued.

The pin out of the 8251A USART is given in Fig. A10.4.1. Apart from the power pins

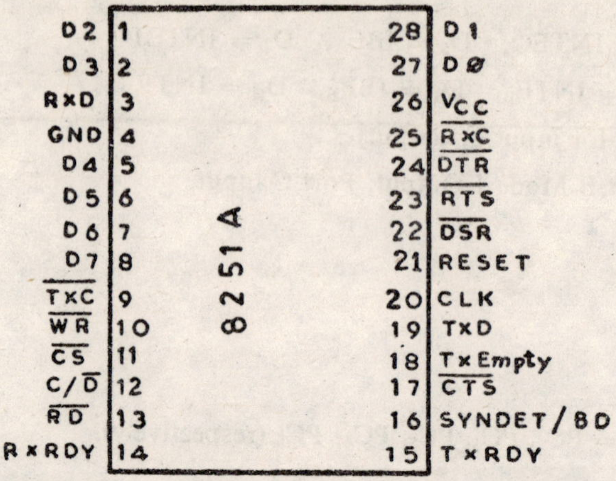

Fig. A10.4.1 Pin-out of Intel's 8251A—Programmable Communication Interface

(V_{cc} = +5V and GND), the pins of the 8251A can be divided into the following categories:

(i) CPU Interface and Control
(ii) Serial Input/Output
(iii) Modem Control.

The various pins of each category are described below.

CPU Interface and Control Pins

DØ–D7 Tristate, bidirectional data bus pins. These connect to the data bus of the microprocessor system. Data to be written into the 8251A is placed on the data bus by the microprocessor. This data can either be commands to the 8251A or data that is to be serially transmitted by the 8251A. Similarly, the 8251A places data on the data bus which is to be read by the microprocessor. This data can either be the 8251A's status information or data that it has received serially.

$\overline{CS}$ Chip Select

$\overline{CS} = Ø$ selects the 8251A for proper READ ($\overline{RD} = Ø$) or WRITE ($\overline{WR} = Ø$) operations. If addressed as two memory locations using memory mapped I/O, the 8251A's chip select should be derived from A15–A1 with $IO/\overline{M} = Ø$. Conversely, if addressed as two I/O ports, chip select should be derived from A7–A1 with $IO/\overline{M} = 1$. (Linear selection can of course be used instead of full decoding.)

$C/\overline{D}$ Command/Data select.

If $\overline{CS}$ is derived as described above, $C/\overline{D}$ should be connected to AØ of the address bus. During WRITE operations, $C/\overline{D}$ (AØ) = 1 writes commands to the 8251A, whereas $C/\overline{D}$ (AØ) = Ø writes data to the 8251A which it is required to transmit serially.

During READ operations, $C/\overline{D}$ (AØ) = 1 reads the status of the 8251A, whereas $C/\overline{D}$ (AØ) = Ø reads data that the 8251A has received serially.

$\overline{RD}$ Read

$\overline{RD}$ = Ø transfers data from the 8251A to the micro-processor's data bus.

$\overline{WR}$ Write

$\overline{WR}$ = Ø writes data from the microprocessor's data bus into the 8251A.

A summary of the operation of the 8251A's $\overline{CS}$, $C/\overline{D}$, $\overline{RD}$ and $\overline{WR}$ inputs are given below.

$\overline{CS}$	$C/\overline{D}$	$\overline{WR}$	$\overline{RD}$	Operation
1	×	×	×	8251A not selected (data bus tristate)
Ø	1	Ø	1	command to 8251A
Ø	Ø	Ø	1	data (for serial transmission) to 8251A
Ø	1	1	Ø	read 8251A's status
Ø	Ø	1	Ø	read data serially received by 8251A
Ø	×	1	1	data bus tristate

RESET This should be connected to the System Reset Line. A high on this input forces the 8251A to an 'idle' state where it is neither receiving nor transmitting but is actually waiting for commands from the CPU. In order to be effective the RESET pulse should be sufficiently wide (at least six cycles of the clock present at the CLK input).

CLK 8251A Clock

This is a clock used to generate internal device timing in the 8251A. Within limits it is independent of the baud rate of operation.

(a) $0.741 \text{ MHz} \leqslant f_{CLK} \leqslant 3.125 \text{ MHz}$

(b) for synchronous operation and ×1 asynchronous operation

$f_{CLK} \geqslant 30$ (Baud Rate)

(c) for asynchronous operation ×16, ×64

$f_{CLK} \geqslant 4.5$ (Transmit or Receive Clocks)

In 8085A systems, the CLK input of the 8251A can be fed from the CLK OUT pin of the 8085A. However, care should be taken to ensure that the 8085A CLK OUT signal does fall within the 0.741 MHz–3.125 MHz frequency range specified earlier.

TxRDY Transmit Ready Output

The 8251A sets TxRDY = 1, when it is ready to accept a new data byte from the microprocessor. It resets TxRDY = Ø, on the leading edge of $\overline{WR}$ when a data character is loaded into the 8251A by the CPU. (It is also reset to Ø if the transmitter is disabled or if the 8251A is reset.) TxRDY output is always Ø if the transmitter is disabled or if $\overline{CTS}$ = 1.

RxRDY Receive Ready Output

The 8251A sets RxRDY = 1, when it has a data byte assembled from the serially received data which it wants the microprocessor to read. It resets RxRDY (=Ø) on the leading edge of $\overline{RD}$ when the microprocessor reads the data bit. Note that the 8251A reports an overrun error if the microprocessor fails to read the assembled data byte (received) by the time the 8251A finishes assembling the next byte from the received serial data stream. In this case, the earlier received character is lost. Disabling the receiver or resetting the 8251A resets RxRDY to Ø. Needless to say, a character is assembled only if the serial data being received is in the same format that the 8251A has been programmed to receive.

The TxRDY and RxRDY outputs can be used if the Transmit/Receive operation of the 8251A is to be controlled via interrupts. Several configurations are possible in this case. Some of them are described below.

(i) Connect TxRDY and RxRDY to two of the three vectored interrupts available on the 8Ø85A. In this case, external hardware for providing the interrupt vector is not needed.

(ii) Generate INT = TxRDY + RxRDY. Connect INT to one of the three vectored interrupts available on the 8085A. In this case, also, no external hardware for interrupt vectoring is needed. Unlike (i), however, the routine servicing the INT interrupt needs to read the Status Register of the 8251A in order to find out if the interrupt was caused by TxRDY or RxRDY. Note that if the microprocessor program knows whether the 8251A is in the transmit or receive mode, it does not need to read the Status Register.

(iii) Use an Interrupt Controller (like the 8259) with two interrupt inputs as TxRDY and RxRDY. The Interrupt Controller can then generate separate interrupt vectors for TxRDY and RxRDY interrupts.

(iv) Generate INT as in (ii) but give that to the INTR input of the 8085A with proper external hardware to generate the interrupt vector. This is, otherwise, similar to (ii) above.

The 8251A can be either interrupt controlled as described above or it could be used in a polled mode. In the latter case, the microprocessor should read the 8251A's Status Register (described later) and take appropriate actions depending on the status of TxRDY and RxRDY.

TxE Transmitter Empty

Tx Empty When the 8251A is in the transmitting mode but does not have a character to send (i.e. the microprocessor did not service the TxRDY signal in time) it can do one of the following:

(a) If it is in the Synchronous mode and is programmed to use one SYNC character, it will send a series of SYNC characters (on the TxD output).

If it is in the Synchronous mode and is programmed to use two SYNC characters—SYNC1 and SYNC2 (SYNC1 and SYNC2 can be same)—it will send a series of this character-pair (on the TxD output).

(b) If it is in the Asynchronous mode, it sends a continuous High signal (on the TxD output). This is referred to as 'Marking'. (An exception to this occurs if the 8251A has been commanded to 'Break' transmission. In this case the TxD output stays Low.)

In both cases (a) and (b) above, the 8251A signals that it has nothing to transmit serially by setting TxE = 1. It resets TxE = Ø, if the microprocessor provides it with new data for serial transmission, if the transmit mode is terminated, or if the 8251A is reset.

SYNDET/
BRKDET SYNC Detect/BRK Detect

This pin serves a number of functions depending on the mode of operation of the 8251A. If it is an output pin and the 8251A is in the synchronous mode, SYNDET is output High if the 8251A detects SYNC characters (one or two, as programmed) in the serially received data. Reading the Status Register of the 8251A or resetting it returns SYNDET output to Ø. (It should be noted that, unlike what we may expect, in the synchronous mode the 8251A does not strip the SYNC characters before assembling the serially received data and passing it to the microprocessor. It, however, provides the SYNDET output to signal that it is passing on SYNC character(s) as the received characters. The responsibility of stripping these characters is left to the microprocessor's software.)

If it is an output and the 8251 is in the asynchronous mode, BRKDET is output High if the 8251 detects a low input on the serially received data line (RxD) for $2(n + 1 + m)$ bit intervals where n = no. of bits/character and m = no. of STOP bits. (Note that n and m values are programmed by software.) This happens if the transmitter at the other end has been commanded to send Break characters—a continuously Low signal.

This pin can also be an input accepting an externally provided SYNC signal. Just as we had described for the TxRDY and RxRDY outputs, the TxE and SYNDET/BRKDET outputs can also be used as interrupts to the microprocessor. One can also use them in various other ways depending on the application.

Modem Control Pins

$\overline{DSR}$ Data Set Ready

$\overline{DTR}$ Data Terminal Ready

$\overline{RTS}$ Request to Send

$\overline{CTS}$ Clear to Send

$\overline{DSR}$, $\overline{DTR}$, $\overline{RTS}$ and $\overline{CTS}$ are standard modem control signals. These pins should be connected to the corresponding signal lines of the modem being used.

The 8251A indicates that it is ready to send data by outputting $\overline{DTR}$ = Ø. The modem replies to this, when it too is ready to transmit, by putting $\overline{DSR}$ = Ø. When the 8251A USART and the modem are both ready for transmission, individual data transmittal sequences are initiated by the USART outputting $\overline{RTS}$ = Ø. This request to send is acknowledged by the modem by its outputting $\overline{CTS}$ = Ø.

The 8251A's transmission logic is enabled by a Low $\overline{CTS}$ input. An ongoing transmission can be stopped if, for some reason, the modem's $\overline{CTS}$ output goes High. In this case, however, the last character output to the 8251A is actually transmitted before the serial data transmission is terminated. (Incidentally, this is the same procedure which is followed, if a 8251A in the act of transmission is ordered to disable its transmitter. Also, as far as the 8251A is concerned, $\overline{DSR}$, $\overline{DTR}$, and $\overline{RTS}$ are essentially general purpose 1-bit I/O ports).

(Note that the 8251A does not have a Carrier Detect input. This may be a disadvantage in certain applications.)

Serial Input/Output Lines

TxD	Transmit Data Output
$\overline{\text{TxC}}$	Transmit Clock Input
RxD	Receive Data Input
$\overline{\text{RxC}}$	Receive Clock Input

TxD is the serial output data line which outputs the binary bits (Ø or 1) which have to be serially transmitted. The rate at which these bits are output is timed by the $\overline{\text{TxC}}$ clock. In the synchronous mode of operation, $\overline{\text{TxC}}$ is required to be identical to the desired baud rate of the serial output. In the asynchronous mode, the 8251A can be programmed to accept either a ×1, ×16 or ×64 (times the baud rate) clock as the $\overline{\text{TxC}}$ input. The 8251A then appropriately divides this clock signal internally (by 1, 16 or 64 respectively) in order to generate the baud clock. Note that the TxD output can change only at the falling edges of $\overline{\text{TxC}}$—which actual edge will be used depends on which factor out of ×1, ×16, ×64 is actually chosen.

RxD is the serial data input to the 8251A. The $\overline{\text{RxC}}$ clock controls the received baud rate by internally providing the 8251A with the time instants at which it should sample the RxD line. In the synchronous mode, the baud rate is the same as the $\overline{\text{RxC}}$ input. In the asynchronous mode, the 8251A can be programmed to accept either a ×1, ×16 or ×64 (times the baud rate) clock as the $\overline{\text{RxC}}$ input. Data is sampled at the appropriate rising edge of $\overline{\text{RxC}}$, depending on the clock option chosen.

In the asynchronous mode, it is better to use the 8251A with the ×16 or ×64 $\overline{\text{RxC}}$ option. In this case, the 8251A uses High-to-Low transitions on the RxD line to synchronize itself so that the instant at which RxD is sampled falls as close as possible to the centre of a data bit. Note that such a transition is guaranteed to be present at the beginning of each character since START bit = Ø, STOP bit = 1 and Marking State = 1. (It will either be a STOP-START or a MARK-START sequence.) Therefore, fresh synchronization is guaranteed for each character at least once. In the ×1 option, (isosynchronous mode) this synchronization feature is absent since the rising edge of the $\overline{\text{RxC}}$ clock provided to the 8251A decides the sampling instant.

In the asynchronous mode, the 8251A also incorporates a Start Bit Detection circuit guarding against a Hi-Lo-Hi *spike* on a 'Marking' line from being interpreted erroneously as a START bit. This circuit first detects a falling edge and, if it expects that this could be the beginning of a START bit, it samples the RxD line 1/[2·(baud rate)] seconds later at what it expects would be the nominal centre of the START bit. It accepts a START bit only if (a) it is expecting a START bit because of the data format, (b) it sees a falling edge on the RxD line, and (c) when it samples RxD as above (at what it expects to be nominal centre of the START bit), it finds RxD = Ø. It is easy to see that it would be difficult to fool this False Start Bit Detection circuit with Hi-Lo-Hi spikes unless these spikes are very fortuitously placed.

It should be noted that in the asynchronous mode, the same ×1, ×16 or ×64 option will have to be chosen for both $\overline{\text{RxC}}$ and $\overline{\text{TxC}}$ (e.g. one cannot have the ×16 option for receive and the ×1 or ×64 option for transmit). It is still possible to have different receive and transmit baud rates by feeding different clocks to the $\overline{\text{RxC}}$ and $\overline{\text{TxC}}$ inputs. However, in most applications, one finds that the receive and transmit baud rates are identical. In such a situation, it is common to tie both the clock inputs together and feed them from a common baud clock generator of the appropriate frequency.

Finally, it should be noted that for the synchronous mode of operation, the 8251A can work at baud rates up to 64 Kbps. For the asynchronous mode of operation, it can only work at baud rates of up to 19.2 Kbps.

Initialising and Controlling the 8251A (Command Code)

We have seen earlier that by issuing a $\overline{WR}$ pulse with $C/\overline{D} = 1$, we can write commands to the 8251A. These commands choose the mode of operation of the 8251A and the various options that can be selected. The 8251A needs a 16-bit command for this purpose which should be issued to it as two Command Code bytes.

From our earlier description of the 8251A, it should be evident that as far as the microprocessor is concerned, the 8251A's command register is accessed as just one location. In order to load two bytes of common code into this one location, the 8251A specifies a (rigid) sequence in which these bytes can be written. Unless this sequence is strictly followed, the 8251A will not function as desired.

The 16-bit command needed by the 8251A is viewed as one Mode Select Command byte and one Control Select Command byte. The Mode Select byte defines the operating environment, e.g. asynchronous or synchronous, number of bits per character, etc. The Control Select byte specifies actions which require specific actions instantly, like enabling or disabling transmission, controlling the Modem Control outputs etc. The sequence in which these command bytes are issued is given below.

On external RESET (through the 8251A's RESET pin) or internal RESET (described later) the 8251A expects a Mode Select Command code byte to be issued to it by the microprocessor. As far as the microprocessor is concerned, when it initially wants to use the 8251A, it should issue a Mode Select Command byte to the device. It should also issue a Mode Select Command byte every time it resets the 8251A via software (internal RESET) or every time the system, as a whole, is reset (external RESET). (The latter assumes that the 8251A's RESET pin is connected to the system's RESET line.)

When we describe the detailed Mode Command later, we will find that, among other things, it will select one of the following three options

Asynchronous Mode

Synchronous Mode with one SYNC character

Synchronous Mode with two SYNC characters—SYNC1 and SYNC2

What follows the Mode Select command depends on which of these options was chosen.

If the Mode Select command chooses the Asynchronous Mode of operation, the 8251A expects that the next command byte ($C/\overline{D} = 1$) issued to it would be a Control Select command byte.

If the Mode Select command specifies the Synchronous Mode of operation with one (or two) SYNC character(s), it expects that the next (or next two, if two SYNC characters are being used) command byte(s) issued to it would be the SYNC character(s) itself. Note that if two SYNC characters are being used, the sequence should be SYNC1 issued first, followed by SYNC2. This is important since the synchronization sequence used by the 8251A will be SYNC1–SYNC2. It should also be noted that, in this case, even if SYNC1 and SYNC2 are identical, they have to be issued twice to the 8251A. After this, the 8251A expects a Control Select command byte to be issued to it. The 8251A does not really become operational until the Control Select command is issued.

Summarising, the sequence is as follows:

Asynchronous: Mode—Control

Synchronous with one SYNC character: Mode—SYNC–Control

Synchronous with two SYNC characters: Mode–SYNC1—SYNC2–Control

It is important to realize that the command code sequence given above has to be followed very rigidly. Otherwise it is likely that the 8251A will misinterpret these commands. Moreover, the sequence is what the 8251A expects and though the microprocessor does not have to issue them in a sequential fashion (it may do other things in between), this is the sequence in which these command bytes should appear to the 8251A. (To clarify this last point, assume that an asynchronous mode of operation is desired. The instructions issuing the Control Select command does not necessarily have to follow the instructions issuing the Mode Select command. The microprocessor can execute other instructions, not meant for the 8251A, in between the two.)

Once a Control Select command byte has been issued, another Control Select command byte can be issued to the 8251A at any time. If we now want to issue a Mode Select command to the 8251A, we need to either reset it externally or reset it internally using the proper Control Select command. (There is a Control Select command *bit* which does this.) If the 8251A is operating in the Synchronous mode and we want to change the number of SYNC characters or the SYNC characters themselves, we will have to first reset the 8251A (as above) and then reprogram the device. This programming sequence will be identical to the sequence described earlier.

Suppose one wants to bring the 8251A to a point where it is expecting a Mode Select Command. One can do this via a Hardware RESET. It can also be done via an internal RESET if one is assured that the 8251A is currently expecting a Control Select command. In case such assurance is not there, one should write ØØ *thrice* into the Command Register and then write a 4ØH in this register; once these have been done, the 8251A will be in a state where it will be expecting a Mode Select command.

Mode Select Command Byte Format

Individual bits of the Mode Select command byte (MS) select the following options.

MS1, MSØ:	MS1 = Ø, MSØ = Ø	Synchronous
	MS1 = Ø, MSØ = 1	Asynchronous (×1 clock)
	MS1 = 1, MSØ = Ø	Asynchronous (×16 clock)
	MS1 = 1, MSØ = 1	Asynchronous (×64 clock)
MS3, MS2:	MS3 = Ø, MS2 = Ø	5 bits/character
	MS3 = Ø, MS2 = 1	6 bits/character
	MS3 = 1, MS2 = Ø	7 bits/character
	MS3 = 1, MS2 = 1	8 bits/character

MS4: MS4 = ∅ parity disable (no parity bit generated or checked)

MS4 = 1 parity enable (parity generated or checked as per MS5)

MS5: MS5 = ∅ odd parity

MS5 = 1 Even parity

(If parity is disabled, MS5 is a 'don't care')

MS7, MS6: The interpretations of these bits are different for the asynchronous and the synchronous modes.

In the *Asynchronous Mode*:

MS7 = ∅, MS6 = ∅ Invalid
MS7 = ∅, MS6 = 1 1 stop bit/character
MS7 = 1, MS6 = ∅ 1.5 stop bit/character
MS7 = 1, MS6 = 1 2 stop bit/character

In the *Synchronous Mode*:

MS6 = ∅ if SYNDET outputs SYNC signal when the 8251A sees the SYNC sequence in the received data (one or two SYNC characters in the SYNCH sequence as per MS7)

MS6 = 1 if SYNDET is an input pin accepting an external SYNC signal

and

MS7 = ∅ 2 SYNC characters (in SYNC sequence)

MS7 = 1 1 SYNC character (in SYNC sequence)

Control Select Command Byte Format

Individual bits of the Control Select command byte (CS) do the operations described below

CS∅: TxEN

CS∅ = 1 enables serial data transmission

= ∅ disables serial data transmission

CS1: DTR

CS1 = 1 forces $\overline{\text{DTR}}$ output of the 8251A to ∅ (recall the modem control signals)

CS2: RxE

CS2 = 1 enables the RxRDY output

= ∅ disables the RxRDY output

Note the CS2 just enables or disables the RxRDY output. The receive logic of a properly initialized 8251A will be active regardless of the value of CS2. If CS2 = 1, the RxRDY output will provide the output described earlier. If CS2 = ∅, RxRDY will always stay low.

CS3: SBRK (Send Break character)

If CS3 (SBRK) = 1, the TxD line is immediately forced to Ø. This continuously low data output is equivalent to sending a BREAK character. The transmitter operates normally if CS3 = Ø.

CS4: ER (Error Reset)

CS4 = 1 resets error flags in Status Register

If the 8251A encounters error conditions (parity error, overflow error, and framing error), it sets appropriate bits in its status register. Issuing a Control Select command with ER = 1 resets any of these bits which may have been set by some previous error.

CS5: RTS

CS5 = 1 forces the $\overline{\text{RTS}}$ output of 8251A to Ø.

(Recall that $\overline{\text{RTS}}$ is a modem control signal.)

CS6: IR (Internal Reset)

CS6 = 1 acts as an Internal Reset

Issuing a Control Select command with IR = 1 acts as an Internal Reset of the 8251A. As discussed earlier, this returns the 8251A to an 'idle' state where it expects the microprocessor to issue a Mode Select command.

CS7: EH (Enter Hunt mode)

CS7 = 1, 8251A enters Hunt mode.

In the synchronous receive mode, data reception does not really start until the leading synchronization sequence (one or two SYNC characters) has been detected. Once this has been detected, assembly of characters from the received serial data stream is started. In the Hunt mode, the USART is commanded to look for the synchronization sequence. After this has been detected, the USART exits the Hunt mode and starts assembling the received serial data.

If the 8251A has been programmed for synchronous operation, issuing a Control Select command with EH = 1 forces the 8251A's receiver to enter the Hunt mode as described above. It exits the Hunt mode only after it detects the synchronization sequence (which it has been programmed with) in the serially received data stream. Note that in synchronous operation, issuing a Control Select command with EH = 1 acts essentially as a Receiver Enable command.

The 8251A takes slightly different kinds of actions to implement the commands of the Control Select bytes issued to it. TxEN, DTR, RxE, SBRK and RTS bits of the Control Select command have an immediate effect on the corresponding signals. IR and EH cause internal IR and EH flip-flops to be set to 1; in effect actions on these command bits are remembered for later use. The ER bit resets error flags till the next error condition is encountered.

The IR flip-flop stays set (in effect disabling the 8251A's serial operation) until a command byte ($C/\overline{D} = 1$) is written to it; it is then reset to Ø. As mentioned earlier, this command byte should be a Mode Select command.

The EH flip-flop stays set until the receiver detects the synchronization sequence (one or two SYNC characters, as programmed) in the received serial data stream; it is then reset to Ø.

8251A's Status Register

We have mentioned earlier that the 8251A also has an internal Status Register which can be read by the microprocessor. Issuing a $\overline{RD} = \emptyset$ signal to the 8251A with its $\overline{CS} = \emptyset$ and $C/\overline{D} = 1$, forces it to output the Status Register contents on the data bus.

Individual bits of the Status Register (SR) should be interpreted as follows:

SRØ: TxRDY

SR1: RxRDY

SR2: TxE

SR3: PE (Parity Error)

SR4: OE (Overflow Error)

SR5: FE (Framing Error)

SR6: SYNDET

SR7: $\overline{DSR}$

Of these, the RxRDY, ΓxE, SYNDET, $\overline{DSR}$ bits reflect the signal levels at the corresponding pins. The TxRDY status bit does not exactly do so. The TxRDY status bit is 1 whenever the transmitter has nothing to transmit (Other things—e.g. whether the transmitter is enabled or not—do not matter). On the contrary, the TxRDY pin output is further masked by $\overline{CTS}$ ($\overline{CTS} = \emptyset$) and TxEN (TxEN = 1).

TxRDY status bit = 1 if there is no data to transmit

TxRDY pin output = TxRDY bit if $\overline{CTS} = \emptyset$ and TxEN = 1; else Ø.

The bits PE, OE, and FE report the three error conditions detected by the 8251A in the receive mode. PE is set to 1 when parity is enabled and a parity error is detected in the received character. OE is set to 1 if the microprocessor fails to read a received data byte by the time the 8251A finishes assembling the next data byte received. FE (asynchronous mode only) is set to 1 if the 8251A fails to see the proper data character format (START bit —n, data bit —1, 1.5, or 2 STOP bits) it was programmed for in the serially received data stream. Recall that PE, OE and FE are reset to Ø by ER = 1 (CS4 = 1) of the Control Select command. It should be noted that 8251A makes no attempt to correct any of these errors; it merely reports them when they are detected.

Before considering an actual example of the use of the 8251A it would be appropriate to consider the buffer arrangement inside the device. The receiver section of the 8251A has two buffers RxA and RxB. As shown in Fig. A10.4.2, the serial data input line RD inputs data to RxB. Once a valid character (as per the way the 8251A has been programmed) has been assembled, the 8251A shifts the contents of RxB to RxA and starts assembling the next character in R×B. If the receiver has been properly enabled (as per our earlier description) the RxRDY bit in the status register is set and the RxRDY output goes to 1. When the microprocessor reads *data* from the 8251A ($\overline{CS} = \emptyset$, $C/\overline{D} = \emptyset$, and $\overline{RD} = \emptyset$), both the RxRDY bit and the RxRDY output are reset to Ø. We had said earlier that an OVERRUN error is reported if the data character (received by the 8251A) is not read in the time interval between when the character was received (and RxRDY was set to 1) and when the next character is assembled. In terms of the receive buffer arrangement described above, the reason for this happening is not difficult to see—the new character assembled by RxB is loaded into RxA before the data previously held by RxA is read. If an OVERRUN error occurs, the earlier character in RxA is lost.

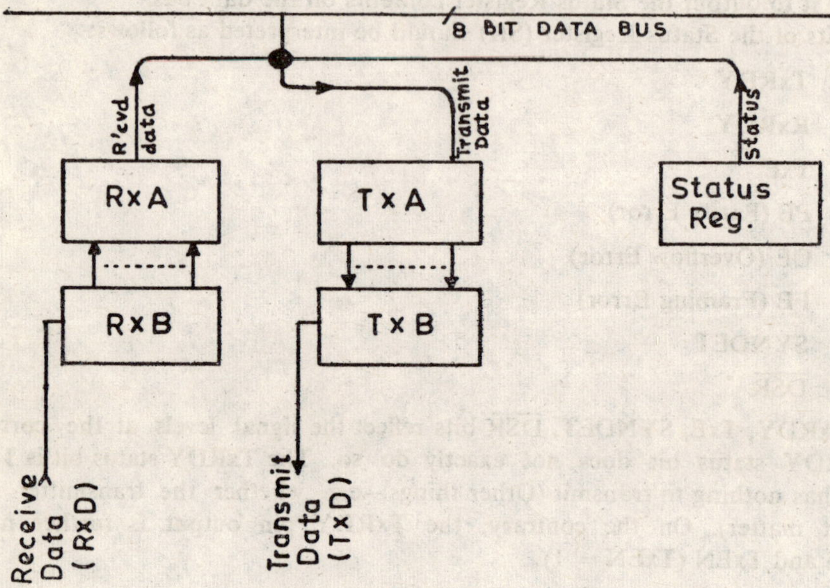

Fig. A10.4.2 Buffer Arrangement Inside the 8251A

The transmit buffer arrangement (Fig. A10.4.2) is very similar to the receive buffer arrangement. There are two transmit buffers—TxA and TxB—in the 8251A device. If the 8251A's transmitter has been properly enabled, the data character that the microprocessor wants to transmit serially, is first loaded into TxA. The buffer TxB holds the earlier character that the 8251A is currently transmitting. Once this character has been transmitted (along with whatever overhead bits are needed, e.g. START, STOP and Parity), the 8251A loads the contents of T×A into TxB and starts transmitting out the contents of TxB. It also simultaneously signals the microprocessor (by setting the TxRDY bit = 1 and the TxRDY output is raised High subject to the conditioning on this output described earlier) that the TxA buffer is empty and that if there is another character to be transmitted it can now be loaded into TxA.

Note that the overhead bits needed in the asynchronous mode of operation is appropriately appended to the data being shifted out to TxB in the transmitter section; the receiver section strips these overhead bits when the character is being assembled in RxB.

A specific example of using the 8251A is shown in the circuit diagram of Fig. A10.4.3. For this circuit, the following specifications are assumed.

(i) Asynchronous Receive/Transmit
(ii) RxC = 16 times baud rate = TxC
(iii) 7 bits/character

(iv) Even parity

(v) 2 STOP bits

(vi) 8085A CLK (out) is acceptable to the 8251A's CLK input

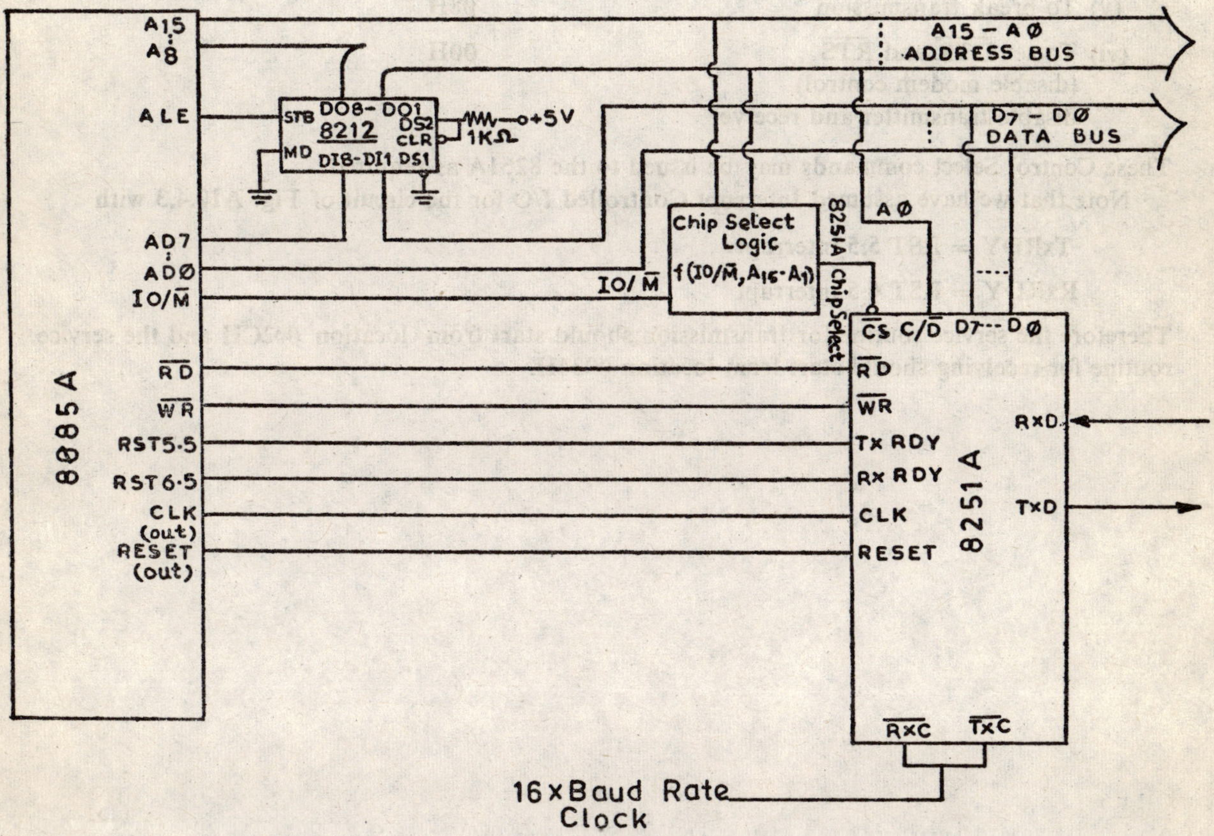

Fig. A10.4.3 Using a 8251A in a 8085A System (Interrupt Controlled)

The Mode Select Command in this case is given as

11111ØØ1B = F9H

We have given below a list of some Control Select commands that can be issued:

(i) To return to Mode Select 4ØH

(ii) To enable transmitter but not receiver

*With modem controls 23H

*Without modem controls Ø1H

(iii) To enable transmitter and receiver

*With modem controls 27H

*Without modem controls Ø5H

(iv) To enable receiver but not transmitter

*With error reset 14H

*Without error reset Ø4H

(v) To break transmission Ø8H

(vi) To set $\overline{DTR}$ and $\overline{RTS}$ ØØH
(disable modem control)
disable transmitter and receiver

These Control Select commands may be issued to the 8251A as required.

Note that we have assumed Interrupt Controlled I/O for the circuit of Fig. A10.4.3 with

TxRDY = RST 5.5 interrupt

RxRDY = RST 6.5 interrupt

Therefore the service routine for transmission should start from location ØØ2CH and the service routine for receiving should start from location ØØ34H.

Chapter 11

Analog Interfacing

11.1 BASIC PRINCIPLES OF ANALOG INTERFACING

It should be evident from the discussion presented in the earlier chapters that the microprocessor—or for that matter any digital device—essentially operates on one or more digital inputs presented via Input I/O Ports and outputs digital data to one or more Output I/O Ports. It is quite possible that the data presented to the microprocessor system or taken from it, is itself digital in nature (e.g. data to or from another processor). However, there will be a large class of applications where either the input or the output, or both, are analog signals. A system of this type is shown in Fig. 11.1.1 where there are two analog inputs and one analog output. The analog inputs are

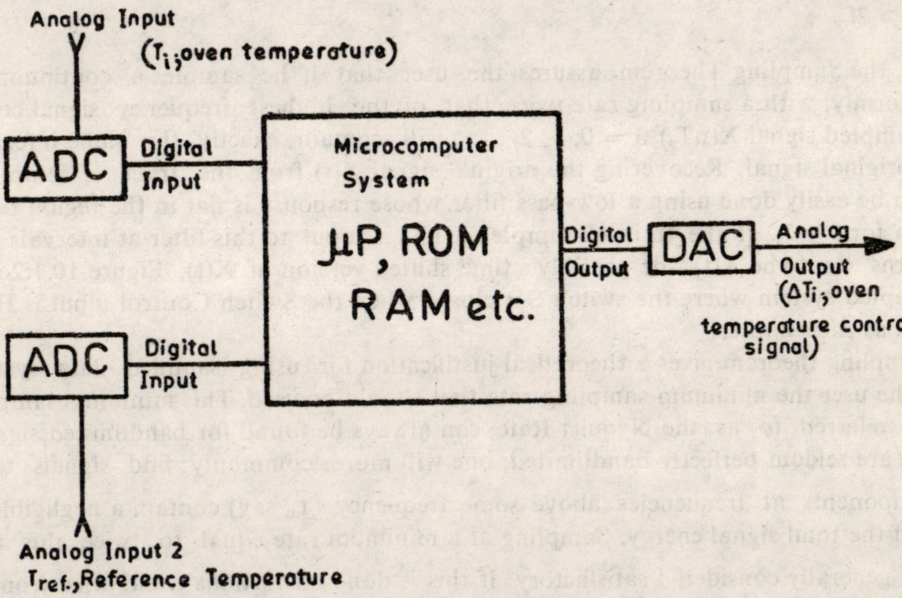

Fig. 11.1.1 A Microcomputer System for Temperature Control

presented to the microprocessor system in digital form where the analog to digital conversion is achieved using two ADCs (Analog to Digital Converters). The microprocessor system outputs digital data which is converted to an analog form by the DAC (Digital to Analog converter). As an example, one can consider the system illustrated in Fig. 11.1.1 to be a controller controlling the temperature of an oven. Let one of the inputs be the current oven temperature T_i and the other input be a reference temperature T_{ref}. Let the system output an analog temperature control signal ΔT_i to the oven; if ΔT_i is positive, it indicates that the oven should be heated up at a rate proportional to ΔT_i; if it is negative, it implies that cooling is required.

There are two basic processes involved in converting an analog signal to a digital form and vice-versa. These are:

(a) Sampling Process,

(b) Quantization Process.

11.1.1 Sampling Theorem and the Sampling Process

The sampling theorem forms the basis for representing a continuous analog signal by a "sampled" analog signal. (This sampled analog signal is then converted into a digital form.) It implies that conversion between a continuous signal and its sampled counterpart can be done without any loss of information provided the sample points are close enough; it also indicates what the minimum sampling rate should be.

Let X(t) represent a continuous analog signal which is band limited to f_m, i.e. its highest non-zero spectral component is at frequency f_m. Let X(t) be sampled at regular intervals T_s (sampling rate $f_s = 1/T_s$) where the sample values are represented by $X(nT_s)$ $n = 0, 1, 2, \ldots$. The sampling theorem states that these sample values uniquely determine the signal X(t) and that they can be used to reconstruct the original signal X(t) without error, provided

$$f_s \geqslant 2f_m$$

In essence, the Sampling Theorem assures the user that if he samples a continuous analog signal uniformly, with a sampling rate twice that of the highest frequency signal component, then his sampled signal $X(nT_s)$ $n = 0, 1, 2, \ldots$, will contain exactly the same information as X(t), the original signal. Recovering the original signal X(t) from the train of sampled signals $X(nT_s)$ can be easily done using a low-pass filter whose response is flat in the region $0 \leqslant f \leqslant f_m$ and is zero for $f \geqslant f_s$. If the train of samples X(nT) is input to this filter at intervals of T_s, the output turns out to be X(t)—or possibly a time shifted version of X(t). Figure 10.1.2 illustrates such a sampled system where the switch S is closed when the Switch Control input is High. This switch acts as the sampler.

The sampling theorem gives a theoretical justification for using sampled data systems and also tells the user the minimum sampling rate that should be used. This minimum sampling rate, sometimes referred to as the Nyquist Rate, can always be found for bandlimited signals. Real life signals are seldom perfectly bandlimited; one will more commonly find signals where the signal components at frequencies above some frequency ($\tilde{f}_m$ say) contain a negligible fraction (say 1%) of the total signal energy. Sampling at a minimum rate equal to twice this frequency (i.e. $2\tilde{f}_m$) is generally considered satisfactory. If this is done, distortions could arise from the fact that the signal components above $\tilde{f}_m$ are not really zero but have a non-zero (albeit small) value. These distortions, if present, are referred to "Aliasing Noise". The Aliasing Noise generated by sampling a non-bandlimited signal can be removed by using a Low Pass Filter prior to sampling. (In the above example one would use a Low Pass Filter whose pass band lies between 0 and $\tilde{f}_m$.) This Low Pass Filter is sometimes referred to as an '*Anti Aliasing Filter*'.

The sampling process converts a continuous analog signal into a sampled form. This sampled form is then converted into a digital form by an 'Analog to Digital Converter'. Normally, this conversion, described later, will take some finite amount of time where the exact duration depends on the converter. It will be desirable to hold the input to the converter fixed until the conversion process is over. In order to do this, the output of the 'Sampling Device' (the switch in our example of Fig. 11.1.2) is held by a '*Sample and Hold*' (S and H) circuit and the output

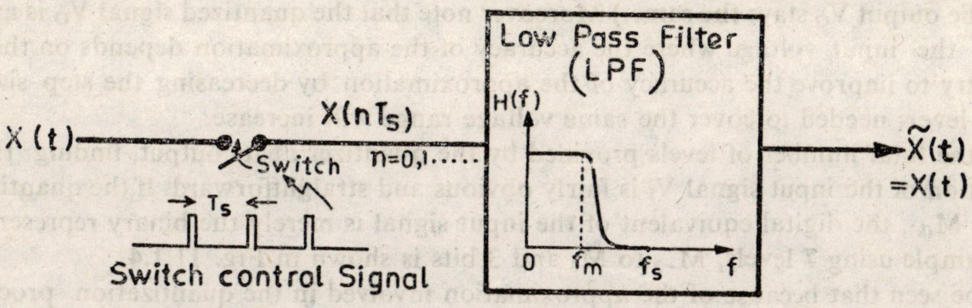

Fig. 11.1.2 Illustration of Sampling Process with Signal X(t) Bandlimited to $f \leqslant f_m$ and Sampled at $f_s \geqslant 2f_m$

of the S and H is fed to the Analog to Digital Converter (ADC). This is illustrated in Fig. 11.1.3. (Note that since the Sample and Hold can also do the sampling, a separate sampler is not needed.)

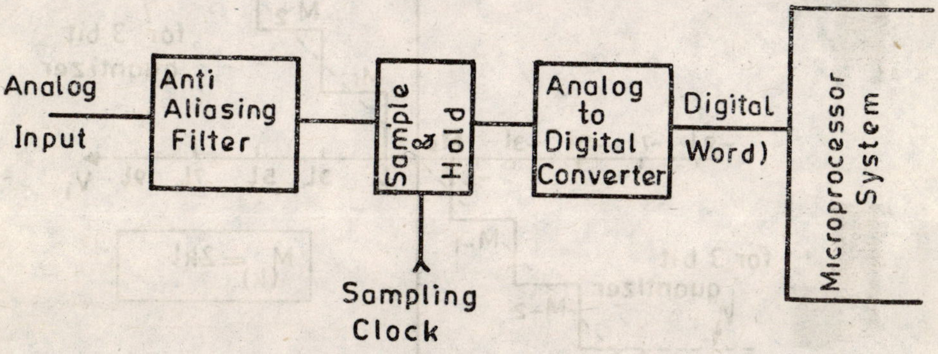

Fig. 11.1.3 Analog to Digital Conversion Incorporating Anti-Aliasing Filter and Sample & Hold

11.1.2 Quantization

The sampling theorem tells us that a continuous analog signal need not be "studied" for all times but that it will suffice if it is observed only at the sampling instants. The Analog to Digital Converter therefore need only examine the signal at these sampling instants in order to evaluate the equivalent digital representation of the signal.

Note that the sample values output are continuous. The digital representation, by its very nature, will, however, be discontinuous. This can be easily seen if one realizes that the 2^n possible digital representations (any n) stand for merely 2^n distinct values whereas the sampled analog signal can have any value, not necessarily equal to any one of these 2^n values. The process of digitizing an analog value is therefore merely making an approximation. This process of approximation is referred to as Quantization.

The input-output relationship of a quantizer, which carries out the quantization process, is shown in Fig. 11.1.4. Note that whereas the input voltage V_i is continuous, the output voltage can take only one of the values . . . M_{-n}, M_{-n+1} . . . , $M_{\emptyset}$ . . . , M_{n-1}, M_n . . . and hence will either be discontinuous or will not change. (The latter might occur if V_i does not change or changes so little that the output V_O stays the same.) Moreover note that the quantized signal V_O is an approximation to the input voltage where the accuracy of the approximation depends on the step size '2l'. If we try to improve the accuracy of the approximation by decreasing the step size 2l, the number of levels needed to cover the same voltage range will increase.

Given the total number of levels provided by the quantizer at its output, finding the digital representation of the input signal V_i is fairly obvious and straightforward. If the quantizer output is given as $M_{(k)}$, the digital equivalent of the input signal is merely the binary representation of (k). An example using 7 levels, M_{-3} to M_3 and 3 bits is shown in Fig. 11.1.4.

It can be seen that because of the approximation involved in the quantization process, some errors will be introduced. This is known as *Quantization Error* and is at most equal to *half the step size*, e.g. 'l' in Fig. 11.1.4. Reconstruction of a signal from quantized samples leads to a

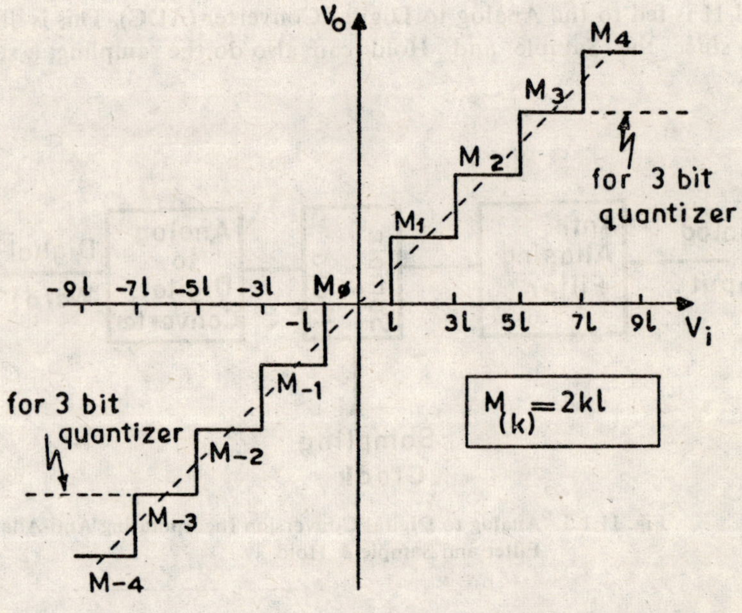

$V_i - V_O$ characteristic of typical quantizer.

For 3 bit (signed output)

$V_i \geq 5l$	$V_O = M_3$	⇒	ø 1 1
$5l > V_i \geq 3l$	$V_O = M_2$	⇒	ø 1 ø
$3l > V_i \geq l$	$V_O = M_1$	⇒	ø ø 1
$l > V_i \geq -l$	$V_O = M_{\emptyset}$	⇒	ø ø ø
$-l > V_i \geq -3l$	$V_O = M_{-1}$	⇒	1 ø 1
$-3l > V_i \geq -5l$	$V_O = M_{-2}$	⇒	1 1 ø
$-5l > V_i$	$V_O = M_{-3}$	⇒	1 1 1

Digital output (in 3 bit Sign Magnitude)

Fig. 11.1.4 Quantizer Operation

signal, which, because of quantization error, is not identical to the original signal. The distortion arising in this fashion from the process of quantization is very often referred to as *Quantization Noise.*

In the example shown in Fig. 11.1.4, we have used sign-magnitude binary representation to represent the quantized voltage output as a digital quantity. The process of assignment of a binary code to each discrete quantizer level is referred to as "*Coding*". Other than the Sign Magnitude Binary shown, a number of other coding techniques can also be used, e.g. BCD, Two's Complement Binary, Offset Binary etc.

In summary, it can be seen that the process of analog to digital conversion is actually achieved in three steps. First, the continuous analog signal is sampled to obtain a sampled analog signal. (For signals which are not bandlimited, Anti-Aliasing Filters may have to be used first in order to reduce Aliasing Noise.) This sampled signal is then quantized by Quantizer. Finally, the quantized levels output by the Quantizer are encoded into a digital representation.

The process of Digital to Analog conversion involves a sequence of steps which are in effect exactly the converse of the steps involved in the Analog to Digital conversion. The digital system (e.g. the microprocessor system) outputs a binary word (of n bits to a Digital to Analog Converter (DAC). The DAC outputs the quantization level corresponding to that digital input. (Note that the conversion from digital word to quantization level should generally follow the converse of the encoding strategy.)

For generating continuous analog signals from digital data, the output of the DAC is fed to a *Reconstruction Filter.* This is typically a Low Pass Filter with a flat (non-zero) response for frequencies 0 to f_m (or $\tilde{f}_m$) and a zero response for frequencies greater than f_s. [Recall f_m = max freq. of bandlimited signal, $\tilde{f}_m$ = max freq. of LPF used for Anti-Aliasing when signal is not bandlimited, f_s = sampling rate.] The output of this filter is the desired continuous analog signal.

One sometimes interposes an *Output Sample and Hold* between the DAC and the Reconstruction Filter. The S and H is used to reduce the output frequency distortion by resampling the output signal using a very narrow sample width. The S and H is also used to store the DAC output when several outputs are being multiplexed during a single sample period and/or when a single DAC is being used to convert several digital outputs. Fig. 11.1.5 illustrates the Digital to Analog conversion process.

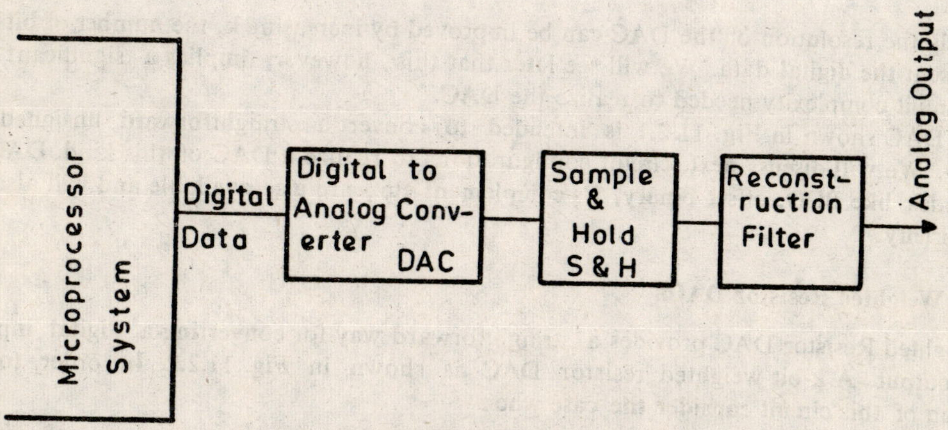

Fig. 11.1.5 Digital to Analog Conversion

11.2 DIGITAL TO ANALOG CONVERSION

The function of a k-bit Digital to Analog Converter is to accept k digital bits (1 or Ø) as input, and output an analog voltage or current. In a completely general situation, the DAC output is a function of the digital input, where the function itself is determined by the coding strategy being used. As an example, consider the k-bit DAC shown in Fig. 11.2.1. Here we want to interpret

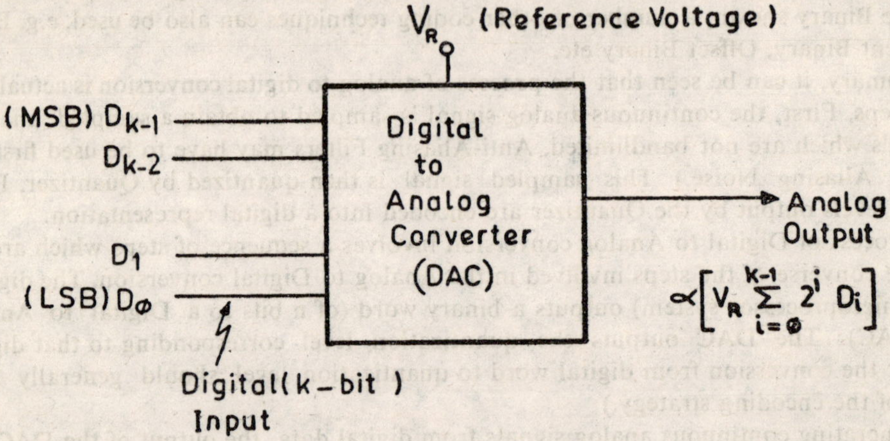

Fig. 11.2.1 DAC Converting a k-bit Unsigned Binary Digital Quantity to an Analog Output

the k-bit data as a straightforward unsigned binary quantity such that the analog output is a voltage or current given by

$$\text{Analog Output Value} = KV_R \sum_{j=0}^{k-1} 2^j D_j \quad (K = \text{constant})$$

The *Resolution* of a DAC is a measure of the output change when the input is changed by 1 LSB. It essentially measures the precision of the DAC. For the k-bit DAC of Fig. 11.2.1

$$\text{Resolution} = \frac{1}{2^k - 1} \times 100\%$$

Obviously the resolution of the DAC can be improved by increasing k, the number of bits needed to represent the digital data. We will see later that this, however, implies a significant increase in the circuit complexity needed to realize the DAC.

The DAC shown in Fig. 11.2.1 is intended to convert a straightforward unsigned binary quantity. We will discuss next, circuit configurations to realize a DAC of this kind. DACs using other codes, like BCD, offset binary, 2's complement etc., are also available and will also be discussed briefly.

11.2.1 Weighted Resistor DACs

The weighted Resistor DAC provides a straightforward way for converting a digital input to an analog output. A k-bit weighted resistor DAC is shown in Fig. 11.2.2. In order to see the operation of this circuit consider the case when

$$V(1) = V_{RH} = V_R \text{ volts}$$

$$V(\emptyset) = V_{RL} = 0$$

(In other words, a reference voltage V_{RH} is used as V(1) whereas the Ø points are shorted to ground.)

In this case, the current I_0 created at the inverting input of the op-amp is proportional to the binary value of the digital input, i.e. $\sum_{i=0}^{k-1} 2^i D_i$. This is achieved by the switches and the weighted resistances shown in Fig. 11.2.2. Consider the i-th digital input D_i. If $D_i = \emptyset$, the switch

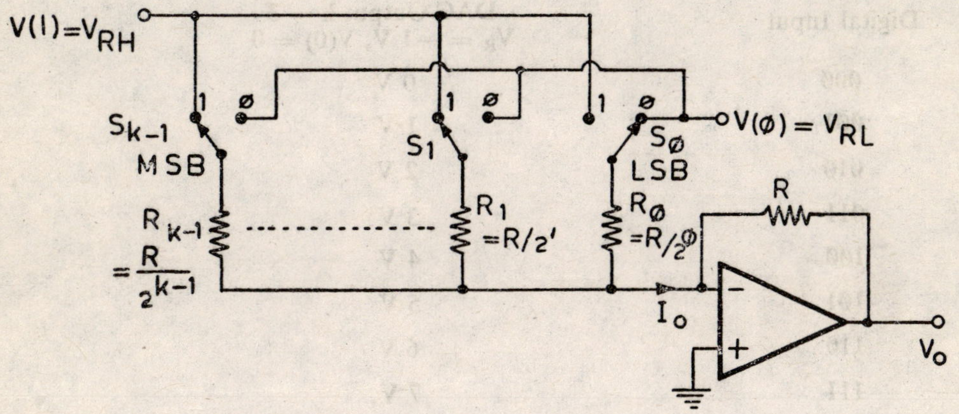

$$\text{if } \begin{Bmatrix} D_i = 1 \Rightarrow S_i = 1 \\ D_i = \emptyset \Rightarrow S_i = \emptyset \end{Bmatrix} \qquad \text{if } \begin{cases} S_i = 1, R_i \text{ connects to } V(1) \\ S_i = \emptyset, R_i \text{ connects to } V(\emptyset) \end{cases}$$

$$I_0 = \frac{V_{RH}}{R} \sum_{i=\emptyset}^{k-1} S_i 2^i + \frac{V_{RL}}{R} \sum_{i-\emptyset}^{k-1} (1 - S_i) 2^i$$

$$V_0 = -V_R \sum_{i=\emptyset}^{k-1} S_i 2^i - V_{RL} \sum_{i=\emptyset}^{k-1} (1 - S_i) 2^i$$

Fig. 11.2.2 Weighted Resistor DAC

for this input is shorted to ground and the current contribution from this bit to I_0 is zero (as it should be!). If $D_i = 1$, the switch connects to $V(1) = V_{RH} = V_R$ volts and the current flowing through the resistor $R/2^i$ for this input is $2^i V_R/R$. This is also the current contribution from this bit to I_0. This yields

$$I_0 = \left[\sum_{i=0}^{k-1} 2^i D_k\right] V_R/R$$

The op. amp itself merely acts as a current summer and a current to voltage converter yielding an output voltage V_0 given by

$$V_0 = -V_R \sum_{i=0}^{k-1} 2^i D_i$$

$$= K \sum_{i=0}^{k-1} 2^i D_i$$

Note that V_0 is Ø volts for an all zero input, ØØ . . . ØØ, and is proportional to the magnitude of the binary digital input. (Remember that the input is being interpreted as an unsigned binary number.) The total voltage swing of the output ranges from 0V to $K(2^k - 1)$V. Obviously, the voltage output of this DAC is either positive ($K > 0$) or negative ($K < 0$) but cannot be both. This type of a DAC is commonly referred to as a Unipolar DAC. For $k = 3$ and $V_R = -1$ volts, the DAC outputs for the various digital inputs are given below

Digital Input	DAC Output k = 3 $V_R = -1$ V, $V(0) = 0$
ØØØ	0 V
ØØ1	1 V
Ø1Ø	2 V
Ø11	3 V
1ØØ	4 V
1Ø1	5 V
11Ø	6 V
111	7 V

The DAC output can be easily made bipolar by choosing V(Ø) to be a polarity opposite to that of V(1). Such a DAC will be a *Bipolar DAC.* An example for k = 3 using $V(1) = -V_{(Ø)} = -0.5$V is given below

Digital Input	DAC Output k = 3, V(1) = −V(Ø) = −0.5V
ØØØ	−3.5 V
ØØ1	−2.5 V
Ø1Ø	−1.5 V
Ø11	−0.5 V
1ØØ	+0.5 V
1Ø1	+1.5 V
11Ø	+2.5 V
111	+3.5 V

For the Bipolar DAC considered in this example (i.e. k = 3, V(1) = −0.5 V, V(Ø) = +0.5V), Fig. 11.2.3 gives a slightly modified circuit where an additional offset of −0.5V is provided at the DAC output. For this circuit the digital input and analog output are related as shown below

Digital Input	DAC Output
ØØØ	−4 V
ØØ1	−3 V

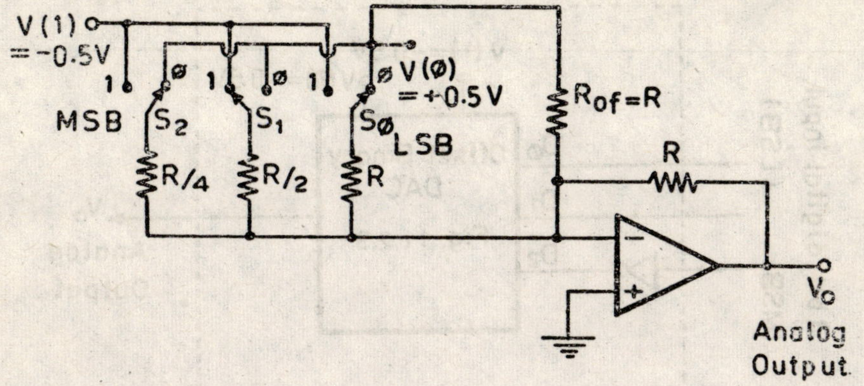

Fig. 11.2.3 Weighted Resistor Bipolar DAC with Offset Adjustment (DAC with Offset Binary Output)

Ø1Ø	−2 V
Ø11	−1 V
1ØØ	0 V
1Ø1	1 V
11Ø	2 V
111	3 V

It would be interesting at this point to digress a little bit and describe one method that can be used to represent digitally, integers in the range of −4 to +3 using 3 bits. This method does the following

(a) Add +4 to the integer (range: −4 to +3)
(b) Using 3 bits, represent in straight binary the result of (a) [range of result of (a): Ø to 7]
Step (b) yields the representation of the original integer

For example, using this scheme −2 = Ø1Ø, 1 = 1Ø1 etc.

This representation of signed numbers is called "*Offset Binary*" representation since the method basically adds an offset to the number before finding its binary equivalent. The DAC of Fig. 11.23 essentially finds the analog output for digital data represented in Offset Binary and is sometimes referred to as an *Offset Binary DAC.*

To generalize, the k-bit offset binary representation of an integer in the range -2^{k-1} to $+(2^{k-1} - 1)$ is found by adding 2^{k-1} to the integer and then finding its binary equivalent. A DAC for this code can be configured by adding more inputs to the inverting input of the op-amp shown in Fig. 11.2.3 keeping $R_{of} = R$.

Examining the offset binary representation of a number, one finds that it is identical to the 2's complement binary representation of the number except that the MSB (the sign bit) is complemented. This is true in general, and can be easily proved. One can, therefore, convert an Offset Binary DAC to one accepting 2's complement data by either (a) Complementing the MSB before giving it to the Offset Binary DAC, or (b) exchanging the 1, Ø switch connections in the MSB switch of an Offset Binary DAC. An illustration of method (a) for a 3-bit DAC is shown in Fig. 11.2.4.

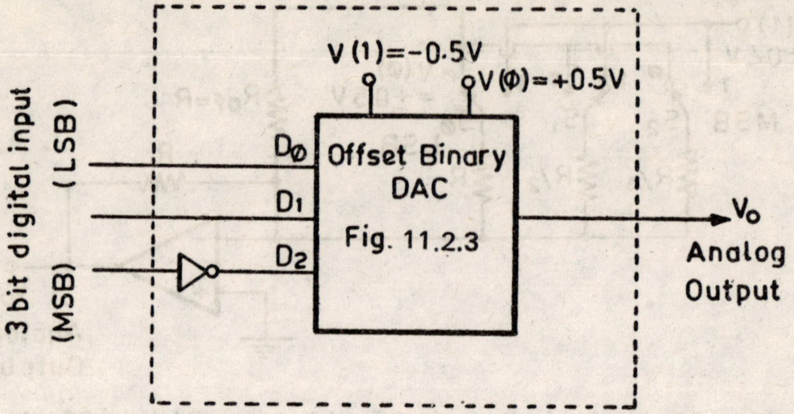

Fig. 11.2.4 DAC for 2's Complement Binary Data

We have discussed above DACs which accept straight binary data, offset binary data or 2's complement binary data and output the corresponding analog output (with, possibly, a scaling factor). It is also possible to build DACs which accept a multidigit BCD code, e.g. 12-bit BCD (3 digits), and output the corresponding analog signal. As a matter of fact, monolithic DAC ICs are commonly available in either complementary binary or complementary BCD form, where the term 'complementary' implies that each individual bit is the complement of what it actually should be (e.g. to find the analog equivalent of Ø1Ø one should input 1Ø1 to the DAC).

There is another major difference in representation between that used by commercial DACs and what we have used here. We have used integer inputs in the DACs we have described, e.g. from ØØØ–111 or Ø-7. Commercial DACs typically represent their data in fractional form *assuming* that the 'binary point' lies to the extreme left of the number, e.g. .ØØØ–.111 or Ø to Ø.875. For this 3-bit example, the maximum possible output of the commercial DAC will only be Ø.875 (Nominal Full Scale value).

11.2.2 R–2R Ladder Network DAC

The precision obtainable in a Digital to Analog Conversion using the Weighted Resistor DAC is a function of the precision of the voltage sources V(1) and V(Ø) and the precision of the various resistances. The latter is a problem, especially in an integrated circuit realization, because of the range of resistances required. For example in an 8-bit DAC the resistances required will range from R to R/128. Obtaining equal precision resistors over such a wide range of values (i.e. 1 to 128) is usually very difficult to achieve. Moreover in an integrated circuit realization, obtaining high value resistances is difficult and is also wasteful from the point of view of chip area. This problem with the Weighted Resistor DAC can be solved by an alternate design, the R–2R Ladder Network DAC, where the only resistance values needed are R and 2R. The circuit of a k-bit R–2R Ladder Network DAC is shown in Fig. 11.2.5. As before, switch S_k connects to the 1-point if $D_k = 1$ and to the Ø point if $D_k = Ø$. We have shown $V(1) = V_R$ and $V(Ø) = 0$ in Fig. 11.2.5 so that the DAC shown is a Unipolar DAC for straight unsigned binary inputs. Modifications for other binary codes—offset binary, 2's complement, etc.—can be made in exactly the same fashion as they were made in the earlier examples. Finally, it should be noted that in

Fig. 11.2.5, if a bit input D_i is $\emptyset$, the corresponding input to the op. amp's non-inverting input is left open. This is not a problem if the op. amp were indeed an ideal op. amp. In an actual

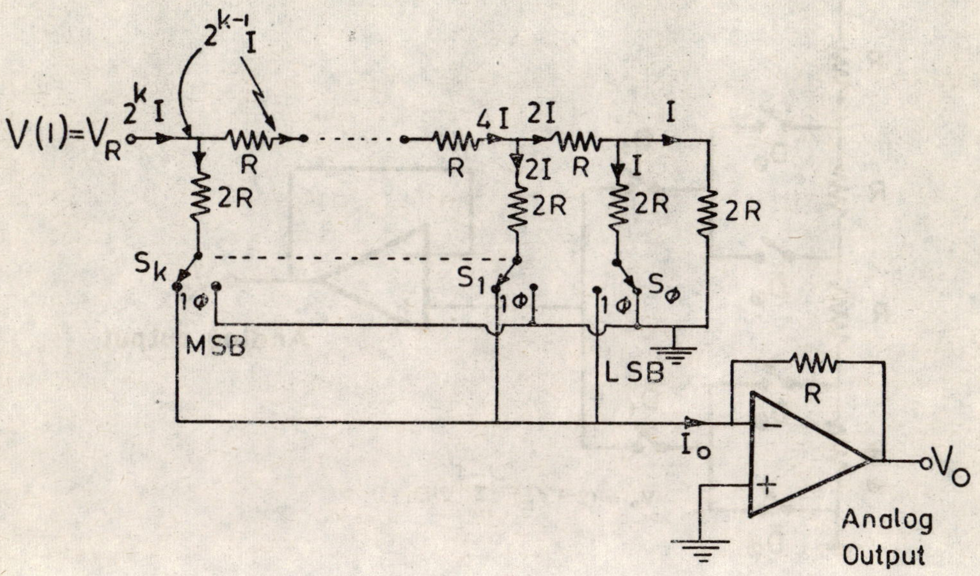

$$I = 2^{-k}/V_R/R$$

$$I_o = (2^{-k}V_R/R)\sum_{i=0}^{k-1} 2^i D_i$$

$$V_o = K\sum_{i=0}^{k-1} 2^i D_i, \quad \text{where } K = -2^{-k}V_R$$

Fig. 11.2.5 R–2R Ladder Network DAC (k-bit, Unipolar)

circuit, though, an open input is not desirable since it might cause 'pick-up' due to unavoidable couplings. It is therefore necessary to provide some means of minimizing pick-ups at such open inputs (not shown in Fig. 11.2.5).

11.2.3 2^k R DAC

In this realization, an n-bit DAC is made using 2^n resistors of equal value and $(2^{n+1} - 2)$ analog switches. Although the number of components are larger than in the earlier circuits, the fact that only a single value resistance is required makes it an attractive option in MOS LSI realization. An example of 2-bit DAC using this circuit is shown in Fig. 11.2.6 where D_1 is the MSB and $D_\emptyset$ the LSB. In this circuit, a particular switch is considered closed only if the indicated binary bit is true (i.e. '1'); otherwise, it is open (i.e. for $\emptyset$). Extending the circuit for n-bits can be easily done.

As can be seen from Fig. 11.2.6, an n-bit DAC of this kind uses potential division to create voltages proportional to each of the 2^n possible input values. The combination of switches is then used to choose one of these 2^n values as indicated by the n-bit digital input. The op. amp is merely a unity gain voltage follower provided for buffering.

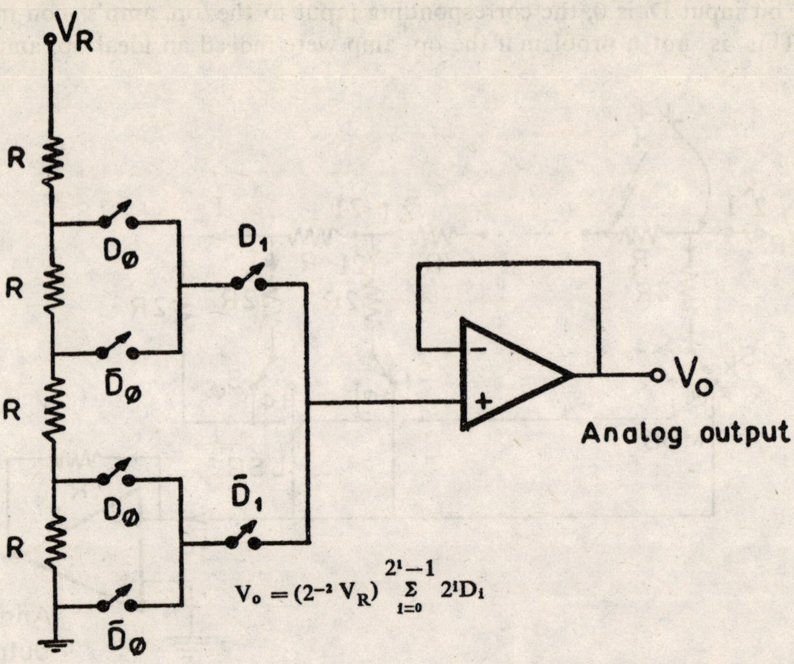

$$V_o = (2^{-2} V_R) \sum_{i=0}^{2^1-1} 2^i D_i$$

Fig. 11.2.5 2^k R DAC (2 bit, Unipolar

11.2.4 Multiplying DACs

We have seen in our earlier examples of DACs with $V(1) = V_R$ and $V(\emptyset) = 0$, that as V_R changes the analog output of the DAC also changes. If the voltage V_R can be provided externally to the DAC by the user and if the DAC is so designed that its output changes linearly with V_R, the device is referred to as a Multiplying DAC.

A Multiplying DAC accepting unsigned binary data and a positive reference voltage always operates in one quadrant. If it can accept bipolar binary data it operates in two quadrants. Two quadrant operation is also obtained if it accepts unipolar binary data but can operate with a bipolar reference voltage. Four quadrant operation requires that the multiplying DAC should accept both bipolar binary data as well as a bipolar reference voltage.

Multiplying DACs are frequently used in microprocessor systems. By changing the reference voltage, one can scale the output analog signal as desired. Another use of multiplying DACs is in building amplifiers with programmable gain. This is illustrated in Fig. 11.2.7. Here the k-bit Multiplying DAC is used in the feedback path of the op-amp where the reference voltage V_R to the DAC is provided by the op. amp's output. The input voltage to this amplifier V_i is fed to the non-inverting input of the op. amp and the output of the DAC V_{OD} is fed to the other op. amp input. Assume that for a given digital input $(D_{k-1} \ldots D_\emptyset)$ and reference voltage V_R, the DAC output is given by

$$V_{OD} = KV_R \sum_{i=\emptyset}^{k-1} 2^i D_i$$

where K = proportionality constant. If the binary input is non-zero, the op. amp would ensure that V_{OD} and V_i are identical, implying that

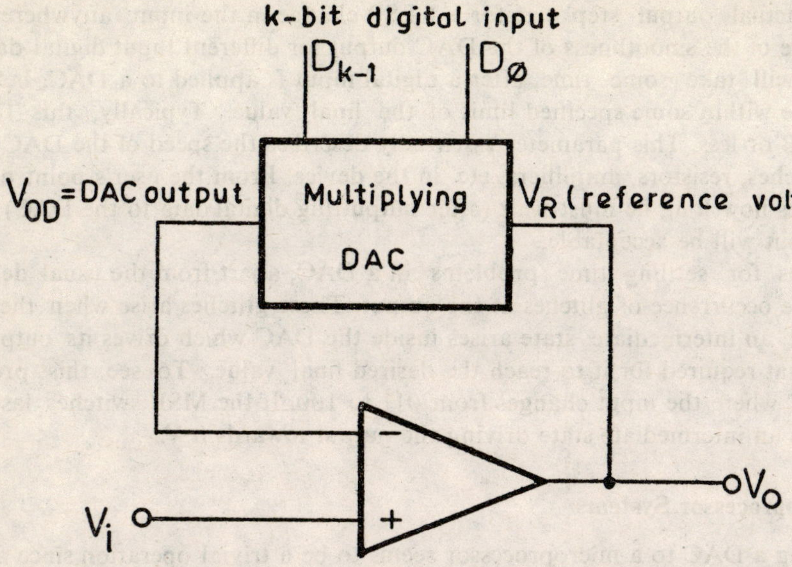

Fig. 11.2.7 Amplifier with Digitally Programmable Gain

$$V_i = KV_R \sum_{i=Ø}^{k-1} 2^i D_i$$

Since $V_0 = V_R$,

$$\frac{V_0}{V_i} = \frac{1}{K \sum_{i=0}^{k-1} 2^i D_i}$$

The range of gain available varies from 1/K to $1/K(2^k - 1)$, a variability of 1 to $(2^k - 1)$.

11.2.5 DAC Specifications

The manufacturer of a commercial DAC usually provides a number of parameters which indicate the quality of performance of the DAC. Some of these parameters are described next.

Resolution: This parameter, introduced earlier, specifies the percentage change in the output compared to the full scale (maximum) output that occurs if the LSB is changed by 1 bit. For a k-bit DAC, this can be calculated to be $1/(2^k - 1)$. Note that this parameter carries information regarding k, the digital input word size of the DAC. It is, therefore, not uncommon to find a k-bit DAC being described as one having a k-bit resolution. This alternative usage of the term 'resolution' should be kept in mind when reading a manufacturer's data sheets.

Linearity: If a DAC were ideal, equal increments (or decrements) in its digital input should effect equal increments (or decrements) in its analog output. The linearity of a converter specifies the accuracy with which this ideal performance is followed. It is important that a DAC's output be *Monotonic*—each step up (down) in the digital input should cause an increase (decrease) in the analog output. This can be guaranteed only if the linearity of the DAC is better than or at least equal to ±1/2 LSB.

Differential linearity of a DAC indicates the difference between the ideal 1 LSB step size (in the output) and the actual output step size for a 1 LSB change in the input anywhere in the range. This is a measure of the smoothness of the DAC output for different input digital data.

Settling Time: It will take some time, after a digital input is applied to a DAC, before its output settles to a value within some specified limit of the final value. Typically, this limit is specified as $\pm 1/2$ LSB or less. This parameter essentially describes the speed of the DAC and is determined by the switches, resistors, amplifiers, etc. in the device. From the user's point of view this parameter specifies how long he must wait (after outputting digital data to the DAC) before the DAC's analog output will be acceptable.

One of the reasons for settling time problems in a DAC, apart from the usual delays in electronic circuits, is the occurrence of glitches at the output. These glitches arise when the DAC input is changed and an intermediate state arises inside the DAC which drives its output in a direction opposite to that required for it to reach the desired final value. To see this problem, consider a 3-bit DAC where the input changes from Ø11 to 1ØØ. If the MSB switches last, it is possible to have ØØ0 as an intermediate state driving the output towards 0 V.

11.2.6 DACs in Microprocessor Systems

At first sight, interfacing a DAC to a microprocessor seems to be a trivial operation since all that the microprocessor is required to do is output the desired data word to the DAC. In practice, the interface might be complicated by one or both of the following problems:

(a) The DAC may not have a latch to store the data word that has to be converted. In this case, the microprocessor would typically have to output its data word to a latch whose outputs drive the DAC inputs.

(b) Commercially, DACs are available with their input data word size varying from 6 to 16 bits. Similarly, microprocessors typically have data word sizes of 4, 8, or 16 bits. It is quite likely that there will be an incompatibility between the word size of the DAC and the word size of the microprocessor. Such an incompatibility, if it exists, will require the system designer to take special care when interfacing the DAC to the microprocessor system.

One typically finds DACs referred to in literature as being *Microprocessor Compatible*. Generally, what this implies is that, to some extent, the DAC takes care of the above problems (a) and (b). Interfacing such a DAC to a microprocessor is relatively easier and requires relatively less extra hardware than other kinds of DACs.

Solving problem (a) is relatively easy since all that is required is to incorporate a latch inside the DAC or incorporate a latch between the microprocessor and the DAC. The microprocessor should be able to address this latch as an Output Port (memory mapped or I/O mapped) and write the data to be converted into this latch. This, of course, assumes that the microprocessor and DAC are word-compatible. Such a latch is almost always provided.

Tackling problem (b)—incompatible word sizes—is also easy if the microprocessor's word size is larger than that of the DAC, e.g. interfacing a 6-bit DAC to the 8085A. In this case, the processor will still treat the DAC's internal latch as an Output Port. The data to be converted is output to this port. The only care that has to be taken is that those bits of the microprocessor's output which are not connected to the DAC's input lines should be regarded as 'don't cares'. In our previous example of a 6-bit DAC to be interfaced to the 8085A, we can conveniently connect only $D_Ø$-D_5 to the DAC input and leave D_6 and D_7 as 'don't cares'.

The solution provided for problem (b) is more complicated if the microprocessor's word size is less than that used by the DAC. Consider an example where a 10-bit DAC is to be interfaced to an 8-bit microprocessor (e.g. 8085A). If the DAC is claimed to be microprocessor compatible, it will typically have a 10-bit latch inside. The lower 8 bits of this latch and the higher 2 bits can be separately loaded. The microprocessor is expected to write two 8-bit words to this DAC. One 8-bit word gets latched into the lower 8 bits of the DAC's latch and two bits of the other 8-bit word gets loaded into the remaining two bits of the latch. The DAC now has the 10-bit data it is supposed to convert. This or a similar approach will typically have to be followed if a microprocessor compatible DAC is to be interfaced to a microprocessor where the latter's word size is less than that of the DAC.

Finally, consider the situation where a DAC which is not claimed to be microprocessor compatible is to be used in a system where the microprocessor's data word size is less than that of the DAC. An example is illustrated in Fig. 11.2.8 where a 10-bit DAC is to be interfaced to a

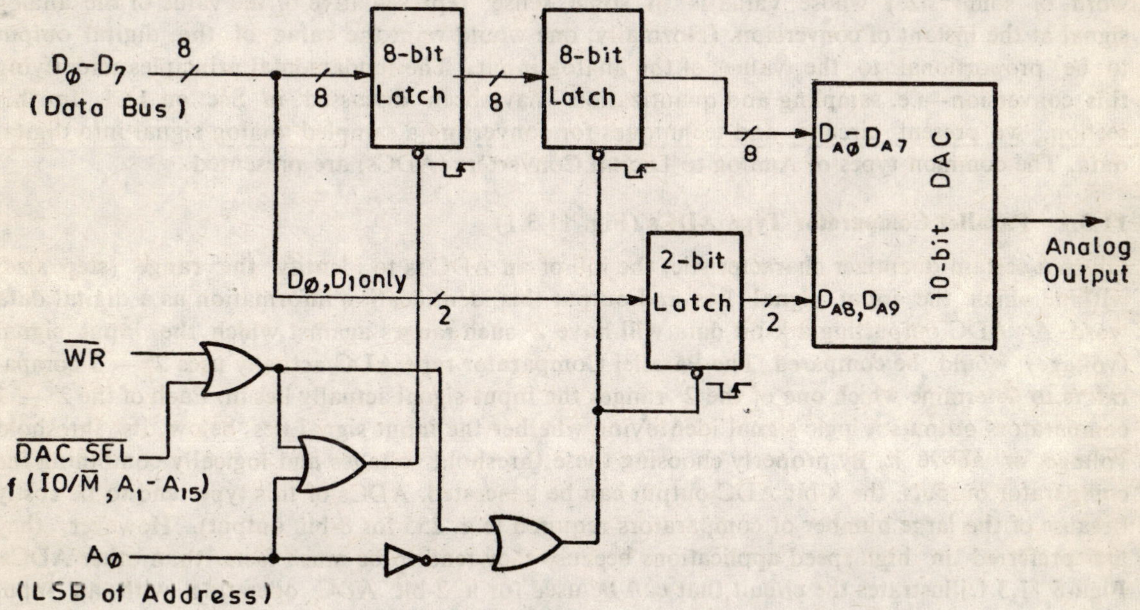

Fig. 11.2.8 Interfacing a 10-bit DAC to a 8-bit Microprocessor (e.g. 8085A)

8-bit microprocessor. In the interface shown, the microprocessor considers the DAC as two output ports where the lower port is addressed with $\overline{\text{DAC SEL}} = \emptyset$ and $A_\emptyset = \emptyset$ and the upper port is addressed with $\overline{\text{DAC SEL}} = \emptyset$, and $A_\emptyset = 1$. (The port can be either I/O mapped or memory mapped.) Note that $\overline{\text{DAC SEL}}$ will have to be appropriately derived using A_1–A_{15} and IO/$\overline{\text{M}}$ in case the 8085A is being used. For the 8085A, assume that the data to be converted is stored in the lower 10 bits of the H-L register pair. If I/O mapped I/O is being used, the following program will load this data into the DAC:

```
; I/O mapped DAC
        MOV     A, L
        OUT     ANLOG; ANLOG = Lower order port's address (8 bits), Aø = 0
        MOV     A, H
        OUT     ANLOG + 1; Aø ≤ 1
```

For memory mapped I/O, the SHLD instruction can be conveniently used. For example,

SHLD ANLGM; ANLGM = 16-bit address of lower order port, $A_\emptyset = \emptyset$

DACs are commonly used in microprocessor systems when one wants to output an analog voltage typically for control or display applications. The accuracy required in the analog output, compatibility considerations and cost typically guide the choice of the DAC. Another common use of DACs is in realizing Analog to Digital Converters. Several examples of this will be seen in the next section.

11.3 ANALOG TO DIGITAL CONVERSION

The analog to digital conversion process typically involves generating a digital output (a digital word of some size) whose value is in some sense representative of the value of the analog signal at the instant of conversion. (Normally, one would want the value of the digital output to be proportional to the value of the analog input.) The fundamental principles underlying this conversion—i.e. sampling and quantization—have been discussed in Section 11.1. In this section, we present circuits and techniques for converting a sampled analog signal into digital data. The common types of Analog to Digital Converters (ADCs) are presented.

11.3.1 Parallel Comparator Type ADCs (Fig. 11.3.1)

Given a certain quantizer characteristic, the job of an ADC is to identify the range (step size) within which the input signal lies and output this identification information as a digital data word. An ADC outputting a k-bit data will have 2^k such ranges against which the input signal (voltage) would be compared. The Parallel Comparator type ADC actually uses $2^k - 1$ comparators to determine which one of the 2^k ranges the input signal actually lies in. Each of the $2^k - 1$ comparators outputs a logic signal identifying whether the input signal lies below its threshold voltage or above it. By properly choosing these threshold voltages and logically combining the comparator outputs, the k-bit ADC output can be generated. ADCs of this type tend to be costly because of the large number of comparators required (e.g. 255 for 8-bit output). However, they are preferred in high speed applications because they tend to be much faster than other ADCs. Figure 11.3.1 illustrates the circuit that can be used for a 2-bit ADC operating with an input signal range of 0 to V_0 volts where V_0 is an externally supplied reference voltage.

The 2-bit ADC of Fig. 11.3.1 can convert input signals lying in the 0 to V_0 voltage range. Its quantizer characteristics are shown in the figure and can be summarised as follows:

$V_i \leqslant V_0/4$ $\quad D_1 = \emptyset \quad D_\emptyset = \emptyset$

$V_0/4 < V_i \leqslant V_0/2$ $\quad D_1 = \emptyset \quad D_\emptyset = 1$

$V_0/2 < V_i \leqslant 3V_0/4$ $\quad D_1 = 1 \quad D_\emptyset = \emptyset$

$3V_0/4 < V_i$ $\quad D_1 = 1 \quad D_\emptyset = 1$

In order to achieve this quantizer behaviour, the threshold voltages for C_2, C_1, and $C_\emptyset$ comparators are chosen as $3V_0/4$, $V_0/2$ and $V_0/4$ respectively. The comparators are assumed to output a logic 1 only if the input signal exceeds its threshold voltage; otherwise, the output is a logic Ø. These outputs are stored in the D-flip-flops at the falling edge of the clock input and the outputs of these flip-flops are combined to produce the digital outputs D_1 and $D_\emptyset$. The circuitry controlling this ADC (e.g. a microprocessor system) issues a Lo-Hi-Lo Start of Conversion Pulse (SOC) to it when it wants the digital equivalent of the analog input. The comparators'

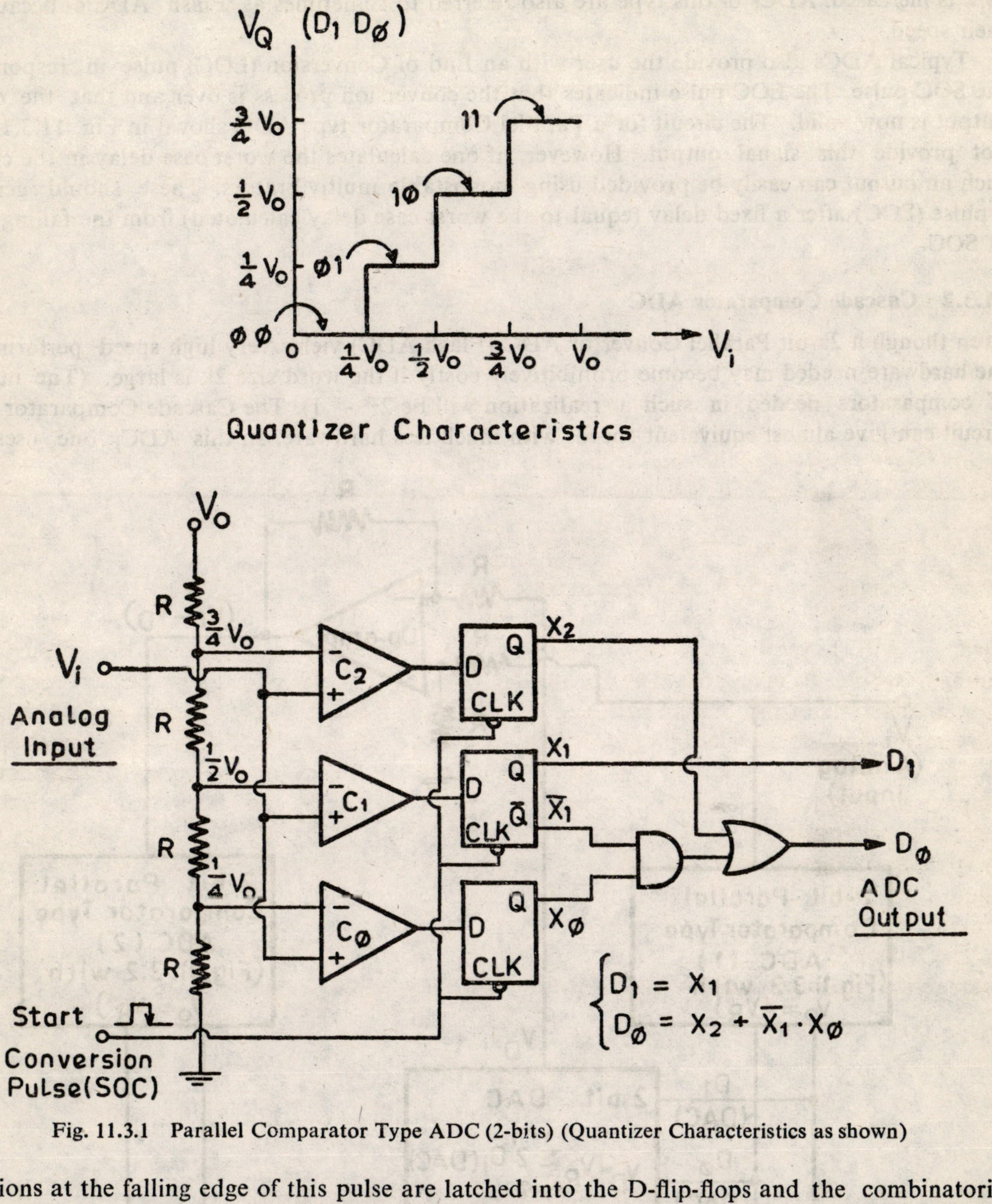

Fig. 11.3.1 Parallel Comparator Type ADC (2-bits) (Quantizer Characteristics as shown)

decisions at the falling edge of this pulse are latched into the D-flip-flops and the combinatorial circuit following it generates D_1 and $D_\emptyset$. Note that the delay between the falling edge of the SOC pulse and the instant when both the outputs D_1 and $D_\emptyset$ are valid is controlled primarily by the flip-flop delay and the delay produced by the combinatorial circuit. Since these contributions will typically be small (e.g. if TTL is used), this ADC will normally operate at a very high speed. Even higher speeds can be obtained by using Schottky TTL or ECL logic. Generalising this approach to a k-bit ADC can also be done except that the level of complexity increases rapidly as k is increased. ADCs of this type are also referred to sometimes as 'Flash ADC's because of their speed.

Typical ADCs also provide the user with an End of Conversion (EOC) pulse in response to the SOC pulse. The EOC pulse indicates that the conversion process is over and that the digital output is now valid. The circuit for a Parallel Comparator type ADC shown in Fig. 11.3.1 does not provide this signal output. However, if one calculates the worst case delay in the circuit, such an output can easily be provided using monostable multivibrators. These should generate a pulse (EOC) after a fixed delay (equal to the worst case delay calculated) from the falling edge of SOC.

11.3.2 Cascade Comparator ADC

Even though a 2k-bit Parallel Convertor ADC (Flash ADC) yields very high speed performance, the hardware needed may become prohibitively costly if the word size 2k is large. (The number of comparators needed in such a realization will be $2^{2k} - 1$). The Cascade Comparator ADC circuit can give almost equivalent speeds with much less hardware. In this ADC, one uses two

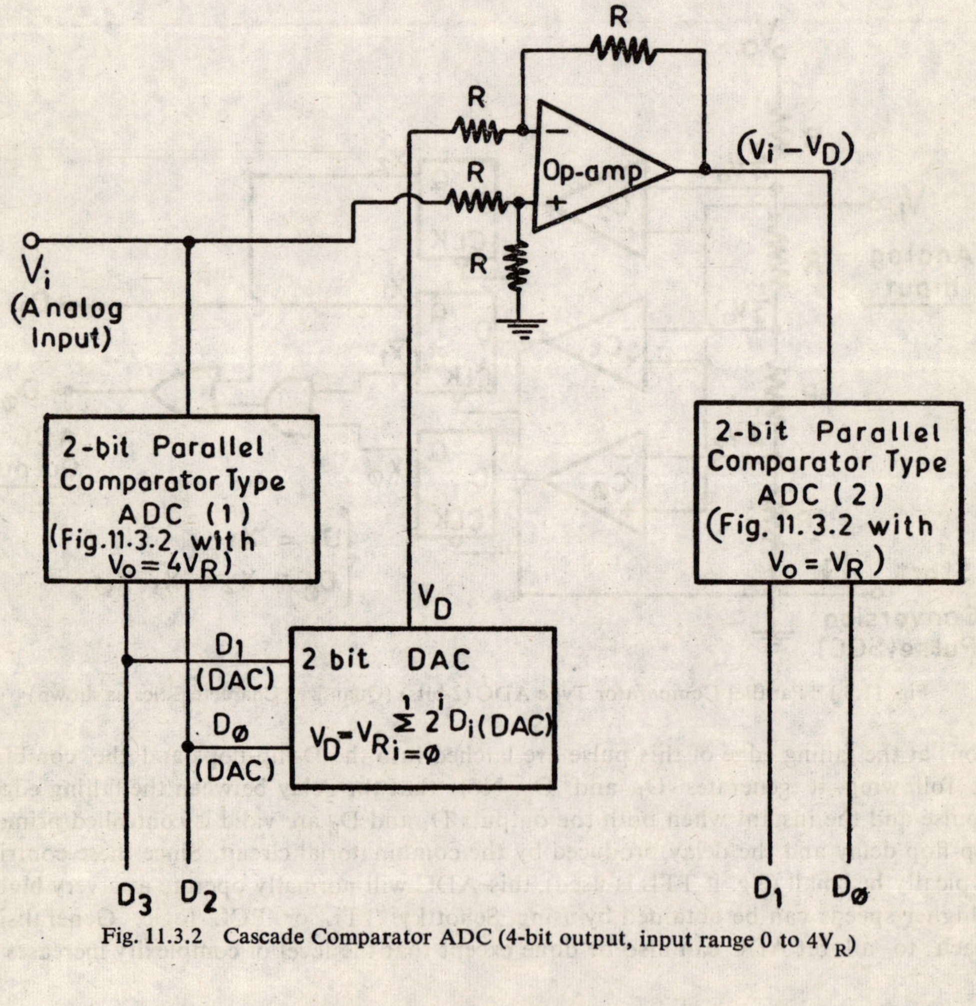

Fig. 11.3.2 Cascade Comparator ADC (4-bit output, input range 0 to $4V_R$)

k-bit ADCs, one k-bit DAC and a high speed op. amp difference amplifier. The total number of comparators needed here is $(2^{k+1} - 2)$ which will be substantially less than the earlier figure of $2^{2k} - 1$, if k is large. (Note that in a general case, for $n < 2k$, one can use one n-bit ADC, one $(2k - n)$-bit ADC and one n-bit DAC to build one 2k-bit cascade comparator ADC from two smaller parallel converter type ADC. In this case, the total number of comparators needed would be $2^n + 2^{2k-n} - 2$. One can show that this number will be minimum when $n = k$). The conversion delay in this type of ADC will be larger than that of an equivalent parallel type ADC. This is primarily because in the cascade comparator ADC, the DAC and the difference amplifier introduce additional delays over and above the delay introduced by the two smaller parallel type ADCs.

The operation of a Cascade Comparator type ADC is illustrated in Fig. 11.3.2 by means of an example. Consider a 4-bit ADC operating in the input voltage range of $0 \leqslant V_i \leqslant 4V_R$. This ADC uses two 2-bit Parallel Converter type ADCs of the type shown in Fig. 11.3.1. The first one of these, ADC(1), works with a reference voltage $V_0 = 4V_R$ (refer to Fig. 11.3.1 also) and outputs D_3 and D_2, the two most significant output bits. The second 2-bit A/D converter, ADC(2), works with a reference voltage $V_0 = V_R$ and outputs D_1 and $D_\emptyset$, the two least significant output bits. The 2-bit output, D_3 and D_2, of ADC(1) is also fed to a 2-bit DAC whose reference voltage is adjusted to yield an output V_D given by

$$V_D = V_R \sum_{i=\emptyset}^{1} 2^i D_{i(DAC)} = V_R[D_{\emptyset(DAC)} + 2D_{1(DAC)}]$$

$$= V_R[D_2 + 2D_3]$$

The op. amp uses the signal input V_i and the DAC output V_D to generate the difference $(V_i - V_D)$ which is fed to ADC(2). The function of ADC(1) is to determine the voltage range (0–V_R, V_R–$2V_R$, $2V_R$–$3V_R$, $3V_R$–$4V_R$ corresponding to D_3, D_2 values of ØØ, Ø1, 1Ø, 11) in steps of V_R within which the input signal lies. This yields the desired data outputs corresponding to the most significant bits D_3 and D_2. These bits are output to the DAC which yields an output voltage V_D related to V_i as follows:

$$0 \leqslant V_i < V_R \qquad V_D = 0\text{ V}$$

$$V_R \leqslant V_i < 2V_R \qquad V_D = V_R$$

$$2V_R \leqslant V_i < 3V_R \qquad V_D = 2V_R$$

$$3V_R \leqslant V_i \qquad V_D = 3V_R$$

(Note that in essence, the combination of ADC(1) with the DAC provides a 4-level quantizer along with the two most significant bits of the result.)

The difference signal $(V_i - V_D)$ is always in the range of

$$0 \leqslant (V_i - V_D) \leqslant V_R$$

This signal can then be converted into a two digit output by ADC(2). These two bits correspond to the less significant two bits D_1 and $D_\emptyset$. The overall circuit, therefore, yields a 4-bit output for signals in the range 0–$4V_R$. (A parallel converter type ADC would have required 15 comparators. In contrast, the circuit of Fig. 11.3.2 will require only 6 comparators.)

11.3.3 Successive Approximation ADC

As the name implies, a Successive Approximation ADC finds the digital equivalent of an analog voltage by comparing it with a sequence of known analog reference voltages where the reference chosen for (n + 1)th comparison depends on the result of the nth comparison. Typically, the sequence of analog reference voltages is obtained by feeding proper digital data words to a DAC. A k-bit ADC using this technique will generally have a k-bit DAC inside where the step size of the ADC's quantization is equal to the change in the analog output of the DAC when its input changes by one LSB. Figure 11.3.3 shows the block diagram of such an ADC. In order to convert a given analog voltage held by a Sample and Hold circuitry, the user issues a Start-of-Conversion (SOC) pulse to the ADC. The SAR/Logic/Control unit then takes over and using the conversion algorithm described later, obtains the digital equivalent of the analog signal. When the conversion is complete, the ADC returns an End-of-Conversion (EOC) pulse to the user indicating that its k-bit digital output is now valid and can be read.

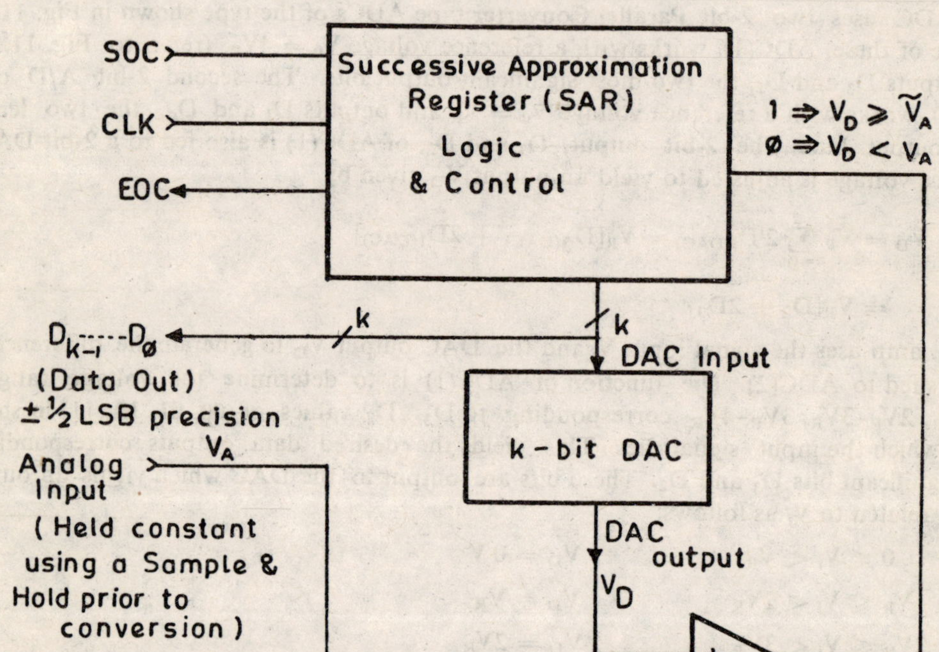

Fig. 11.3.3 Block Diagram of k-bit Successive Approximation ADC (±½ LSB precision)

For the ADC shown in Fig. 11.3.3, the Successive Approximation Algorithm can be summarised as follows:

(i) Add an analog voltage equal to 1/2 LSB of DAC to the analog input voltage V_A to be converted: (see note below)

$$\tilde{V}_A = V_A + V_{1/2\ LSB(DAC)}$$

(ii) Reset all SAR bits (k-bit register)

$$(SAR) = \emptyset\emptyset\emptyset \ldots \emptyset\emptyset$$

(iii) Initialize $N = k$

(iv) Set Bit $(N - 1)$ of SAR and output SAR contents to the k-bit DAC

(v) Compare V_D with $\tilde{V}_A$. If $V_D < \tilde{V}_A$ go to step (vii); else, continue

(vi) Reset Bit $(N - 1)$ of SAR

(vii) $N = N - 1$

(viii) If $N \neq \emptyset$, go to step (iv); else, continue

(ix) Output SAR contents as the conversion result

(x) Generate EOC pulse

(*Note*: Step (i) of the above algorithm is only needed to ensure the accuracy of the result to $\pm\frac{1}{2}$ LSB. If this step is omitted, the accuracy of the result will only be $\pm$ LSB. Also note that if step (i) is omitted, there is no need to check the SAR bit ∅ position. Instead, one can arbitrarily set or reset that bit without affecting accuracy. Moreover step (viii) can be modified as

'step (viii). If $N \neq 1$, go to step (iv); else, continue'

in which case bit ∅ of the output is always ∅ and is never tested.)

The ADC shown in Fig. 11.3.3 can be easily interfaced to a microprocessor system via an I/O port. The microprocessor would ask the ADC to convert the 'sampled and held' analog input by issuing an SOC pulse to it. After the conversion is over, the ADC will issue the EOC pulse which should load the data output into the I/O port and should also place an interrupt request to the microprocessor. This interrupt request informs the microprocessor that the Analog to Digital conversion is over and that the conversion result is waiting to be read from the I/O port.

An alternative approach to using a Successive Approximation type of ADC in microprocessor systems uses the microprocessor and an Output Port to implement the SAR/Logic/Control block of Fig. 11.3.3. In this case, the Successive Approximation Algorithm is implemented by the microprocessor itself via software. A circuit to do this is shown in Fig. 11.3.4. In this circuit the input to the 8-bit DAC is provided through the Output Port shown, which is addressed by the microprocessor as PORTA. (The microprocessor is assumed to be the 8085A). The result of comparing the DAC's output (V_D) with the analog input voltage (actually offset by $+\frac{1}{2}$ LSB from precision considerations) $\tilde{V}_A$ is monitored by the 8085A via its SID input where

$$\begin{aligned} SID &= 1 \quad \text{if} \quad V_D \geqslant \tilde{V}_A \\ &= \emptyset \quad \text{if} \quad V_D < \tilde{V}_A \end{aligned}$$

The 8085A subroutine (ADCON) which does the conversion is given below.

```
; ADCON converts analog input to 8-bit digital form using the successive approximation
;           technique. Refer Fig. 11.3.4.
;
```

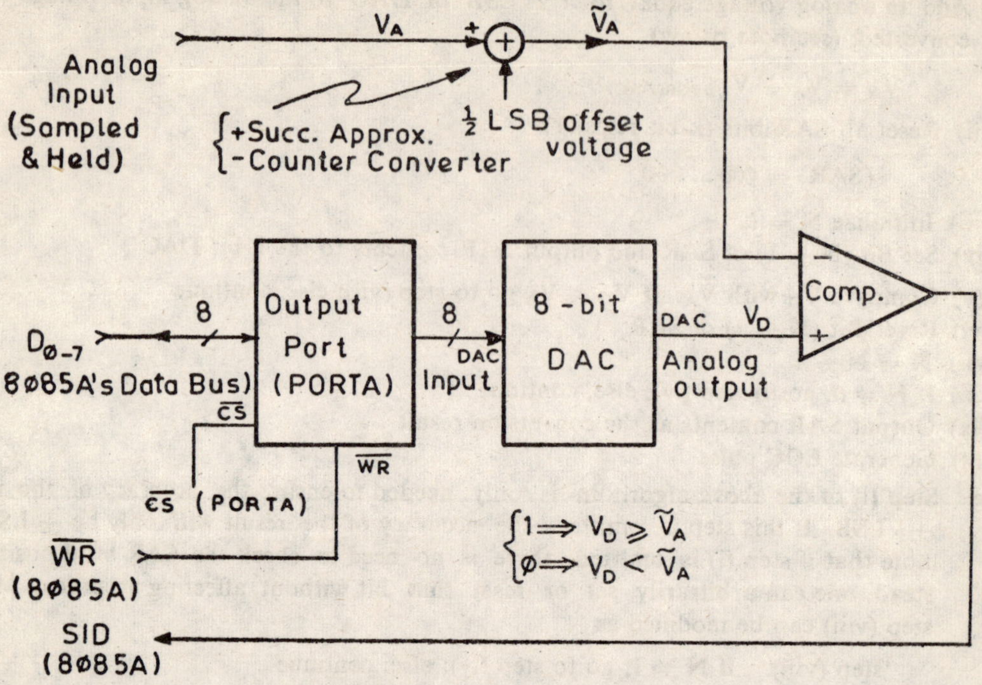

Fig. 11.3.4 Hardware for Microprocessor (8085A) Controlled 8-bit ADC. (Successive Approximation and Counter Converter types)

```
ADCON:  LXI D, 8∅∅∅H    ;  (D) = bit to be tested, (E) = result till now
LOOP :  MOV  A, E
        ORA  D          ;  new test output
        MOV  B, A       ;  also saved in B
        OUT  PORTA      ;  output to DAC to generate
        RIM             ;  read SID input
        ANI  8∅∅∅H      ;  check SID
        JNZ HSUB        ;  jump to HSUB if SID = 1
        MOV E, B        ;  update result register E if SID = ∅
HSUB:   MOV A, D        ;  update bit
        RRC             ;     to be tested
        RC              ;  if (CY) = 1, conversion over
        MOV D, A        ;  store new (D)
        JMP LOOP
;  Result of Conversion in E-reg on return from ADCON
```

If this routine is used, care should be taken to ensure that the DAC and the comparator output should be *valid* in the time between the 'OUT PORTA' and the RIM instructions; otherwise a sufficient number of NOPs will have to be interposed between the two instructions.

Finally, it should be noted that the time taken by a Successive Approximation conversion for a k-bit output is k-times the time taken for one approximation cycle. Therefore, this type of ADCs requires longer conversion times than the Parallel Comparator and Cascade Comparator ADCs discussed earlier. However, the hardware used by this type of ADC is substantially less.

11.3.4 Counting ADC (Counter Converter)

A Counting ADC uses a DAC and a counter to find the digital equivalent of an analog input. The counter output is fed to a DAC whose output is compared with the analog input. The counter, which is initially cleared, is made to count up bit by bit until the output of the comparator changes state. At that point, the counter is 'frozen' (stops counting) and its output gives the result of the analog to digital conversion. A hardware and software realization of this type of ADC is given in Fig. 11.3.5 and 11.3.4 respectively.

In the circuit shown in Fig. 11.3.5, a k-bit unipolar ADC is implemented using a k-bit DAC and a k-bit counter. The input analog voltage is offset by 1/2 LSB (voltage) in order to improve the precision of the ADC from $\pm$LSB to $\pm 1/2$ LSB. The conversion process is started by a SOC pulse which clears both the k-bit counter and the D-flip-flop shown in Fig. 11.3.5. The counter then counts up with the clock until the analog value of its output becomes high enough so that the comparator output becomes 1. At that point, the counter stops incrementing and EOC = 1 is output signalling end of conversion. The output of the counter at this point is the digital value of the analog input voltage. In order to interface this ADC in a microprocessor system, we need an I/O port which will be used by the microprocessor to read the output of the k-bit counter once the conversion is over. The EOC output can be used as an interrupt to inform the microprocessor that the counter output is valid and can now be read. Note that, in contrast to the Succ. Approx. ADC which always does the conversion in k cycles for a k-bit output, the ADC described here may require 0 to $2^k - 1$ cycles depending on the input level. This is a disadvantage with this type of ADC since the user will never actually know how long the conversion will take until the conversion itself is over.

The circuit shown in Fig. 11.3.4 can also be used to implement a Counting ADC with the microprocessor replacing the k-bit counter and the associated logic. In this case, the counting algorithm is implemented by the microprocessor itself via software. If an 8085A is being used to implement an 8-bit Counting ADC, the following program can do the conversion:

```
;  ADCNT   uses the hardware of Fig. 11.3.4 to implement on
;          8-bit counting type ADC

 ADCNT:    MVI   B, ØØH    ;  initialize counter B
 LOOP:     MOV   A, B      ;  output current
           OUT   PORTA     ;  count to DAC
           RIM             ;  read SID
           ANI   8ØH
           RNZ             ;  if SID = 1, conversion over
```

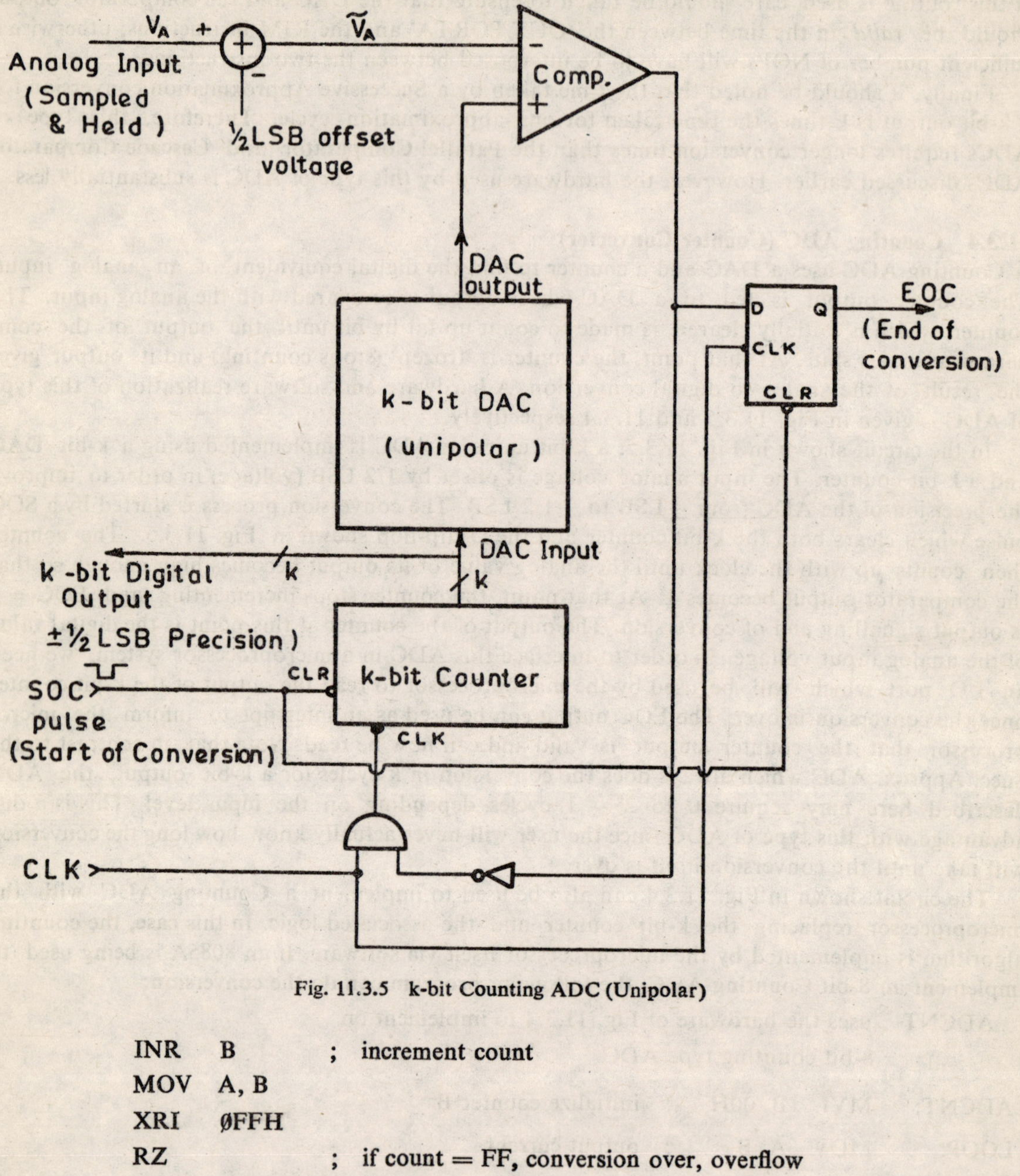

Fig. 11.3.5 k-bit Counting ADC (Unipolar)

```
        INR   B      ; increment count
        MOV   A, B
        XRI   ØFFH
        RZ           ; if count = FF, conversion over, overflow
        JMP   LOOP
; ADCNT  routine returns result of conversion in B reg.
```

As in the ADCON routine given earlier for a Successive Approximation ADC, care should be taken to ensure that the comparator output should be valid when the RIM instruction is issued

(All mnemonics copyright Intel Corporation 1985).

to read the SID line, If necessary, one or more NOPs should be placed between 'OUT PORTA' and 'RIM' in order to ensure this condition.

11.3.5 Dual Slope Integrating ADC

Consider a situation where an unknown analog input voltage (sampled and held) is used to charge up a capacitor (initially discharged) for a fixed time T. Now, this capacitor is disconnected

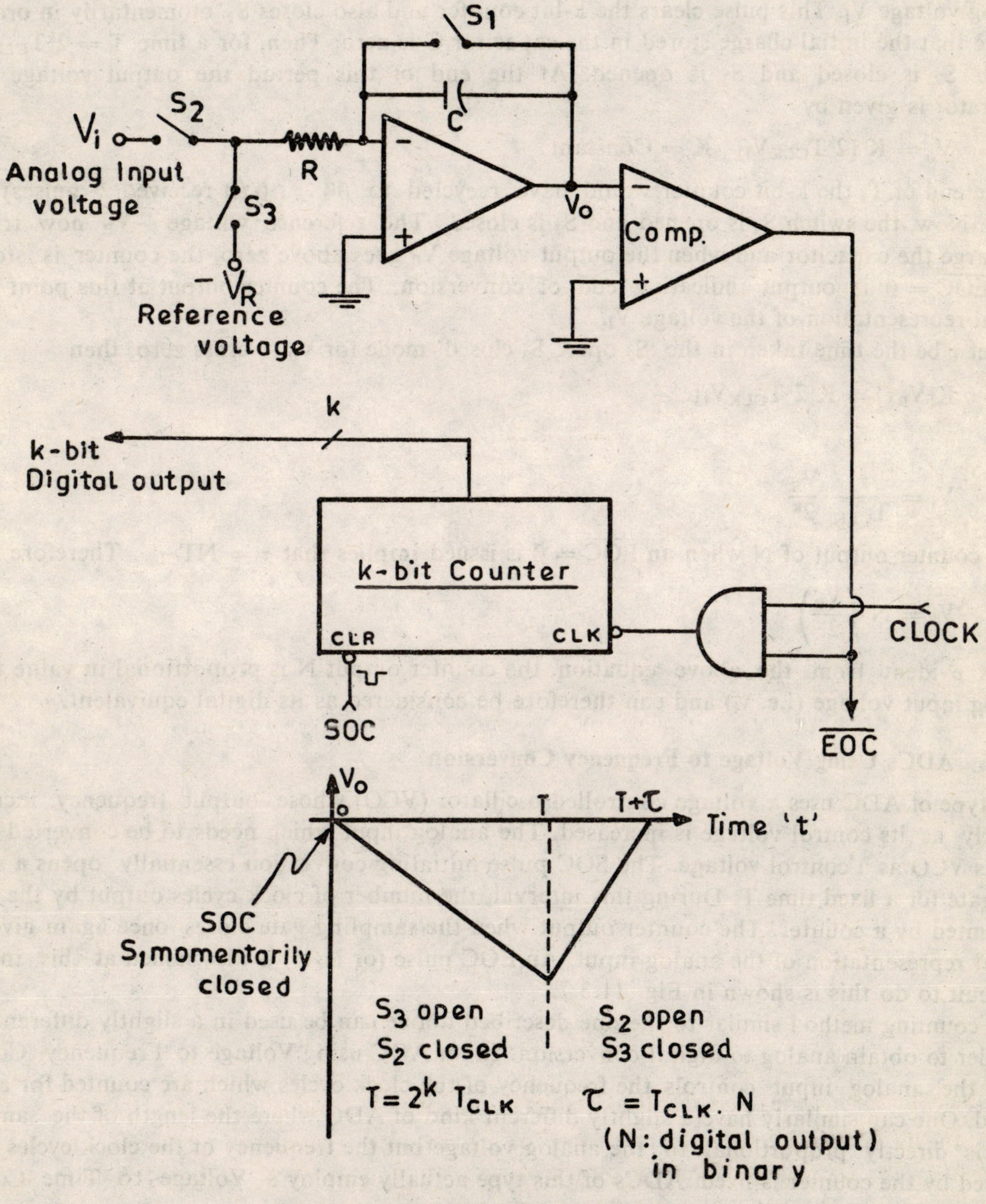

Fig. 11.3.6 Dual Slope Integrating ADC

from the unknown analog voltage and is, instead, discharged by a known reference voltage. The time taken to discharge the capacitor will then be proportional to the magnitude of the analog input. If a clock signal and a counter are used to measure this time, then the counter output will give a digital representation of the analog signal. Figure 11.3.6 gives a simplified block diagram representation of a Dual Slope Integrating ADC which uses this approach.

Consider the operation of the circuit shown in Fig. 11.3.6. At $t = 0$, the controlling hardware (e.g. a microprocessor) issues on SOC pulse to initiate conversion of the sampled and held analog voltage V_i. This pulse clears the k-bit counter and also closes S_1 momentarily in order to ensure that the initial charge stored in the capacitor C is zero. Then, for a time $T = 2^k T_{CLK}$, the switch S_2 is closed and S_3 is opened. At the end of this period the output voltage of the integrator is given by

$$V_0 = K\,[2^k T_{CLK} V_i] \quad K = \text{Constant}$$

At the end of T, the k-bit counter would have recycled to ØØ . . . Ø (it received 2^k pulses) once again. Now, the switch S_2 is opened and S_3 is closed. The reference voltage $-V_R$ now tries to discharge the capacitor and when the output voltage V_0 goes above zero, the counter is stopped and $\overline{EOC} = \emptyset$ is output indicating end of conversion. The counter output at this point is the digital representation of the voltage V_i.

Let τ be the time taken in the 'S_2 open, S_3 closed' mode for V_0 to cross zero; then

$$K[V_R \tau] = K[2^k T_{CLK} V_i]$$

or

$$V_i = \frac{\tau}{T_{CLK}} \cdot \frac{V_R}{2^k}$$

A counter output of N when an $\overline{EOC} = \emptyset$ is issued implies that $\tau = N T_{CLK}$. Therefore

$$V_i = N \cdot \left(\frac{V_R}{2^k}\right)$$

As is evident from the above equation, the counter output N is proportional in value to the analog input voltage (i.e. v_i) and can therefore be considered as its digital equivalent.

11.3.6 ADCs Using Voltage to Frequency Conversion

This type of ADC uses a voltage controlled oscillator (VCO) whose output frequency increases linearly as its control voltage is increased. The analog input which needs to be converted is fed to this VCO as a control voltage. The SOC pulse initiating conversion essentially opens a sampling gate for a fixed time T. During this interval, the number of clock cycles output by the VCO is counted by a counter. The counter output when the sampling gate closes once again gives the digital representation of the analog input. An EOC pulse (or level) is also issued at this instant. A circuit to do this is shown in Fig. 11.3.7.

A counting method similar to the one described above can be used in a slightly different way in order to obtain analog to digital conversion. In an ADC using Voltage to Frequency Conversion, the analog input controls the frequency of the clock cycles which are counted for a fixed period. One can similarly have a slightly different kind of ADC where the length of the sampling gate is directly proportional to the analog voltage but the frequency of the clock cycles being counted by the counter is fixed. ADCs of this type actually employ a Voltage to Time Conversion and can also be used.

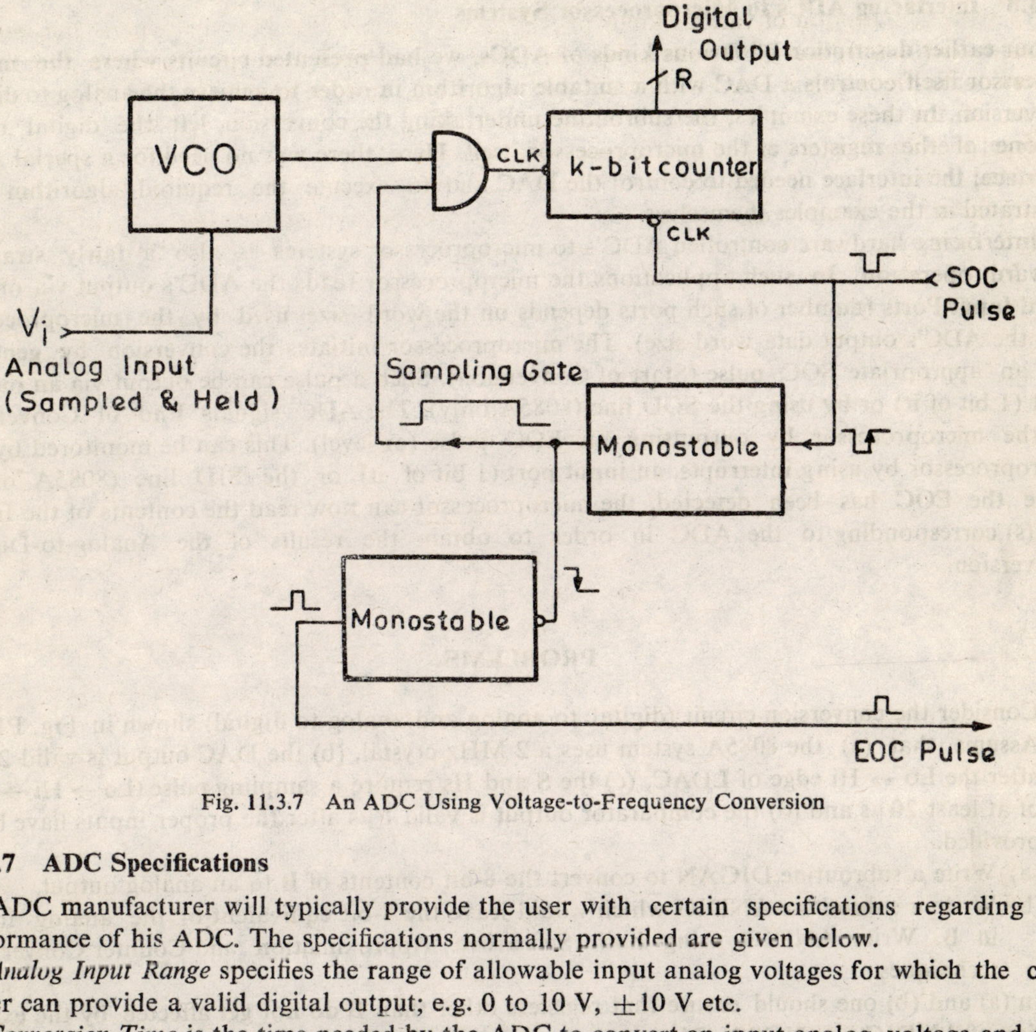

Fig. 11.3.7 An ADC Using Voltage-to-Frequency Conversion

11.3.7 ADC Specifications

An ADC manufacturer will typically provide the user with certain specifications regarding the performance of his ADC. The specifications normally provided are given below.

Analog Input Range specifies the range of allowable input analog voltages for which the converter can provide a valid digital output; e.g. 0 to 10 V, ± 10 V etc.

Conversion Time is the time needed by the ADC to convert an input analog voltage and output the corresponding digital results. It can vary from a few μs to a few tens of nanoseconds. The lower the conversion time, the higher will be the speed of the ADC.

Format represents the form of the digital output presented by the ADC, e.g. unipolar binary, 1's complement, 2's complement, offset binary, BCD etc.

Accuracy: Apart from the quantization errors inherent in all digital-analog conversions, an ADC will generally also have other sources of errors, e.g. noise, drift, errors in reference voltage, etc. The quantization error is usually specified as $\pm 1/2$ LSB. (Recall the addition of a fixed offset voltage in our circuits in order to achieve this when needed.) Apart from this error, the manufacturer also specifies a worst case bound. For example, consider an ADC with an input range specified to be ± 10 V. If this is a 10-bit ADC, the quantization error will be about 5 mV ($= 1/2$ LSB). The manufacturer may specify an additional error of 0.01% of the Full Scale Range (FSR = 10 V) implying that the cumulative errors from all other sources will not exceed ± 1 mV.

11.3.8 Interfacing ADCs to Microprocessor Systems

In our earlier description of various kinds of ADCs, we had presented circuits where the microprocessor itself controls a DAC with a suitable algorithm in order to achieve the analog to digital conversion. In these examples, the subroutine undertaking the conversion, left the digital result in one of the registers of the microprocessor itself. Here, there was no need for a special ADC interface; the interface needed to control the DAC and to execute the required algorithm was illustrated in the examples themselves.

Interfacing hardware controlled ADC's to microprocessor systems is also a fairly straightforward operation. In such applications the microprocessor reads the ADC's output via one or more Input Ports (number of such ports depends on the word size used by the microprocessor and the ADC's output data word size). The microprocessor initiates the conversion by generating an appropriate SOC pulse (Start of Conversion). Such a pulse can be output via an output port (1 bit of it) or by using the SOD line (8085A only). The ADC signals End of Conversion to the microprocessor by outputting an EOC pulse (or level). This can be monitored by the microprocessor by using interrupts, an input port (1 bit of it) or the SID line (8085A only). Once the EOC has been detected, the microprocessor can now read the contents of the Input Port(s) corresponding to the ADC in order to obtain the results of the Analog-to-Digital Conversion.

PROBLEMS

(1) Consider the conversion circuit (digital to analog and analog to digital) shown in Fig. P11.1. Assume that (a) the 8085A system uses a 2 MHz crystal, (b) the DAC output is valid 20 μs after the Lo → Hi edge of LDAC, (c) the S and Hs require a sampling pulse (Lo → Hi → Lo) of at least 20 μs and (d) the comparator output is valid 1 μs after the proper inputs have been provided.

(a) Write a subroutine DIGAN to convert the 8-bit contents of B to an analog output.

(b) Write a subroutine ANDIG which would leave the 8-bit equivalent of the analog input in B. Write ANDIG using both Successive Approximation and Counter Conversion techniques.

In (a) and (b) one should ensure that registers other than B do not get affected by the execution of ANDIG or DIGAN. Moreover, executing ANDIG should not change the analog output.

(2) Using a 16-bit DAC and 8-bit output ports modify the circuit of (1), i.e. Fig. P11.1, so that one can do 16-bit conversions rather than merely 8-bit ones. Write the software so that the modified circuit will give the analog equivalent of (BC) in Analog Out via DIGAN and convert the Analog In to a 16-bit quantity in B-C via ANDIG.

(3) Redo (2) using 8-bit DACs only. You can use op. amps etc. for whatever other circuitry is needed.

(4) A Voltage-To-Time Converter is available in IC form. It has an Analog Input (AI) and a Control Input (SOC or Start of Conversion). A Lo → Hi → Lo pulse at the SOC input causes the converter to output a pulse (Lo → Hi → Lo) of width T_w seconds to appear at the Pulse Out output where T_w is proportional to the voltage level of AI (always positive or zero) at the falling edge of SOC.

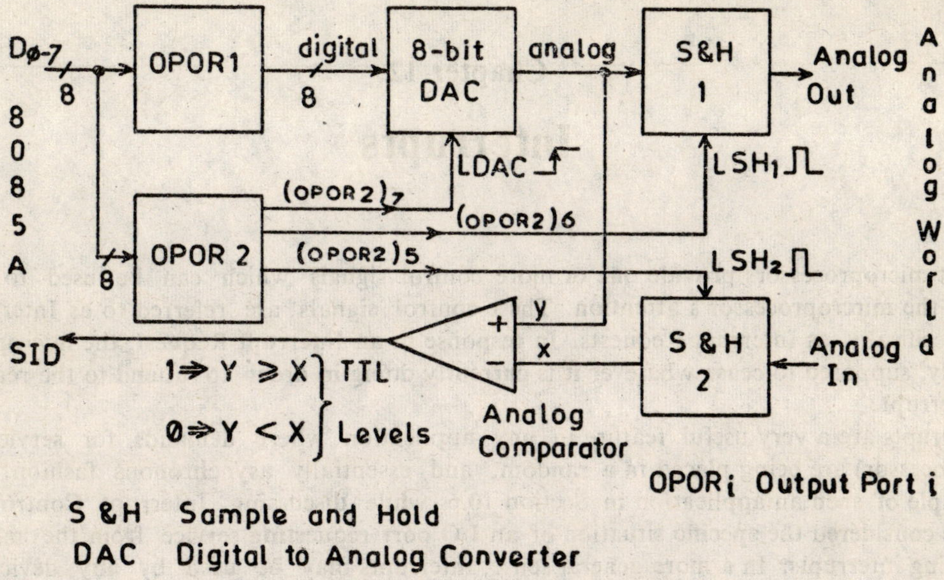

Fig. P11.1 Digital to Analog Conversion and Vice-versa

Using an 8253 Timer/Counter and appropriate software, design a system which will load the 8085A's Accumulator with a digital value proportional to the level of AI whenever a conversion routine DIGIN is called.

(5) Redo (4) without using an 8253, i.e. measuring T_w by a software delay loop. Mention any approximation or assumption that need to be made.

Chapter 12

Interrupts

Most microprocessors provide one or more control signals which can be used to directly demand the mircroprocessor's attention. These control signals are referred to as Interrupts, or more specifically, as Interrupt Requests. In response to an Interrupt Request, the microprocessor is, ideally, supposed to cease whatever it is currently doing in order to attend to the reasons for that interrupt.

Interrupts are a very useful feature in any application where demands for services (to a microprocessor) are being placed in a random, and essentially asynchronous fashion. We saw an example of such an application in Section 10.6 while discussing Interrupt Controlled I/O; there we considered the specific situation of an I/O port requesting service from the microcomputer using interrupts. In a more general sense, interrupts may be used by any device in the microcomputer system when it wants to get attention from the microcomputer. It is not uncommon to find applications where the microcomputer system is completely interrupt driven. On RESET or START, the microprocessor initializes the system properly; it then waits for interrupts in a WAIT loop. Specific tasks are executed only when the corresponding interrupt is requested; moreover, having finished the task, the microcomputer goes back to the WAIT loop.

The advantage of using interrupts (rather than using a polled status approach) is that the execution of the main program is not interfered with except when it is necessary. When a service is requested from the microcomputer via an interrupt, the microcomputer would temporarily leave whatever it is doing to service that interrupt; once the required service has been rendered, the microcomputer would resume its original task. It is possible, in this case, to divide the microcomputer's work into a background task, continually executed, and a few foreground tasks which are executed only when requested via interrupts.

12.1 INTERRUPT RECOGNITION

On recognizing an interrupt request, the CPU should break away from whatever program sequence it is currently executing and instead execute the required interrupt service routine. The following considerations are important and should be taken into account in the interrupt recognition process.

(i) It is not possible to respond to an interrupt request immediately after it is made. Specifically, the interrupt request might come in the middle of an instruction; in this case, the CPU should acknowledge that interrupt and start servicing it only after the current instruction has been executed. (Chaotic situations might happen if this strategy is not followed). The 8085A CPU ensures this by monitoring the status of the various interrupt requests only in the last-but-one clock state (T-state) of the last machine cycle of each instruction. It then completes the remaining clock state of the instruction even if an active interrupt request is found; only then does it respond to the interrupt request. (Note that, in essence, this is an effort on the CPU designer's part to integrate the reaction to an asynchronous input in a synchronous framework.) Other microprocessors also follow a similar strategy.

The point made above may not sound very important. It, however, has two major consequences in the design of a system. Firstly, it shows that the effect of an interrupt request is not instantaneous; that there is always a finite, but random, time delay between requesting the CPU for an interrupt and the CPU actually responding to that request. Secondly, it shows that in order to place an interrupt request effectively, it is necessary to keep the request 'alive' long enough for the CPU to monitor it. If the requesting device cannot ensure this, separate latching, either external or internal to the CPU will have to be provided for proper operation.

(ii) A microcomputer may not always like to be interrupted. Specifically, it may at times be executing (time) critical code sequences which it does not want interrupted. Moreover, the microcomputer should not be interrupted following a RESET (when it is initializing the system) or immediately after it responds to an interrupt (and has not yet reached the stage where the interrupt has been serviced). In the 8085A CPU, this is ensured by conditioning the external interrupt request lines (all but one of them, see (iii) next) on the output of an INTE flip-flop, internal to the CPU. (The INTE output masks the external interrupt requests before passing them on to the CPU). This INTE flip-flop can be set or reset under software control; i.e. it is reset by executing the DI (disable interrupts) instruction and set by executing the EI (enable interrupts) instruction. It is also reset following a RESET or after an interrupt request has been responded to and stays that way until an EI is executed. Most other microprocessors provide a similar feature. (The 8085A also provides a facility for selectively enabling/disabling some specific interrupt request lines. This is over and above the global enabling/disabling feature provided via the INTE flip-flop).

(iii) In spite of (ii) above, there is some need in microcomputers to provide interrupt request lines which can never be disabled. These are generally referred to as Non-maskable Interrupts (NMI) and are dedicated to detecting and servicing catastrophic conditions in the system (e.g. impending power failure or a run-away condition in the system). In the 8085A, this facility is provided by the TRAP interrupt which is not masked by the INTE flip-flop and can never be disabled. Some other microprocessors, like the 8086, 8088 or Z80 etc., also provide a similar facility.

In a system where the non-maskable interrupt is not being used, the system designer should always terminate it properly to ground or V_{cc} to prevent spurious activation of this interrupt. Even when it is needed and is being used, great care is recommended. Note that unless it is properly taken care of at the design stage itself, the NMI input might get activated on power-up, before the system even has time to properly initialize itself. If this problem exists, the designer will have to take care of this either through appropriate software or proper external hardware.

As mentioned in (i) above, the interrupt request has to stay active as long as the CPU does not respond to it. This is true for the interrupt request lines of most microprocessors; they are generally level sensitive. Some microprocessors provide interrupt lines which are edge sensitive, e.g. the RST 7.5 interrupt in the 8085A is activated by a LO→HI transition. In this case, the microprocessor provides internal latching to store this interrupt request freeing the external device from the responsibility of keeping the request active till it is serviced. The reader should take care of the following when using edge sensitive interrupts. Firstly, it is now possible for the external device to place multiple interrupt requests even when a prior interrupt request has not yet been serviced. Note that in such situations, the microprocessor has no way of knowing that multiple interrupt requests were made. Secondly, such an interrupt request line will be very sensitive to noise spikes. For the TRAP interrupt, the 8085A takes care of this problem by making this input both level and edge sensitive. The TRAP request is activated at the LO → HI edge but must also stay HI till the CPU samples it. (For further details regarding this feature, the reader is referred to Fig. 12.1.1.) As long as the noise spike itself does not behave in this

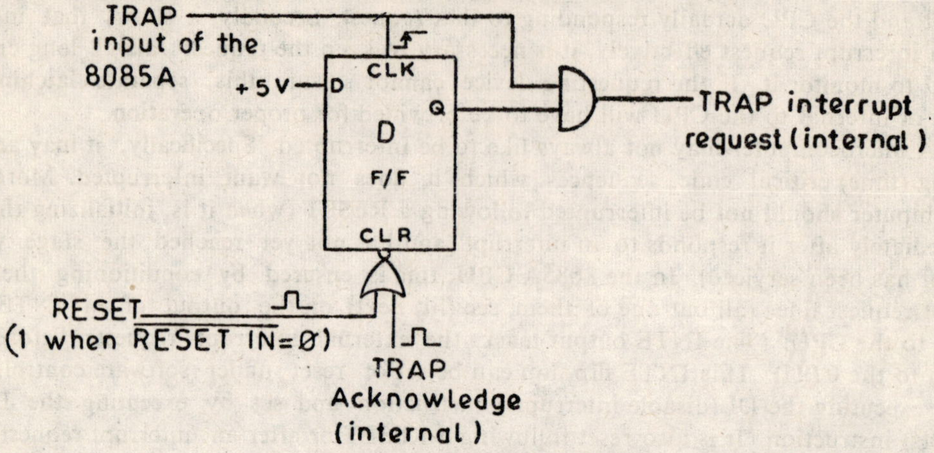

Fig. 12.1.1 TRAP Input of the 8085A

fashion, it will not cause a spurious TRAP interrupt request. Finally, it should be noted that as shown in Fig. 12.1.2, one can always use external circuitry to convert a level sensitive interrupt

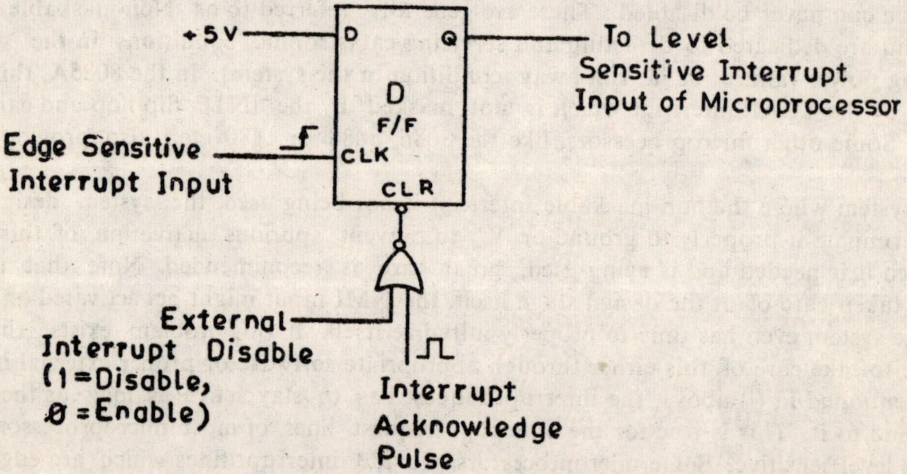

Fig. 12.1.2 Converting Level Sensitive Interrupts to Edge Sensitive Ones

input into an effectively edge sensitive one. The interrupt recognition procedure itself is summarised by the flow-chart of Fig. 12.1.3.

12.2 INTERRUPT RESPONSE AND INTERRUPT VECTORS

The operations undertaken by a microprocessor in response to an interrupt request not only differ from microprocessor to microprocessor but may also be different for the different interrupt

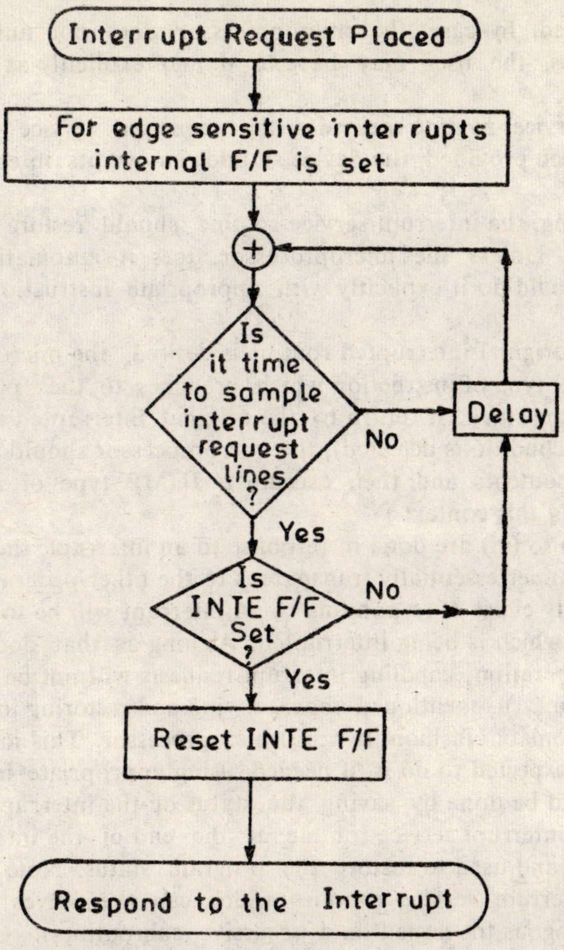

Fig. 12.1.3 CPUs Response to an Interrupt Request

inputs of the same microprocessor. However, in general terms, some or all of the following actions are taken in response to an interrupt request.

(i) The microprocessor branches to the program memory location containing the beginning of the interrupt service routine (for the interrupt) after saving the Program Counter contents in the stack. Essentially, the interrupt causes a CALL ADDR type instruction to be executed, where ADDR is the address of the first instruction of the interrupt service routine. The address ADDR to which program execution is directed is also commonly referred to as the Interrupt Vector corresponding to that interrupt request.

(ii) The INTE flip-flop, internal to the microprocessor, is reset disabling all further interrupts until the appropriate interrupt enabling instruction (e.g. EI in 8085A) is executed. (This is generally done at the end of the interrupt service routine).

(iii) The program status—status of various flags and register contents—just prior to the interrupt branching is saved. This saving could be done using the stack or by providing an alternate bank of registers in the CPU. In the latter case, this alternate bank of registers is used by the interrupt service routine so that the pre-interrupt program status can be exactly restored later, if so needed. In case, the microprocessor does not automatically save the pre-interrupt program status, the user may have to do so explicitly at the very beginning of the interrupt service routine.

(iv) The interrupt service routine provides the requested service to the requesting device. Once this service has been provided, the device should remove its interrupt request (if it has not done so before).

(v) Before terminating, the interrupt service routine should restore the program status to its pre-interrupt conditions. Unless the microprocessor does it automatically on interrupt service termination, the user should do it explicitly with appropriate instructions. (See (ii) above in this context.)

(vi) If return to the original interrupted routine is desired, the microprocessor should be able to do so with a RETURN type of instruction which branches to the previous Program Counter (PC) contents saved in the stack. If return to the original interrupted routine is not desired, for some reason (e.g. error conditions detected), the microprocessor should be able to pop the stack to lose the stored PC contents and then execute a JUMP type of instruction to the desired address. (See (i) again in this context.)

Note that if all of (i) to (vi) are done in response to an interrupt, the interrupt service routine will be executed in a manner essentially transparent to the other programs being executed by the microcomputer. The only effect of responding to an interrupt will be to extend the time required to execute the program which is being interrupted. As long as that does not have a detrimental effect on the system's operation, handling interrupt requests will not be a problem.

Of the actions—(i) to (vi)—mentioned above, saving and restoring interrupted program status is rarely done in an automatic fashion by the microprocessor. This job is generally left for the system designer who is expected to do it, if needed, using appropriate instructions. For example, in the 8085A, this should be done by saving the status of the interrupted program in the stack at the beginning of the interrupt service routine; at the end of the interrupt service routine this information is retrieved and used to restore the program status. Note that with this approach, it is possible to have interrupt service routines which can themselves be interrupted by other interrupt requests (analogous to nested and recursive subroutines). Other processors, like the Z80 or 8048, provide alternate register banks which may also be used for this purpose. Here, the programmer should ensure that different register banks are used by the interrupted program and by the interrupt service routine (sometimes referred to as 'context change'). By properly choosing the appropriate register banks one can ensure that the interrupted program's status is restored on return from the interrupt service routine. If this technique is used by itself, it is quite restrictive. This is because it then becomes difficult to have interrupt service routines which can themselves be interrupted. However, used in conjunction with the earlier approach using the stack, (as is, for example, normally done in Z80 based systems) one can develop a very powerful and efficient way of handling responses to an interrupt request.

12.3 VECTORED INTERRUPTS

In the previous section, we saw that the microprocessor's response to a particular interrupt was to direct program execution to branch to an address pointed to by the Interrupt Vector. Most microprocessors provide a general purpose interrupt input (e.g. INTR in the 8085A) for which the interrupt vector has to be provided by external circuitry associated with the interrupting device. (For the specific case of the 8085A, this is described in greater detail in Sec. 12.5). Sometimes, other interrupt inputs, referred to as Vectored Interrupts may also be provided where the required interrupt vector is generated internally by the microprocessor. (In the 8085A, these are the RST 7.5, 6.5 and 5.5 interrupts described in greater detail in Sec. 12.5). When a number of such interrupt inputs are available on a microprocessor, the entire set of these interrupts is sometimes collectively referred to as Vectored Interrupts.

Vectored interrupts provide an easy and efficient way of using the interrupt approach without the cost of designing additional hardware to generate interrupt vectors. Their use presents problems if design considerations force more than one device to share the same vectored interrupt line. In this case, the interrupt vector will be the same, whatever be the source of the interrupt. Hence, the interrupt service routine needs to poll the status of each device sharing that interrupt line in order to determine the exact source of the interrupt. It is possible to service the interrupting device only after this determination has been made. Sharing the same vectored interrupt line between several devices, therefore, not only requires more software but also implies that more time is needed to respond to an interrupt request. An INTR (8085A) type of general purpose interrupt may also be shared by more than one device. One can, of course, also use a polling approach here, as for vectored interrupts. However, it is more common to provide sufficient external circuitry so that each device sharing that general purpose interrupt line provides a unique interrupt vector to the microprocessor. This avoids the disadvantages of more software and greater response delays associated with the polled approach; however, the cost of the external hardware is now greater. In the final analysis, which kind of interrupt is actually used, is usually decided based on cost considerations.

12.4 INTERRUPT PRIORITY

Consider a microprocessor like the 8085A which provides multiple interrupt inputs to the user. It is possible that more than one such inputs have simultaneously active interrupt requests (e.g., more than one of these inputs get interrupt requests when the interrupts are disabled).

In order to ensure proper operation, the microprocessor will be forced to choose one of these requests as the one to respond to first. This choice is guided by the Interrupt Priority Structure of the microprocessor. Following this priority structure, the microprocessor responds to an interrupt request at the highest priority interrupt input currently active, whenever it is in a position to respond to an interrupt request at that input. An interrupt request of lower priority is responded to only when higher priority interrupts have been taken care of or are, for some reason, currently disabled. Summarising the above one can say that for microprocessors with more than one interrupt inputs, if at any instant, there exist two or more interrupt requests to which the microprocessor can respond, it will always respond to the one having higher priority in its interrupt priority structure. For the 8085A, the interrupts arranged in order of decreasing priority are TRAP, RST 7.5, RST 6.5, RST 5.5 and INTR. The non-maskable TRAP interrupt has the highest priority whereas the general purpose interrupt input INTR has the lowest priority. It should be noted that the general purpose interrupt input of a typical microprocessor may be expanded using external hardware to provide more than one interrupt request inputs, (see Sec. 12.5). In such situations, the priority structure of these (expanded) interrupts also have to be decided by the external hardware.

From the CPU's point of view, the Interrupt Priority Structure ensures unambiguous operation when it is faced with multiple interrupt requests at the same time. The microcomputer system designer using interrupts can use this priority structure to decide which interrupt line should be used by a particular interrupting device. The rational approach in this case would be to assign higher priority interrupts to devices which have critical timing requirements or are considered more important by the system designer. In applications where several interrupt driven tasks are to be performed, it would be reasonable to have the highest priority interrupt request the highest priority task.

As an example of an interrupt driven system using an interrupt priority structure, consider a computer terminal incorporating a keyboard, a CRT monitor, a high speed serial data link to a host computer and a low speed serial data link to a cassette interface. For purposes of illustration assume that all the tasks (i.e. reading the keyboard, refreshing the CRT display, data transfer to and from the host computer and the cassette interface) are interrupt driven. Since human reaction times are typically slow, it would be logical to service the keyboard with the lowest priority interrupt. (Moreover, even if once in a while a key depression is not read, the operator will probably just need to press the key again. Even though this might irritate the operator, especially if it happens often enough, the operation of the equipment will not be jeopardized). The most time critical job, as well as the most important one from the viewpoint of proper operation, will be servicing the serial link for communicating data and commands with the host. Accordingly, this should probably be given the highest priority interrupt for requesting service. The other two devices will then use interrupts with priorities in between the highest and the lowest.

A question generally left unanswered by the interrupt priority structure of the CPU is that of higher priority interrupts interrupting the service routine of a lower priority interrupt. The interrupt priority structure of the CPU typically does not provide this feature, leaving it to the user to do this, if needed, using additional hardware and software. In typical CPUs, all interrupts become disabled when the processor starts executing a particular interrupt service routine. If interrupts are re-enabled at any point in this routine, any interrupt, whether it has higher or lower priority, can interrupt the ongoing interrupt service routine.

The importance of this feature (of allowing higher priority interrupts to interrupt a lower priority interrupt service routine) can be better appreciated by referring to the earlier example of a microprocessor based computer terminal. Consider a situation where the terminal is transferring a file to the cassette interface when it gets a service request interrupt from the host serial link. It would be proper at this point to interrupt the cassette interface routine, service the host serial link and then resume the cassette interface routine. In the opposite situation, when the terminal is involved in a transaction with the host and a transaction with the cassette interface is requested, it will not be desirable to interrupt the ongoing service routine. In this case, it would be more proper to wait until the transaction with the host is complete before servicing the cassette interface.

The interrupt structures of most CPUs utilize the priority information on interrupts only at that point in time where an interrupt service routine is initiated. As per our earlier discussion, the interrupt priorities do not hold during the interrupt service routine itself when typically all interrupts are disabled. The designer generally has to make an extra effort (more hardware and software), if his application demands that particular interrupt service routines should allow themselves to be interrupted by higher priority interrupts. Priority Interrupt Controller Units or PICUs (e.g. INTEL 8214 or 8259) do allow this feature with proper programming (see Sec. 12.6). In the 8085A, one can provide this feature via software, if only the vectored interrupts—TRAP, RST 7.5, RST 6.5 and RST 5.5—are being used (see Prob. 12.6). (Note that, in this case, INTR should not be used in the system, i.e. it should be grounded). It should be noted that in the 8085A, this feature is provided by using the SIM instruction of the 8085A to properly mask (and hence disable) lower priority interrupts while an interrupt service routine is being executed. Such a feature can also be provided for interrupts sharing the INTR input by using additional software. In this case, the software should store the 'current interrupt service status' in memory and should also implement (also in memory) a 'current interrupt service queue'. A particular interrupt service routine (say for interrupt n) should do the following (apart from appropriately providing the requested service).

(i) Check current interrupt service status at the beginning to see if interrupt n is interrupting the interrupt service routine of a higher priority interrupt. If this is the case, interrupt n should place itself in the current interrupt service queue and return to the interrupt service routine. Otherwise, the interrupt service routine for interrupt n should be executed.

(ii) After concluding the service routine for interrupt n, the CPU should examine the interrupt service queue to see if a lower priority interrupt request is waiting in this queue. This interrupt should now be serviced. If its previous execution was interrupted, the software should ensure that reexecution of this routine is appropriately resumed.

[The reader should be aware of the fact that the complexity of the above software may be such that using a PICU or other external hardware for the same purpose may be more efficient].

12.5 INTERRUPT STRUCTURE OF THE 8085A

In the earlier sections, we had highlighted some features of the interrupts available in the 8085A CPU. This section presents a detailed description of the interrupt structure of this microprocessor.

The 8085A provides five interrupt inputs to the user; these, in order of decreasing priority, are TRAP, RST 7.5, RST 6.5, RST 5.5 and INTR. The TRAP interrupt is also referred to as a Non-maskable Vectored Interrupt—it can never be disabled. The other interrupts can be disabled or enabled as a group by executing the DI and EI instructions respectively. These interrupts are also disabled following RESET and when the CPU acknowledges an interrupt request. (It is therefore important to put EIs in appropriate places in the software so that after an interrupt service routine is executed, the interrupts are re-enabled. The user may also use EIs and DIs to write specific sections of code which he does not want interrupted by these interrupts). Over and above this global enable/disable, the 8085 also allows the user to mask (disable) or unmask (enable subject to whether an EI has been appropriately executed) specific vectored interrupts—i.e. the RST 7.5, RST 6.5 and RST 5.5 interrupts. This is done by executing the SIM (Set Interrupt Masks) instruction with appropriate contents in the accumulator (register A). The current masking/unmasking status of these interrupts can be read by using the RIM (Read Interrupt Masks) instruction. (See Chapter 8 for details). All the vectored interrupts are masked following RESET (except TRAP).

The 8085A has an INTE (Interrupt Enable) flip-flop internal to the CPU. The output of this flip-flop indicates the current (global) enable/disable status of the interrupts. If interrupts are disabled (e.g. following a DI), the INTE flip-flop is reset; after an EI instruction is executed to enable interrupts, the INTE flip-flop is set. The current status of this INTE flip-flop can also be read using the RIM instruction.

The RST 6.5, RST 5.5 and the INTR inputs are level sensitive. An interrupt request for these is placed by providing a HI (1) input at the corresponding pin of the CPU; moreover the HI input has to be maintained until the CPU acknowledges the interrupt request. The RST 7.5 interrupt, on the contrary, is edge-sensitive. A Lo → Hi transition on the corresponding input places a RST 7.5 interrupt request to the 8085A CPU. Internally, the 8085A has a RST 7.5 flip-flop which is set by a Lo → Hi transition at the RST 7.5 input. The RST 7.5 interrupt request to the CPU stays active as long as the RST 7.5 flip-flop is set. This flip-flop is reset when the CPU responds to the RST 7.5 interrupt. It may also be reset through software by executing the SIM instruction appropriately. The TRAP interrupt is both level and edge sensitive. (See Sec. 12.1 Fig. 12.1.1.)

For the TRAP, RST 7.5, RST 6.5 and RST 5.5 interrupts, the interrupt vectors are generated internally by the CPU. When the CPU acknowledges one of these interrupts, it executes a CALL ADDR instruction internally, where the branch address ADDR is as follows:

TRAP	ØØ24H
RST 5.5	ØØ2CH
RST 6.5	ØØ34H
RST 7.5	ØØ3CH

It should be noted that for these interrupts, the corresponding CALL ADDR instruction is generated completely internal to the CPU; nothing has to be provided by the external hardware. As stated earlier, the CPU checks in the last but one T-state of every instruction (except EI and DI), whether there is an active interrupt request on these interrupts. The highest priority interrupt which is currently enabled (note TRAP will always be enabled) and has an active interrupt request will be the one to which the CPU will respond. In order to respond to such an interrupt request, the CPU executes the appropriate CALL ADDR instruction internally during a six-cycle Bus Idle (BI) Machine Cycle (MC) as shown in Fig. 12.5.1. This is followed by two normal

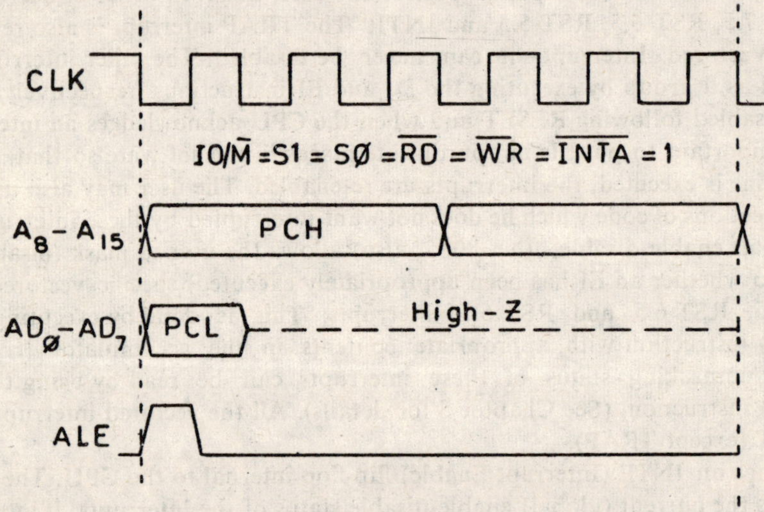

Fig. 12.5.1 RST/TRAP Bus Idle Machine Cycle

Memory Write (MW) machine cycles which are used by the CPU to push the contents of the Program Counter (PC) onto the stack. These last two machine cycles are exactly the same as those used by a normal CALL type of instruction to save the Return Address on the stack. After the 6-cycle BI-MC, the operation of the CPU is exactly the same as that for a normal CALL type of instruction. Just as for such instructions, the CPU resumes instruction execution from the location ADDR which should contain the first instruction of the corresponding interrupt service routine. As in a normal subroutine, terminating the interrupt service routine with a RETurn type of instruction will allow the CPU to resume the execution of the interrupted routine from the point of interruption. It should also be noted that the 8085A disables all interrupts (except TRAP) once it responds to a particular interrupt request. Therefore one should execute an EI instruction as part of the interrupt service routine. Normally, though not always, the EI instruction is placed just before the RETURN instruction. When the branching to the interrupt service routine takes place, the 8085A does not, by itself, save the registers and the FLAGs, so that their contents can be restored on RETURN. If such saving and restoration is required, the user must do this with his own software (see Sec. 12.2).

[There is one major advantage in not sampling the status of the interrupt requests during the EI instruction. Show that if all the interrupt service routines end with EI followed by a RETURN type of instruction and if this feature were not there, then there is a possibility that the stack will begin to grow without limit. Similarly, interrupts are not monitored during the execution of DI as it would be improper any way to respond to an interrupt immediately after a DI is executed.]

The INTR input is the one general purpose interrupt input available on the 8085A CPU. Unlike the vectored interrupts, here the interrupt vector is not generated internally but has to be provided by external hardware. Therefore, this interrupt input is more difficult (and costlier) to use than the vectored interrupts. However, by providing the facility of generating different interrupt vectors for different interrupting devices, a number of devices can share this interrupt input. In other words, this interrupt input is somewhat more difficult to use than the vectored interrupts but allows the very desirable feature of expandibility.

Like the vectored interrupts, the 8085A CPU tests the INTR input for an active interrupt request in the last but one clock state of each instruction (except EI and DI). If interrupts are enabled (INTE = 1) and an active INTR request is found, the CPU generates an INA (Interrupt Acknowledge) cycle as the next machine cycle. As described in Sec. 9.2.6, this INA machine cycle is similar to an Opcode Fetch cycle, except that instead of pulsing $\overline{RD}$ Hi→Lo→Hi, the CPU generates a Hi→Lo→Hi pulse at its $\overline{INTA}$ output. On getting this pulse, the interrupting device, or the hardware associated with it, places on the data bus the opcode of the instruction it wants executed in response to this interrupt. If the opcode happens to be one for a multibyte instruction, the CPU will follow the previous INA-MC with one or two more INA-MCs (for 2-byte and 3-byte instructions respectively). The interrupting device, or the external hardware associated with it, should place the required byte (or bytes) on the data bus during these machine cycles. These latter INA-MCs are similar to MR-MCs except that the CPU pulses its $\overline{INTA}$ output rather than its $\overline{RD}$ output. Once the INA-MCs are over, the CPU executes the instruction provided to it by the interrupting device (or the associated hardware). Such execution is done in exactly the same fashion as if the instruction has been fetched from program memory.

Apart from the fact that EI and DI should not be used, the 8085A does not place any other restrictions on what instruction should be given to the CPU during the INA machine cycle(s). Normally, however, the instruction provided is a RST_n or CALL ADDR instruction. (Recall that RST_n is really a one byte subroutine call instruction. The CALL ADDR instruction, on the other hand, requires three bytes.) The first instruction of the corresponding interrupt service routine is, in that case, located at the branch address 8n for RSTn (n = 0, . . . , 7) and at ADDR for the CALL ADDR. In accordance with the terminology introduced earlier, these addresses are the interrupt vectors provided by hardware external to the CPU. It should also be noted that only eight different interrupt vectors are possible if the RSTn instructions are used for vectoring; if CALL ADDR type of vectoring is provided, there is no limit on the number of interrupt vectors possible.

For most applications, eight interrupt lines on the INTR input are sufficient. In such situations, interrupt vectoring is usually done using the one-byte RSTn instruction rather than the 3-byte CALL ADDR type of instruction. (The external hardware is usually simpler in the latter case.) A circuit to do this is shown in Fig. 12.5.2. Here the interrupting device raises INT = INTR (8085A) Hi when it needs service. The 8085A pulses $\overline{\text{INTA}}$ (Hi→Lo→Hi) when it acknowledges the interrupt request. When $\overline{\text{INTA}}$ is Lo, the 8212 puts the opcode (FF_H) for the RST7 instruction on the data bus. The 8085A interprets this as the next instruction to be executed and hence performs a subroutine call to location $3ØØ8_H$; this location should have the first instruction of the service routine for serving this interrupt. The interrupting device should return its INT output to Lo once it has obtained the appropriate service from the CPU; the service routine should reenable its interrupts (execute EI) only after this happens. The circuit shown in Fig. 12.5.2 can

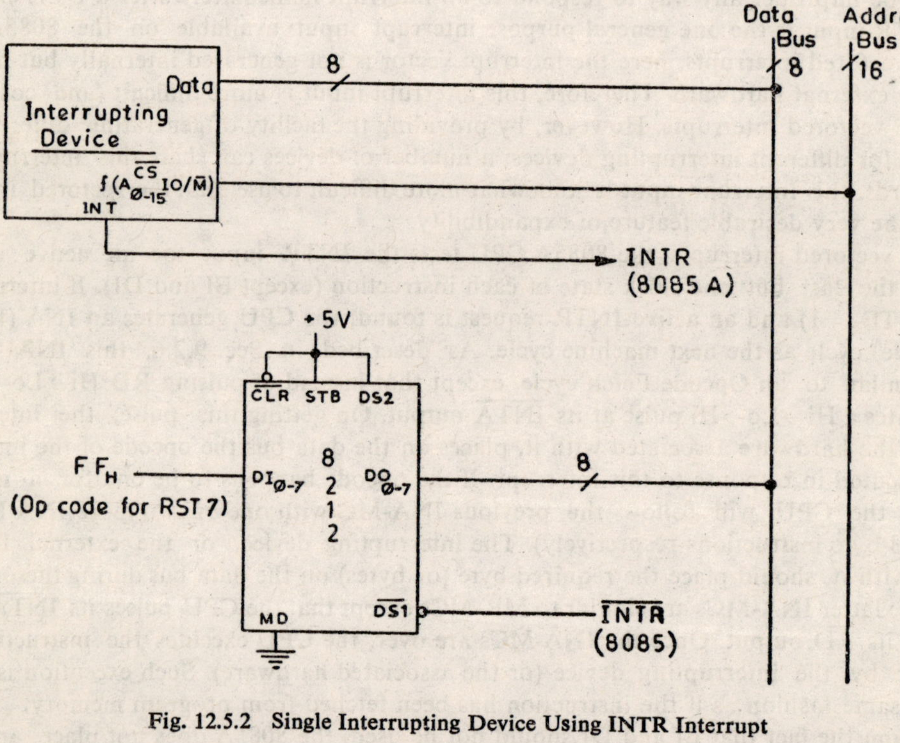

Fig. 12.5.2 Single Interrupting Device Using INTR Interrupt

be used in situations where only one interrupt vector is needed for an interrupt on the 8085A's INTR input.

Consider a situation where eight interrupt lines (INTØ–INT7) share the INTR interrupt input and where the application demands that separate interrupt vectors (RSTØ–RST7) be generated for each interrupt. In this situation, it is quite possible that more than one interrupt requests are simultaneously present. In that case, the interrupt acknowledgement circuit should ensure that amongst these, the interrupt with the highest priority should be handled first. In our example, suppose Device 7 has the highest priority and Device Ø the lowest. This implies that if all device interrupts were simultaneously present, Device 7 would be served first. An Interrupt Acknowledgement Circuit incorporating these features and serving 8 interrupting devices is given in

Fig. 12.5.3. This circuit uses a *74148 priority encoder* (*8 line to 3 line*) whose relevant partial truth table is given below.

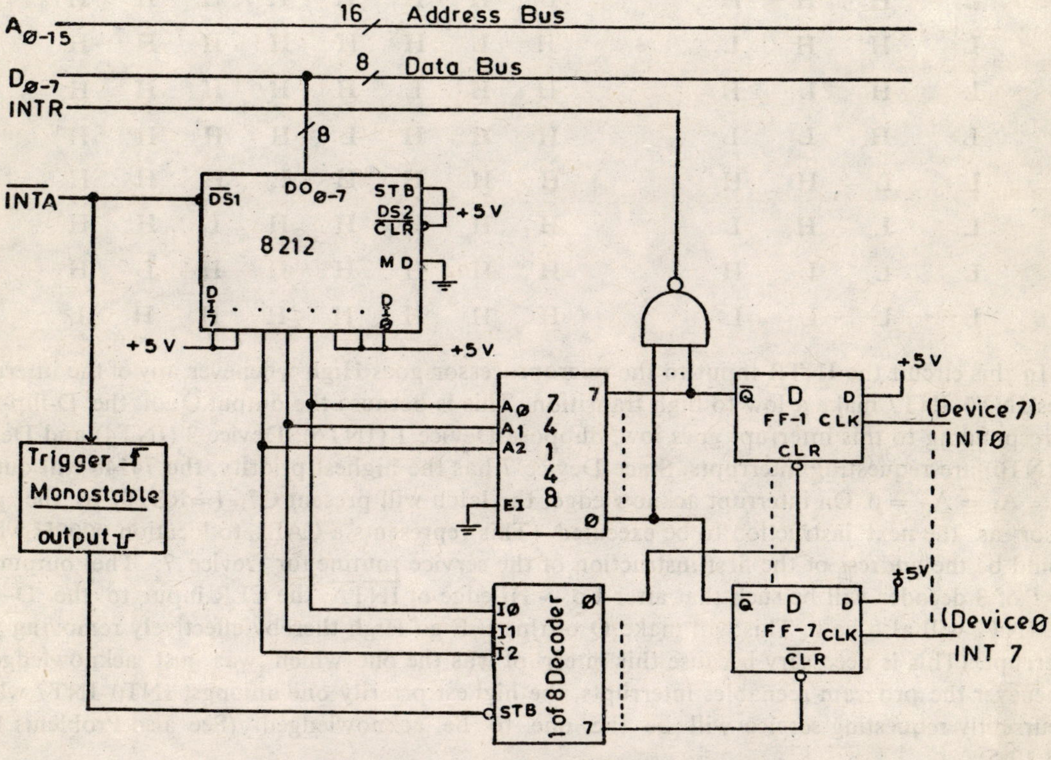

Fig. 12.5.3 Multiple Interrupting Devices with Separate Interrupt Vectors

Inputs (with EI = Ø)								*Output*		
0	1	2	3	4	5	6	7	A_2	A_2	$A_Ø$
H	H	H	H	H	H	H	H	H	H	H
×	×	×	×	×	×	×	L	L	L	L
×	×	×	×	×	×	L	H	L	L	H
×	×	×	×	×	L	H	H	L	H	L
×	×	×	×	L	H	H	H	L	H	H
×	×	×	L	H	H	H	H	H	L	L
×	×	L	H	H	H	H	H	H	L	H
×	L	H	H	H	H	H	H	H	H	L
L	H	H	H	H	H	H	H	H	H	H

The truth table of the 1 of 8 decoder needed here is shown below.

$\overline{STB}$	I_2	I_1	$I_\emptyset$	7	6	5	4	3	2	1	∅
H	×	×	×	H	H	H	H	H	H	H	H
L	H	H	H	L	H	H	H	H	H	H	H
L	H	H	L	H	L	H	H	H	H	H	H
L	H	L	H	H	H	L	H	H	H	H	H
L	H	L	L	H	H	H	L	H	H	H	H
L	L	H	H	H	H	H	H	L	H	H	H
L	L	H	L	H	H	H	H	H	L	H	H
L	L	L	H	H	H	H	H	H	H	L	H
L	L	L	L	H	H	H	H	H	H	H	L

In this circuit the INTR input to the microprocessor goes High whenever any of the interrupt lines INT∅–INT7 make a low to high transition. This is because the output $\overline{Q}$ of the D-flip-flop corresponding to this interrupt goes low. Suppose Device 1 (INT6), Device 3 (INT4) and Device 7 (INT∅) are requesting interrupts. Since Device 7 has the highest priority, the 74148 will output $A_2 = A_1 = A_\emptyset = \emptyset$. On interrupt acknowledge, the latch will present $C7_H$ (=RST∅) to the processor as the next instruction to be executed. (This represents a CALL to location ∅∅∅∅H which should be the address of the first instruction of the service routine for Device 7. The output of the 1 of 8 decoder will be such that after Lo → Hi edge of $\overline{INTA}$, the CLR input to the D–F/F for INT7 will also be ∅. This will make $\overline{Q}$ of this F/F go High thereby effectively removing this interrupt. (This is necessary because this interrupt was the one which was just acknowledged!) Whenever the program reenables interrupts, the highest priority one amongst INT∅–INT7 which is currently requesting service will be the one to be acknowledged. (See also Problems 12.3 and 12.5).

12.6 LSI INTERRUPT CONTROLLERS

In the previous section, we presented some circuits using SSI and MSI chips for providing interrupt control hardware for the 8085A's general purpose interrupt input INTR. Today microprocessor manufacturers may provide LSI chips which essentially serve the same function for the corresponding microprocessor(s). Some of the features typically provided by Programmable Interrupt Controller Unit LSI are given below:

(a) Expanding the INTR-type Interrupt Input: Using these, the general purpose interrupt input of the microprocessor can be shared by a number of interrupt lines coming from different interrupting devices.

(b) Providing Interrupt Vectors: These devices, properly programmed, provide an easy way of generating appropriate interrupt vectors for each interrupt line sharing the general purpose interrupt.

(c) Prioritizing the Shared Interrupt Lines: As mentioned in (a) above, these devices may be used to allow several interrupt lines to share the general purpose INTR-type interrupt. These may then be used to impose a priority structure on these interrupt lines. Depending on the particular device, it may also be possible to implement different priority schemes like fixed priority, rotating priority, dynamically assigned priority, etc.

(d) Interrupt Masking: These may also provide means of masking (i.e. selectively disabling) one or more of the interrupt lines sharing the interrupt input of the microprocessor. This is a very powerful feature which allows the user to disable selected interrupt lines under software control without disabling others.

(e) Interrupting an interrupt service routine with a higher priority interrupt: As indicated in Sec. 12.4, this is a very useful feature which is required in some applications. These devices may allow the feature of disabling interrupts below a certain priority level where the priority level can be specified by the CPU. The CPU does this by issuing appropriate commands to it (i.e. programming it) and can do this dynamically under software control. Exploiting this feature with appropriate software, the user can write interrupt service routines which will allow themselves to be interrupted if a higher priority interrupt is requested while another such routine is under execution.

INTEL provides a powerful PICU, the 8259, which is intended for use with 8085 or 8086/88 based systems. The reader is referred to the manufacturer's data sheets and application notes for a description of this complex, yet extremely powerful product.

12.7 PROS AND CONS OF INTERRUPT DRIVEN SYSTEMS

A microcomputer may be viewed as a combination of software and hardware, whose objective is to execute one or more tasks as required by the application and the operating environment. These tasks may be interdependent and a task may also be required to run concurrently with other tasks. For tasks initiated by external events, configuring these tasks to be interrupt driven offers several important advantages. It also creates some problems in system design and implementation.

In most applications, the greatest advantage in having an interrupt driven system comes from the fact that the service routines for each interrupt can then be written and tested more or less independent from each other. It is then possible to write the software required in terms of independent (or at least semi-independent) routines and then combine them appropriately at a later date. This, very often, provides an immense advantage in the design and development of microcomputer systems. This advantage is offset by a number of problems some of which are described next. The system designer should take these into account while designing interrupt driven systems.

(a) It is almost impossible to completely debug a system which uses interrupts. This is primarily because interrupts being asynchronous, it is never quite possible to find out the worst case combination of events and check the system's performance for that situation. This statement is actually a little more pessimistic than what happens in reality. In real life it is possible to design interrupt driven systems, which take into account most of the event combinations that might occur. Such a carefully designed system will generally work pretty well in practice; however, the designer should be aware of the fact that a peculiar combination of events may exist which will make the system malfunction.

(b) Consider a situation where a particular subroutine, say SUB, is used by a number of interrupt service routines at different priority levels or by one or more such service routines and the main program. In that case, it is possible that a service routine (or main program) which uses SUB gets initiated while SUB is being executed by another service routine. This situation is referred to as reentering a subroutine; a subroutine like SUB which allows reentry is also referred to as a Reentrant Subroutine. Unless proper care is taken, the effect of reentering a subroutine can be the loss of data, loss of intermediate results, loss of the proper sequence of execution, erroneous I/O operations and other effects leading to failure of the system. (As an example, consider what would happen if SUB uses a certain fixed RAM area as a scratch pad area.)

In interrupt driven systems, problems of reentry can be avoided in many ways. For example by appropriately duplicating a subroutine like SUB, i.e. one for each routine using SUB, one can completely avoid reentry. Reentry can also be avoided if all interrupts are disabled when a reentrant (potentially!) subroutine like SUB is being executed. One can also write a proper reentrant subroutine which can be reentered and reexecuted without any of the problems mentioned earlier. This, however, makes the task of writing these routines more difficult and also tends to make them longer and hence less efficient in terms of memory space and execution times.

(c) Consider an interrupt driven system which has interrupts at different priority levels. Unless due care is taken, the low priority interrupts in such a system may find it very difficult to get the desired service. This may lead to system failure; after all, the fact that a task has low priority does not imply that it is redundant.

(d) If software delays are being used in an interrupt driven system, one must take care that the delay loops (generating the delays) are not interrupted. It is easy to see that if such interruptions do indeed occur, the delays generated will become unpredictable and hence unreliable.

(e) In interrupt driven systems, disabling interrupts should be resorted to in the software only when that is absolutely necessary. This is because the time intervals when interrupts are disabled are essentially time periods during which the system cannot respond to requests for a particular task; such periods should be as few and as small as possible. In any case, these time intervals should be taken into account while calculating worst-case response times.

(f) Most LSI peripherals need to be properly initialized by the microcomputer before they can be used. The designer should ensure that interrupts do not have a detrimental effect on such initialization sequences.

(g) Special care should be taken when using non-maskable interrupts (NMIs) in the system. Typically every microcomputer system must do a certain amount of initialization before it can respond to an interrupt. A NMI can potentially wreak havoc if it occurs before this basic initialization is complete. The system designer should be aware of this problem and take it into account in his hardware and software design.

The reader should note that the intent of the above was not to discourage him/her from using interrupts or from designing interrupt driven systems. It is true that at the practical level, such systems are easy to design and, constructed with proper care, can be made to function satisfactorily. The purpose of the above discussion was to warn the reader of the potential problems of such a design so that appropriate precautions can be taken at the design and development stage itself.

PROBLEMS

1. Two contiguous bytes of memory store a 16-bit count of the number of times an external event occurs. A routine starting at location ØØ38H increments this 16-bit counter each time it is called. Write this interrupt service routine. Assume that the external event generates a hardware signal and draw a diagram of the external hardware required to interrupt an 8085A and vector to the above service routine. Make appropriate assumptions. Use the INTR input and an 8212 to provide the interrupt vector. Repeat using SID.

2. A device generates a train of equally spaced pulses which are fed to the RST 7.5 interrupt of the 8085A. Write a subroutine SPEED which when called by the main program does the following:

   —Counts the number of times a software loop is executed between two successive pulses. The 16-bit count should be left in the HL register pair.

   —Having obtained the count between two pulses, it should return to the main program with the count in H-L.

   Write a software loop which will serve the above counting objective.

3. Design a hardware scheme to put the instruction CALL 2ØØØH on the data bus when an INTR interrupt is responded to by the 8085A.

4. Write an interrupt service routine which does the following: It reads a byte in sign magnitude from PORT 4ØH, converts it to 2's complement and stores it in the memory location whose address is given in ADDRH and ADDRL = ADDRH + 1. This address is to be incremented each time and a new I/O operation initiated by pulsing Low the normally High SOD output for at least 10 μs. Assume $f_{crystal}$ = 2 MHz. What will be the total time required to execute this interrupt service routine?

5. Three devices A, B and C are to be connected to an 8085A. Each has a RQ output which goes Hi when an interrupt is required to service the device. When the interrupt is acknowledged, the appropriate service routines are to be found starting respectively from locations ØØØ8H, ØØ18H and ØØ28H. Device A should have the highest priority and Device C the lowest.

   (a) Suggest a way to implement the above. You should clearly specify whatever software or hardware you need. You may assume that each of the interrupt service routines executes an EI before returning. You may use 8212s and other gates and flip-flops in your design but not a Priority Encoder IC.

   (b) Compare (a) with the situation where RST 5.5, RST 6.5 and RST 7.5 are available for the above.

   (c) It is possible to do (a) with either one 8212 and some software or three 8212s and less software. If you choose one such technique in (a), try doing the same thing with the other technique.

6. A certain 8085A-based microcomputer system only uses its RST 7.5, 6.5 and 5.5 interrupts. The system designer wants to provide a prioritized interrupt structure such that a higher priority interrupt can interrupt the interrupt service routine of a comparatively lower priority interrupt and, if two or more interrupts are awaiting service, the higher priority one will get serviced first.

   Describe, in detail, how such an interrupt priority structure can be implemented for this case. How can one implement a rotating priority structure using these vectored interrupts?

7.
```
LOOP : DCX B      ; execute
       MOV A, B   ; loop
       ORA C      ; (BC)
       JNZ LOOP   ; times
```

   Write a program using LOOP to measure the time interval between the RST 7.5 and 6.5 interrupts initializing BC appropriately.

Chapter 13

Single-Chip Microcomputers

Examination of the components inside a microcomputer reveals that any such system will usually have one or more of the following devices in some form—i.e. ROM, RAM, I/O ports, Clock Generator and may be a Timer/Counter. Though these are functionally distinct, it has been the trend amongst semiconductor manufacturers to integrate them on the same chip. For example, Intel Corp. provides the 8155 with RAM, I/O ports and a timer or the 8755 with EPROM and I/O. Similarly, the trend among microprocessor manufacturers is to integrate the Clock Generator with the CPU on the same chip, requiring only an external crystal for proper clock generation. It is easy to see that in the extreme limit, one can integrate all the components of a microcomputer in one IC package; what we get as a result of such integration is a 'Single-Chip Microcomputer'.

Apart from the CPU, such a single-chip microcomputer will therefore contain its own clock generator and some amount of ROM (or EPROM), RAM and I/O ports on the same chip. (It may also have other features like Timer/Counter, USART/UART, Analog I/O channels etc. on that chip. These, however, are not basic requirements like ROM, RAM or I/O ports.) The amount of ROM, RAM or I/O available on a given single-chip microcomputer is essentially a decision made by the IC designer and the semiconductor manufacturer. The user is constrained to stay within these limits in designing with a particular single-chip microcomputer. In case the design requires more ROM, RAM or I/O, there are usually only two choices open. One may, of course, choose a different single-chip microcomputer—one which better satisfies the requirements. Alternatively, one may expand the internal resources of the single-chip microcomputer by adding additional resources (ROM, RAM or I/O) outside the chip; this, however, is only possible if the particular single-chip microcomputer does allow such expansion. This latter solution is not generally preferred because it increases system complexity as well as goes against the very philosophy which leads one from multi-chip microcomputers to single-chip microcomputers.

Even though the design is limited by the restrictions imposed by the manufacturer, a designer sometimes prefers to use a single-chip microcomputer for a number of reasons. A single-chip microcomputer-based design leads to a lower chip count in the system than what would be possible with a general purpose microprocessor like the 8085A. This is especially true if the designer does not expand ROM, RAM or I/O beyond what is already available on the chip. Single-chip microcomputers are also very attractive in special purpose applications where once the system has been configured, it is unlikely that further augmentation of that will be needed. Some special dedicated interfaces may also be built using single-chip microcomputers. Examples of such applications are interfaces for a digital cassette storage system, floppy disk controllers, controllers for a GPIB interface, etc. In these and other similar applications, one can think of the single-chip device as being a slave processor to a larger microcomputer system. The latter system, working as the master, relegates tasks (like special interfacing) to the slave processor whose program is specialized to handle such tasks. Single-chip microcomputers are extremely popular as specialized slave processors. So much so, that one finds some special single-chip microcomputers whose internal architectures have been modified to facilitate such usage.

A number of single-chip microcomputers are available in the market today. Of the various manufacturers supplying such devices, Intel Corp. probably offers one of the more comprehensive product lines ranging from a device with 1K×8 ROM, 64×8 RAM and 13 I/O lines to a device with 4K×8 ROM, 256×8 RAM and 27 I/O lines. All these devices offer a 8-bit Timer/Counter, a RESET feature, Interrupts, and on-chip Clock Generator. In addition, some of them provide Serial I/O ports and ADCs on the same chip. Typical single-chip microcomputers available from Intel Corp. are 8050H, 8749H, 8049H, 8748, 8048H, 8022, 8021H, 8020H, 8041A, 8741A, 8042 and 8742. (A chip with number XOYZ has an internal ROM programmed by the manufacturer as per the user's specifications. A chip with number X7YZ is identical to the XOYZ chip except that it has a user programmable/erasable EPROM instead of a ROM). In this chapter, we will describe in detail the 8048/8748 single-chip microcomputer as its features fall somewhat midway between the range offered. In addition, we will give a brief description of the 8041/8741 device which is really a 8048/8748 modified to be used as a slave processor. It should also be noted, that the architecture, assembly language and other features of these chips are very similar to each other. Knowing one of these chips well, the reader can easily pick up the modifications needed to handle another chip of this family.

13.1 THE ARCHITECTURE OF THE INTEL 8048/8748

The Intel 8048/8748 single-chip microcomputer is a device widely available and used in many applications. In terms of its features and capabilities, it falls somewhat midway in the range of single-chip microcomputers available in the market today. It is an 8-bit microcomputer chip with 1K bytes of ROM (8048) or EPROM (8748), 64 bytes of (data) RAM, an 8-bit Timer/Counter, an interrupt input, a reset input, 24 I/O lines and two test inputs. Using additional hardware one can expand the microcomputer to a maximum configuration of 4K bytes of ROM (or EPROM), 256 bytes of (data) RAM and 36 I/O lines. In Fig. 13.1.1, we have given a simplified block diagram showing the important details of the internal architecture of this device. A description of this, the pin-out and the assembly language of the 8048/8748 device is given in this and subsequent sections. The reader should note that our discussion of the 8048/8748 device will be limited to its use without expanding ROM, RAM or I/O. We will also not describe the programming aspects of the 8748, assuming that an appropriate EPROM programmer is available to do the necessary programming.

13.1.1 The CPU

The Intel 8048/8748 single-chip microcomputer features a 8-bit CPU on the chip. This contains a 8-bit ALU, a 8-bit Accumulator (A or Acc), a 8-bit Program Status Word (PSW) register, a 12-bit Program Counter (PC) and other (user invisible) storage elements to fetch and execute instructions, to fetch and store variable data or to fetch constants. The CPU can address up to 4K bytes of ROM (EPROM) and up to 256 bytes of RAM. An interesting feature of this CPU is the fact that the ROM address space and RAM address space are different and non-overlapping (i.e. orthogonal). The ROM is used to get program instructions and constants. The RAM is used to store and read variable quantities, to implement register banks for the CPU and to implement the CPU's stack.

The ALU has a fairly standard instruction set which allows arithmetic and logical manipulations of the accumulator with another 8-bit quantity. It has instructions to do additions, AND,

OR, and EX-OR operations but does not have instructions to do subtractions. One can also do some amount of BCD arithmetic by using a DAA instruction which decimal adjusts the accumulator just like the DAA instruction for the 8085A. It also has some instructions which operate

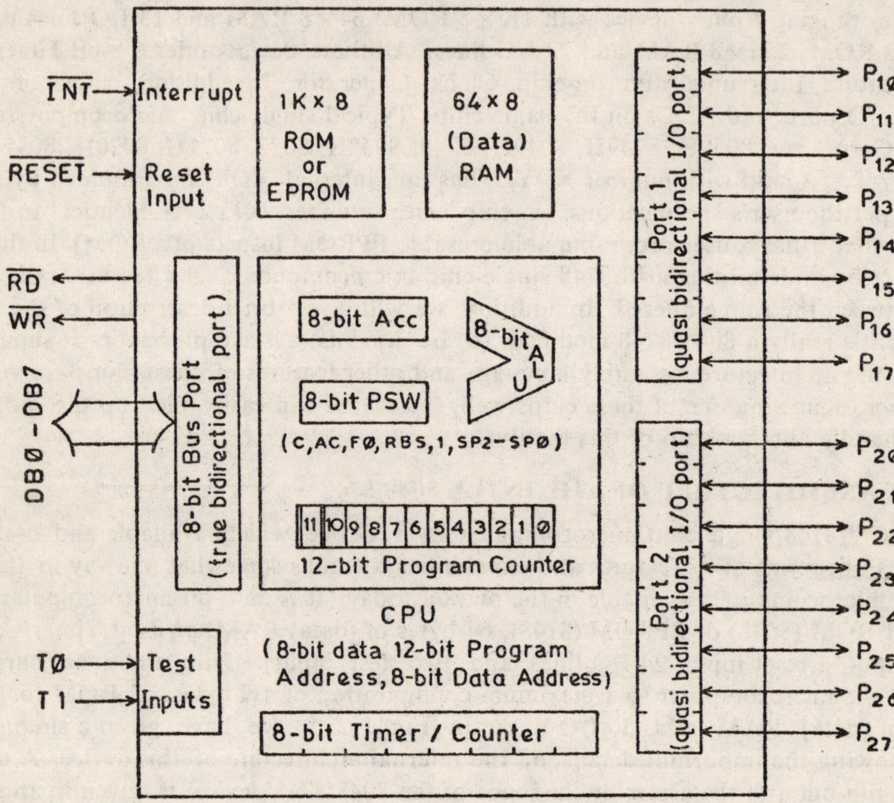

Fig. 13.1.1 A Simplified Block Diagram of the Intel 8048/8748 Single-Chip Microcomputer

on nibbles (i.e. 4-bit data), e.g. SWAP A exchanges the upper and lower nibbles of the accumulator. These make it easier to do some nibble-based operations.

As shown in Fig. 13.1.1 and Fig. 13.1.2, the lower three bits of the PSW register implement a modulo-8 stack pointer (SP). The stack itself occupies sixteen bytes (address $Ø8_H$–17_H) of the internal RAM with two bytes of the stack associated each value of SP. Pushing data into the stack increments SP by 1 whereas popping data decrements it by 1. The 8048/8748 does not have any PUSH or POP instruction but uses the stack only for subroutine linkage. A subroutine CALL saves the 12-bit PC and bits 4–7 of PSW (described subsequently) on stack; on RETURN, PC is restored from the stack but the programmer may exercise his/her choice regarding restoration of bits 4–7 of PSW. (There are two RETURN instructions, RET and RETR. The former does not restore bits 4–7 of PSW which the latter does).

The upper 4-bits of PSW store flag information. (Bit 3 of PSW has not been used and is always '1'.) Bit 7 and bit 6 store the carry (C) and the Auxiliary Carry (AC) flags. The carry (C) flag is set when an arithmetic operation causes an overflow out of bit 7 of the accumulator.

Similarly, the AC flag is set if there is an overflow from bit 3 to bit 4 of the accumulator. Bit 5 of the PSW register stores the value of a software flag FØ which can be set or reset via instructions. Other (conditional) instructions also exist which test the status of this flag. (The 8048/8748 also has another software flag F1 which is similar to FØ except that it does not form a part of PSW.) Bit 4 of PSW contains a Register Bank Select (RBS) bit which indicates the current Register Bank that is being used by the CPU. This bit is needed because the 8048/8748 contains two register banks, Register Banks Ø and 1, of eight 8-bit registers each (all of them implemented in the 64×8 internal RAM). RBS = Ø indicates that register bank Ø is currently in use and vice-versa for RBS = 1. Note that these four bits (bits 4–7) of PSW are the only flags which get stored in the stack and can be restored later. A schematic diagram of the PSW register is shown in Fig. 13.1.2.

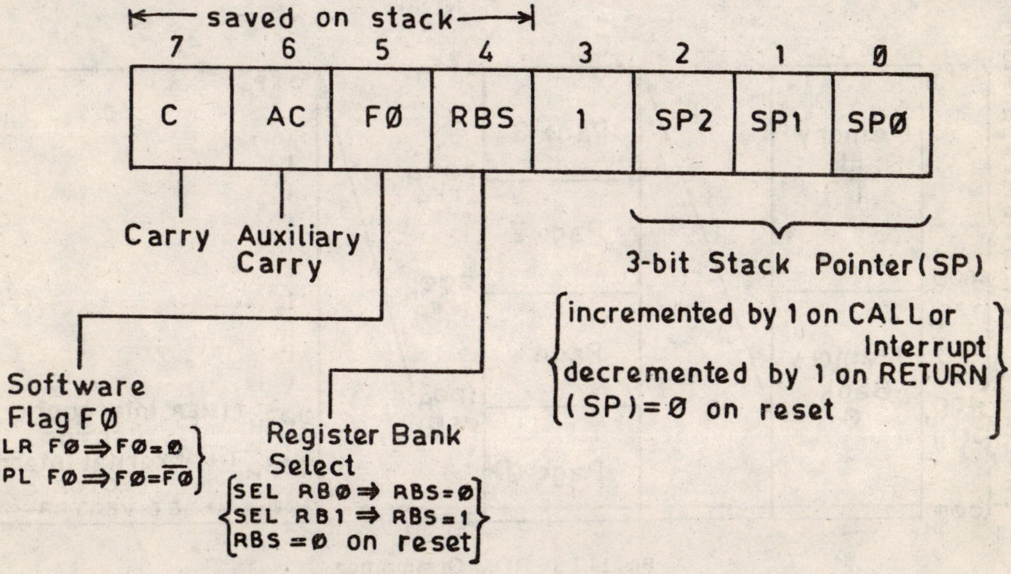

Fig. 13.1.2 Bits of the Program Status Word (PSW) Register

Comparing the 8048/8748 with the 8085, we find that the 8048/8748 has no provisions for storing the Zero, Parity or Sign flags. As far as the Sign flag is concerned, it is not relevant for the 8048/8748 because it does not have any instruction for subtraction. The 8048/8748 checks for a Zero (Z) or Non-Zero (NZ) condition by directly testing the current contents of the accumulator. This makes the Zero flag redundant. No provision is made for the parity flag since it is not often used in writing assembly language programs anyway.

Referring to Fig. 13.1.1, we find that the 8048/8748 has a 12-bit Program Counter (PC) allowing it a program memory address range of ØØØH–FFFH (4K). Of this, the lowest 1K block (ØØØ–3FFH) is available on-chip; the rest, if needed, must be implemented externally. The Program Counter of the 8048/8748 itself, shows a somewhat anomalous behaviour. PCØ–PC1Ø behave normally, but PC11 is not affected by incrementing or decrementing PC. As a matter of fact, the only way PC11 can be changed is through an internal DBF flip-flop. This DBF is Ø on reset and can be made Ø or 1 by executing, the instructions SEL MBØ and SEL MB1, respectively. When a CALL or JMP instruction is executed PC11 is loaded with DBF; this is virtually the only way PC11 can be changed from Ø to 1 or vice-versa. The effect of this behaviour on branch instructions will be discussed in Sec. 13.1.2.

13.1.2 The Organization of the ROM (EPROM)

The ROM/EPROM (internal or external) of the 8048/8748 is used to store the program code and data constants. With a 12-bit PC, the total address range for this is 4K from ØØØH to FFFH. The organization of this ROM is shown in Fig. 13.1.3. As shown in the figure, the 4K of addressable ROM is divided into two memory banks Ø and 1 of 2K bytes each.

ØØØH–7FFH Memory Bank Ø (Accessible with $PC_{11} = Ø$)

8ØØH–FFFH Memory Bank 1 (Accessible with $PC_{11} = 1$)

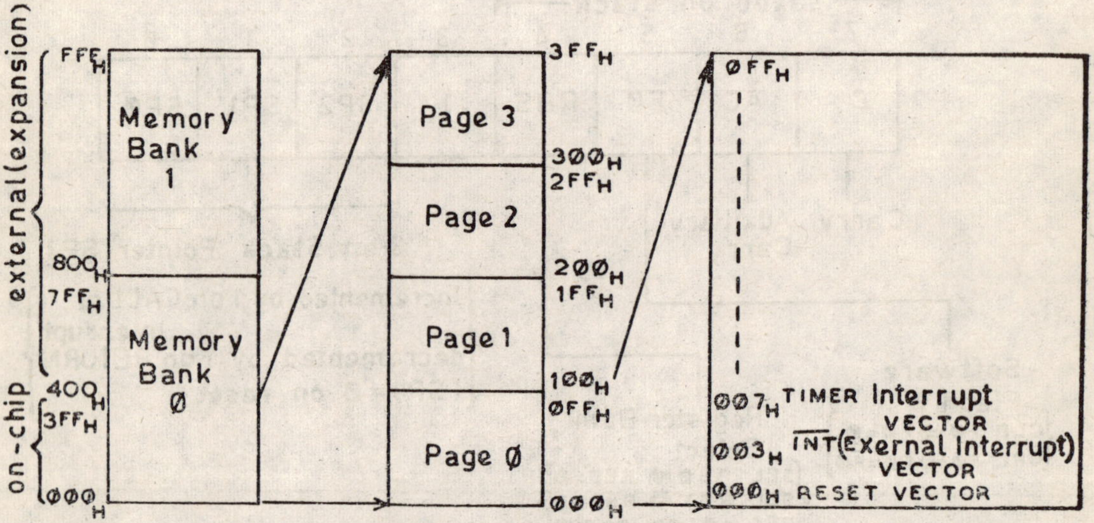

Fig. 13.1.3 ROM Organization

The lower 1K bytes of Memory Bank Ø are provided by the on-chip ROM. More ROM, if needed, must be added outside the chip. Every 1K block of ROM is further considered to be 'paged' into four pages of 256 bytes each. This page-division is shown for the lowest 1K block of ROM in Fig. 13.1.3. (Note that, on any given page, the upper 4-bits of address do not change).

The 8048/8748 provides one unconditional jump (JMP) instruction, one unconditional subroutine call (CALL) instruction, twenty-eight conditional jump instructions and two returns (RET and RETR) instructions for program branching. These instructions are such that—

(i) Program execution can branch to anywhere in the current memory bank by means of the CALL and the JMP instructions.

(ii) Virtually the only way of going from one memory bank to the other is by executing a CALL or a JMP properly preceded by a SEL MBØ or a SEL MB1 instruction, as needed.

(iii) All the conditional jump instructions are limited to branching within the current page (i.e. only last 8-bits of PC can be changed by these instructions).

(iv) Conditional jumps which test individual bits of the accumulator and branch as per (iii) are also available.

(v) For returning from a subroutine, RETR restores PSW whereas RET does not.

(vi) Both the RETR and the RET instructions, recover the 12 bits of stored PC from the stack and can therefore return to the calling program across the memory bank boundary. Such a return across the bank boundary does not however change DBF. Therefore any subsequent CALL or JMP will once again cause a bank switch. The user should take this into account.

The 4K bytes of ROM may also be used to keep the constants needed by the program. Two instructions are available to load the accumulator with constants from the program memory space using register indirect addressing. These are—

(i) MOVP which loads ACC with the contents of the program memory location addressed by ACC on the current page.

(ii) MOVP3 which loads ACC with the contents of the program memory location addressed by ACC on Page 3 of the internal ROM (3ØØH–3FFH).

Of course, one can also load the accumulator, registers, or data memory locations with constants by using a 'Move Immediate' kind of instruction.

13.1.3 The Organization of the RAM

The 8048/8748 has 64 bytes of internal RAM which can be further augmented by adding 256 bytes of external RAM. The internal RAM, organized as shown in Fig. 13.1.4, provides the Register Banks and the stack locations needed by the CPU. The remaining RAM locations may be used for data storage. The only addressing mode available for this in the 8048/8748 is 'Register Indirect Addressing'.

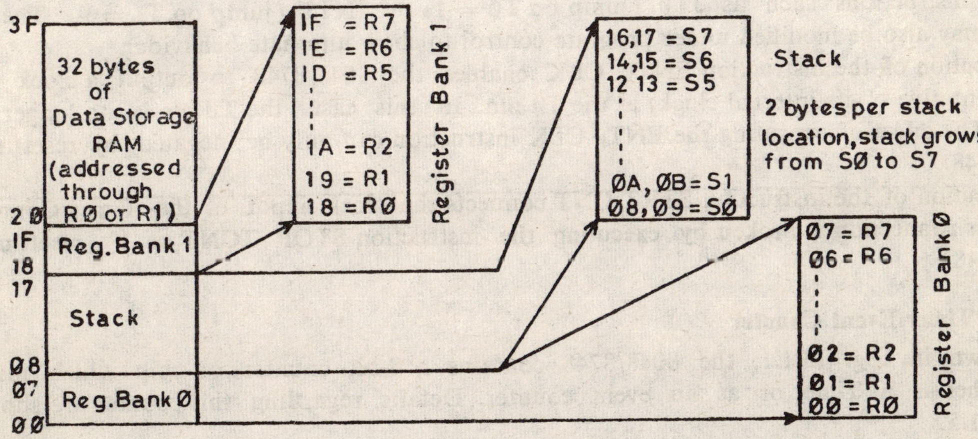

Fig. 13.1.4 Organization of the Internal RAM

As shown in Fig. 13.1.4, locations ØØ–Ø7 and 18–1F of the internal RAM act as Register Banks Ø and 1, respectively. The actual Register Bank currently in use is determined by the RBS bit in PSW. The availability of two Register Banks is a very useful feature if interrupts or subroutines are being used. In this case, the main program (calling program or interrupted program) may use one Register Bank whereas the subroutine or interrupt service routine uses the other Register Bank. By switching the RBS bit properly, one can then achieve the equivalent or saving all the registers (RØ–R7) while branching away from the main program; of course, RBS should be properly restored on or before returning. (Note that this technique will only work for a single

(All mnemonics copyright Intel Corporation 1985).

level nesting of subroutines and interrupt service routines.) The registers RØ–R7 of the current Register Bank are directly addressable by means of instructions such as MOV A, R7 (which loads the accumulator with the contents of register R7).

Locations Ø8–17 (16 bytes) of the internal RAM are used by the 8048/8748 to implement the 8-position (SØ–S7) stack needed for subroutine linkage (i.e. 2 bytes per stack location). This stack is addressable by the SPØ–SP2 bits of PSW. Note that because of this limited stack, one can have only eight levels of subroutine (and interrupt) nesting on the 8048/8748. Moreover, the stack cannot be used for data storage/retrieval. (PUSH or POP type instructions are not available on the 8048/8748.)

All the 64 bytes of internal RAM are addressable by means of 'Register Indirect Addressing' using registers RØ or R1 of the current Register Bank. Thus, one has instructions like MOV @ RØ,A which stores into the internal RAM location addressed by RØ the contents of the accumulator. Note that for instructions which use such addressing, the 6 less significant bits of the address register (RØ or R1) are used (i.e. bits 0–5).

The 8048/8748's RAM space may be augmented by adding 256 bytes of external RAM. The 8048/8748 provides only four instructions MOVX @ RØ,A. MOVX @ R1,A MOVX A, @ RØ and MOVX A, @ R1 to access this external RAM. Note that even in this case, the addressing mode used is register-indirect through RØ or R1 of the current Register Bank.

13.1.4 The Test Inputs

The 8048/8748 provides two test inputs TØ and T1 which can be directly tested by conditional branch instructions such as JTØ (jump on TØ = 1) or JNT1 (jump on T1 = Ø). These test inputs may also be modified under program control to show alternate behaviour.

Execution of the instruction ENTØ CLK enables the 8048/8748 to output a clock signal (representative of its internal clock) at the TØ pin. In this case, the TØ pin is no longer a test input. The effect of executing the ENTØ CLK instruction can only be negated by resetting the 8048/8748.

Execution of the instruction STRT CNT connects the clock input of the Timer/Counter to T1. This connection is broken by executing the instruction STOP TCNT or by resetting the 8048/8748.

13.1.5 Timer/Event Counter

As shown in Fig. 13.1.1, the 8048/8748 contains a 8-bit counter on-chip which may be used either as a Timer or as an event counter. Details regarding this counter are shown in Fig. 13.1.5.

The 8-bit internal timer/counter of the 8048/8748 counts up in binary at the Hi → Lo edges of its CLK clock input. This is a modulo-FF counter which recycles from FF to ØØ setting the TF flag at this transition. The counter is cleared to ØØH on reset. It can be loaded or read through the accumulator of the 8048/8748 (MOV A,T loads the accumulator with the current contents of the counter. MOV T,A loads the counter with the contents of the accumulator). The mod-FF counting up feature of the counter is disabled on reset or by doing STOP TCNT; this can be enabled by doing either the STRT T or the STRT CNT instruction. The STRT T instruction connects the CLK input of the counter to an internal clock source which has requency $f = \frac{1}{480} f_{crystal}$. The STRT CNT instruction connects the CLK input of the counter

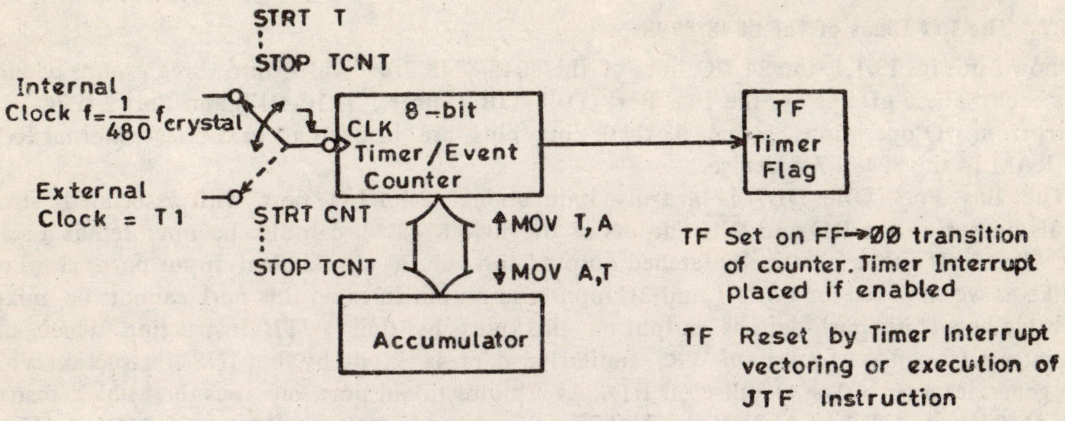

Fig. 1.31.5 The Intel Timer/Event Counter

to the T1 input of the 8048/8748 as mentioned in Sec. 13.1.4 thereby providing an external clock source to the counter. Operated as a timer, one usually chooses the clock source to be internal via the STRT T instruction. For an event counter type of operation, one typically uses an external clock whose 1 → Ø edge indicates one occurrence of the event. For this kind of use, the STRT CNT instruction should be executed after connecting the external event counting clock source to the T1 input of the 8048/8748.

The TF flag gets set on the FF → ØØ transition of the counter indicating a counter overflow situation. The 8048/8748 provides an internal Timer Interrupt for responding to this event. If the Timer Interrupt is enabled, TF = 1 causes an internal CALL ØØ7H to be executed vectoring program execution to location ØØ7H of the (internal) ROM; this vectoring also resets TF. The TF flag may also be tested via the conditional branch instruction JTF which causes a jump if TF = 1; once again, execution of this instruction resets TF. TF is also reset on RESET.

13.1.6 Interrupts

The 8048/8748 provides two interrupts to the user. One of them is the Timer Interrupt described in Sec. 13.1.5. This is a vectored interrupt vectoring to location ØØ7H of the ROM when TF = 1. This interrupt is enabled by the instruction EN TCNTI and disabled by either DIS TCNTI or on reset. The other interrupt input is external through the $\overline{\text{INT}}$ input of the 8048/8748. This interrupt is enabled by the EN I instruction and disabled either by the DIS I instruction or on reset. If this interrupt is enabled and $\overline{\text{INT}}$ input is Ø, the processor does an internal CALL ØØ3H vectoring program execution to location ØØ3 of the (internal) ROM (i.e. this is also a vectored interrupt). If the $\overline{\text{INT}}$ interrupt and the Timer Interrupt are both enabled and both of them are currently being requested, then the $\overline{\text{INT}}$ interrupt is attended to first, i.e. $\overline{\text{INT}}$ has a higher priority than the Timer Interrupt.

The 8048/8748 disables both interrupts on being reset. Both the interrupts are also disabled when the processor enters an Interrupt Service Routine (ISR). The RETR instruction should terminate the ISR as apart from restoring bits 4–7 of PSW to the prior-to-interrupt values, it also restores the interrupt enable/disable status to whatever it was before the interrupt. Note that the 8048/8748 does not produce any specific signal to show that the external interrupt has been acknowledged. If such a signal is needed, it must be generated under program control by one of the I/O lines of the 8048/8748.

13.1.7 The I/O Lines of the 8048/8748

As shown in Fig. 13.1.1, the 24 I/O lines of the 8048/8748 are divided into three groups of eight lines each. These groups are the Bus Port (DB0–DB7), Port 1 (P10–P17) and Port 2 (P20–P27). Apart from I/O operations, some of these port pins are also used to connect external ROM and RAM to the 8048/8748 device.

The Bus Port (DB0–DB7) is a truly bidirectional 8-bit I/O port with associated strobe signals output on the $\overline{RD}$ and $\overline{WR}$ outputs of the 8048/8748. It can also be operated as a static port where it is either a statically latched output port or a non-latched input port. However, (unlike as we shall see for Ports 1 and 2), input and output lines on this port cannot be mixed. Operated as a static port, data is output on this port by the OUTL instruction which also generates a (1 → 0 → 1) pulse on $\overline{WR}$; similarly, data is input by the INS instruction which also generates a (1 → 0 → 1) pulse on $\overline{RD}$. As a bidirectional port, one uses the MOVX instructions (MOVX A, @R0 or MOVX A, @R1 for input and MOVX @R0, A or MOVX @R1,A for output), which also generates a $\overline{RD}$ pulse for input and a $\overline{WR}$ pulse for output operations. One can also do direct logical manipulation of the Bus Port as outputs by using ANL or ORL instructions for directly ANDing or ORing the bus contents with a logical mask. This port may also be used to interface the 8048/8748 to other microprocessors.

Ports 1 and 2 are both 8-bit ports and show identical characteristics. The I/O lines of these ports may be used for either input or output even though these lines are not truly bidirectional. These I/O lines are, instead, referred to as 'Quasibidirectional' I/O lines. This quasibidirectionality implies that any of these lines may be used as an input, a latched (and always active) output or both. This is achieved by a special output circuit structure which decides the nature of an I/O line on the basis of external circuitry connected to it and the instructions executed by the processor. To use one of the port lines as an output, it should be connected to the input of some other device. Data can be output on this (data latched), in that case, by means of an OUTL instruction; moreover, logical manipulation of this output line is also possible using ANL or ORL instructions with an appropriate logical mask (ANL for ANDing, ORL for ORing). For use as an input line, the port line should be driven by the output of some other device. To input from that line, 8048/8748's software should first write (output) a '1' to that line via the OUTL or ORL instruction and then issue an IN instruction. Note that any mix of input or output lines is possible with these 16 port lines, provided one ensures proper hardware connections and software. Used properly with the proper software, these lines can also be connected to a bus.

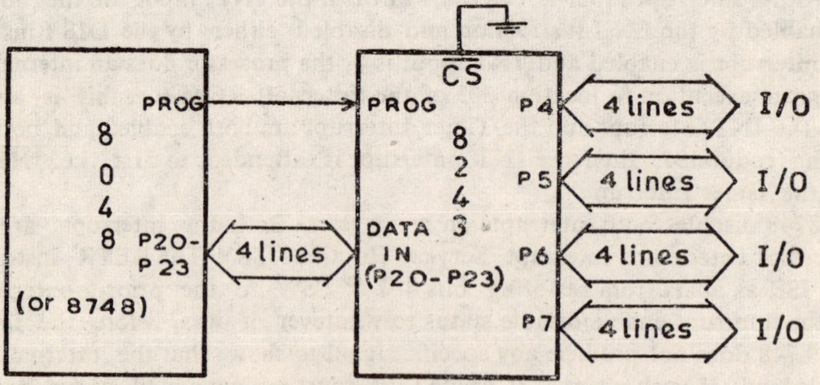

Fig. 13.1.6 I/O Expansion with the 8243

The I/O lines available with the 8048/8748 may be expanded with the use of an 8243 I/O Expander as shown in Fig. 13.1.6. Using the circuit shown, one gets four 4-bit bidirectional I/O ports P4, P5, P6, and P7 sacrificing four bits P20–P23 of the 8048/8748's. As with Ports 1 and 2, we can output on these ports using commands MOVD Pp,A ANLD Pp,A or ORLD Pp,A (p = 4, 5, 6 or 7); similarly input operations can be done with the instructions MOVD A,Pp (p = 4, 5, 6, 7).

The Bus Port lines are also used to interface external ROM and RAM to the 8048/8748. In addition, external ROM access also uses the port lines P20–P23, the lower four lines of Port 2.

13.1.8 The RESET Input

A low pulse (1 → Ø → 1) at the $\overline{\text{RESET}}$ input resets the 8048/8748. Specifically, resetting the 8048/8748 does the following:

(i) (PC) = ØØØH
(ii) (SP) = ØØØB
(iii) Ports 1 and 2 are set to input and Bus Port is tristated.
(iv) Register Bank Ø and Memory Bank Ø chosen (i.e. RBS = Ø and DBF = Ø)
(v) Both interrupts disabled
(vi) Timer is stopped and counting is disabled
(vii) Timer Flag (TF), and software flags F1 and FØ are cleared to Ø
(viii) Clock output from TØ disabled (effect of ENTO CLK instruction is negated)

13.2 THE PIN-OUT OF THE 8048/8748

The pin-out of the 8048/8748 is shown in Fig. 13.2.1. A brief description of these pins is given below.

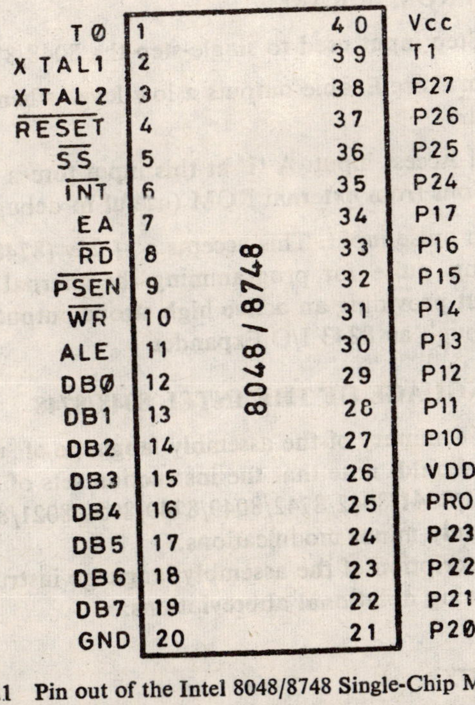

Fig. 13.2.1 Pin out of the Intel 8048/8748 Single-Chip Microcomputer

V_{CC}, V_{DD}, GND	:	Power supply pins, V_{CC} and V_{DD} are connected to +5v and GND is connected to the circuit ground. V_{CC} is the main power supply and V_{DD} is the low power stand by pin. In the 8048, even if V_{CC} power fails, keeping V_{DD} at +5v (possibly from a battery or back-up supply) will keep the RAM contents undisturbed.
P10–P17	:	8-bit quasibidirectional Port 1 (See Sec. 13.1.7)
P20–P27	:	8-bit quasibidirectional Port 2 (See Sec. 13.1.7)
DBØ–DB7	:	8-bit bidirectional Bus Port (See Sec. 13.1.7)
$\overline{RD}$, $\overline{WR}$	:	Read (output strobe) and write (input strobe) outputs respectively for Bus Port and external data memory accesses.
$\overline{INT}$	:	External, active low, level-triggered vectored interrupt (See Sec. 13.1.6).
TØ	:	Test input TØ. It can also output a clock signal (See Sec. 13.1.4).
T1	:	Test input T1. It may also provide clock signal to the internal timer/counter operated as an event counter (See Sec. 13.1.4 and 13.1.5).
XTAL1, XTAL2	:	Crystal inputs. Connect crystal, external clock source ($XTAL2 = \overline{XTAL1}$) or LC circuit for generating internal clock reference.
$\overline{RESET}$	:	Active low Reset input. A low pulse at this input resets the 8048/8748 (See Sec. 13.1.8).
$\overline{ALE}$	:	Address Latch Enable. This pin outputs an active low pulse in every machine cycle. It is useful to demultiplex address and data while accessing external ROM or RAM.
$\overline{SS}$	:	Single Step input used to single-step the 8048/8748 (useful in debugging).
$\overline{PSEN}$	:	Program Store Enable outputs a low level when external ROM is being accessed.
EA	:	External Access input. A '1' at this input forces the 8048/8748 to fetch all instructions from external ROM (useful in debugging).
PROG	:	Program pulse input. This accepts a +23v (8748) or +18v (8748H) programming pulse for programming the internal EPROM. This also acts as an output providing an active high strobe output for accessing Ports 4, 5, 6 or 7 through an 8243 I/O Expander

13.3 THE ASSEMBLY LANGUAGE OF THE INTEL 8048/8748

In this section, we give a brief summary of the assembly language of the 8048/8748 single-chip microcomputer. The reader should note that the instruction sets of a number of Intel's single-chip microcomputers (e.g. 8041/8741/8042/8742/8049/8749/8050/8021/8020/8035 etc.) are similar to that of the 8048/8748 with only minor modifications.

In order to facilitate our description of the assembly language instruction set of the 8048/8748 we would like to use the following notational abbreviations:

A	Accumulator
AC	Auxiliary Carry

addr	Program Memory Address, 11-bit for JMP and CALL and 8-bit for other (conditional) jump instructions
Bb	Bit designator for bit b of accumulator (b = 0–7), e.g. BØ, B1 etc.
BUS	BUS Port
C	Carry
CLK	Clock
data	8-bit
DBF	Memory Bank Flip-Flop
Pp	Port p (p = 1, 2 or p = 4, 5, 6, 7), e.g. P1, P2 etc.
R_r	Register r, e.g. RØ, R1 etc. of current bank
T	Timer
TF	Timer Flag
#	prefix for immediate data
@	prefix for indirect address
(X)	Contents of X
((X))	Contents of location addressed by X

The instructions of the 8048/8748 can be derived into groups depending on their functions. We give below a group by group description of these instructions.

13.3.1 Data Moves

Instructions in this group move data between registers, between registers and data memory, or from program ROM to the accumulator.

MOV	A,R_r	;	(A) ← (R_r)	r = 0–7	
MOV	A,@R_r	;	(A) ← ((R_r))	r = 0, 1	internal RAM access
MOV	A,# data	;	(A) ← data		move immediate
MOV	R_r,A	;	(R_r) ← (A)	r = 0–7	
MOV	@R_r,A	;	((R_r)) ← (A)	r = 0, 1	internal RAM access
MOV	R_r,# data	;	(R_r) ← data	r = 0–7	move immediate
MOV	@R_r,# data	;	((R_r)) ← data	r = 0, 1	internal RAM access
MOV	A,PSW	;	(A) ← (PSW)		
MOV	PSW,A	;	(PSW) ← (A)		
XCH	A,R_r	;	(A) ↔ (R_r)	r = 0–7	internal RAM access
XCH	A, @R_r	;	(A) ↔ ((R_r))	r = 0, 1	internal RAM access
XCHD	A,@R_r	;	($A_{Ø-3}$) ↔ ((R_r))$_{Ø-3}$	r = 0, 1	internal RAM access

MOVX	A,@R_r	; (A) ← ((R_r))	r = 0, 1 external RAM access
MOVX	@R_r,A	; ((R_r)) ← (A)	r = 0, 1 external RAM access
MOVP	A,@A	; (A) ← (XXXX(A)) XXXX 4 bits for current ROM Page	
MOVP3	A,@A	; (A) ← (ØØ11(A)) Move from Page 3 of lowest 1K of ROM	

13.3.2 Input/Output

These instructions allow input or output operations on the I/O ports of the 8048/8748, possibly expanded with an 8243 device.

IN	A,Pp	; (A) ← (P_p)	p = 1, 2
OUTL	Pp,A	; (Pp) ← (A)	p = 1, 2
ANL	Pp,# data	; (Pp) ← (Pp) (AND) data	p = 1, 2
ORL	Pp,# data	; (Pp) ← (Pp)(OR) data	p = 1, 2
INS	A, BUS	; (A) ← (BUS), $\overline{RD}$ pulse generated.	
OUTL	BUS, A	; (BUS) ← (A) latched and $\overline{WR}$ pulse generated	
ANL	BUS, # data	; (BUS) ← (BUS)(AND) data	
ORL	BUS,# data	; (BUS) ← (BUS)(OR) data ANL BUS/ORL BUS assume prior execution of OUTL BUS	
MOVD	A,Pp	; ($A_{Ø-3}$) ← (Pp), (A_{4-7}) = $Ø_H$	p = 4–7
MOVD	Pp,A	; (Pp) ← ($A_{Ø-3}$)	p = 4–7
ANLD	Pp,A	; (Pp) ← (Pp)(OR)($A_{Ø-3}$)	p = 4–7
ORLD	Pp,A	; (Pp) ← (Pp)(OR)($A_{Ø-3}$)	p = 4–7

Pp, p = 4–7 refers to the expander using 8243. This is connected to the 8048/8748 using the latter's PROG output and port pins $P_{20\text{-}23}$ as shown in Fig. 13.1.6.

13.3.3 Accumulator Instructions (Arithmetic and Logical)

These instructions perform arithmetic and logical manipulation of the contents of the accumulator.

ADD	A,R_r	; (A) ← (A) + (R_r)	r = 0–7
ADD	A,@R_r	; (A) ← (A) + ((R_r))	r = 0, 1 internal RAM access
ADD	A,# data	; (A) ← (A) + data	add immediate
ADDC	A,R_r	; (A) ← (A) + (R_r) + (C)	r = 0–7, add with carry
ADDC	A,@R_r	; (A) ← (A) + ((R_r)) + (C)	r = 0, 1 internal RAM access
ADDC	A,# data	; (A) ← (A) + data + (C)	add immediate with carry
ANL	A,R_r	; (A) ← (A)(AND)(R_r)	r = 0–7
AND	A,@R_r	; (A) ← (A)(AND)((R_r))	r = 0, 1 internal RAM access
ANL	A,# data	; (A) ← (A)(AND) data	AND immediate

ORL	A,R_r	;	$(A) \leftarrow (A)(OR)(R_r)$	$r = 0$–7
ORL	A,@R_r	;	$(A) \leftarrow (A)(OR)((R_r))$	$r = 0$, 1 internal RAM access
ORL	A,# data	;	$(A) \leftarrow (A)(OR)$ data	OR immediate
XRL	A,R_r	;	$(A) \leftarrow (A)(EX\text{–}OR)(R_r)$	$r = 0$–7
XRL	A,@R_r	;	$(A) \leftarrow (A)(EX\text{–}OR)((R_r))$	$r = 0$, 1 internal RAM access
XRL	A,# R_r	;	$(A) \leftarrow (A)(EX\text{–}OR)$ data	EX-OR immediate
INC	A	;	$(A) \leftarrow (A) + 1$	increment
DEC	A	;	$(A) \leftarrow (A) - 1$	decrement
CLR	A	;	$(A_{Ø-7}) = ØØ_H$	clear
CPL	A	;	$(A) \leftarrow (\bar{A})$	complement
DA	A	;	decimal adjust A, just like DAA of 8085A	
SWAP	A	;	Swap nibbles of A	
RL	A	;	$(A_Ø) \leftarrow (A_7)$, $(A_{n+1}) \leftarrow (A_n)$	$n = 0$–6 } rotate
RLC	A	;	$(C) \leftarrow (A_7)$, $(A_Ø) \leftarrow (C)$, $(A_{n+1}) \leftarrow (A_n)$	$n = 0$–6 } rotate
RR	A	;	$(A_7) \leftarrow (A_Ø)$, $(A_n) \leftarrow (A_{n+1})$	$n = 0$–6 } rotate
RRC	A	;	$(A_7) \leftarrow (C)$, $(C) \leftarrow (A_Ø)$, $(A_n) \leftarrow (A_{n+1})$	$n = 0$–6 } rotate

13.3.4 Increment/Decrements of Registers and Memory

Instructions in this group permit incrementing/decrementing of registers RØ–R7 and incrementing data memory location contents.

INC	R_r	;	$(R_r) \leftarrow (R_r) + 1$	$r = 0$–7
INC	@R_r	;	$((R_r)) \leftarrow ((R_r)) + 1$	$r = 0$, 1 internal RAM access
DEC	R_r	;	$(R_r) \leftarrow (R_r) - 1$	$r = 0$–7

(N.B. No DEC @R_r instruction)

13.3.5 Branch (Jump) Instructions

This group comprises of one unconditional direct, one unconditional indirect and several conditional jump instructions.

JMP	addr	;	$(PC_{Ø-10}) \leftarrow (Addr_{Ø-10})$, $(PC_{11}) \leftarrow$ DBF, direct jump
JMPP	@ A	;	$(PC_{Ø-7}) \leftarrow ((A))$, indirect jump
DJNZ	R_r, addr	;	$(R_r) \leftarrow (R_r) - 1$, if $(R_r) \neq ØØ_H$, $R = Ø$–7 then $(PC_{Ø-7}) \leftarrow addr_{Ø-7}$, if $(R_r) = ØØ_H$, continue sequentially.
JC	addr	;	$(PC_{Ø-7}) \leftarrow$ addr if $(C) = 1$
JNC	addr	;	$(PC_{Ø-7}) \leftarrow$ addr if $(C) = Ø$
JZ	addr	;	$(PC_{Ø-7}) \leftarrow$ addr if $(A) = ØØ$

JNZ	addr	;	$(PC_{\emptyset-7}) \leftarrow$ addr if $(A) \neq \emptyset\emptyset$
JTØ	addr	;	$(PC_{\emptyset-7}) \leftarrow$ addr if $T\emptyset = 1$
JNTØ	addr	;	$(PC_{\emptyset-7}) \leftarrow$ addr if $T\emptyset = \emptyset$
JT1	addr	;	$(PC_{\emptyset-7}) \leftarrow$ addr if $T1 = 1$
JNT1	addr	;	$(PC_{\emptyset-7}) \leftarrow$ addr if $T1 = \emptyset$
JFØ	addr	;	$(PC_{\emptyset-7}) \leftarrow$ addr if $F\emptyset = 1$
JF1	addr	;	$(PC_{\emptyset-7}) \leftarrow$ addr if $F1 = 1$

(N.B. No JNFØ & JNF1 instruction)

JTF	addr	;	$(PC_{\emptyset-7}) \leftarrow$ addr if $TF = 1$
JNI	addr	;	$(PC_{\emptyset-7}) \leftarrow$ addr if $\overline{INT} = \emptyset$

(N.B. (a) execution of JTF resets TF to Ø
(b) JNI can be executed only if interrupt is disabled).

JB_b	addr	;	$(PC_{\emptyset-7}) \leftarrow$ addr if bit b of Accumulator is 1.

Except JMP all other jumps are short and are limited to within the current page.

13.3.6 Subroutine Linkage Instructions

CALL	addr	;	return address saved on stack with PSW_{4-7}
			$(PC_{\emptyset-10}) \leftarrow addr_{\emptyset-10}$
			$(PC_{11}) \leftarrow DBF$
RET		;	return without PSW restore and without changing interrupt enable/disable status
RETR		;	return with PSW restore and interrupts enabled to prior status.

13.3.7 Instructions Manipulating FLAGs

The carry (C), FØ and F1 flags can be directly manipulated by these instructions.

CLR	C	;	$(C) \leftarrow \emptyset$
CPL	C	;	$(C) \leftarrow (\bar{C})$
CLR	FØ	;	$(F\emptyset) \leftarrow \emptyset$
CPL	FØ	;	$(F\emptyset) \leftarrow (\overline{F\emptyset})$ Note FØ part of PSW but not F1
CLR	F1	;	$(F1) \leftarrow \emptyset$
CPL	F1	;	$(F1) \leftarrow (\overline{F1})$

13.3.8 Timer/Counter Instructions

These instructions manipulate the internal 8-bit Timer/Counter.

MOV	A,T	;	$(A) \leftarrow (T)$
MOV	T,A	;	$(T) \leftarrow (A)$

STRT	T	;	Start timer, internal clock, $f = f_{crystal}/480$
STRT	CNT	;	Start counter, clock external from T1
STOP	TCNT	;	Stop timer/counter
EN	TCNTI	;	enable timer/counter interrupt
DIS	TCNTI	;	disable timer/counter interrupt

13.3.9 Control Instructions

Various machine control instructions fall in this group.

EN	I	;	enable external interrupt
DIS	I	;	disable external interrupt
SEL	RBØ	;	select Reg. Bank Ø
SEL	RB1	;	select Reg. Bank 1
SEL	MBØ	;	select Memory Bank Ø
SEL	MB1	;	select Memory Bank 1
ENTØ	CLK	;	enable clock output on TØ
NOP		;	no operation

13.4 THE INTEL 8041A/8741A/8042/8742 UNIVERSAL PERIPHERAL INTERFACE (UPI), SINGLE-CHIP MICROCOMPUTER

As mentioned earlier, a very common use for single-chip microcomputers is as a slave processor for a master microcomputer in a multiple processor system. In this kind of an application, the master microcomputer considers the slave processor as another peripheral device, relegating special processing or interfacing tasks to it, whenever the need arises. A device like the 8048/8748 can be used as the slave processor in such an application by adding some special hardware for interfacing purposes. To simplify this interfacing problem, Intel Corp. manufactures a series of single-chip microcomputers 8041A/8741A/8042/8742 which have been specially designed to act as a slave microcomputer to another master microcomputer without any additional interface hardware. (These devices can also act as stand-alone microcomputers, if needed.) A close examination of these devices reveals that these devices are actually modified versions of the 8048/8748 which have been specialized to make interfacing with a master microcomputer easier.

In this section, we will present a brief description of the 8041A/8741A/8042/8742 Universal Peripheral Interface Single-Chip Microcomputer. We will describe these UPI devices by emphasizing their differences with each other and with the 8048/8748 single-chip microcomputers. This is specially desirable as these devices are otherwise very similar to each other.

13.4.1 The Pin-Out of the UPI Devices (8041A, 8741A etc.)

In Fig. 13.4.1 we have shown the pin-out of the 8041/8741A etc. UPI devices. Comparing this with Fig. 13.2.1, we find that the pin-out of these devices is virtually identical to that of the 8048/8748. The functions of the various pins of the 8048/8748 and the UPI devices are virtually identical. The important differences are highlighted below.

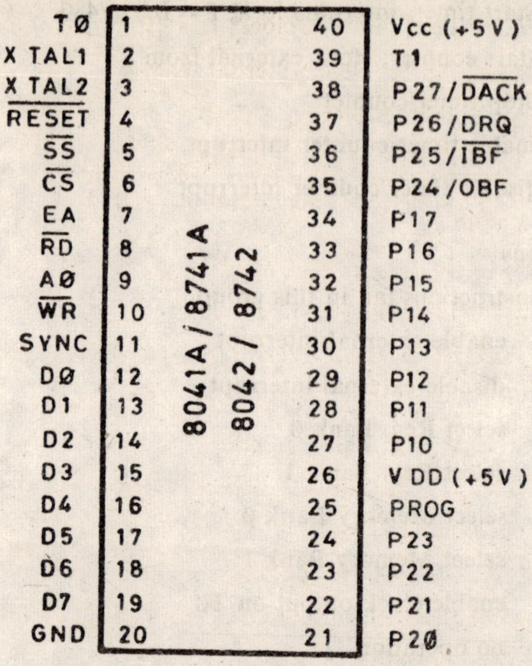

Fig. 13.4.1 Pin-Out of the Intel 8041A/8741A/8042/8742 Single-Chip Microcomputer (Universal Peripheral Interface)

Differences between the pins of the 8041 etc. with those of the 8048/8748

(a) TØ can no longer output system clock.

(b) $\overline{CS} + \overline{RD} = \emptyset$ to read from output buffer (DBBOUT) of the 8041A etc. or from its STATUS register.

(c) $\overline{CS} + \overline{WR} = \emptyset$ to write into DBBIN input buffer of 8041A etc. Data written may be interpreted as data or command byte. Internal flag F1 is set to value of AØ during $\overline{CS} + \overline{WR} = \emptyset$. (F1 is also modifiable under program control.)

(d) AØ is the Select Input (for the internal registers of the 8041A etc.) as per table given below.

$\overline{CS}$	$\overline{RD}$	$\overline{WR}$	AØ	Operation
Ø	Ø	1	Ø	Read DBBOUT of 8741A
Ø	Ø	1	1	Read STATUS of 8741A
Ø	1	Ø	Ø	Write to DBBIN data register
Ø	1	Ø	1	Write to DBBIN command register

(e) The SYNC output outputs a clock pulse for every instruction.

(f) $D_Ø$–D_7 is like the BUS Port of the 8048/8748. It is connected to the Data Bus of Master CPU in applications where the 8041A/8741A is the slave CPU.

(g) P_{24}–P_{27} (i.e. upper half of Port 2) may be programmed to show functional behaviour other than their normal function as quasi-bidirectional port lines. If EN FLAGS is executed,

P_{24} = OBF = Full/$\overline{\text{Empty}}$ status of DBBOUT register.

P_{25} = $\overline{\text{IBF}}$ = $\overline{\text{Full}}$/Empty status of DBBIN register.

provided a '1' is written to these port lines.

(i.e. pin $P_{24} = P_{24} \cdot \text{OBF}$; pin $P_{25} = P_{25} \cdot \overline{\text{IBF}}$)

If EN DMA is executed,

P_{26} = DRQ, DMA Request Output

P_{27} = $\overline{\text{DACK}}$, DMA Acknowledge Input

Data Transfers between DBBOUT (or DBBIN) and memory of Master CPU may be done under DMA control. ($\overline{\text{DACK}}$ acts as the Chip Select Signal during DMA transfers.)

13.4.2 The Buffer Structure of the UPI

The UPI devices have a special buffer structure between their internal data bus and the data lines $D_{\emptyset}$–D_7 which connect to the data bus of the master microcomputer. This buffer structure is shown in Fig. 13.4.2. The UPI devices have a double buffer arrangement for data transactions

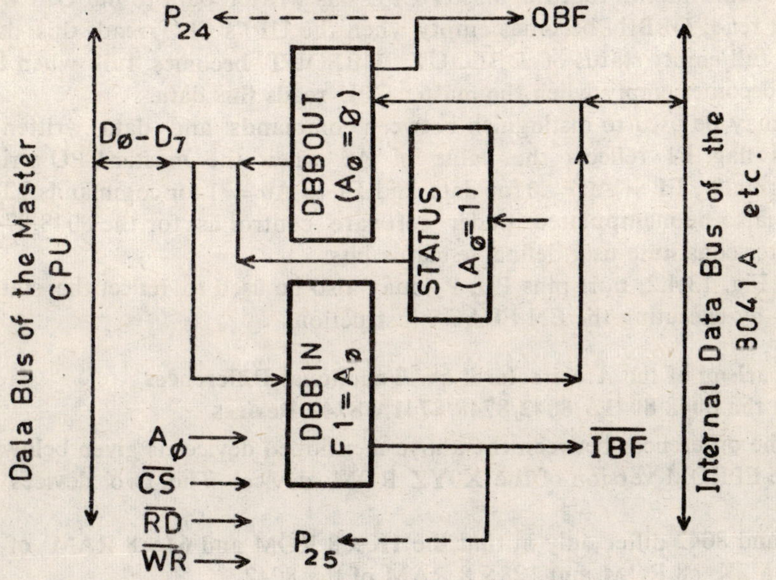

Fig. 13.4.2 The $D_{\emptyset}$–D_7 Buffer Structure of the Intel 8041A/8741A/8042/8742 UPI

between the internal and external data bus. From the UPI's point of view, it has two registers—an input register DBBIN and an output register DBBOUT. The master CPU writes to DBBIN with $\overline{\text{CS}} = \emptyset$, $\overline{\text{RD}} = 1$, $\overline{\text{WR}} = \emptyset$, A∅ = ∅ or 1, which can then be read by the UPI's CPU by the instruction IN A, DBB. Similarly, to send data to the master CPU, the UPI's CPU writes data to DBBOUT by the instruction OUT DBB, A; the master CPU can read this data with $\overline{\text{CS}} = \emptyset$, $\overline{\text{RD}} = \emptyset$, $\overline{\text{WR}} = 1$, A∅ = ∅.

The condition (full of empty status) or these DBBIN and DBBOUT registers is reflected by the contents of the STATUS register. The contents of this STATUS register is defined as shown in Fig. 13.4.3. The Input Buffer Flag (IBF) indicates when DBBIN is full.

IBF = 1 for DBBIN full

= Ø for DBBIN empty

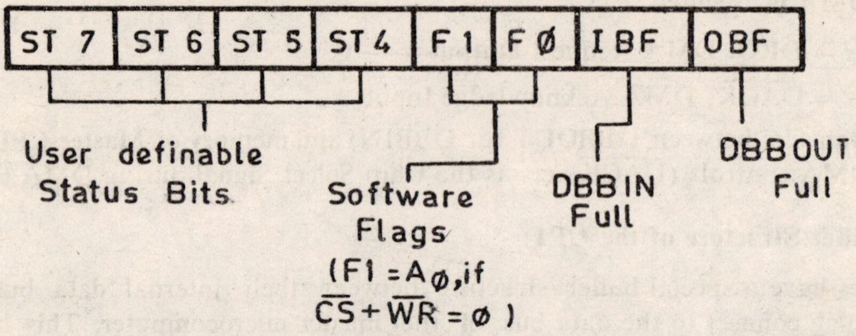

Fig. 13.4.3 The Status Register of the Intel 8041A/8741A/8042/8742 UPI

Note that DBBIN full implies that the master CPU has written data to the UPI which the UPI's CPU has not yet read; DBBIN becomes empty when the UPI's CPU reads this data. Similarly, OBF reflects the full/empty status of DBBOUT; DBBOUT becomes full when the UPI's CPU writes to it and becomes empty when the master CPU reads this data.

The flag F1 may be used to distinguish between commands and data written to the UPI by the master. This flag F1 reflects the value of AØ when the master CPU writes to DBBIN of the UPI. (Normally, F1 = AØ = Ø for data and F1 = AØ = 1 for commands.) This flag F1 and the flag FØ may also be manipulated under software control as for the 8048/8748. Bits 4–7 of the Status Register constitute user definable status bits.

As shown in Fig. 13.4.2, port pins P_{24}, P_{25} may also be used to reflect the status of OBF and $\overline{\text{IBF}}$ respectively by executing the EN FLAGS instruction.

13.4.3 A Comparison of the Architectural and Functional Differences between the 8048/8041A/8042/8748/8741A/8742 Devices

A summary of the differences between the above mentioned devices is given below—

(a) X7YZ is the EPROM version of the X0YZ ROM device. The two devices are otherwise identical.

(b) The 8041A and 8042 differ only in that the 1K×8 ROM and 64×8 RAM of the 8041A is replaced by a 2K×8 ROM and 128×8 RAM of the 8042.

(c) The pins of the 8041A/8741A/8042/8742 are exactly identical and differ from those of the 8048/8748 as described earlier.

(d) The BUS port of the 8041A/8741A/8042/8742 is specialized as an interface between a slave 8041A etc. and a Master CPU with the buffer arrangement as shown (Sec. 13.4.2 and Fig. 13.4.2–13.4.3).

(e) The 8041A etc. cannot be expanded with external program or data memory. (I/O expansion with an 8243 I/O Expander is however possible as with the 8048/8748).

(f) Interrupts: No external interrupt (like $\overline{\text{INT}}$ in 8048/8748) is available in the 8041A etc. The

two interrupts available are:
(i) Timer Overflow Interrupt forcing a CALL ØØ7H
(ii) Input Buffer Full interrupt (IBFINT) forcing a CALL ØØ3H if DBBIN is full.
An external interrupt, if needed, can be 'faked' on the 8041A etc. by programming the internal Timer/Counter in the Event Counter mode (input from T1) with the Counter preset to FF_H. In that case, T1 is an edge sensitive interrupt input forcing a CALL ØØ7H on the $1 \rightarrow Ø$ edge of T1, provided the Timer Overflow Interrupt is enabled.

(g) An IBF or OBF interrupt can be generated by the 8041A etc. for the Master CPU, by using P_{25} and P_{24} respectively, following an EN FLAGS instruction.

(h) Reading from DBBOUT or writing to DBBIN of the 8041A etc. can be done under DMA Control by executing the EN DMA instruction. This feature (not present in the 8048/8748) allows P_{26} to generate DRQ to the DMA Controller and P_{27} to accept $\overline{DACK}$ from the DMA Controller. $\overline{DACK} = Ø$ selects the 8041A etc. for a Read/Write operation bypassing its $\overline{CS}$ input.

13.4.4 Differences Between the Instruction Sets of the 8041A etc. and the 8048/8748

The two instruction sets are basically similar. However, the *8041A etc. does not have the following instructions—*

INS	A, BUS	OUTL	BUS, A	ANL	BUS, # data
ORL	BUS, # data	JNI	addr		
MOVX	A, @ R_r	MOVX	@ R_r, A	ENTØ	CLK
SEL	MBØ	SEL	MB1		
EN	I/DIS I (different function in the 8041A etc.)				

Compared to the instruction set of the 8048/8748, the instruction set of the *8041A etc. has the following new instructions—*

EN	DMA ;	enable DRQ and $\overline{DACK}$ on P_{26} and P_{27}
EN	FLAGS;	enables $\overline{IBF}$ and OBF flag outputs on P_{25} and P_{24} *provided a '1' is written to these port pins before this instruction*; P_{24} and P_{25} can then be used as interrupts to the master CPU, if so needed.
IN	A, DBB;	(A) ← (DBBIN), clear IBF
OUT	DBB, A;	(DBBOUT) ← (A), Set OBF
MOV	STS, A ;	$(ST_{4-7}) \leftarrow (A_{4-7})$, ST4–ST7 are additional bit information for the Master CPU (other than F1, FØ, IBF, OBF)
EN	I ;	enable IBF interrupt to 8041A etc.
DIS	I ;	disable IBF interrupt to 8041A etc.
JNIBF	addr ;	jump to addr, $(PC_{Ø\ 7}) \leftarrow addr_{Ø-7}$, if IBF = Ø
JOBF	addr ;	jump to addr, $(PC_{Ø-7}) \leftarrow addr_{Ø-7}$, if OBF = 1

13.4.5 Interfacing a UPI—An Example

We have given in Fig. 13.4.4 an example of interfacing a UPI to a master microcomputer on one side and a LSI peripheral—a 8251A USART—on the other side. The UPI is considered to be a

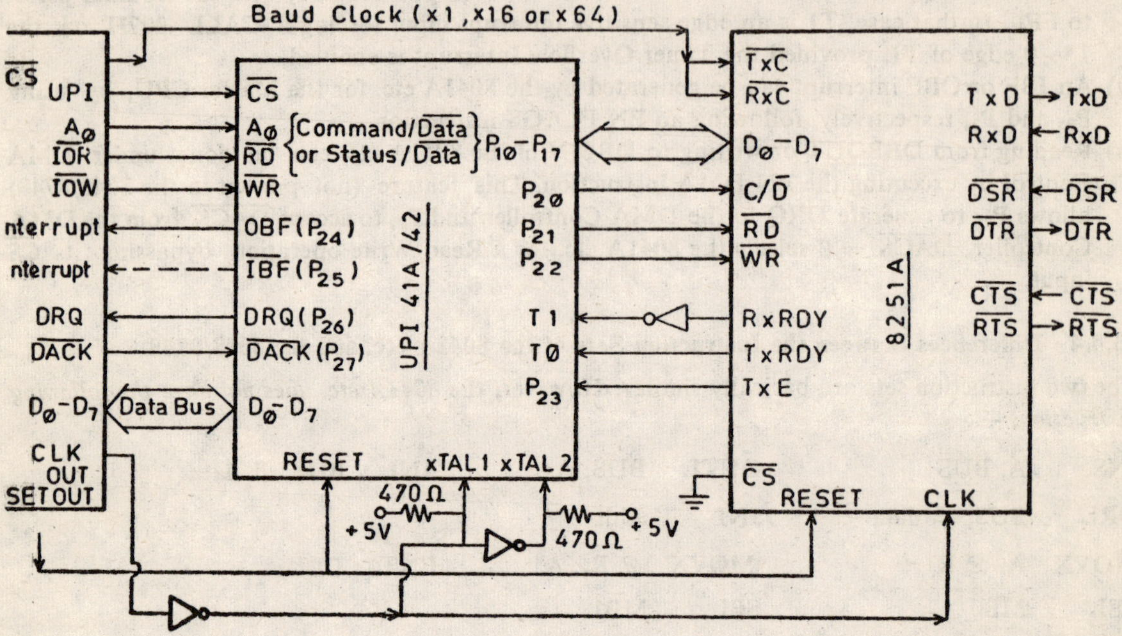

Fig. 13.4.4 Interfacing a UPI to a Peripheral Device (8251A) and a Master Microcomputer

slave controlling the 8251A to give an intelligent serial interface. The hardware features of this interface are given below.

Features of the Master-UPI Interface

(i) UPI functions as two I/O ports (I/O Mapped) of Master.

(ii) UPI is interrupted by IBFINT when Master writes command or data to it. Command (F1 = 1) or Data (F1 = ∅) may be distinguished by UPI on the basis of F1

(iii) UPI interrupts Master by OBF output when it has data for the latter to read in its DBBOUT.

(iv) UPI may also interrupt Master by its $\overline{\text{IBF}}$ output to signal that DBBIN is empty and it wants more data from the Master.

(v) No provision is made for interrupting the UPI when its DBBOUT is empty. UPI must achieve this by monitoring the OBF flag under program control.

(vi) Even with the given hardware connection, any one or more of (ii), (iii) or (iv) can be made software controlled (Master's side or UPI's side, as appropriate) instead of interrupt controlled.

(vii) The Master can monitor the UPI's internal status by reading the latter's STATUS register ($\overline{\text{CS}} + \overline{\text{RD}} = \emptyset$, $A_\emptyset = 1$). The UPI can define ST4–ST7 as it wants, in order to pass on four special bits of control information to the Master. This is apart from the usual status information regarding F1, FØ, IBF and OBF.

(viii) The UPI may also occupy one DMA channel of the Master Microcomputer by executing the EN DMA instruction. This one DMA channel, if needed, must be shared between UPI DBBOUT-to-Master or Master-to-DBBIN of UPI (with F1 = Ø regardless of AØ) transfers. Note that with DRQ = 1 and $\overline{DACK}$ = Ø, the CS input of the UPI is a 'don't care'. The direction of transfer in this case is decided by the $\overline{RD}$ and $\overline{WR}$ inputs of the UPI. If the DMA feature is not required, one can use $\overline{RD}$ and $\overline{WR}$ of the Master Microcomputer directly instead of its decoded $\overline{IOR}$ and $\overline{IOW}$ outputs.

Features of the 8251A–UPI Interface

(i) Port 1 ($P_{1Ø}$–P_{17}) which is quasi-bidirectional in nature is used to read data/status from the 8251A and write data/commands to it. Port 2 pins $P_{2Ø}$, P_{21}, P_{22} are used to control these reading and writing operations by properly driving the C/$\overline{D}$, $\overline{RD}$ and $\overline{WR}$ inputs of the 8251A. Note that $\overline{CS}$ of the 8251A is always enabled.

(ii) The Master Microcomputer provides the Baud Clock ($\overline{TxC}$, $\overline{RxC}$) to the 8251A probably through some Timer/Counter IC like the 8253. The slave UPI can still have its say about what it wants the Baud Clock to be, by passing appropriate data to the Master Microcomputer.

(iii) RxRDY, TxRDY and TxE of the 8251A is monitored by the UPI via its T1 and TØ test inputs and the P_{23} port pin respectively. Note that by programming the internal Timer/Counter of the UPI in the event counter mode (STRT CNT), loading FF_H into the counter (MOV A,#FF; MOV T,A) and enabling the Timer Overflow Interrupt (EN TCNTI), the T1 input becomes an edge triggered interrupt input. A 1→Ø transition at the T1 input will cause a CALL $ØØ7_H$ to be executed in that case. If the internal Timer/Counter of the UPI is otherwise required (e.g. for generating time-out signals), the UPI can still operate the 8251A easily by monitoring RxRDY, TxRDY and TxE under program control.

Note that both the UPI and the 8251A USART are reset by the RESET OUT output of the Master Microcomputer. Similarly, the clock signal needed by both the UPI and the 8251A for their internal operations is provided by the CLK OUT output of the Master Microcomputer.

PROBLEMS

1. Write a subroutine to subtract RØ from the accumulator leaving the result in the accumulator.
2. Write a subroutine MULT to do the following:

   (A) ← (A) * (RØ)
3. How can one implement multi-byte addition with the 8048 etc.?
4. The instruction MOVP3 A, @A provides an easy way of doing code conversions. Investigate its use by writing subroutines to convert 2-digit Hex numbers into the corresponding ASCII characters stored in two successive RAM locations, and for the reverse conversion.

5. The 8253's three counters may be operated in six modes, Mode 0–5. Can you make the internal counter of the 8048 etc. emulate these modes to some extent by using some pins of Port 1 and additional software?
6. For the 8085A-UPI-8251A interface example discussed in the text, write all the 8085A and UPI software needed to handle the interface.

Chapter 14

The Zilog Z-80 Microprocessor

For users familiar with the 8085A (or its predecessor, the 8080A), the Z-80 CPU from Zilog Inc. is a fairly easy CPU to learn and use. Just as the 8085A represents Intel's approach to enhancing the 8080A CPU, Z-80 represents Zilog's approach to this. It is therefore not surprising to find a substantial degree of similarity between the Z80 and the 8085A. Apart from the RIM and the SIM instructions, the 8085A's instruction set is a subset of the Z80's instructions. For this common subset of instructions, even the op-codes are the same for the two processors implying that (barring RIM and SIM) a ROM having instructions for the 8085A may be used as such in a Z80 system. Even from the point of view of their internal architecture, the two CPUs are quite similar. The major differences here are that the Z80 has Index Registers to allow Indexed Addressing and that some of the 8085A's registers have been provided in duplicate form in the Z80.

Even though both the 8085A and the Z80 are enhancements of the same device, their features are different enough to give the impression that the enhancements went in somewhat different directions for the two. The reader is referred to Sec. 9.3 and Table 9.1 for a quick comparison of the two devices. From this comparison, it is hard not to get the impression that apart from the 8085A's features of Serial I/O and maskable vectored interrupts, the Z80 shows more advanced features than the 8085A. As mentioned in Sec. 9.3, this would imply that the current popularity of the 8085A can be explained on the basis of the support available for it and its features of on-chip Serial I/O and its maskable vectored interrupts.

Compared to the 8085A, the biggest advantage of the Z80 is a much more powerful instruction set. This enhancement has been done in such a fashion that all software developed for the 8080A (and the 8085A if RIM and SIM are not used) may be used with the Z80 without change. Moreover, the Z80 provides some facilities for refreshing dynamic memory devices and has instructions which essentially mimic DMA-like operations by moving blocks of data between memory and I/O ports or between two areas of memory. It provides better memory and I/O addressing modes, more ways of transferring data between registers, memory and I/O locations, simpler interrupt vectoring, bit manipulation commands, normal address and data busses (not multiplexed), simple instruction timing and other desirable features compared to the 8080A and the 8085A.

In this chapter, we present a brief description of the Z80 CPU in terms of its internal architecture, pin-out, instruction set, interrupts and other important features. For the sake of brevity, we will rely heavily on a comparative approach indicating similarities between the Z80 and the 8085A whenever possible. Even though our description of the Z80 CPU is fairly complete, the reader is advised to refer to the manufacturer's data sheets and manuals prior to actual use.

14.1 INTERNAL ARCHITECTURE OF THE Z80 CPU

The Z80 is a 8-bit microprocessor with a 16-bit address bus whose internal organization (as far as the user is concerned) resembles very closely that of the 8080A and the 8085A CPU. As a matter of fact, except for the differences mentioned next, the reader may regard the Z80 to have the architecture shown in Fig. 6.3.1 for the 8085A device. In this section, therefore, we will limit ourselves to indicating the differences between the two devices.

14.1.1 Program Status Word (PSW) of the Z80

Recall that the 8085A used a special register, referred to as the PSW, to store the FLAG bits Carry (CY), Auxiliary Carry (AC), Zero (Z), Parity (P) and Sign (S) as described in Sec. 8.1. The Z80 also uses a PSW register to store the FLAG bits as they are affected by various instructions. Figure 14.1.1 shows the FLAGs stored in the Z80's PSW register and the positions they occupy.

S	Z	x	AC	x	P/O	N	C

X : Not used
Z : Zero
P/O : Parity/Overflow (Parity only in 8085A)
C : Carry (same as CY for 8085A)
S : Sign
AC : Auxiliary Carry
N : Subtract (used for BCD subtraction, not present in 8085A)

Fig. 14.1.1 Program Status Word (PSW of the Z80

Comparing this with the 8085A (or the 8080A) CPU, one finds that the PSW registers of the two devices are essentially very similar. The FLAGs S, Z, AC and C (referred to as CY in the 8085A) occupy the same positions, as the corresponding FLAGs of the 8085A and denote the same conditions, i.e. Sign, Zero, Auxiliary Carry and Carry, respectively. (Note that as in the 8085A, the Z80's Carry FLAG indicates a carry-out on addition but a borrow-in on subtraction.) The Parity FLAG of the 8085A is modified to show both Parity and Overflow conditions appropriately. For signed binary arithmetic operations, the P/O FLAG indicates overflow in the result (i.e. it is set to 1, if there is an overflow but stays reset at Ø, otherwise). For other instructions, this bit indicates the parity of the result just as in the case of the 8085A. The bit N shows a Subtract status which is used by the Z80 internally (in conjunction with AC) for properly implementing the Decimal Adjust instruction DAA. The combination of N and AC allows the Z80 to properly adjust the result for both BCD additions and subtractions. (Recall that the 8085A can do this only for BCD additions by issuing DAA following an addition.) The Z80, like the 8085A, provides conditional branching instructions which test the status of the S, Z, P/O, and C FLAGs. The FLAGs AC and N, however, cannot be used for such conditional branchings.

14.1.2 Programmable Registers of the Z80

The Z80 CPU has more programmable registers than the 8085A. (Referring to Fig. 6.3.1 and Sec. 8.1, recall that the programmable registers of the 8085A are PSW, A, B, C, D, E, H, L, PC and SP). The Z80 has all the registers of the 8085A and more. The additional registers in the Z80 are of two types—one set of registers merely provide an alternate register bank duplicating some other registers of the Z80, the other set of registers (i.e. IX, IY, IV and R) provide some additional functional power to the Z80. Fig. 14.1.2 shows the programmable registers of the Z80; this figure also shows which registers of the Z80 CPU are common to both this and the 8085A.

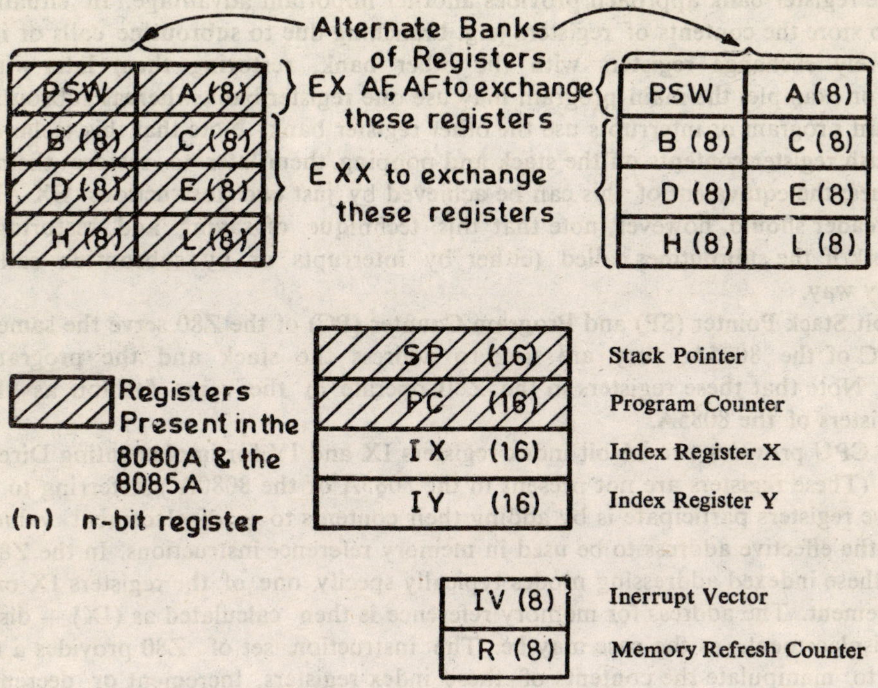

Fig. 14.1.2 Programmable Registers of the Z80

Consider any one of the two register banks of the Z80. In this bank, the registers PSW, A, B, C, D, E, H and L are functionally similar to the corresponding registers of the 8085A. As for the 8085A, the register A is the primary accumulator whereas B, C, D, E, H, L are the secondary accumulators which may be concatenated to give 16-bit registers B, D and H; the PSW register has been described earlier in Sec. 14.1.1. All the instructions involving these registers present in the 8085A (except RIM and SIM) are also present in the Z80. In addition, they participate in many more instructions in the Z80, i.e. instructions not there in the 8085A. In the Z80, registers B and C serve as counters in some specialized instructions like block moves from memory to I/O or memory, or from I/O to memory, and block search operations. All these registers of the Z80 CPU (except PSW) may be rotated left or right, individual register bits may be set or reset or transferred in complemented form to the Zero flag. Note that, in the 8085A, one could only rotate the accumulator A and that direct manipulation of the individual register bits was not at all possible.

The registers PSW A, B, C, D, E, H and L have been duplicated in the Z80 in the form of two alternate register banks. On power-up or RESET, the user starts using one of these register banks (any one of the two may be used, the actual choice is immaterial). In order to use registers in the other bank, the user must exchange registers of the current bank with the other by means of appropriate instructions; the original registers may be brought back later by doing another exchange operation. The Z80 provides two instructions for doing such exchanges; the EX AF, AF instruction exchanges only A and PSW (leaving the others unchanged) whereas EXX exchanges the B, C, D, E, H and L with the corresponding registers of the other bank.

Apart from providing the user with additional storage locations for operands inside the CPU, this alternate register bank approach provides another important advantage. In situations where one wants to store the contents of registers (e.g. branching due to subroutine calls or interrupts), one can merely exchange registers with the other bank, restoring them later with another exchange. For example, the main program may use one register bank whereas subroutines called from the main program or interrupts use the other register bank. Note that this technique avoids having to push register contents on the stack and popping them later to restore them to their original values; the equivalent of this can be achieved by just two instructions EX AF, AF and EXX. The reader should, however, note that this technique of saving and restoring registers will only work if the subroutines called (either by interrupts or by subroutine calls) are not nested in any way.

The 16-bit Stack Pointer (SP) and Program Counter (PC) of the Z80 serve the same functions as SP and PC of the 8085A; they are used to address the stack and the program memory respectively. Note that these registers in the Z80 operate in the same fashion as the corresponding registers of the 8085A.

The Z80 CPU provides two 16-bit index registers IX and IY for implementing Direct Indexed Addressing. (These registers are not present in the 8085A or the 8080A.) Referring to Sec. 7.4.5, the way these registers participate is by adding their contents to a displacement value in order to calculate the effective address to be used in memory reference instructions. In the Z80, instructions using these indexed addressing modes typically specify one of the registers IX or IY and a 8-bit displacement. The address for memory reference is then calculated as (IX) + displacement or (IY) + displacement, as the case may be. The instruction set of Z80 provides a number of instructions to manipulate the contents of these index registers, increment or decrement them, load them from other registers or memory locations or store their contents in other registers or memory locations.

Another new register of the Z80 is its 8-bit Interrupt Vector Register (IV). The use of this register is closely tied with the interrupt vectoring technique used by the Z80 and will be discussed in detail in Sec. 14.4 while discussing the Interrupts of the Z80. The Z80 provides the following instructions to load IV and to read its contents

LD A,IV; (A) ← (IV), load contents of IV into A, read IV

LD IV,A; (IV) ← (A), load IV from A, load IV

The 8-bit Memory Refresh Counter (R) of the Z80 is useful in applications where dynamic memories are being used. (This is not present in the 8085A.) Just as for IV, R may be loaded or read via the accumulator by the following instructions:

LD A, R; (A) ← (R), load contents of R into A, read R

LD R, A; (R) ← (A), load contents of A into R, load R

Referring to Sec. 3.5.2, recall that dynamic RAMs require to be refreshed so that they do not lose their contents. This typically requires a counter to keep track of the progress of the refreshing action. This counter is provided internally in the Z80 by the Memory Refresh Counter R (also referred to as the Refresh Register). In clock states where the Z80 does not require the system bus, the lower 7 bits of the Refresh Register are output on the lower 7 lines of the Address Bus. It also outputs a control signal $\overline{RFSH}$ Low at these times to signal that a memory refresh operation is in progress. It is easy to see that with proper use, the bulk of the logic needed to refresh dynamic memories is provided by the Z80 itself.

From the user's point of view, the differences between the internal architectures of the Z80 and the 8085A basically amount to some changes in the definition of the PSW bits and a much greater number of registers in the Z80. The more powerful instruction set of the Z80 is essentially a corollary to this abundance of registers inside the CPU. This will be evident from the Z80's instruction set given in Sec. 14.5.

14.2 The Pin-Out of the Z80 CPU

The Z80 is available in a 40-pin DIP (Dual In Line Package) like the 8085A. The actual pin-outs of the two devices are however completely different. A major difference between the two devices is that the Z80, unlike the 8085A, does not multiplex its data and address busses. The control bus of the Z80 is also differently organized, even though it performs essentially similar functions. The Z80 does not have serial I/O lines like the 8085A and has fewer interrupt inputs. However, it provides appropriate signals for refreshing dynamic memory devices in the system. In the following, we describe the pins of the Z80 relying heavily, wherever possible, on comparisons with the pins of the 8085A. Note that the numbers given in brackets indicate the actual pin numbers of the corresponding lines in the 40-pin DIP package of the Z80.

V_{CC} (+5V) (11), GND (29) — Power Connections

Like the 8085A (and unlike the 8080A), the Z80 operates from a single +5 V power supply. All inputs and outputs of the Z80 are TTL compatible in terms of their voltage levels

Φ (6) — CPU Clock Input

Unlike the 8085A, the Z80 does not have an on-chip clock generator. For Z80 systems, one would require a clock generator outputting a clock signal to the Z80's Φ input.

A0–A15 — Address Bus, Tristate Outputs, A0–A10 (30–40), A11–A15 (1–5)

Unlike the 8085A, the Z80's address-data busses are not multiplexed. The address bus outputs a 16-bit memory address for memory accesses, i.e. instruction fetch, memory read or memory write. During I/O accesses (read or write), only A0–A7 carry the 8-bit address of the I/O port. (A8–A15 contain other information. This is unlike the 8085A, where for I/O operations, A0–A7 are equal to A8–A15.) During memory refresh times, indicated by $\overline{\text{RFSH}} = \emptyset$, A0–A6 contain the lower 7-bits of register R.

The address buses A0–A15 are tristate outputs which may be floated by the Z80 if external logic, e.g. a DMA Controller, requests the system bus. The Z80 can be made to drive its A0–A15 lines tristate by inputting $\overline{\text{BUSRQ}} = \emptyset$.

D0–D7 — Data Bus, Bidirectional and Tristate D0(14), D1(15), D2(12), D3(8), D4(7), D5(9), D6(10), D7(13)

This is a normal data bus (unlike that of the 8085A which is multiplexed with A0–A7). It floats in situations where the Z80 relinquishes the system bus to other devices, i.e. by applying $\overline{\text{BUSRQ}} = \emptyset$, or when memory refresh is in progress, i.e. $\overline{RFSH} = Ø$. The data bus lines are inputs during instruction fetches, memory or I/O read operations and during interrupt acknowledge vectoring; they are outputs during write operations to memory or I/O.

$\overline{INT}$ (16) General Purpose Interrupt Request Input

This is the Z80's counterpart of the INTR input of the 8085A. It is a general purpose interrupt input which can be enabled or disabled (i.e. masked) and can be shared by a number of interrupting devices. The operation of this interrupt input differs considerably from the INTR input of the 8085A and will be discussed in detail in Sec. 14.4. Note that $\overline{INT} = Ø$ is needed to place an interrupt request to the Z80.

$\overline{NMI}$ (17) Non-maskable Interrupt Request Input

This interrupt cannot be disabled (like TRAP of the 8085A) and must be pulled Low, i.e. $\overline{NMI} = Ø$, to place the corresponding interrupt request to the Z80. It will be considered in detail in Sec. 14.4.

$\overline{WAIT}$ (24) Wait State Request Input

Applying $\overline{WAIT} = Ø$, forces the CPU to interpose Wait states in its machine cycles until $\overline{WAIT}$ is returned to 1. Functionally, $\overline{WAIT}$ is equivalent to the READY input of the 8085A and is used to interface slow memory or I/O devices to the system, i.e. it extends the width of the READ or WRITE operations.

$\overline{RESET}$ (26) Reset Input.

A Hi→Lo→Hi pulse at this input resets the Z80 (Unlike the 8085A, no RESET OUT is generated.) While $\overline{RESET} = Ø$, the system bus lines (A0–A15, D0–D7, $\overline{MREQ}$, $\overline{IORQ}$, $\overline{RD}$ and $\overline{WR}$) are floated. Following the $\overline{RESET}$ pulse, the interrupt requests on $\overline{INT}$ are disabled and the registers IV and R are cleared to ØØ. The Program Counter (PC) is reset to ØØØØ, so that following a reset the next instruction to be executed is fetched from the memory location ØØØØ.

$\overline{M1}$ (27) Output identifying an Instruction Fetch machine cycle (OF–MC)

$\overline{M1} = Ø$ indicates that the Z80 is currently fetching an op-code from memory, i.e. executing an op-code Fetch machine cycle (OF–MC). (*Note*: For the Z80, the OF–MC is referred to as an Instruction Fetch machine cycle. We will, however, retain our earlier nomenclature of referring to this as the OF–MC, as introduced in Chaps. 6, 8, and 9.)

$\overline{MREQ}$ (19) Memory Request Tristate Output

$\overline{MREQ} = Ø$ indicates that a memory access operation is in progress. This could be for fetching instructions or for reading from memory or writing to it.

$\overline{IORQ}$ (20) I/O Request, Tristate Output

$\overline{IORQ} = \emptyset$ indicates that an I/O access operation (I/O read or I/O write) is in progress. Moreover, an Interrupt Acknowledge cycle is indicated by the combination $\overline{IORQ} = \overline{M1} = \emptyset$.

$\overline{RD}$ (21) Read Control, Tristate Output

$\overline{RD} = \emptyset$ indicates that in the current machine cycle, the Z80 CPU wants to read data from I/O port or memory (data or instruction op code in case of reading from memory). Note that $\overline{RD}$ actually outputs an active low pulse in such machine cycles.

$\overline{WR}$ (22) Write Control, Tristate Output

$\overline{WR} = \emptyset$ implies that in the current machine cycle, the Z80 wants to write data to an I/O port or memory location. Note that $\overline{WR}$ actually outputs an active low pulse in such machine cycles.

The signals $\overline{MREQ}$, $\overline{IORQ}$, $\overline{RD}$ and $\overline{WR}$ form the control bus of the Z80 system. These signals are floated (along with A0–A15 and D0–D7) whenever the Z80 relinquishes the system bus. This is done whenever $\overline{RESET} = \emptyset$ or when the Z80 vacates the system bus in response to a request to that effect on $\overline{BUSRQ}$.

The control signals $\overline{M1}$, $\overline{RD}$, $\overline{WR}$, $\overline{IORQ}$ and $\overline{MREQ}$ of the Z80 carry out more or less the same functions as the S1, SØ, IO/$\overline{M}$, $\overline{RD}$ and $\overline{WR}$ outputs of the 8085A. The logical interpretation of the Z80's control signals is, however, different from that of the corresponding signals of the 8085A. The way the control signals of the Z80 behave in the various machine cycles is summarized in the following:

	Machine Cycle	$\overline{M1}$	$\overline{MREQ}$	$\overline{IORQ}$	$\overline{RD}$	$\overline{WR}$
1	OF–MC	Ø	Ø	1	Ø	1
2	MR–MC	1	Ø	1	Ø	1
3	MW–MC	1	Ø	1	1	Ø
4	IOR–MC	1	1	Ø	Ø	1
5	IOW–MC	1	1	Ø	1	Ø
6	INA–MC (for the $\overline{INT}$ interrupt)	Ø	1	Ø	1	1
7	INA–MC (for the $\overline{NMI}$ interrupt)	Ø	Ø	1	Ø	1

$\overline{HALT}$ (18) Halt State Output

$\overline{HALT}$ is output Low by the CPU whenever it executes the Halt instruction. $\overline{HALT} = \emptyset$ is output by the CPU as long as no interrupt request is placed or the Z80 is not reset.

$\overline{\text{RFSH}}$ (28) — Memory Refresh Output

The CPU outputs the signal $\overline{\text{RFSH}}$ = Ø to refresh dynamic memories using register R as the memory refresh counter. Specifically, $\overline{\text{RFSH}}$ = Ø is output whenever the Z80 does not require the system bus in its normal operation. During these times, AØ–A6 output the lower 7 bits of R with $\overline{\text{MREQ}}$ = Ø and $\overline{\text{M1}}$ = $\overline{\text{IORQ}}$ = $\overline{\text{RD}}$ = $\overline{\text{WR}}$ = 1 and the data bus is floated by the Z80.

$\overline{\text{BUSRQ}}$ (25) — Request for Z80 to Vacate the System Bus, Input

External logic (e.g. DMA Controller) requests the Z80 to vacate the system bus (by floating A0–A15, D0–D7, $\overline{\text{MREQ}}$, $\overline{\text{IORQ}}$, $\overline{\text{RD}}$ and $\overline{\text{WR}}$) by inputting $\overline{\text{BUSRQ}}$ = Ø. This is the functional equivalent of the 8085A's HOLD input.

$\overline{\text{BUSAK}}$ (23) — Acknowledge for $\overline{\text{BUSRQ}}$, Output

The Z80 acknowledges a bus request with $\overline{\text{BUSRQ}}$ = Ø by means of its $\overline{\text{BUSAK}}$ output. It outputs $\overline{\text{BUSAK}}$ = Ø when it is actually able to relinquish the system bus. Functionally, this is equivalent to the HLDA output of the 8085A. Note that unlike the 8085A, $\overline{\text{BUSRQ}}$ and $\overline{\text{BUSAK}}$ are both active low signals.

14.3 EXECUTION TIMING OF THE Z80 CPU

Compared to the 8085A, the Z80 is certainly simpler in one respect. Its address and data bus are not multiplexed implying that, in a machine cycle, the timing complexity introduced by such multiplexing is not there. The control signals output by the Z80 (i.e. $\overline{\text{M1}}$, $\overline{\text{MREQ}}$, $\overline{\text{IORQ}}$, $\overline{\text{RD}}$, $\overline{\text{WR}}$) are different from those output by the 8085A (i.e. IO/$\overline{\text{M}}$, S1, SØ, $\overline{\text{RD}}$, $\overline{\text{WR}}$, $\overline{\text{INTA}}$), though taken together they still serve the same purpose. Note that this implies that given a 8085A based system, to convert to a Z80 system one needs to make changes in the signals selecting memory and I/O devices and the signals actually doing the reading or writing.

As the timing diagrams given subsequently show, for Z80 systems one may derive chip select ($\overline{\text{CS}}$s) in the following fashion:

$\overline{\text{CS}}$ = f(A0–A15, $\overline{\text{MREQ}}$) for memory and memory mapped I/O

$\overline{\text{CS}}$ = f(A0–A7, $\overline{\text{IORQ}}$) for I/O mapped I/O

where the function f(·) is decided by where in the address space the memory or I/O device should be placed. Here, the $\overline{\text{RD}}$ and $\overline{\text{WR}}$ outputs of the Z80 should be connected directly to the $\overline{\text{RD}}$ and $\overline{\text{WR}}$ inputs of the memory or I/O device, as appropriate.

Like the 8085A, the instructions of the Z80 get executed by executing a sequence of different kinds of Machine Cycles where each machine cycle consists of a number of T-states. A T-state corresponds to one cycle of the clock Φ input to the Z80 and, therefore, the number of T-states required for each instruction is directly indicative of the time taken to execute it. The machine cycles required by the Z80 are:

(i) Instruction Fetch or Op-code Fetch (OF–MC)

(ii) Memory Read (MR–MC)
(iii) Memory Write (MW–MC)
(iv) I/O Read (IOR–MC)
(v) I/O Write (IOW–MC)
(vi) Interrupt Acknowledge (INA–MC)

[The reader is urged to review Secs. 6.2 and 9.2 in this context.]

Figure 14.3.1 shows the number of T-states required for the different kinds of machine cycles without introducing additional Wait states (see Sec. 9.2.8–9.2.9) by pulling $\overline{\text{WAIT}}$ Low. Note that the machine cycles of the Z80 take between 3 to 6 T-states for execution.

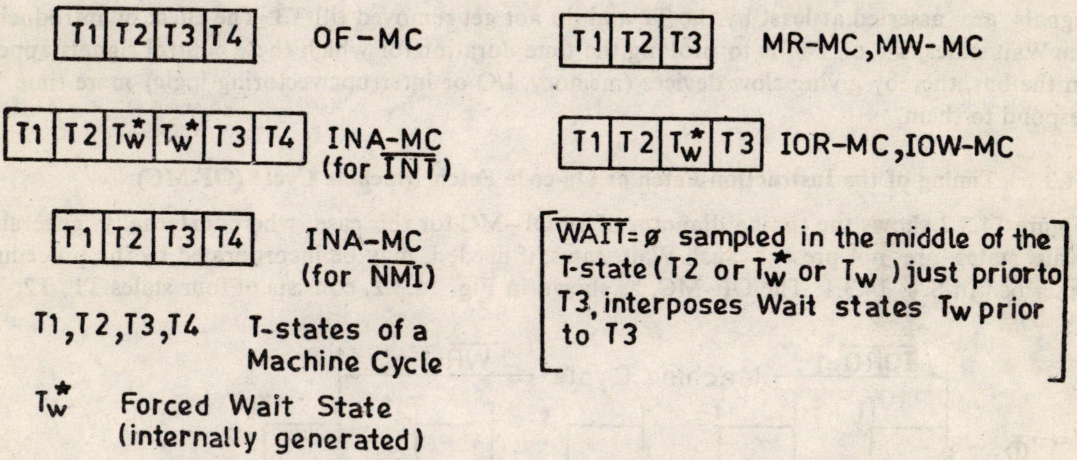

Fig. 14.3.1 T-states of the Z80's Machine Cycles without External Wait States

Referring to Fig. 14.3.1, note that whereas states T1, T2, T3 and T4 are the normal states occurring in sequence, the states labelled T_W^* are internally generated Forced Wait states which the Z80 interposes between states T2 and T3 without regard to the $\overline{\text{WAIT}}$ input. The various signals output by the Z80 stay unchanged over the state T_W^*, i.e. from the end of T2 to the beginning of T3. This reflects the Z80 designer's belief that whereas the system memory (ROM or RAM) will be fast enough to operate without Wait states, I/O devices will be slower and may require an additional state T_W^* for proper operation. (A designer concerned with the resultant lowering of the speed of I/O operations may limit himself to memory mapped I/O for higher speeds.) Similarly, the INA–MC for the $\overline{\text{NMI}}$ interrupt does not have forced Wait states since in this case the vectoring is completely internal. On the other hand, the $\overline{\text{INT}}$ interrupt may be shared between a number of users through 'Daisy Chaining' or Priority Encoding and the INA–MC for this may require more time for external logic to respond with an interrupt vector; for this INA–MC, this additional time is provided by the two forced Wait states T_W^* between T2 and T3.

14.3.1 Externally Generated Wait States

Apart from the internally generated forced Wait states T_W^*, one can also introduce externally generated Wait states in all the machine cycles of the Z80 except the INA–MC following the $\overline{\text{NMI}}$ interrupt. These externally generated Wait states, referred to as T_W, are introduced by making $\overline{\text{WAIT}} = \emptyset$ of the Z80 CPU. The Z80 samples the $\overline{\text{WAIT}}$ input in the middle of the state preceding T3 (T_W^* for INA, IOR and IOW and T2 for OF, MR and MW). It proceeds in the normal fashion to T3 if $\overline{WAIT}$ is sampled to be '1' at this instant; if $\overline{WAIT}$ is sampled to be 'Ø', it introduces a Wait state T_W instead of going to T3. In the latter case, $\overline{WAIT}$ is again sampled at the middle of T_W with the Z80 introducing another T_W state if at the sampling instant $\overline{WAIT} = Ø$; on the other hand, it goes to T3 if $\overline{WAIT} = 1$ at that instant. This procedure is repeated as long as $\overline{WAIT}$ is Ø, thereby allowing the user to introduce an integral number of externally generated wait states in the machine cycle by keeping the $\overline{WAIT}$ input Low for a sufficiently long time.

The function of both types of Wait states, externally generated (T_W) or internally generated (T_W^*) is to allow slow peripheral logic or devices to be used in the system. With this function in mind, all the Z80's machine cycles are such that the relevant $\overline{CS}$s and $\overline{RD}$, $\overline{WR}$ and other control signals are asserted at least by the T2 and do not get removed till T3. The effect of introducing the Wait states, T_W or T_W^*, is to prolong the time duration for which these control signals appear on the bus, thereby giving slow devices (memory, I/O or interrupt vectoring logic) more time to respond to them.

14.3.2 Timing of the Instruction Fetch or Op-code Fetch Machine Cycle (OF-MC)

Figure 14.3.2 shows the timing diagram of an OF–MC for the case where externally generated Wait states are not present. Such Wait states, if needed, may be incorporated by the procedure described in Sec. 14.3.1. The OF–MC, as shown in Fig. 14.3.2, consists of four states T1, T2, T3

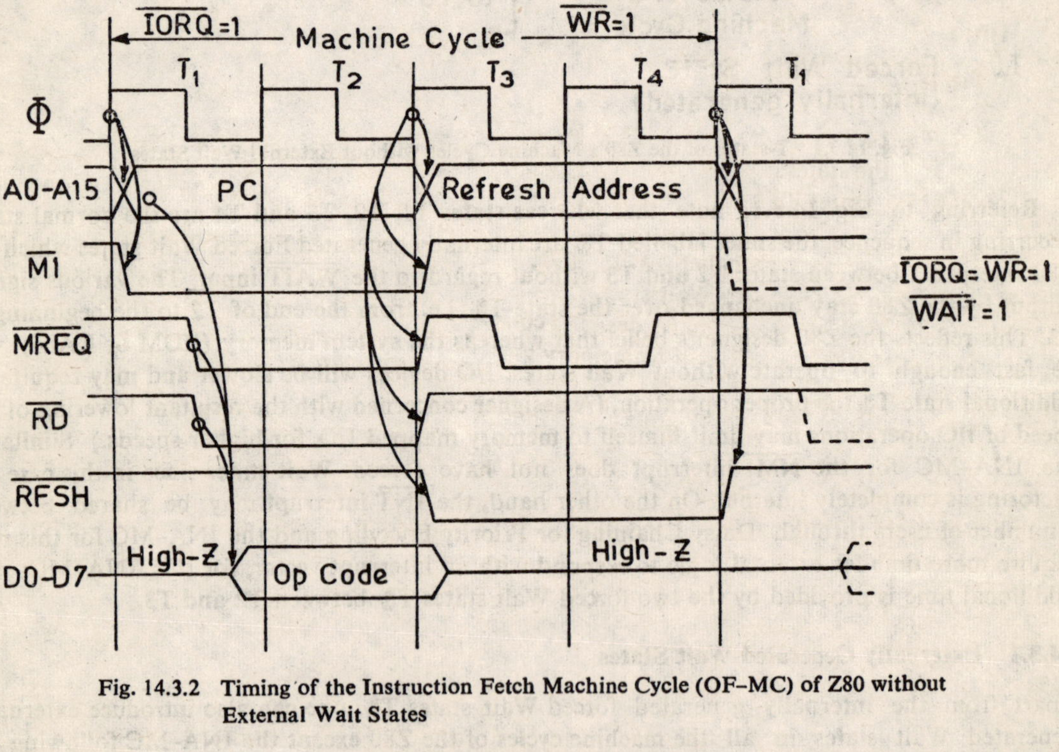

Fig. 14.3.2 Timing of the Instruction Fetch Machine Cycle (OF–MC) of Z80 without External Wait States

and T4 of which states T1 and T2 are used to read the instruction op-code from memory and states T3 and T4 are used to refresh dynamic memories by granting a refresh cycle on the system bus. The contents of the Program Counter (PC) are output on the address bus A0–A15 during T1 and T2 pointing to the memory location containing the instruction op code. $\overline{\text{M1}}$ is output Low during these states to signal that it is an OF–MC. Moreover $\overline{\text{MREQ}}$ and $\overline{\text{RD}}$ are also output active Low from the middle of T1 to the end of T2 so that the selected memory device can drive the data bus with the contents of the addressed location during this time; the Z80 interprets the data on the data bus as the op code (or part of the op code, see later).

During T3 and T4 of the OF–MC, the Z80 outputs $\overline{\text{RFSH}} = \emptyset$ to indicate that these T-states may be used, if so desired, for refreshing dynamic memories in the system. The memory area to be refreshed is indicated by A0–A6 which output the lowest seven bits of the Memory Refresh Counter as the 'Refresh Address' during these states. Since this is a memory operation $\overline{\text{MREQ}}$ goes Low (from the middle of T3 to the middle of T4) but since no actual reading or writing is involved; $\overline{\text{RD}}$ and $\overline{\text{WR}}$ stay at '1' during these states. (Actually, as expected, $\overline{\text{IORQ}} = \overline{\text{WR}} = 1$ during the entire OF–MC.)

One major distinction between the 8085A and the Z80 is the fact that whereas all 8085A op codes have only one byte, Z80 op codes may have one byte or two bytes. (Note that these are the bytes required to specify an instruction, not including what may be additionally specified as immediate data or address.) This implies that, whereas all 8085A instructions will have only one OF–MC each, a Z80 instruction may have either one or two OF–MCs depending on whether it has a one-byte or two-byte op code.

14.3.3 Timing of the Memory Read Machine Cycle (MR–MC)

Figure 14.3.3 shows the timing of the Z80's Memory Read machine cycle without externally generated Wait states. Note that this is a three T-state machine cycle where A0–A15 outputs

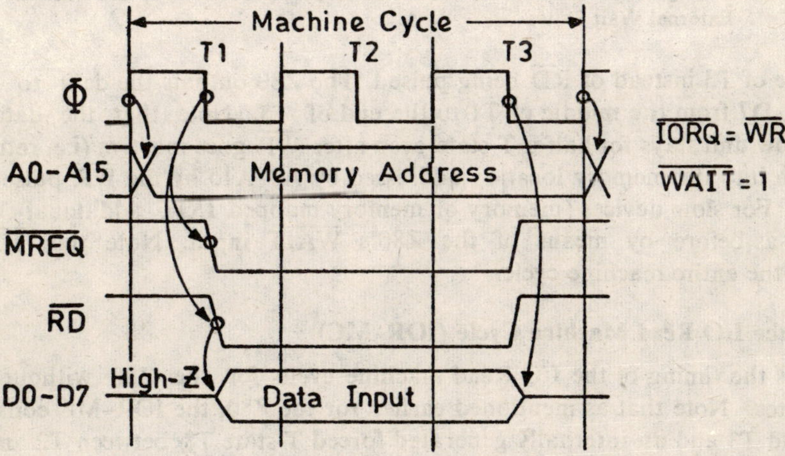

Fig. 14.3.3 Timing of the Memory Read Machine Cycle (MR–MC) of Z80 without External Wait States

the address of the memory location from which data is to be read and $\overline{\text{MREQ}}$ and $\overline{\text{RD}}$ are Low from the middle of T1 to the middle of T3. With proper chip selects, the selected memory device outputs on the data bus the contents of the addressed memory location for the Z80 to read. This data would be present on the data bus while $\overline{\text{RD}}$ is Low. Additional Wait states, T_W, may be interposed between T2 and T3 by pulling $\overline{WAIT}$ Low for an appropriate duration; this extends the chip select, $\overline{MREQ}$ and $\overline{RD}$ allowing slower devices (memory or memory mapped I/O) more time to respond. Note that $\overline{IORQ} = \overline{WR} = \overline{M1} = 1$ throughout the machine cycle.

14.3.4 Timing of the Memory Write Machine Cycle (MW–MC)

The timing of this machine cycle without external Wait states is shown in Fig. 14.3.4. This is a three T-state machine cycle similar to MR–MC except that $\overline{WR}$ is pulsed Low from the middle

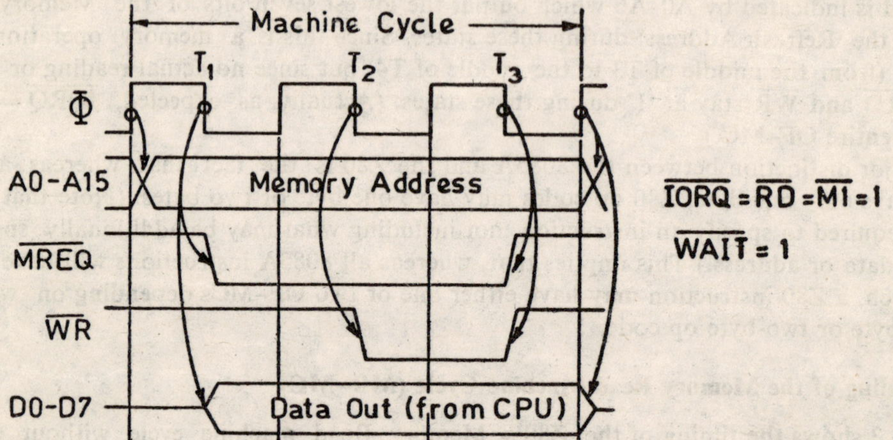

Fig. 14.3.4 Timing of the Memory Write Machine Cycle (MW–MC) of Z80 without External Wait States

of T2 to the middle of T3 instead of $\overline{RD}$ being pulsed. The Z80 outputs the data to be written on its data bus D0–D7 from the middle of T1 to the end of T3. Note that the data is stable before $\overline{WR}$ goes Low and stays for half a T-state even after $\overline{WR}$ goes inactive (i.e. returns to '1'). This data is written into the memory location addressed by A0–A15 by the $\overline{WR}$ pulse conditioned by $\overline{MREQ} = \emptyset$. For slow devices (memory or memory mapped I/O), additional Wait states may be introduced as before by means of the Z80's $\overline{WAIT}$ input. Note that $\overline{IORQ} = \overline{RD} = \overline{M1} = 1$ during the entire machine cycle.

14.3.5 Timing of the I/O Read Machine Cycle (IOR–MC)

Figure 14.3.5 shows the timing of the I/O Read machine cycle for the Z80 without externally generated Wait states. Note that as mentioned earlier for the Z80, the IOR–MC consists of four T-states T1, T2, and T3 and an internally generated forced T-state T_W^* between T2 and T3. The address of the I/O port is output only on A0–A7. (In the 8085A, A0–A7 = A8–A15 = port address). $\overline{IORQ}$ is pulsed Low from the beginning of T2 to the middle of T3 indicating the fact that this is a machine cycle for I/O access. $\overline{RD}$ is pulsed Low with the same timing as $\overline{IORQ}$. The port selected by $\overline{IORQ} = \emptyset$ and by decoding chip select from A0–A7 is read by the $\overline{RD}$ signal and outputs its contents on D0–D7 for the Z80 to read. For even slower devices, one can introduce additional Wait states T_W by manipulating the $\overline{WAIT}$ input. Note that $\overline{M1} = \overline{MREQ} = \overline{WR} = 1$ for the entire machine cycle.

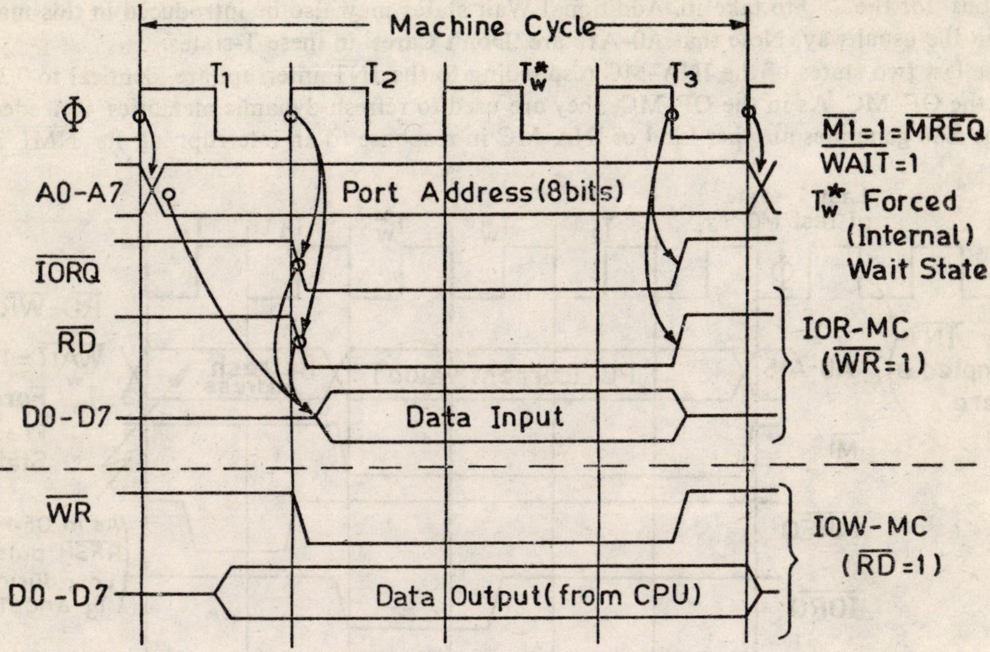

Fig. 14.3.5 Timing of the I/O Read (IOR–MC) and I/O Write (IOW–MC) Machine Cycles of Z80 without External Wait States

14.3.6 Timing of the I/O Write Machine Cycle (IOW–MC)

As shown in Fig. 14.3.5, the timing of the I/O Write machine cycle and the IOR–MC are very similar to each other. In the IOW–MC, $\overline{WR}$ is pulsed Low in the same position as $\overline{RD}$ of IOR–MC; this writes the contents of the data bus, output by Z80, into the addressed port. As usual, additional Wait states T_W may be generated by manipulating $\overline{WAIT}$ properly. Note that $\overline{M1} = \overline{MREQ} = \overline{RD} = 1$ throughout IOW–MC.

14.3.7 Timing of the Interrupt Acknowledge Machine Cycles (INA–MC)

In response to an interrupt at its $\overline{INT}$ input, the Z80 may require external logic to place a byte on the data bus which would be used by it to handle the interrupt. (When this is needed and how this byte will be interpreted is discussed in Sec. 14.4.2.) This is done by the Z80 by means of a six T-state INA–MC whose timing diagram is shown in Fig. 14.3.6. The Z80 samples the $\overline{INT}$ input at the beginning of the last T-state of the last machine cycle of each instruction; if, at this instant, the $\overline{INT}$ interrupt is enabled and $\overline{INT} = \emptyset$, then the Z80 executes an INA–MC to get the interrupt vector byte needed to handle the interrupt.

Comparing the INA–MC (Fig. 14.3.6) with the OF–MC (Fig. 14.3.2), one finds the two MCs to be very similar to each other. The two MCs both have states T1–T4 with the INA–MC having two additional forced Wait states T_W^*. The Z80 provides these additional T_W^* states to allow more time to the vectoring logic to do interrupt arbitration between competing devices, e.g. priority encoding or daisy chaining. Both the INA–MC and OF–MC pulse $\overline{M1}$ Low as shown; they can however be distinguished by the fact that $\overline{IORQ}$ is pulsed Low in INA–MC. While $\overline{IORQ}$ is Low in INA–MC, external logic should place the interrupt vector byte on the data bus for the Z80 to take in. Additional Wait states may also be introduced in this machine cycle in the usual way. Note that A0–A15 are 'Don't Cares' in these T-states.

The last two states of the INA–MC responding to the $\overline{INT}$ interrupt are identical to T3 and T4 of the OF–MC. As in the OF–MC, they are used to refresh dynamic memories, if needed.

The Z80 generates another kind of INA–MC in response to an interrupt at its $\overline{NMI}$ input.

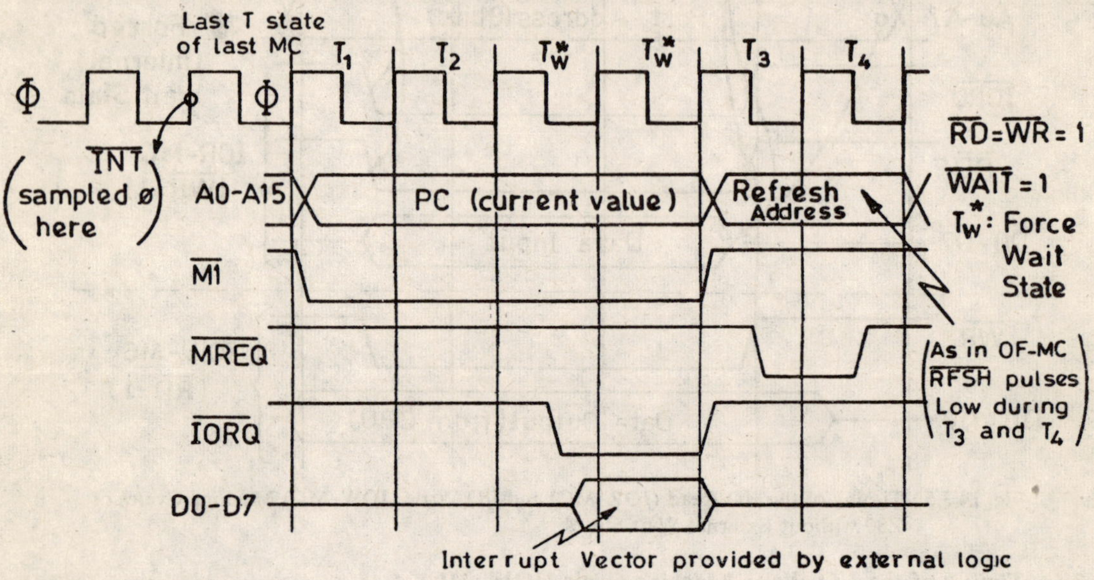

Fig. 14.3.6 Timing of the Interrupt Acknowledge Machine Cycle (INA–MC) of Z80 in response to interrupt at $\overline{INT}$ (without External Wait States)

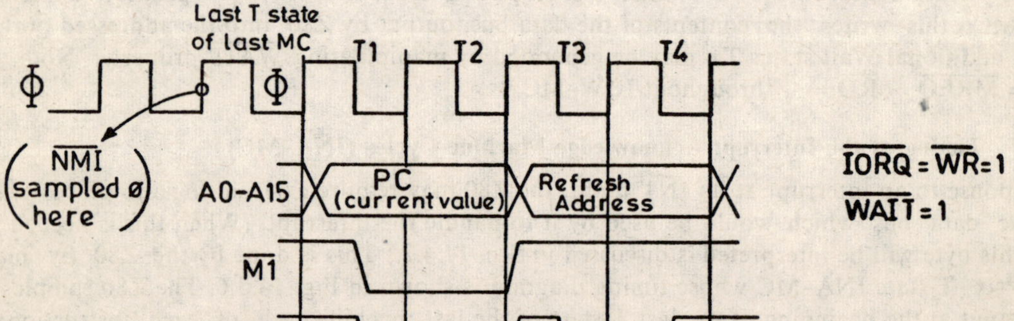

Fig. 14.3.7 Timing of the Interrupt Acknowledge Machine Cycle (INA–MC) of Z80 in response to interrupt at NMI (Wait States cannot be introduced)

The Z80 samples the $\overline{NMI}$ input at the beginning of the last T-state of the last machine cycle of each instruction; if $\overline{NMI}$ = Ø at this instant, then the Z80 executes an INA–MC of the type shown in Fig. 14.3.7. Note that this machine cycle is virtually identical to the OF–MC shown in Fig. 14.3.2. The only differences are that Wait states T_W cannot be introduced in this INA–MC (because $\overline{WAIT}$ is not sampled) and the contents of the data bus is irrelevant as far as the Z80 is concerned (in OF–MC, this will be the op code). The reason for the latter is that, in this INA–MC, the Z80 generates the necessary interrupt vector internally, without regard to the contents of the data bus.

14.3.8 Bus Requests with the Z80

The Z80 may be made to float its system bus (A0–A15, D0–D7, $\overline{MREQ}$, $\overline{IORQ}$, $\overline{RD}$ and $\overline{WR}$) by applying a Ø at its $\overline{BUSRQ}$ input. This is sampled at the beginning of the last T-state of every machine cycle and at the rising edge of Φ in each T-state when the bus is floating. The Z80 continues to float the bus in the next T-state if $\overline{BUSRQ}$ is sampled Low at these instants; otherwise, normal operation continues with the next machine cycle after returning the bus to its active state. The Z80 outputs $\overline{BUSAK}$ = Ø whenever it is floating the system bus in response to $\overline{BUSRQ}$ = Ø. Note that this is analogous to introducing HOLD states in the 8085A. As a matter of fact, except for the fact that they are active Low instead of being active High, $\overline{BUSRQ}$ and $\overline{BUSAK}$ correspond functionally to HOLD and HLDA, respectively, of the 8085A. Refresh cycles are not generated while $\overline{BUSAK}$ = Ø, i.e. the system bus is floating.

14.3.9 Execution of the HALT Instruction

Following the execution of the HALT instruction (analogous to HLT of the 8085A), the Z80 continues to execute a sequence of NOP instructions (no operation), each with four T-states, while outputting $\overline{HALT}$ = Ø. The four T-states T1, T2, T3, and T4 are such that $\overline{IORQ}$ = $\overline{MREQ}$ = $\overline{RD}$ = $\overline{WR}$ = 1 but the interrupts are sampled at the beginning of T4 in each NOP instruction. One can exit from this condition only through an $\overline{INT}$ (if enabled) or $\overline{NMI}$ interrupt or by a RESET (Low pulse at $\overline{RESET}$). Note during T3 and T4 of these NOP instructions, the Z80 will allow refresh cycles for refreshing dynamic memories.

14.4 INTERRUPTS OF THE Z80

Unlike the five interrupt inputs present in the 8085A, the Z80 provides only two interrupt inputs, $\overline{INT}$ and $\overline{NMI}$, to the user. The $\overline{NMI}$ interrupt has the higher priority and is a non-maskable interrupt like TRAP of the 8085A. The $\overline{INT}$ interrupt corresponds to the general purpose INTR interrupt of the 8085A; in some respects, the Z80's response to this interrupt is, however, superior to the 8085A's response to INTR, i.e. in ease of vectoring. Note that the $\overline{INT}$ interrupt is the one to be shared by a number of users, whenever such sharing is needed.

The Z80 suffers from one definite drawback compared to the 8085A, i.e. it has less interrupts than the 8085A. Specifically, it does not have anything corresponding to the maskable vectored interrupts, RST 5.5, RST 6.5 and RST 7.5 of the 8085A.

The Z80 samples its interrupts $\overline{NMI}$ and $\overline{INT}$ at the beginning (rising edge of Φ) of the last T-state of the last machine cycle of each instruction (including the NOP instructions following the HALT instruction's execution). It responds to the $\overline{NMI}$ interrupt if $\overline{NMI}$ = Ø. Otherwise, if $\overline{NMI}$ = 1 but $\overline{INT}$ = Ø and the $\overline{INT}$ interrupt is enabled, the Z80 CPU will respond to the $\overline{INT}$ interrupt. These two interrupts are described in detail next.

14.4.1 The $\overline{\text{NMI}}$ Interrupt

The $\overline{\text{NMI}}$ interrupt of the Z80 is a non-maskable interrupt which cannot be disabled. It has a higher priority than $\overline{\text{INT}}$ and will be responded to by the CPU if the $\overline{\text{NMI}}$ input is sampled Low as described earlier. It is a level sensitive interrupt and is vectored internally. This implies that the INA–MC generated in response to the $\overline{\text{NMI}}$ interrupt (Fig. 14.3.7) does not require any data from external logic but is used by the Z80 to do the vectoring internally.

In response to an $\overline{\text{NMI}}$ interrupt, the Z80 forces internally the instruction CALL ØØ66$_H$ to be executed; the first instruction of the corresponding interrupt service routine should be at memory location ØØ66$_H$. It, moreover, disables other interrupts while this service routine is being executed by clearing the Interrupt Enable Flip-Flop (IFF) inside the CPU. Before doing this, however, it stores internally the value of IFF prior to the $\overline{\text{NMI}}$ interrupt. On return from the interrupt service routine with the RETN instruction, IFF is restored to its earlier (stored) value. This restores the interrupt enable/disable status to its earlier value while keeping interrupts disabled during $\overline{\text{NMI}}$'s service routine. One can, of course, reenable interrupts explicitly by executing the EI instruction.

The $\overline{\text{NMI}}$ interrupt, like TRAP of the 8085A, is used to handle catastrophic events like impending power-failure. Like TRAP it should be used judiciously and with caution because it cannot be disabled (and may come at a time when the microcomputer may not like to be interrupted). If not being used, it would be desirable to tie the $\overline{\text{NMI}}$ input to V_{CC} with a pull-up resistor.

14.4.2 The General Purpose $\overline{\text{INT}}$ Interrupt

The Z80 provides a general purpose $\overline{\text{INT}}$ interrupt which is intended to be shared between a number of interrupt controlled peripherals. This has a lower priority than the $\overline{\text{NMI}}$ interrupt and generally (with one exception) requires the interrupting device (or associated logic) to provide vectoring information to identify the Interrupt Service Routine (ISR) to be executed. This is a level sensitive interrupt which must be sampled Low, as described earlier, to be acknowledged. The $\overline{\text{INT}}$ interrupt is disabled following RESET or by executing the DI (Disable Interrupt) instruction. It is also disabled when the Z80 responds to an $\overline{\text{INT}}$ or $\overline{\text{NMI}}$ interrupt. An ISR for the $\overline{\text{INT}}$ interrupt should normally be terminated by the RETI (Return from Interrupt) instruction; this is similar to a subroutine-return instruction except that it reenables the $\overline{\text{INT}}$ interrupt. This interrupt may also be explicitly enabled by executing the EI (Enable Interrupt) instruction.

In response to the $\overline{\text{INT}}$ interrupt, when it is acknowledged, the Z80 CPU generates an INA–MC as shown in Fig. 14.3.6. This is distinguished by the fact that both $\overline{\text{M1}}$ and $\overline{\text{IORQ}}$ are pulsed Low. (Note that the Z80 does not output any signal like the $\overline{\text{INTA}}$ of the 8085A). Peripherals or external logic doing the interrupt vectoring must get activated by $\overline{\text{M1}} + \overline{\text{IORQ}} = \emptyset$ and (after may be interrupt arbitration between contending devices) place on the data bus, the interrupt vector to be supplied to the Z80 for the $\overline{\text{INT}}$ interrupt. The nature (and interpretation) of the interrupt vector depends on the mode of operation of the $\overline{\text{INT}}$ interrupt.

The $\overline{\text{INT}}$ interrupt of the Z80 may be programmed to operate in one of Modes 0, 1 and 2. On reset, the Z80 sets the mode to Mode 0. Subsequently, the mode may be set to any of Modes 0, 1, or 2 by executing the instructions 'IM 0', 'IM 1' or 'IM 2', respectively.

In Mode 0, the interrupt vector is interpreted as a single-byte op code which will be the first instruction to be executed following the INA–MC. Typically, the op code provided is that of the RST n instruction (equivalent to 'RST n' of the 8085A) causing essentially a CALL 8 * n operation. This implies that the corresponding ISR starts from locatton 8n in memory. This mode of operation is very similar to the 8085A's treatment of its INTR interrupt when the interrupt vector provided is the op code of the one-byte 'RST n' instruction.

In Mode 1, the Z80 disregards the data on D0–D7 during INA–MC and does the necessary vectoring internally (like $\overline{\text{NMI}}$ of Z80 or RST 5.5, RST 6.5, RST 7.5 or TRAP of 8085A). In Mode 1, the Z80 executes the RST7 instruction (equivalent to CALL ØØ38$_H$) following the INA–MC for this interrupt. Note that this implies that the corresponding ISR starts from location ØØ38$_H$. This mode of operation is useful if one wants a vectored interrupt which does not require any additional interrupt vector generation logic.

The Mode 2 operation of the Z80's $\overline{\text{INT}}$ interrupt does not have any 8085A equivalent. This mode allows the system designer to provide more interrupts than the maximum of eight allowed by Mode 0 operation (corresponding to RST0–RST7). In this mode, the actual vectoring is done through an Interrupt Vector Table. The interrupt vector byte obtained through the INA–MC points to this vector table with the help of the IV register. The contents of this vector table (2 bytes) provides the actual branching address to which the Z80 places a subroutine call. This address gives the starting location of the corresponding ISR.

The Interrupt Vector Register IV can be read and loaded through the accumulator A; the instruction LD A,IV loads the contents of IV into A and LD IV,A loads IV with the contents of A. The register IV gives the upper byte of address for the interrupt vector table. The vector provided through INA–MC provides the lower byte of this address except that the lowest bit of this vector byte is a Don't Care bit; this is because the address INADDR generated by concatenating IV with this vector byte has its LSB reset to Ø. This procedure is illustrated in Fig. 14.4.1. Note that INADDR therefore points to a memory location with an even-numbered address. As shown in Fig. 14.4.1, for a Mode 2 $\overline{\text{INT}}$ interrupt, the Z80 places a subroutine call to location (INADDR + 1)(INADDR) where, as usual, (X) means 'contents of X'. In essence, the pointer INADDR points to the memory location containing the lower address byte of the

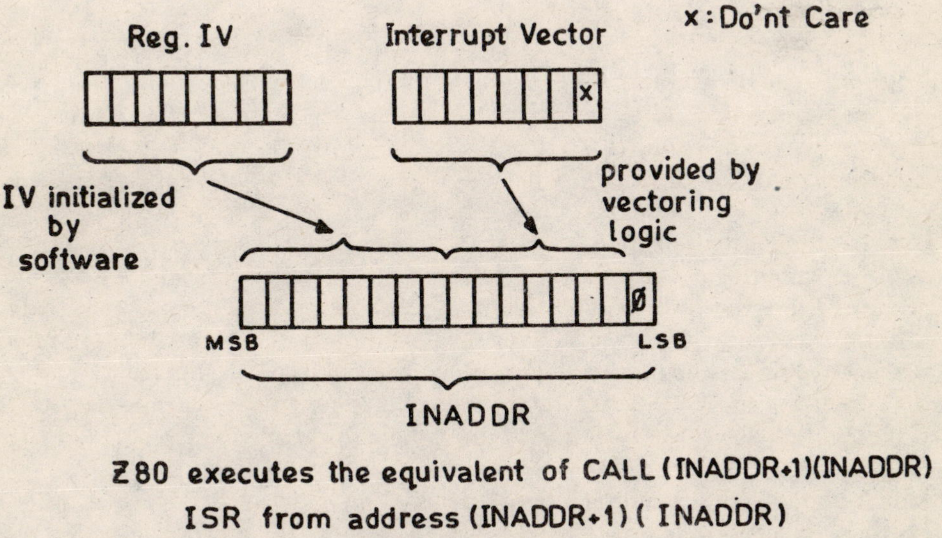

Fig. 14.4.1 Mode 2 Operation of the $\overline{\text{INT}}$ Interrupt

ISR's starting address, whereas INADDR+1 points to the memory location containing the higher order address byte.

In the Z80, we cannot directly vector for interrupts by providing a CALL Addr instruction through three INA–MCs (as is the case with the 8085A). The interrupt vector table approach for the Mode 2 $\overline{\text{INT}}$ interrupt is Z80's way of achieving the same thing. Since only one vector byte has to be provided through an INA–MC, this approach is simpler. It is not restrictive either, because with the same value of IV one can directly accommodate up to 128 interrupts with unique vectoring. This is much more than the number of interrupts a microcomputer system is likely to have.

14.5 THE INSTRUCTION SET OF THE Z80 CPU

As mentioned earlier, the Z80's instruction set is upward compatible with that of the 8085A (except RIM and SIM). Except RIM and SIM, all other 8085A instructions are present in the Z80's instruction set with the same one-byte op code. In addition, the Z80 has a large number of other instructions not present in the 8085A. As indicated earlier, the Z80's instructions may have either one-byte or two-byte op codes depending on the instruction; in addition, the instruction may require one or two bytes of immediate data. For instructions with two-byte op codes, the first byte is generally found to be one of the unused op codes of the 8085A.

In the following pages, we present a brief summary of the instruction set of the Z80 using its assembly language mnemonics (reproduced with permission from Zilog Inc.). Note that for the same instruction, the Z80 and the 8085A may use different mnemonics. We leave it to the reader to infer the common instructions of the Z80 and the 8085A and, wherever possible, identify the 8085A counterpart of the Z80's instruction set.

Z80, Z80A Instruction Set

The following is a summary of the Z80, Z80A instruction set showing the assembly language mnemonic and the symbolic operation performed by the instruction. A more detailed listing appears in the Z80-CPU technical manual, and assembly language programming manual. The instructions are divided into the following categories:

8-bit loads	Miscellaneous Group
16-bit loads	Rotates and Shifts
Exchanges	Bit Set, Reset and Test
Memory Block Moves	Input and Output
Memory Block Searches	Jumps
8-bit arithmetic and logic	Calls
16-bit arithmetic	Restarts
General purpose Accumulator & Flag Operations	Returns

In the table the following terminology is used.

b ≡ a bit number in any 8-bit register or memory location

cc ≡ flag condition code
- NZ ≡ non zero
- Z ≡ zero
- NC ≡ non carry
- C ≡ carry
- PO ≡ Parity odd or no over flow
- PE ≡ Parity even or over flow
- P ≡ Positive
- M ≡ Negative (minus)

d ≡ any 8-bit destination register or memory location

dd ≡ any 16-bit destination register or memory location

e ≡ 8-bit signed 2's complement displacement used in relative jumps and indexed addressing

L ≡ 8 special call locations in page zero. In decimal notation these are 0, 8, 16, 24, 32, 40, 48 and 56

n ≡ any 8-bit binary number

nn ≡ any 16-bit binary number

r ≡ any 8-bit general purpose register (A, B, C, D, E, H, or L)

s ≡ any 8-bit source register or memory location

s_b ≡ a bit in a specific 8-bit register or memory location

ss ≡ any 16-bit source register or memory location

subscript "L" ≡ the low order 8 bits of a 16-bit register

subscript "H" ≡ the high order 8 bits of a 16-bit register

( ) ≡ the contents within the ( ) are to be used as a pointer to a memory location or I/O port number

8-bit registers are A, B, C, D, E, H, L, I and R

16-bit register pairs are AF, BC, DE and HL

16-bit registers are SP, PC, IX and IY

Addressing Modes implemented include combinations of the following:

Immediate	Indexed
Immediate extended	Register
Modified Page Zero	Implied
Relative	Register Indirect
Extended	Bit

	Mnemonic	Symbolic Operation	Comments
8-BIT LOADS	LD r, s	r ← s	s ≡ r, n, (HL), (IX+e), (IY+e)
	LD d, r	d ← r	d ≡ (HL), r (IX+e), (IY+e)
	LD d, n	d ← n	d ≡ (HL), (IX+e), (IY+e)
	LD A, s	A ← s	s ≡ (BC), (DE), (nn), I, R
	LD d, A	d ← A	d ≡ (BC), (DE), (nn), I, R
16-BIT LOADS	LD dd, nn	dd ← nn	dd ≡ BC, DE, HL, SP, IX, IY
	LD dd, (nn)	dd ← (nn)	dd ≡ BC, DE, HL, SP, IX, IY
	LD (nn), ss	(nn) ← ss	ss ≡ BC, DE, HL, SP, IX, IY
	LD SP, ss	SP ← ss	ss = HL, IX, IY
	PUSH ss	(SP-1) ← ss_H; (SP-2) ← ss_L	ss = BC, DE, HL, AF, IX, IY
	POP dd	dd_L ← (SP); dd_H ← (SP+1)	dd = BC, DE, HL, AF, IX, IY
EXCHANGES	EX DE, HL	DE ↔ HL	
	EX AF, AF'	AF ↔ AF'	
	EXX	(BC, DE, HL) ↔ (BC', DE', HL')	
	EX (SP), ss	(SP) ↔ ss_L, (SP+1) ↔ ss_H	ss ≡ HL, IX, IY

	Mnemonic	Symbolic Operation	Comments
MEMORY BLOCK MOVES	LDI	(DE) ← (HL), DE ← DE+1 HL ← HL+1, BC ← BC-1	
	LDIR	(DE) ← (HL), DE ← DE+1 HL ← HL+1, BC ← BC-1 Repeat until BC = 0	
	LDD	(DE) ← (HL), DE ← DE-1 HL ← HL-1, BC ← BC-1	
	LDDR	(DE) ← (HL), DE ← DE-1 HL ← HL-1, BC ← BC-1 Repeat until BC = 0	
MEMORY BLOCK SEARCHES	CPI	A-(HL), HL ← HL+1 BC ← BC-1	
	CPIR	A-(HL), HL ← HL+1 BC ← BC-1, Repeat until BC = 0 or A = (HL)	A-(HL) sets the flags only. A is not affected
	CPD	A-(HL), HL ← HL-1 BC ← BC-1	
	CPDR	A-(HL), HL ← HL-1 BC ← BC-1, Repeat until BC= 0 or A = (HL)	
8-BIT ALU	ADD s	A ← A + s	
	ADC s	A ← A + s + CY	CY is the carry flag
	SUB s	A ← A - s	
	SBC s	A ← A - s - CY	s ≡ r, n, (HL) (IX+e), (IY+e)
	AND s	A ← A ∧ s	
	OR s	A ← A ∨ s	
	XOR s	A ← A ⊕ s	

Group	Mnemonic	Symbolic Operation	Comments
8-BIT ALU	CP s	A - s	s = r, n (HL) (IX+e), (IY+e)
	INC d	d ← d + 1	d = r, (HL) (IX+e), (IY+e)
	DEC d	d ← d - 1	
16-BIT ARITHMETIC	ADD HL, ss	HL ← HL + ss	ss ≡ BC, DE HL, SP (ADD HL, ADC HL, SBC HL)
	ADC HL, ss	HL ← HL + ss + CY	
	SBC HL, ss	HL ← HL - ss - CY	
	ADD IX, ss	IX ← IX + ss	ss ≡ BC, DE, IX, SP
	ADD IY, ss	IY ← IY + ss	ss ≡ BC, DE, IY, SP
	INC dd	dd ← dd + 1	dd ≡ BC, DE, HL, SP, IX, IY
	DEC dd	dd ← dd - 1	dd ≡ BC, DE, HL, SP, IX, IY
GP ACC. & FLAG	DAA	Converts A contents into packed BCD following add or subtract.	Operands must be in packed BCD format
	CPL	A ← $\overline{A}$	
	NEG	A ← 00 – A	
	CCF	CY ← $\overline{CY}$	
	SCF	CY ← 1	
MISCELLANEOUS	NOP	No operation	
	HALT	Halt CPU	
	DI	Disable Interrupts	
	EI	Enable Interrupts	
	IM 0	Set interrupt mode 0	8080A mode
	IM 1	Set interrupt mode 1	Call to 0038_H
	IM 2	Set interrupt mode 2	Indirect Call
ROTATES AND SHIFTS	RLC s	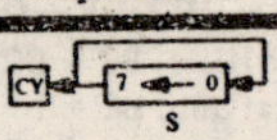	s ≡ r, (HL) (IX+e), (IY+e)
	RL s	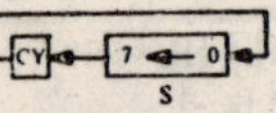	
	RRC s	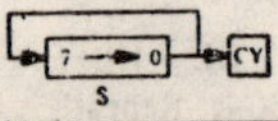	
	RR s	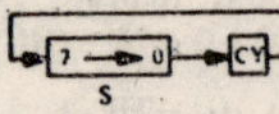	
	SLA s	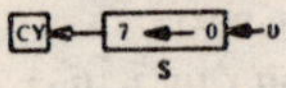	
	SRA s	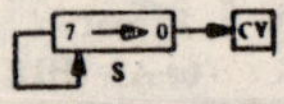	
	SRL s		
	RLD	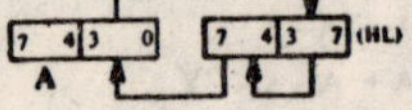	
	RRD	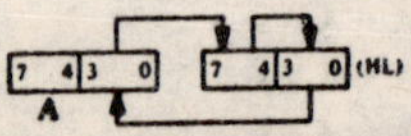	

Group	Mnemonic	Symbolic Operation	Comments
BIT S, R, & T	BIT b, s	Z ← $\overline{s_b}$	Z is zero flag
	SET b, s	s_b ← 1	s ≡ r, (HL) (IX+e), (IY+e)
	RES b, s	s_b ← 0	
INPUT AND OUTPUT	IN A, (n)	A ← (n)	
	IN r, (C)	r ← (C)	Set flags
	INI	(HL) ← (C), HL ← HL + 1 B ← B - 1	
	INIR	(HL) ← (C), HL ← HL + 1 B ← B - 1 Repeat until B = 0	
	IND	(HL) ← (C), HL ← HL - 1 B ← B - 1	
	INDR	(HL) ← (C), HL ← HL - 1 B ← B - 1 Repeat until B = 0	
	OUT(n), A	(n) ← A	
	OUT(C), r	(C) ← r	
	OUTI	(C) ← (HL), HL ← HL + 1 B ← B - 1	
	OTIR	(C) ← (HL), HL ← HL + 1 B ← B - 1 Repeat until B = 0	
	OUTD	(C) ← (HL), HL ← HL - 1 B ← B - 1	
	OTDR	(C) ← (HL), HL ← HL - 1 B ← B - 1 Repeat until B = 0	
JUMPS	JP nn	PC ← nn	cc: NZ PO, Z PE, NC P, C M
	JP cc, nn	If condition cc is true PC ← nn, else continue	
	JR e	PC ← PC + e	
	JR kk, e	If condition kk is true PC ← PC + e, else continue	kk: NZ NC, Z C
	JP (ss)	PC ← ss	ss = HL, IX, IY
	DJNZ e	B ← B - 1, if B = 0 continue, else PC ← PC + e	
CALLS	CALL nn	(SP-1) ← PC_H (SP-2) ← PC_L, PC ← nn	cc: NZ PO, Z PE, NC P, C M
	CALL cc, nn	If condition cc is false continue, else same as CALL nn	
RESTARTS	RST L	(SP-1) ← PC_H (SP-2) ← PC_L, PC_H ← 0 PC_L ← L	
RETURNS	RET	PC_L ← (SP), PC_H ← (SP+1)	
	RET cc	If condition cc is false continue, else same as RET	cc: NZ PO, Z PE, NC P, C M
	RETI	Return from interrupt, same as RET	
	RETN	Return from non-maskable interrupt	

Chapter 15

Microprocessor System Development Aids

Traditionally, designers of electronic systems have used a variety of instruments for designing, developing and maintaining such systems. For example, for analog systems, signal generators and oscilloscopes were the most popular instruments used; the signal generator was used (if needed) to provide an input to the system, whereas the oscilloscope allowed the user to see the response waveforms at the output and other intermediate points.

Even though oscilloscopes (especially multi-channel ones) may still be used for examining the behaviour of digital system, they are not as useful here as for the analog case. After all, in digital systems, one is typically only curious to know whether a logic variable is 1 or Ø; one is interested in the exact waveform of the associated variable as shown by an oscilloscope, only when a study of rise and fall times is needed. For this, one typically used 'Logic Probes' to see quickly if a particular variable was 1 or Ø. These probes have an LED which glows brightly if the probe-tip touches a point which is at logic '1'; the LED stays dark if the probe-tip touches a point which is at logic 'Ø' or glows dimly if the point is floating.

Such logic probes are usually sufficient to examine combinatorial circuits or even simple, slow sequential circuits. For more complex circuits, one typically needs a Logic State Analyzer (LSA) to study the behaviour (1 or Ø) over time of a number of logic variables. The LSA, at least conceptually, is very much like a multi-channel oscilloscope with a large number of channels (typically 16 or more) which displays only in terms of 1-Øs or Hi-Lo levels. A good LSA typically simplifies its usage by allowing a large variety of triggering modes and storing logic patterns for comparison with a subsequent pattern or a reference pattern. These instruments are invaluable for testing or studying the behaviour of complex digital systems, whether they are microprocessor based or not.

Even though they may be used, instruments like oscilloscopes, logic probes or LSAs, provide rather limited facilities to the designer of microprocessor based systems. These instruments are basically used to test the hardware of the system and are not helpful (except in trivial cases) in testing the software. From the earlier chapters, it should be evident that hardware and software are the two components of a microprocessor based design. Apart from whatever development and testing one has to do separately for the two, one must also usually carry out a similar exercise while integrating them. This integrated testing, unless it is very limited in scope, is also usually beyond the capability of instruments like LSAs. Another difficulty with these instruments is that they can typically be used more or less only after the system has been designed; they are of little use in the development process itself. Finally, even when LSAs or oscilloscopes may be used to study or monitor the performance of a microprocessor-based digital system, they are usually not very 'user-friendly' in terms and their usage or their presentation or interpretation of data.

As microprocessor applications grew in volume, one felt the need for a system that could support all that a user needs to develop designs (hardware and software), test them as well as maintain and repair them, if needed. All this (and more) are provided by the Microprocessor Development Systems available from various manufacturers. A brief description of these systems is given in Sec. 15.1. In Secs. 15.2–15.5, we give a summary of some of the software and hardware tools available on these to help in developing microprocessor based systems. In the other sections, we present some other development tools.

15.1 MICROPROCESSOR DEVELOPMENT SYSTEMS

Any computer system which supports the software and hardware needed to make designing with microprocessors easier may be classified as a Microprocessor Development System. Some microprocessor manufacturers (e.g. INTEL Corp.) make their own development systems which are primarily intended to support design and development activities with their own range of processors. Others, typically equipment or instrumentation manufacturing firms (e.g. Hewlett-Packard, Tektronix, Phillips etc.), make more general purpose development systems which, with suitable hardware and software augmentation, can support most of the popular microprocessors available today. It is also possible to add hardware or software to other, more general purpose computers (e.g. Personal Computers), in order to make them work as microprocessor development systems when the need arises. To confuse the issue, an equipment marketed as a microprocessor development systems may, with proper software, also be used for general purpose computing in languages like BASIC, FORTRAN or PASCAL.

A development system supporting a particular microprocessor provides a whole range of support facilities to help the designer to design and develop a digital system using that processor. It will typically provide an Assembler to allow the designer to write software using assembly language mnemonics rather than in machine code. It may also allow the designer to write the software in high-level languages (e.g. PASCAL or FORTRAN). In both cases, the development system takes the responsibility of translating the software to the machine code for that CPU. It may also provide ready-made application software packages which the designer may incorporate directly in his own software. Typically, the development system will support a Simulator which allows the user to interactively execute his own software to verify that it is working properly and as desired. The machine code may then be 'burned into' an EPROM using an EPROM Programmer controlled by the development system. The user may also store his source program and object program (machine code) in various storage media supported by the system for later retrieval; typical storage media supported are floppy disks, hard disks, Winchester disks, and may be even paper tapes, cassette tapes or data cartridges. For testing the hardware designed by the user, one will also find the development system supporting an emulator for the processor. This emulator allows the user to read/write into the memory and I/O locations of the target in order to isolate and correct hardware faults in the target system. It also allows integrated hardware/software testing of the target system being designed. The emulator also finds use as a powerful tool to debug and repair faults in the final system for maintenance purposes. These facilities are described in greater detail in Secs. 14.2–14.5.

In terms of their own operating software, development systems are very much like general-purpose microcomputers. They usually do not have Real-Time Operating Systems and range from systems which can support only single-users to ones capable of supporting multiple-users. Typically, a single-user system is built around a 8-bit CPU with a single-tasking operating system using floppy disks as the storage medium. The more complex systems usually work with a 16-bit CPU and generally support multi-user, multi-tasking operating systems (very often UNIX or a

UNIX-like operating system); these, generally, also support Winchesters or hard disks apart from the usual floppy disks.

15.2 ASSEMBLERS

A typical development system will usually provide an Assembler Program which will translate the user-software (source program), written in the assembly language of a particular CPU, into the machine code (object program) of that CPU. Typical functions of an assembler have been discussed in Sec. 8.8. Note that with proper software it is possible to have a development system (or general-purpose computer) running on one CPU to generate machine code for another CPU from a source program written in the assembly language of the latter; these special types of assemblers are referred to as 'Cross-Assemblers'. An assembler which is able to support a macro-facility is referred to as a 'Macro-Assembler'. (Macros are instruction sequences which are written only once and referred to by a name. The macro-assembler inserts the corresponding instructions whenever it finds the macro's name in the assembly language program.) A good assembler will also be able to provide 'Relocatable Object Code'. This implies that the object code generated by the assembler can, with appropriate software, be moved to a different location in the address space of the target system without changing the nature of its execution. (Code which cannot be relocated is referred to as Absolute Code.)

15.3 APPLICATION SOFTWARE PACKAGES AND HIGH LEVEL LANGUAGE SUPPORT

With a sufficiently powerful development system, it is possible for the user to write his software in a high level language like PASCAL or FORTRAN rather than in assembly language. A compiler program on the development system may then be used to translate that program into executable object code for the target CPU. Writing correct software in a high-level language is generally a much easier and faster process than writing them in assembly language. This facility, therefore, saves a lot of development time. However, the object code generated by this process is generally larger and runs more slowly than the object code generated by a relatively good assembly language program. In the author's opinion, it is probably better to write programs in assembly language for 8-bit CPUs. For 16-bit CPUs, with their more complex instruction set, it is probably advisable to write programs in a modular fashion using a high-level language. Later, time-critical portions of the software may be rewritten using assembly language.

A good development system will generally provide a Linker program to connect different program modules developed in assembly language or one or more high level languages. One can then load the linked-program to generate machine-code which will run from any address specified by the user. These support programs encourage the use of modular programming techniques.

The development system may also provide ready-made software which the designer can directly incorporate in his program and thus reduce the development time. In this category, one typically finds software packages for arithmetic operations and other complex calculations. One can even get software to help a system designer write an operating system for his own application.

A development system may also provide a Simulator program for a microprocessor. This allows the user to execute interactively a program written for that microprocessor and verify its proper operation. Note that this simulation process is independent of the target system hardware (which may not even exist.)

15.4 EPROM PROGRAMMERS

At least at the developmental stage, if not later, the program for the microprocessor based system will be in EPROMs. An EPROM Programmer provides the means for transferring the machine code developed in the development system to the desired EPROM. The programmer is a piece of hardware appropriately controlled by the software of the development system to 'burn-in' the required machine code into the EPROM. The details of the programming (e.g. how to present the data to be burned-in and the address on the EPROM where it should go, voltage-levels, pulse-widths, etc.) are taken care of by the hardware-software combination of the EPROM programmer and the development system.

15.5 EMULATORS

An emulator is useful in developing and testing the hardware and software of a microprocessor based systems. It is a piece of hardware functioning typically as a peripheral of the development system. A pod from the emulator replaces the microprocessor of the target system (i.e. the system under development or being tested). With this connection between the host and the emulator, the typical emulator functions that may be carried out are given below. Note that even though the CPU of the target system is replaced by the emulator pod, some part of the emulator hardware will generally have a CPU which is functionally the CPU of the target.

(i) Read memory locations or I/O ports of target.
(ii) Write into memory locations or I/O ports of target.
(iii) Display CPU status (i.e. CPU registers).
(iv) Change CPU status.
(v) Map selectively memory or I/O ports of the target system into resources of the development system (useful if the target system is not complete).
(vi) Execute software on the target system with specified starting address and break-points. The software may be resident either on the host or on the target system or a combination thereof.
(vii) Execute software as in (vi) but with a single-stepping or trace option with the CPU status displayed at each step.
(viii) Execute software as in (vi) or (vii) but with a dumping facility to display to the user the execution of the instructions prior to termination.

The emulator can be used to verify that the target system's hardware is functioning properly by using (i)–(iv) above. Proper operation of the software or the hardware-software combination may be subsequently verified using (vi)–(viii). Interestingly enough, the latter can be done even when the entire hardware is not functional by using (v) judiciously. These features are extremely helpful in developing, testing or repairing microprocessor based digital systems.

15.6 SINGLE-BOARD CARDS

From Chaps. 6–12, it should be evident to the reader that the hardware design by microprocessor based systems is fairly standard. Apart from the CPU, they typically contain EPROMs, RAMs and Parallel I/O ports along with some more specialized devices like Serial I/O devices and Interrupt Controllers, if they are needed. Some manufacturers sell single-board CPU cards which contain more or less the above components. Usually the CPU's system bus is also brought out on these cards (along with other I/O lines) and may be used to expand the facilities available on the card itself or to add more features.

There is a trend among many designers to use these cards (properly augmented, if needed) to build microprocessor based systems. Since hardware development is then virtually not needed, total development time is considerably reduced. Moreover, one can also buy various kinds of packaged software which run on these systems and by properly incorporating them in one's application, software development is also reduced. Over and above these advantages, one also stands to gain on serviceability by using these standard off-the-shelf single board cards in the system.

15.7 MICROPROCESSOR KITS

These are basically single-board cards meant to train and familiarise people in the use of microprocessors but are also used (especially in a lab environment) for designing microprocessor based systems. Apart from the CPU, the hardware consists of EPROM, RAM, some parallel I/O ports, a HEX keyboard, LED display, keyboard and display controller, and possibly serial I/O lines for connecting to a teletype. As for the single-board cards of Sec. 15.6, the CPU's system bus and the I/O port lines are also brought out on edge-connectors and may be used to augment the system. Data entry is done as HEX characters from the keyboard and the 7-segment LEDs are used for display purposes. These kits typically run a small monitor program which apart from interfacing with the keyboard and the display allows the following functions:

(i) Read memory locations, input ports or CPU registers of the kit.
(ii) Write into RAM locations, output ports or CPU registers of the kit.
(iii) Execute programs resident in the memory (EPROM or RAM) from a specified initial address.
(iv) Execute programs as in (iii) but in a single-step fashion.

(Apart from the keys on the HEX keyboard, other keys are provided to let the user choose functions such as the above). Usually, these kits can also be interfaced to a teletype for data entry and display purposes. Some kit manufacturers also provide additional cards which may be interfaced to their kits to provide additional features like Analog I/O or EPROM programming.

These kits typically have to be programmed directly in machine code through the keyboard (or teletype). This makes them somewhat difficult to use. Even in spite of this and other disadvantages, kits like these are heavily used in a lab environment for training purposes

15.8 THE IIT/K MICROPROCESSOR WORKSTATION

As indicated above, a major problem with using microprocessor kits is that the programming has to be done in machine code. To circumvent this problem, IIT/K designed and fabricated augmented versions of these kits, referred to as workstations, which have a resident assembler allowing the users to program directly in the assembly language of the 8085A. Equipped with a full ASCII keyboard and a CRT monitor, these workstations fall somewhat half-way between kits and development systems as an aid in developing 8085A based microprocessor system. Apart from an Assembler and a Disassembler, these workstations also provide an Editor to make program entry and correction easier and a number of other utility and arithmetic routines that the user can call from his software. It is equipped with an EPROM programmer to burn object code into EPROMs and an I/O card which provides Parallel and Serial I/O, Analog I/O and Timer/Counters. A dual processor architecture allows the user to execute his program at full CPU speed while also handling the keyboard and the display.

A reasonably sophisticated monitor allows the user to read or modify memory locations, I/O ports or CPU registers. Software on the workstation's ROM or RAM may be executed under full user control (i.e. starting point, break-points, single-step and trace options). In addition, various other peripherals, such as an emulator, extension memory, network interfaces etc., have also been developed for these workstations. Apart from its use in a lab environment for training purposes, equipped with the proper peripherals, these workstations can help substantially in the integrated hardware-software development needed for designing microprocessor-based systems (8085A based, in this case).

Index